Cross-Border Cooperation in the Age of Climate Change

Rui Alexandre Castanho
https://orcid.org/0000-0003-1882-4801
WSB University, Poland & Polytechnic Institute of Portalegre, Portugal

IGI Global
Scientific Publishing
Publishing Tomorrow's Research Today

Vice President of Editorial Melissa Wagner
Director of Acquisitions Mikaela Felty
Director of Book Development Jocelynn Hessler
Production Manager Mike Brehm
Cover Design Jose Rosado

Published in the United States of America by
IGI Global Scientific Publishing
701 East Chocolate Avenue
Hershey, PA, 17033, USA
Tel: 717-533-8845 | Fax: 717-533-7115
Website: https://www.igi-global.com E-mail: cust@igi-global.com

Library of Congress Cataloging-in-Publication Data

Names: Castanho, Rui Alexandre editor
Title: Cross-border cooperation in the age of climate change / edited by Rui Alexandre Castanho.
Description: Hershey, PA : IGI Global Scientific Publishing, 2026. | Includes bibliographical references and index. | Summary: "This book investigates the role of cross-border cooperation as a strategic response to climate change. It highlights how collaborative governance, joint planning, and integrated territorial approaches can enhance resilience in border regions"-- Provided by publisher.
Identifiers: LCCN 2025028212 (print) | LCCN 2025028213 (ebook) | ISBN 9798337339726 hardcover | ISBN 9798337339733 paperback | ISBN 9798337339740 ebook
Subjects: LCSH: Environmental management--International cooperation | Climatic changes--International cooperation | Borderlands--International cooperation
Classification: LCC GE300 .C764 2026 (print) | LCC GE300 (ebook)
LC record available at https://lccn.loc.gov/2025028212
LC ebook record available at https://lccn.loc.gov/2025028213
British Cataloguing in Publication Data
A Cataloguing in Publication record for this book is available from the British Library.

Table of Contents

Detailed Table of Contents

socio-economic conditions, media influence, and political contexts. Advancements in data collection, machine learning, and predictive models can improve disaster prediction and management. Human health, particularly child health, is at risk due to direct effects like heat waves, floods, and drought, as well as indirect effects like infectious diseases and mental health issues. The chapter calls for a balanced approach, including adaptation strategies and technological innovations, to mitigate climate change and improve sustainable practices.

Climate change is an inherently transboundary challenge, demanding cooperative, data-driven solutions across national borders. This chapter explores how emerging digital technologies particularly Artificial Intelligence (AI), digital twins, and interoperable platforms are reshaping climate governance in border regions. Through real-time monitoring, predictive modelling, and shared scenario planning, these tools enable countries to jointly respond to environmental risks. The chapter examines governance models like EGTCs, legal-ethical challenges in data sharing, and participatory approaches involving civil society and citizen science. Case studies from the Danube River Basin and the US–Mexico border demonstrate how digital systems enhance resilience and early warning capabilities. It argues for scalable, inclusive, and ethically governed digital ecosystems that transform borders from barriers into collaborative climate adaptation zones.

As climate change crosses national boundaries, the urgency for cross-border cooperation in South Asia increases. This chapter explores eco-tourism as a tool of climate diplomacy, using shared ecological assets to foster environmental resilience and peaceful regional engagement. It focuses on the Eastern Himalayas, the Terai Arc, and the Sundarbans Delta to illustrate how ecologically sensitive tourism can support biodiversity conservation, climate adaptation, and trust-building. These transboundary landscapes, rich in ecosystems and cultural ties, offer potential for co-managed tourism aligned with integration goals. The chapter reviews frameworks like SAARC, BIMSTEC, and bilateral environmental agreements, while identifying

local communities but also for ensuring regional stability, security, and sustainable development.

This chapter explores cross-border air pollution as a serious challenge that affects both the environment and human health. It looks at how different regions, such as Europe, Southeast Asia, North America, and East Asia, have built legal and institutional systems to deal with pollution that crosses national borders. Using a doctrinal and comparative method, it examines key agreements, policy tools, and the role of regional cooperation. The chapter also discusses how uneven capacity and lack of enforcement affect outcomes. It highlights the need for stronger legal frameworks and inclusive cooperation to protect air quality, support public health, and promote sustainable development across regions.

In an era marked by escalating climate variability and increasingly frequent extreme weather events, the importance of robust and coordinated climate monitoring cannot be overstated. Climate change knows no borders; its effects transcend political, cultural, and geographical boundaries, impacting ecosystems, economies, and communities across nations. As such, the responsibility for observing, analyzing, and responding to climatic shifts must extend beyond the confines of individual states. Climate monitoring in neighboring territories plays a pivotal role in constructing a comprehensive understanding of regional and global climate systems. By pooling data and resources across borders, countries are better equipped to predict, adapt to, and mitigate the multifaceted impacts of climate change. Neighboring territories often share interconnected ecosystems, river basins, coastlines, and atmospheric systems, making it crucial that climate data is collected and shared across these areas.

Environmental diplomacy signifies the collective effort of nations to tackle pressing environmental challenges that transcend individual borders. As the global community confronts increasingly complex and interconnected issues like climate change, pollution, and deforestation, the importance of fostering international cooperation has grown exponentially. Unlike traditional diplomacy, which primarily centers on matters of politics, economics, and security, environmental diplomacy strives to encourage collaboration to safeguard and sustain the natural world. The health of the environment directly impacts global stability and security, making it imperative for nations to engage in multilateral approaches to address these concerns. Through agreements, knowledge-sharing initiatives, and resource allocation, environmental diplomacy aims to develop sustainable solutions that benefit humanity as a whole. The interconnected nature of environmental challenges demands a collaborative approach that takes into account diverse cultures, perspectives, and interests.

India and the United Arab Emirates (UAE) have evolved their long-standing bilateral relationship into a strategic climate partnership. Both countries are committed to combating climate change through cooperation in renewable energy, green hydrogen, and climate finance, aligned with their nationally determined contributions under the

Paris Agreement. India's Green Hydrogen Mission and the UAE's Net Zero 2050 strategy form the foundation for collaborative energy transition goals. Joint initiatives, MOUs, and investment agreements have expanded clean energy development, grid connectivity, and support for developing nations. This study assesses the effectiveness of these bilateral efforts, their influence on global climate negotiations, and the shared vision for a sustainable, low-carbon future.

This research explores the critical role of cross-border renewable energy integration in facilitating a sustainable and just green transition. It highlights how regional collaboration across legal, technical, and policy domains enhances energy security, grid stability, and climate resilience. The manuscript evaluates successful models like the North Sea Wind Power Hub and India-Bhutan Hydropower Cooperation, illustrating the benefits of shared infrastructure and joint governance. Emphasis is placed on harmonized regulations, stakeholder inclusion, smart grid technologies, and robust environmental safeguards. The work concludes that cross-border systems are not only technologically feasible but geopolitically essential for achieving net-zero goals, economic integration, and equitable energy access.

This research highlights the impact and repercussions of so-called "environmental risks" on the future of territories marked by the cross-border impacts generated by the evolutionary processes of the renowned climate change. In this way, a concise study of the most northwestern region of Spain is carried out, based on the analysis of extreme meteorological phenomena -such as extratropical cyclonic low pressure systems-, on the physical, human and economic environment in the coastal areas of the Province of Pontevedra, focus attention on how these atmospheric processes constitute a key element in the configuration of the climate regime, where their frequency and intensity generate episodes of high disturbance, with significant implications for territorial dynamics and the safety and activity of its inhabitants.

Likewise, it emphasizes the need for climate risk planning tools, based on action and performance measures by local governments that promote cooperation to mitigate or eradicate cross-border risks and impacts, thus strengthening resilience in border regions.

This chapter examines the role of regional climate cooperation in the European Union's Green Transition process, with a particular focus on subnational governance mechanisms. Drawing on multi-level and polycentric governance theories as well as transnational municipal networks, the study examines the strategic importance of local and regional actors in shaping and implementing EU climate policies. In this study we examine climate governance frameworks and four empirical case studies (Vienna, Transdanube.Pearls, Eurometropolis Lille-Kortrijk and Ghent). The chapter assesses how regional initiatives contribute to innovation, participatory governance and cross-border policy coherence. With IPCC recommendations, we explain that effective climate governance in the EU requires institutional coherence, democratic legitimacy and the integration of resilience-sensitive strategies in line.

With climate change increasingly crossing political boundaries, the importance of local and regional governments (LRGs) in promoting cross-border climate cooperation has grown significantly. LRGs are often better positioned than national governments to tackle localized climate challenges through context-specific collaboration. This chapter examines the growing importance of LRGs in transboundary climate governance, emphasizing their ability to innovate, involve communities, and carry out adaptive solutions. It investigates mechanisms like cross-border networks, inter-municipal forums, joint environmental initiatives, and digital platforms that facilitate collaboration among subnational actors across borders. Through the examination of global case studies and policy scrutiny, this chapter pinpoints essential obstacles—like financial shortfalls and legal constraints—and proposes tactical policy suggestions, such as bolstering legal acknowledgment, establishing regional climate funds, and fostering knowledge sharing.

This chapter is not a critique of failures; it is a reflection on why those failures persist despite repeated attempts. International climate governance has not lacked diplomacy, nor has it suffered from ignorance of science or absence of public concern. What it lacks, we argue, is intelligibility across worlds. The frameworks imposed in the name of global cooperation presume a world populated by rational actors, seeking utility, governed by secular institutions, and pursuing infinite growth. But such assumptions do not travel well. They do not describe most human communities; they describe a particular cultural experience made universal. Drawing on comparative insights and case studies, we show how this epistemological imposition produces alienation, not alignment; mistrust, not participation. Compliance becomes performative. To move beyond this impasse, climate governance must be reimagined—not as a global consensus on Western terms, but as a dialogue between diverse lifeworlds. Without plural epistemologies, transformation remains a metaphor.

Foreword

As someone who has long witnessed the growing convergence of environmental, social, and economic challenges across borders, I am convinced that climate change represents the defining issue of our time. It is a crisis that transcends nations, ideologies, and generations, demanding responses that are both global in vision and local in execution. In this context, I welcome *Cross-Border Cooperation in the Age of Climate Change* as a timely and necessary contribution to one of the greatest collective challenges humanity has ever faced.

Borders are among humanity's most paradoxical creations. They are drawn by men, formalized in treaties, and defended by laws, yet the forces that shape life on Earth remain indifferent to such boundaries. Rising global temperatures, increasingly frequent extreme weather events, the degradation of ecosystems, and the accelerated loss of biodiversity are not hypothetical risks. They are lived realities, disrupting communities, eroding economies, and threatening human security, generally affecting those that live in poorer conditions, strengthening social inequities and promoting even more segregation between the ones that have it all, and the other which have nothing.

Crucially, this is not a distant problem for future generations; it is a crisis of *this* generation. Climate impacts are already reshaping where and how people live, and they will inevitably give rise to a new category of refugees: those forced from their homes not by conflict or persecution, but by drought, desertification, floods, and fires. These displacements will test not only the resilience of infrastructures but also the moral responsibility of the international community to respond with compassion and foresight.

Climate change has no borders, and for this reason our responses must be equally borderless. This is why strategic planning becomes indispensable. To confront a crisis of such magnitude, societies must move beyond piecemeal solutions toward comprehensive, long-term strategies that integrate climate resilience into every dimension of governance. Strategic planning allows us to anticipate risks, allocate resources efficiently, and align local, national, and international efforts in ways that maximise impact. Without such planning, our actions will remain fragmented, reactive, and insufficient against the scale of the challenge.

The complexity of this task requires not only robust science and sound policy but also the active participation of governments, international organisations, businesses, and civil society. Cooperation must be multidimensional: strengthening renewable energy transitions, advancing circular economies, restoring ecosystems, and creating inclusive governance frameworks that embed justice and equity. Sharing knowledge and best practices across borders can accelerate these transformations, while reducing inequalities and fostering stability.

This book rightly places emphasis on the concept of circular landscapes — re-imagining abandoned or degraded areas not as problems to conceal, but as opportunities for innovation and regeneration. By embedding circularity into planning and design, we can minimise resource loss, repurpose obsolete infrastructures, and create adaptive environments aligned with the Sustainable Development Goals. Importantly, the volume also demonstrates, through the analysis of case studies, that these ideas are not abstract theories but practical pathways already being tested in diverse contexts.

The lessons are clear. Recovery must mean more than remediation; it must mean transformation. Borders should not be barriers but bridges, spaces of dialogue and solidarity where collective challenges become opportunities for shared progress. Climate change, for all its risks, also offers us the possibility of building a new kind of cooperation — one that is more inclusive, more forward-looking, and more attuned to the realities of our interdependence.

The fight against climate change is, at its core, a test of our collective imagination and determination. If we act with urgency, guided by strategic vision and strengthened by international cooperation, we can turn this crisis into an unprecedented opportunity for sustainable development. The responsibility is ours, and the time to act is now.

Ana Loures

Portalegre Polytechnic University

Preface

Climate change has become one of the greatest shared challenges of our time. Its impacts transcend political and administrative boundaries, affecting communities, ecosystems, and economies in ways that remind us of our interconnectedness. No country or region can address these challenges alone. Instead, cooperation, dialogue, and shared responsibility are essential to building resilience and advancing sustainable development. This book, *Cross-Border Cooperation in the Age of Climate Change*, emerges from the conviction that collaboration across borders is not only desirable but also fundamental to achieving meaningful climate action.

Over the past decade, my research and editorial work have explored the nexus between borderlands, sustainable planning, and regional development (Castanho *et al.*, 2019; Vulevic *et al.*, 2020; Castanho, 2023). Through projects and publications, I have seen how areas often considered peripheral can become laboratories of innovation in governance, cooperation, and climate adaptation. These insights align with broader contributions in the literature, which emphasize how cross-border regions can promote territorial cohesion (Perkmann, 2003), sustainable governance (Oliveira, 2015), and collaborative responses to global challenges (Schmitt & Van Well, 2016). As editor of related volumes such as *Urban Agglomeration: Extracting Lessons for Sustainable Development* (Castanho, 2024) and *ESG and Ecosystem Services for Sustainability* (Castanho, 2024), I have sought to curate platforms where diverse voices can contribute to advancing knowledge and practice. The chapters included in this book reflect the diversity and complexity of cross-border cooperation under climate change pressures. They address topics ranging from energy transition and resource management to governance mechanisms and community resilience. Importantly, they also acknowledge the barriers that persist—such as regulatory asymmetries, political tensions, and disparities in resources—while offering pathways for transforming these obstacles into opportunities.

CHAPTER OVERVIEW

Chapter 1: Arbitration as a Tool for Cross-Border Climate Cooperation: Bridging Legal Gaps in Environmental Governance

The opening chapter positions arbitration as a valuable mechanism for addressing cross-border climate disputes. It explores how conflicts over natural resources, environmental damages, and treaty commitments can be managed through state and investor arbitration. By analyzing recent institutional developments and case law, the chapter demonstrates the potential of arbitration to foster cooperation, strengthen governance, and create legally binding solutions in transboundary environmental contexts.

Chapter 2: Climate Change Risks – An Overview

This chapter provides a comprehensive overview of climate change risks across sectors such as agriculture, biodiversity, human health, and ecosystems. It outlines how uneven rainfall, rising temperatures, and limited adaptation strategies amplify vulnerabilities, while also emphasizing how data-driven technologies and predictive models can support better risk management. The discussion highlights both direct and indirect effects on human health and stresses the importance of balanced adaptation strategies and technological innovation.

Chapter 3: Climate Data Without Borders: AI, Digital Twins, and Collaborative Adaptation in Border Regions

This chapter turns attention to digital cooperation across borders, focusing on how artificial intelligence, digital twins, and interoperable data platforms can reshape climate governance. Through real-time monitoring and predictive modeling, these technologies facilitate shared responses to climate risks in border regions. Case studies from the Danube Basin and the US–Mexico border illustrate the promise of data-driven collaboration, while the chapter also addresses the ethical and governance challenges of building inclusive, participatory digital ecosystems.

Chapter 4: Climate Diplomacy through Tourism: Leveraging Cross-Border Eco-Tourism for Peace and Environmental Resilience

Tourism emerges as a tool of climate diplomacy in this chapter, which examines how cross-border eco-tourism can foster environmental resilience and regional peace. Using examples from the Eastern Himalayas, Terai Arc, and Sundarbans Delta, the chapter illustrates how co-managed tourism initiatives can strengthen biodiversity conservation and trust-building. By engaging with frameworks such as SAARC and BIMSTEC, the analysis emphasizes tourism's potential to transform borders into ecological continuums while promoting inclusive, sustainable development.

Chapter 5: Climate-Induced Displacement and Cross-Border Cooperation

This chapter examines climate-induced displacement and the shortcomings of current international frameworks in addressing the needs of displaced populations. Drawing on practices from regions such as the Pacific Islands, South Asia, and Sub-Saharan Africa, it explores emerging solutions like humanitarian visas, planned relocation, and regional protocols. The discussion underscores the importance of cooperative frameworks rooted in human rights and climate justice, aiming to ensure protection, equity, and resilience for vulnerable communities.

Chapter 6: Climate Risk and Resilience Planning in Border Regions

Border regions often experience unique vulnerabilities due to fragile ecosystems, high population densities, and limited cross-border coordination. This chapter highlights the role of collaborative governance, data sharing, and community-based adaptation in enhancing resilience. It emphasizes the importance of inclusive, multi-level stakeholder engagement and cross-border partnerships in ensuring stability, security, and sustainable development.

Chapter 7: Cross-Border Air Pollution Control and Regional Environmental Health Cooperation

Air pollution is the focus of this chapter, which investigates its cross-border impacts on public health and environmental stability. By comparing legal and institutional approaches across regions such as Europe, Southeast Asia, and North America, the chapter evaluates the effectiveness of international agreements and policy tools. It underscores the unevenness of enforcement capacity while making the case for stronger legal frameworks and inclusive cooperation to ensure clean air and sustainable development.

Chapter 8: Data Sharing and Climate Monitoring Between Neighboring Territories

This chapter highlights the importance of cross-border data sharing for effective climate monitoring. Recognizing that climate change knows no borders, the chapter shows how pooling resources and integrating monitoring systems across neighboring territories enhances prediction and adaptation. Shared ecosystems and atmospheric systems necessitate such cooperation, which ultimately strengthens collective resilience and the global understanding of climate dynamics.

Chapter 9: Environmental Diplomacy and Cross-Border Negotiations

This chapter frames environmental diplomacy as an essential response to transboundary environmental challenges. Unlike traditional diplomacy focused on politics or security, environmental diplomacy emphasizes collaboration to safeguard natural systems. The chapter discusses multilateral agreements, knowledge-sharing, and joint resource management as critical components of building sustainable solutions that serve the collective interests of humanity.

Chapter 10: India–UAE Bilateral Cooperation to Combat Climate Change: Strategic Synergies and Future Pathways

This chapter explores the bilateral partnership between India and the United Arab Emirates in advancing renewable energy, green hydrogen, and climate finance. Anchored by India's Green Hydrogen Mission and the UAE's Net Zero 2050 strategy, their cooperation illustrates how joint initiatives and investment agreements strengthen climate commitments. The chapter situates these efforts within global climate negotiations and emphasizes the role of strategic partnerships in creating a low-carbon future.

Chapter 11: Integrated Cross-Border Renewable Energy Infrastructure for a Green Transition

This chapter emphasizes the role of renewable energy integration in driving a just and sustainable transition. Case studies such as the North Sea Wind Power Hub and India-Bhutan hydropower cooperation highlight the benefits of shared infrastructure and harmonized regulations. The discussion demonstrates that cross-border renewable energy systems are technologically feasible and geopolitically essential for achieving net-zero goals and equitable energy access.

Chapter 12: The Cross-Border Influence of Climate Change on the Current Reality of Environmental Risks in Spain

Spain's experience with climate risks is explored in this chapter, with a focus on the coastal Province of Pontevedra. Extreme meteorological events, including cyclonic low-pressure systems, are shown to disrupt local safety, livelihoods, and territorial stability. The chapter stresses the importance of climate risk planning tools and local government cooperation to mitigate cross-border risks and strengthen resilience in vulnerable regions.

Chapter 13: The Impact of Regional Climate Cooperation in the EU in the Green Transition

This chapter examines the European Union's governance frameworks for climate cooperation, emphasizing the role of local and regional actors in the Green Transition. Through case studies such as Vienna and Eurometropolis Lille-Kortrijk, it shows how subnational governance and cross-border networks foster participatory climate action. The analysis underscores the need for institutional coherence, democratic legitimacy, and resilience-sensitive strategies in EU climate governance.

Chapter 14: The Role of Local and Regional Governments in Fostering Cross-Border Climate Cooperation

This chapter highlights the importance of local and regional governments in addressing transboundary climate challenges. Positioned to innovate and implement context-specific solutions, these actors foster cooperation through networks, joint initiatives, and digital platforms. While financial and legal barriers persist, the chapter proposes solutions such as regional climate funds and stronger legal recognition to empower local governments in global climate governance.

Chapter 15: Why the World Fails to Act? The Silent Collapse of International Climate Governance

The final chapter reflects on the deeper reasons behind the repeated shortcomings of international climate governance. It argues that global frameworks often assume a universal cultural and institutional logic that fails to resonate with diverse human communities. This mismatch leads to alienation, mistrust, and performative compliance. The chapter calls for reimagining governance as a plural dialogue between lifeworlds, where diverse epistemologies and values are recognized as essential for genuine transformation.

It is my hope that this book will serve as a resource not only for researchers and students but also for policymakers, municipal leaders, and practitioners. By sharing experiences from multiple contexts, the volume demonstrates that border regions can shift from being perceived as barriers to being understood as spaces of opportunity, where cooperation transforms vulnerability into a source of strength.

I extend my gratitude to all contributors for their commitment to this collective endeavor. Contextually, I invite readers to approach this book not only as an academic reference, but as a call to rethink the way we work together in the age of climate change.

Rui Alexandre Castanho
Editor
2025

REFERENCES

Castanho, R. A. (2019). Identifying Processes of Smart Planning, Governance and Management in European Border Cities. Learning from City- to- City Cooperation (C2C). *Sustainability (Basel), 2019*(11), 5476. DOI: 10.3390/su11195476

Castanho, R. A. (2023). The Fourteen Critical Factors for Regional Development in Borderlands: Focusing on European Cross-Border Cooperation (CBC) Areas. Chapter in the book Borderology. Ed. Springer. ISBN 978-3-031-29719-9 pp. 135–145 DOI: DOI: 10.1007/978-3-031-29720-5_11

Castanho, R. A. (2024). Introductory Chapter: Urban Agglomeration – Extracting Lessons for Sustainable Development – The Opening Act. Chapter in the Book: Urban Agglomeration - Extracting Lessons for Sustainable Development (Ed) IntechOpen DOI: pp.1-7DOI: 10.5772/intechopen.1004204

Castanho, R. A. (2024). *Preface. In the Book: ESG and Ecosystem Services for Sustainability*. IGIGLOBAL. DOI: 10.4018/979-8-3693-3771-4

Oliveira, E. (2015). Constructing regional advantage in branding the cross-border Euroregion Galicia–northern Portugal. *Regional Studies, Regional Science*, *2*(1), 341–349. DOI: 10.1080/21681376.2015.1044020

Perkmann, M. (2003). Cross-Border Regions in Europe. *European Urban and Regional Studies*, *10*(2), 153–171. Advance online publication. DOI: 10.1177/0969776403010002004

Schmitt, Peter & Van Well, Lisa. (2015). Territorial governance across Europe: Pathways, practices and prospects. .DOI: 10.4324/9781315716220

Vulevic, A., Obradovic, V., Castanho, R.A. and Djordjevic, D. (2020). Cross-Border Cooperation (CBC) in a Multi-Level Governance System in Southeastern Europe Territories: How to Manage Territorial Governance Processes in Serbia-Romania Border Space. Chapter in the book: Cross-Border Cooperation (CBC) Strategies for Sustainable Development. Editor of the Book. IGI GLOBAL. pp 70-85. ISBN13: 9781799825135. DOI: DOI: 10.4018/978-1-7998-2513-5.ch004

Chapter 1

Arbitration as a Tool for Cross-Border Climate Cooperation: Bridging Legal Gaps in Environmental Governance

Shubhang Gomasta
https://orcid.org/0009-0000-7834-2084
Amity University, Gwalior, India

Aakash Gupta
https://orcid.org/0009-0008-3920-9991
Amity University, Gwalior, India

ABSTRACT

Cross-border disputes have evolved from political and sovereignty disputes to issues such as climate change, natural resource distribution, water-sharing treaties, and environmental pollution. These disputes often affect communities living on the other side of the border, particularly those relying on natural resources or the environment for survival. In such cases, arbitration is a neutral, fair, and acceptable dispute resolution mechanism. State- and investor-based arbitration can help solve transboundary climate governance disputes. Arbitration facilitates transboundary climate governance agreements and public-private partnerships, as countries often use treaties or contracts to make promises legally binding. The paper explores the possibilities and restrictions of using arbitration to settle disputes involving resource sharing, environmental damages, and infrastructure promises made. It also explores recent institutional developments, arbitral case law, and the international legal framework in cross-border environmental conflicts.

DOI: 10.4018/979-8-3373-3972-6.ch001

INTRODUCTION

Climate change is an urgent issue that crosses scientific, political, and legal boundaries. The climate crisis creates global environmental, economic, and human effects, articulated in terms of glacial melting, sea-level rise, desertification, and weather extremes. These effects cross national boundaries; they disrupt ecosystems, compel the displacement of human beings, and imperil the security and development agenda of many states at the same time. With the intensity and frequency of climate change conflicts escalating, there is an increasingly recognized recognition that existing legal instruments are not effective in meeting the nature, magnitude, and immediacy involved in transboundary environmental harm.

The institutional design of multilateral treaties encompassing the United Nations Framework Convention on Climate Change (UNFCCC), the Kyoto Protocol, and the Paris Agreement is the central building block of international climate regime. Yet these systems are heavily dependent on non-binding legal commitments and the goodwill of political leaders. Their mechanisms for resolving disputes are weakly developed, substantially non-binding, and inaccessible to private parties, civil society, or the affected communities afflicted by climate harms. This gap between articulated international norms and enforceable remedies leaves unaddressed climate grievances and a lack of complete environmental responsibility.

At the same time, climate disputes have come to engage not only nation-states but also foreign investors, private corporations, and development donors. Climate disputes have grown in size, tackling matters ranging from the carbon pricing and subsidies of renewable energy to grievances against the lack of proper environmental protection in resource extraction. However, international judicial institutions such as the International Court of Justice (ICJ), the International Tribunal for the Law of the Sea (ITLOS), and the regional tribunals of human rights have lacked climate litigation. Procedurally, in most cases, restrictions, standing requirements, and jurisdictional barriers render it impossible for them to effectively adjudicate climate disputes.

Application of arbitration to climate disputes has never been a novel aspect. Early cases in the 1990s and the 2000s, particularly those that were submitted to the Permanent Court of Arbitration (PCA), showed arbitration's potential to resolve public interest cases. PCA adopted the Optional Rules for Arbitration of Disputes Relating to the Environment and Natural Resources in 2001, and it was an acknowledgment of environmental concerns in transboundary conflict resolution. However, application of such mechanisms has never been common, mainly because of lack of institutional acknowledgment and fear of traditional investors about the expansion of arbitral jurisdiction.

Furthermore, the United Nations Environment Programme (UNEP) and UNCITRAL have also discussed the use of arbitration in environmental regulation but stopped short of creating climate-specific regimes. These nascent developments indicate that the building blocks for climate arbitration have been in place for more than two decades, but were at the periphery because of doctrinal rigidity and institutional conservatism. As climate litigation gains momentum globally, the necessity to draw upon and capitalize on these nascent efforts is increasingly important.

Against this background, international arbitration historically the preserve of commercial and investment disputes—can now potentially be the forum to decide some forms of climate disputes. Arbitration does have some procedural strengths: party control, cross-border enforceability in treaties such as the New York Convention (1958), and the possibility of tailoring proceedings to suit technical or scientific subject matter. Arbitration's flexibility also means that institutions such as the ICC, SIAC, or PCA can, in theory, design specialized climate arbitration structures or rulesets incorporating climate law principles and public interest factors.

However, the current arbitration framework is poorly positioned to address climate change disputes effectively. Doctrinally, arbitration is mostly focused on commercial considerations and bilateral investor-state matters. Procedurally, it systematically excludes important stakeholders, such as affected communities, environmental non-governmental organizations, and indigenous peoples. Substantively, it lacks the ability to offer non-monetary remedies such as ecological restoration, injunctive relief, or remediation—all of which are critical to resolving climate-related conflicts. Politically, additionally, it can exacerbate inequalities by enabling capital-exporting parties to challenge climate policies taken by Global South nations on expropriation or discriminatory treatment grounds.

In recent times, many arbitration claims have raised serious concerns about regulatory chill, a situation where states fear to undertake progressive climate action because of the risk of expensive arbitration litigation. The Vattenfall v. Germany and Rockhopper v. Italy cases demonstrate how arbitration can unintentionally impose sanctions on sovereign attempts to cancel fossil fuel projects. This turn of events raises serious ethical, legal, and governance questions on the place of arbitration in the environmental law framework.

1. OBJECTIVE OF THE CHAPTER

This chapter examines the possible structural and procedural development of international arbitration in dealing with the unique needs of climate governance. It contends that, with appropriate adjustment, arbitration could be a method of resolving transnational climate disputes in enforceable, accessible, and responsive to global equity terms. Specifically, the chapter aims to:

1. Determine the existing doctrinal and procedural constraints in applying arbitration in climate conflict.
2. Explore opportunities for integrating climate law norms and stakeholder participation into arbitral practice;
3. Recommend institutional reforms and legal innovations to promote a more equitable and sustainable climate arbitration system.

The paper is structured into eight distinct sections. Section 2 gives a doctrinal and historical context for climate law and international arbitration. Section 3 gives an overview of the literature, pointing to current scholarly gaps and introducing arbitration as an enormously underleveraged instrument of climate governance. Section 4 outlines the principal doctrinal and procedural defects that permeate current arbitral regimes. Section 5 outlines specific reforms—such as the creation of climate-specific arbitration protocols, mechanisms for stakeholder inclusion, and hybrid remedies—to drive the realignment of arbitration with climate justice principles forward. Section 6 outlines the specific challenges of the Global South, i.e., access, equity, and unevenness of legal power. Section 7 outlines practical future directions and new policy interventions, and Section 8 concludes with a reflection on the potential of arbitration as a transformative instrument within the architecture of global environmental law.

The chapter thus contributes to an emerging literature viewing arbitration not merely as a tool for the resolution of disputes, but as an instrument of regulation that can shape international climate norms, promote compliance, and offer restorative environmental justice. It challenges the view that arbitration is necessarily ill-suited for public interest disputes, arguing that with proper doctrinal ingenuity and institutional innovation, it can grow up to address the urgent needs of global climate governance.

2. BACKGROUND: CLIMATE LAW, DISPUTE RESOLUTION, AND THE ROLE OF ARBITRATION

One of the most important legal and governance issues of the twenty-first century is climate change. Traditional legal institutions have been stretched to their limitations by the global scope of climate harm, its disproportionate effect on vulnerable groups, and the intricate web of public and private parties involved. A framework that can settle claims from several jurisdictions, incorporate scientific data, and create legally binding remedies that are enforceable internationally is needed to address climate issues. The doctrinal and institutional framework required to comprehend the potential function of arbitration in this changing climate governance environment is provided in this section.

2.1 Emergence of International Climate Governance

The Paris Agreement (UNFCCC, 2015) introduced nationally determined contributions but left dispute resolution largely facilitative. UN reports (United Nations, Year) emphasize the need for dispute resolution tools within the global climate regime.

• The United Nations Framework Convention on Climate Change (UNFCCC), which was ratified at the Rio Earth Summit in 1992, established the framework for contemporary climate governance. The UNFCCC established a framework for state collaboration and acknowledged climate change as a universal human problem. It established the common but differentiated responsibilities (CBDR) idea, which still has an impact on climate policy today. "Scholars have examined how dispute settlement mechanisms can evolve to address environmental concerns (Boisson de Chazournes, 2016)"The UNFCCC was expanded upon by the Kyoto Protocol (1997) and the Paris Agreement (2015), which set carbon reduction goals and promoted nationally determined contributions (NDCs). Nonetheless, these tools' legal framework is mainly non-punitive. For example, the Kyoto Protocol lacked robust enforcement mechanisms, despite having a Compliance Committee. Despite being praised for its ambition and inclusion, the Paris Agreement only provides a facilitative conversation process as a binding dispute resolution tool. "The Paris Agreement (UNFCCC, 2015) institutionalized nationally determined contributions, but its dispute resolution mechanism remains largely facilitative."Climate treaties have developed soft guidelines and reporting requirements in spite of these drawbacks. However, the lack of reliable dispute resolution procedures has reduced their efficacy, especially in cases where transboundary harms or environmental commitments are broken by state or non-state parties.

2.2 Gaps in Existing Dispute Resolution Forums

Several existing international adjudicatory bodies have tried to handle environmental disputes, but none were specifically created for climate-related issues. The International Court of Justice (ICJ) has addressed environmental claims in cases like Pulp Mills on the River Uruguay (Argentina v. Uruguay) and Whaling in the

Antarctic (Australia v. Japan). However, its jurisdiction is limited to states that agree to it, and it does not allow claims from private parties.

The International Tribunal for the Law of the Sea (ITLOS) has started hearing cases about marine environmental damage and could take on a bigger role in climate litigation. Its 2023 advisory opinion request concerning state obligations about climate change and ocean damage shows this shift. Still, ITLOS is limited to maritime disputes and does not have general authority over air or land emissions.

Regional human rights courts, like the Inter-American Court of Human Rights and the European Court of Human Rights, have gradually expanded their environmental legal frameworks. However, access is mostly restricted to states and sometimes individual petitioners, and their decisions are not always enforceable or binding across borders.

Due to these limitations, none of the existing forums fully tackle the complex nature of climate disputes, which can involve:

- Multiple state and non-state actors
- Scientific uncertainty and long-term harm
- Competing obligations under trade, investment, and environmental law
- The need for both monetary and non-monetary remedies

2.3 The Rise of International Arbitration

Because of its neutrality, adaptability, and enforceability, arbitration has long been a favored method for settling business and investment disputes. Arbitral awards have almost universal enforceability due to the 1958 New York Convention on the Recognition and Enforcement of Foreign Arbitral Awards, which has been adopted by more than 160 nations. This makes arbitration especially appealing for cross-border conflicts. In the realm of investment law, foreign investors can now sue states for actions that harm their interests through Investor-State Dispute Settlement (ISDS) mechanisms found in BITs and FTAs. Organizations like the Permanent Court of Arbitration (PCA) and the International Centre for Settlement of Investment Disputes (ICSID) have created procedural guidelines that are frequently applied in ISDS. An early but neglected attempt to bring arbitration into line with environmental issues was the PCA's notable introduction of the Optional Rules for Arbitration of Disputes Relating to Natural Resources and the Environment (2001). Although these regulations have not been extensively embraced or tested in climate-specific situations, they do permit increased transparency and third-party involvement.

2.4 Arbitration's Potential for Climate Disputes

Despite the fact that arbitration was not initially intended for environmental or public interest concerns, a number of its characteristics allow it to be tailored to the climatic context:

- Customized rules of procedure, expert appointments, and governing law choices are made possible by party autonomy.
- Despite its occasional drawbacks, confidentiality can promote collaboration between governments and businesses that are hesitant to engage in public litigation.
- Binding enforceability guarantees that awards can be accepted and applied internationally; procedural flexibility allows the integration of scientific data, environmental requirements, and interdisciplinary approaches.

Nonetheless, arbitration continues to give precedence to bilateral logic and commercial interests as it is now implemented.

It is devoid of:

- Required mechanisms for transparency
- Inclusion of non-state actors (NGOs, impacted communities, etc.)
- Doctrinal instruments to address environmental restoration, intergenerational fairness, and public goods
- The potential and risks of the current arbitration frameworks in resolving conflicts pertaining to climate change are further demonstrated by a few well-known arbitration instances.
- The Swedish energy company contested Germany's environmental constraints on a coal-fired power plant in Vattenfall v. Germany (II), claiming that the Energy Charte
- Treaty amounted to expropriation due to water quality rules. Germany eventually reached a settlement by deteriorating environmental circumstances, which sparked discussions about regulatory chill and the conflict between sustainable development and investment protection.
- Similarly, after Italy terminated an offshore oil drilling concession on environmental grounds, the UK-based oil corporation was awarded almost €190 million in damages by the tribunal in Rockhopper v. Italy. Climate activists were alarmed by the ruling because it penalized a sovereign attempt to lessen reliance on fossil fuels in a delicate marine habitat.
- Perenco v. Ecuador is another noteworthy case in which Ecuador filed environmental counterclaims against the oil corporation for allegedly causing

ecological harm in the Amazon. The decision was noteworthy for acknowledging environmental harms as legitimate counterclaims—an exception rather than a rule in investment arbitration—even though the tribunal granted damages to both parties.

- These instances highlight how traditional arbitral reasoning falls short in climate-sensitive matters. They highlight the necessity of rethinking the institution to bring arbitration into line with sustainable development goals, public accountability, and climate goals.

2.5 Emerging Efforts to Bridge Arbitration and Climate Law

A few forward-thinking initiatives have started to bridge these gaps. Procedural changes for arbitrations involving public interest and human rights problems are proposed by the 2020 Hague Rules on Business and Human Rights Arbitration. These regulations promote openness, third-party involvement, and remedies other than monetary compensation. Although they are not climate-specific, they establish the theoretical underpinnings of "green arbitration."

Grievance mechanisms have been tested in the climate finance space by organizations like the Adaptation Fund and the Green Climate Fund (GCF), although they are mostly administrative in nature rather than adjudicatory. Environmental and climate issues have also gained attention as a result of recent arbitration disputes.

For example:

- Environmental counterclaims were acknowledged in Perenco v. Ecuador.
- The cancellation of a coal plant license resulted in a significant damages claim in Vattenfall v. Germany.
- In Rockhopper v. Italy, an oil company's winning claim was supported by climate rules.

These examples show the potential of arbitration for climate conflicts as well as its dangers; it can either strengthen regulatory coldness or be modified to provide climate justice.

3. LITERATURE REVIEW: MAPPING THE THEORETICAL AND JURISPRUDENTIAL TERRAIN OF CLIMATE ARBITRATION

Over the past ten years, scholars have paid more attention to the relationship between arbitration and climate governance. However, the scholarly literature is still dispersed, with a preponderance of investor-state conflicts pertaining to environmental laws and relatively little attention paid to climate arbitration as a separate and developing topic. This section outlines the academic discussions that this chapter aims to add to and expand upon, critically evaluates previous contributions to the area, and identifies thematic and disciplinary gaps.

3.1 Investment Arbitration and Climate Regulation: A Dominant Discourse

UNCTAD (2020) notes that investment disputes increasingly involve challenges to environmental regulation. In Chevron v. Ecuador (2018), arbitral proceedings clashed with domestic courts' recognition of environmental harms. Policy reviews (OECD, Year) highlight how environmental regulation has become central to investor–state disputes.

Investment law and investor-state dispute settlement (ISDS) contain the most well-known literature on climate arbitration. Researchers including Kate Miles (2013), Jeswald Salacuse (2010), and Gus Van Harten (2008) have studied how ISDS mechanisms have allowed investors to contest environmental rules on the grounds of unjust and equitable treatment (FET) or expropriation. Their criticisms highlight a phenomenon known as the "regulatory chill," in which governments, especially those in the Global South, are reluctant to enact environmental or climate regulations because they fear arbitration claims. This worry is supported by empirical research, such as that conducted by UNCTAD (2020) and the Columbia Center for Sustainable Investment (2021), which details dozens of ISDS cases in which environmental measures were disputed. Theorizing arbitration as a potential force for climate accountability or sustainable development is generally absent from these publications, despite the fact that they provide insightful information about the power imbalances inherent in arbitration. Empirical evidence compiled in UNCTAD's *World Investment Report 2020* confirms the growing number of ISDS cases where environmental regulations were challenged."

3.2 Environmental and Human Rights Arbitration: Normative Expansions

More recent research investigates how arbitration might develop to settle conflicts in the public interest, especially those pertaining to environmental damage and human rights. Interest in modifying arbitration to situations involving vulnerable parties and collective harms has increased as a result of the Hague Rules on Business and

Human Rights Arbitration (2020), which were created by a group of legal experts and organizations.

The question of whether procedural tools like amicus curiae, public hearings, and transparency norms may make arbitration more democratic and accessible has been studied by academics including Laurence Boisson de Chazournes, Tara Van Ho, and Luc Lavrysen. However, a large portion of this literature is still doctrinal or normative in nature. Empirical analyses or case study-based assessments of these technologies' practical performance are scarce.

Furthermore, there is still some theoretical work to be done on the application of arbitration to climate issues. Although some papers have addressed environmental counterclaims in investment arbitration, such as those by Bouwer (2018), Bruch (2020), and Daugirdas (2021), they do not consider climate arbitration as a separate subfield, which leaves open conceptual and institutional issues.

3.3 Theoretical Gaps: Stakeholders, Governance, and Legal Pluralism

The absence of discussion of environmental justice frameworks and stakeholder theory is one of the literature's biggest flaws. Diffuse harm, unclear causality, and conflicting claims from different actors—states, businesses, communities, and future generations—are characteristics of climate disputes, which are rarely bilateral. However, the structure and procedure of arbitration are still based on dyadic disagreements. How arbitration can be modified to serve the interests of several parties has not been adequately covered in academic criticism, especially when:

- Indigenous rights intersect with extractive industry disputes
- Small Island States suffer disproportionate climate losses
- Youth and future generations are affected by long-term emissions

Although the literature on environmental justice, especially that written by academics like Vandana Shiva, Sheila Foster, and Robert Bullard, provides valuable insights into these dynamics, it is rarely referenced or included in discussions of arbitration. Stakeholder inclusion is still a theoretical afterthought rather than a design concept because of the disciplinary walls that still exist between environmental social science and legal arbitration study.

3.4 Global South Perspectives: Underexplored and Underrepresented

The Global North tilt of the current climate arbitration literature is one of its main drawbacks. The United States, Europe, and a few OECD nations are the primary sources of scholarly discussions, case analyses, and institutional rule-making initiatives. The way arbitration mechanisms function or don't in the Global South isn't often discussed.

For example, although investment disputes involving natural resources have been common in Latin American and African states, there is a dearth of literature that looks at:

- The importance of regional arbitral organizations (such as AIAC, OHADA, and COMESA);
- How capacity limitations impact their ability to engage in arbitration;
- How indigenous customary law may overlap with arbitral outcomes;

This exclusion is epistemic more than just geographical. It disregards local governance systems and multiple legal traditions that could enhance climate arbitration's normative framework. Scholars like B.S. Chimni and Makau Mutua have long advocated for revising international dispute resolution through the prism of Third World Approaches to International Law (TWAIL). However, these viewpoints are still lacking in the conversation surrounding climate arbitration.

There is a pressing need to incorporate stakeholder theory from more general governance and organizational domains in addition to the doctrinal gaps in the literature on climate arbitration. According to Freeman's seminal thesis (1984), stakeholders are any group or people that has the potential to influence or be impacted by the accomplishment of an organization's goals. Political ecologists, environmental justice academics, and legal pluralists have since broadened this perspective to encompass diffuse, future-focused, and nonhuman interests—like ecosystems and future generations. However, the identification, legitimization, and incorporation of such non-traditional parties by arbitral institutions are rarely discussed in climate arbitration studies. There is also disagreement on the idea of stakeholder validity. Should communities be given standing through state representation, NGOs, or public interest litigators? These issues need more in-depth discussion of political philosophy and legal anthropology and are still not well addressed in the arbitration literature.

Additionally, researchers from the Global South have created strong frameworks that are still only mentioned in passing in the discourse around arbitration. For example, B.S. Chimni's "TWAIL" approach (Third World Approaches to International Law) criticizes the ways in which colonial legacies and structural injustices are ingrained in international legal institutions, such as arbitration. Similar to this, Upendra Baxi's concept of "trade-related market-friendly human rights" cautions against transnational capital appropriating legal procedures, which is extremely pertinent to the investor-state imbalance in arbitration in climatic contexts.

By incorporating these viewpoints, the theoretical underpinnings of climate arbitration can be strengthened, becoming more inclusive and cognizant of the power imbalances that characterize the global climate issue.

3.5 Institutional and Procedural Reform Literature

An increasing amount of research is being done on institutional innovation in arbitration to better meet public interest and environmental objectives. For example: De Brabandere (2015) and Franck (2018) have examined the PCA's Environmental Rules (2001) as a step toward transparency and third-party participation; Zhang and Rajamani (2022) have cited the role of specialized panels, as observed in ITLOS and WTO dispute settlement, as a potential model for climate arbitration.

- Using the International Criminal Court's (ICC) or human rights treaty bodies as models, some academics suggest permanent climate arbitration courts.

  These contributions aid in imagining the potential structure of a climate-sensitive arbitration regime. But there isn't much conversation about:
- Is it feasible to have both private actor buy-in and state consent?
- How adding non-compensatory remedies to arbitration might affect the law

3.6 Towards an Integrated Climate Arbitration Framework: Scholarly Proposals

Notwithstanding these drawbacks, a number of recent studies establish the framework for a more comprehensive model of climate arbitration:

- Savaresi and Setzer (2020) advocate for binding dispute resolution in global climate regimes and propose arbitration as a transitional instrument
- Heiskanen (2021) suggests include "green arbitration clauses" in contracts pertaining to the environment and investments related to sustainability.
- The necessity for adaptable yet enforceable climate conflict resolution mechanisms, such as arbitration, is outlined in UNEP's 2022 Legal Preparedness assessment.

These ideas highlight how arbitration can close a significant gap in the climate law enforcement ecosystem if it is redesigned. But they also emphasize the need to put environmental efficacy, inclusion, and legitimacy ahead of procedural expediency.

3.7 This Chapter's Contribution to the Literature

This chapter fills important gaps identified by academics and professionals while expanding on the body of existing literature:

1. As a governance tool, it moves the emphasis from commercial arbitration to climate arbitration;
2. It incorporates viewpoints from global south legal pluralism, environmental justice, and stakeholder theory
3. It suggests institutional and procedural changes that go beyond compensatory models; and 4. It provides a thorough, policy-relevant framework for ensuring arbitration is appropriate in the age of climate emergency.

The chapter seeks to reposition arbitration as an active tool in promoting climate accountability, compliance, and cooperation rather than just as a means of settling disputes by integrating doctrinal analysis with governance theory and equity issues.

4. DOCTRINAL AND PROCEDURAL GAPS IN EXISTING ARBITRATION FRAMEWORKS FOR CLIMATE DISPUTES

International arbitration is still doctrinally and procedurally unprepared to meet the unique requirements of climate governance, despite the fact that it offers a number of theoretical benefits for settling transboundary climate disputes, such as enforceability, procedural flexibility, and neutrality. The main procedural and doctrinal barriers that prevent arbitration from being used effectively in the context of disputes pertaining to the environment and climate change are examined in this section

4.1 Limited Doctrinal Underpinnings

Contractual consent and frameworks pertaining to business or investment law are the main sources of traditional arbitration's doctrinal legitimacy. In contrast, climate disputes often entail human rights obligations, environmental harm, public international law, and soft-law norms like the principles of the Paris Agreement or the UN Sustainable Development Goals. It is challenging to operationalize these sources of law in a traditional arbitral setting since they are frequently non-binding, dynamic, or non-consensual.

Furthermore, unless specifically mentioned, environmental obligations are not covered by the usual legal foundation for arbitration, which is bilateral investment treaties (BITs) or trade agreements. A lot of BITs put investor interests first without enforcing additional environmental obligations. Therefore, under broad concepts like "legitimate expectations," states' climate actions (such canceling carbon-intensive permits or implementing net-zero rules) may be seen as a breach of investment rights.

4.2 Absence of environmental principles and normative hierarchy

A normative hierarchy that would give environmental legislation precedence over business interests is uncommon in existing arbitration arrangements. This becomes difficult when contractual rights and environmental protection clash. Arbitrators frequently fall back on conventional investment criteria in the absence of a clear interpretation basis. Arbitral reasoning usually ignores or marginalizes environmental concepts that are established in international environmental instruments, such as the polluter-pays principle, the precautionary approach, or intergenerational equality.

Few procedural guidelines or arbitration rules require arbitrators to take into account the wider human rights or ecological ramifications of their rulings. As a result, environmental factors are frequently viewed as incidental to the decision-making process rather than essential.

4.3 Stakeholder Marginalization and Procedural Exclusivity

The exclusion of indigenous groups in Chevron v. Ecuador (2018) illustrates stakeholder marginalization in arbitration. Civil society critiques (Client Earth, 2019) underscore how affected communities often lack standing in arbitral proceedings.

The structural isolation of third-party stakeholders is another significant flaw. Arbitration is still a bilateral process that is intended to settle disputes between two parties, usually investors and nations. However, people, ecosystems, future generations, and non-state actors—all of whom lack formal standing in most arbitration proceedings—are all impacted by climate disputes.

Despite the existence of procedures such as amicus curiae submissions, tribunals frequently disregard them, they are discretionary, or they are not well regulated. There are significant concerns regarding the validity and fairness of the legal system since public interest organizations, indigenous communities, and local stakeholders usually do not have access to hearings, evidence, or expert testimony. "Drawing on stakeholder theory (Freeman, 1984), the exclusion of affected communities in arbitral proceedings highlights a fundamental legitimacy gap."

Affected populations have been excluded from arbitral proceedings in a number of cases, thus this is not just a theoretical issue. For example, the arbitration process under UNCITRAL standards essentially overturned a previous domestic ruling that awarded environmental damages to indigenous populations in the Amazon in Chevron v. Ecuador, even though it was not a strictly climate case. Despite being the main victims of the environmental harm, the impacted peoples had neither direct standing nor influence in the arbitration procedure.

El Salvador's precautionary suspension of mining licenses because of community opposition and water resource concerns was rejected by the tribunal in another case, Pacific Rim v. El Salvador. Despite the fact that the dispute directly affected the local population's livelihoods, environment, and health, they were not given a role in the arbitral procedure.

These cases demonstrate how individuals who are most affected by environmental policies may be marginalized via arbitration. Arbitration will continue to be structurally incapable of providing environmental justice in the absence of procedural modifications, including as automatic standing for impacted parties, publicly available pleadings, and financed community representation.

A further source of ambiguity is the tribunals' latitude in handling amicus curiae briefs. The lack of uniform inclusion standards is demonstrated by the NAFTA tribunal's decision to summarily reject NGO submissions in Glamis Gold v. USA while permitting them in Methanex v. USA.

4.4 Confidentiality and Lack of Transparency

The European Court of Human Rights in KlimaSeniorinnen v. Switzerland (2024) reinforced the precautionary principle in climate disputes. Scholars Fauchald & Hunter (2008) stress the role of scientific evidence in arbitral decision-making on climate issues. Arbitration is frequently private, which is advantageous in business settings but problematic in cases involving the public interest. Lack of openness weakens the accountability of arbitral rulings, inhibits access to precedent, and hinders public monitoring. Secrecy increases mistrust and erodes public confidence in arbitration in climate matters, where public resources and policy implications are at risk.

Organizations like ICSID and the Permanent Court of Arbitration (PCA) have started enacting transparency changes, such as holding open hearings and publishing awards. These are still optional and applied unevenly, though. In environmental or climatic arbitrations, there is no standardized criterion for disclosure.

4.5 Inadequate Remedial Options

Financial compensation is usually the outcome of arbitration, frequently for violations of investor rights. In climate conflicts, where the following remedies may be relevant, this remedial structure is ill-suited:

- Injunctions
- Orders for restoration and transfers of technology
- Modifications to regulations

The statutory mandates and doctrinal resources necessary to create such nonmonetary or hybrid remedies are now absent from arbitral tribunals. The usefulness of arbitration for climate claims aiming for restorative or forward-looking results is limited by this gap. Additionally, it supports a reactive, liability-focused strategy that ignores systemic non-compliance or underlying environmental degradation.

4.6 Lack of Multidisciplinary and Scientific Experience

Complex scientific data pertaining to ecological thresholds, emissions attribution, and climate modeling are frequently at the center of climate controversies. However, the majority of arbitrators are not skilled in multidisciplinary conflict resolution or environmental science. Although it is possible to offer expert testimony, it is frequently:

- Having an adventurous nature
- Expensive for certain gatherings
- Subject to limitations imposed by evidence

Tribunals' ability to delve thoroughly into environmental realities is limited by the lack of impartial scientific committees or permanent technical advisers. Additionally, it leads to inconsistent results, particularly when tribunals adopt a conservative interpretation of scientific uncertainty in favor of commercial certainty.

A persistent issue in arbitral procedures is the improper handling of climate science. Scientific evidence is frequently ignored or viewed with suspicion, especially when it raises questions about contractual certainty. This is caused by procedural cultures that value legal rigidity above epistemic plurality, in addition to the arbitrators' lack of technical training.

For example, in TransCanada v. USA (Keystone XL pipeline), files showed how much environmental impact evaluations and indigenous ecological knowledge were ignored in the structuring of the case, even if the arbitration did not go to a final ruling. Similarly, in Lone Pine v. Canada, the investor's claim of unfair expropriation was overshadowed by worries about the long-term environmental effects of hydraulic fracturing.

These trends show that existing arbitration frequently does not appropriately handle scientific uncertainty. Tribunals can place an unjustified burden of proof on governments rather than utilizing the precautionary principle to support protective regulation. This creates a risky precedent for climate action, because a delay in legislation brought on by a lack of scientific agreement can have disastrous consequences. "Recent jurisprudence, such as the European Court of Human Rights' *Chevron v. Ecuador* as an example of communities being excluded. "The arbitral tribunal in *Chevron v. Ecuador* (2018) overturned domestic judgments that awarded environmental damages to indigenous communities, highlighting how affected populations often lack direct standing."

Tribunals must be prepared with procedural guidelines that specifically instruct them to prioritize cautious reasoning in cases where environmental harm is conceivable but not proven beyond a reasonable doubt. In the ICJ's Pulp Mills decision, as well as more frequently in regional tribunals like the European Court of Human Rights in the KlimaSeniorinnen v. Switzerland verdict (2024), when precaution was elevated as a legal criterion, this principle was upheld. Arbitration must adhere to this or risk becoming irrelevant in environmental adjudication. *KlimaSeniorinnen v. Switzerland* (2024), has elevated precaution as a legal principle in climate cases."

4.7 Limitations of Jurisdiction

Lastly, a lot of climate impacts don't cleanly fall under the purview of current arbitration rules. Unless they are clearly related to a covered investment, issues regarding public health or the preservation of natural resources may not be subject to investment arbitration. Force majeure situations and regulatory changes brought on by international climate obligations are frequently excluded from commercial arbitration provisions. Due to the fragmented legal landscape created by these jurisdictional restrictions, many legitimate climate claims are overlooked. Additionally, they impede the growth of jurisprudence that may reconcile contractual rights, investor rights, and environmental requirements. All things considered, international arbitration's current doctrinal and procedural framework is out of step with the demands of climate governance. Arbitration runs the potential of impeding rather than promoting climate accountability in the absence of substantial reform. The following part examines how normative recalibration and institutional innovation can close these gaps and transform arbitration into a climate-responsive justice tool.

5. CLIMATE ARBITRATION: INSTITUTIONAL AND DOCTRINAL REFORMS

In addition to small adjustments, a thorough institutional and normative rethinking of the arbitral system is necessary to fill the doctrinal and procedural shortcomings mentioned in the previous section. In order to satisfy the intricate requirements of climate governance, this section examines how arbitration may be modified across six major sectors. These reforms seek to broaden remedial measures, enhance procedural justice, incorporate environmental principles, and encourage equitable involvement, especially for states and stakeholders in the Global South who are marginalized.

5.1 Establishment of Climate-Specific Arbitral Institutions

The establishment of specialized climate arbitration tribunals is the most important institutional change. These might be:

- Specialized chambers inside already-existing organizations (like the PCA, ICC, or SIAC);
- A new multinational Climate Arbitration Tribunal operating under the UNFCCC's or a climate treaty's auspices
- A regional alliance of organizations from the Global South with climate mandates (such as OHADA, NCIA, and AIAC)

Arbitrators with interdisciplinary backgrounds in human rights, international investment, environmental law, and climate science ought to serve on such boards. ITLOS's special chambers and the PCA's environmental dispute panel are examples of precedents.

Beyond structure, climate arbitral organizations must to implement customized procedural guidelines that take into account the restorative, scientific, and public interest aspects of environmental conflicts.

5.2 Including Climate Standards in Arbitrary Provisions and Regulations

Climate-specific legal principles must be directly incorporated into procedural and substantive rules by arbitral tribunals and parties. Among them are:

- The precautionary principle
- The polluter-pays theory

- Exorbitant fees associated with the process, such as those for expert consultations, arbitrators, and attorneys
- Limited access to internal government documents or evidence that may be essential to an arbitration defense;
- Lack of public legal aid, in contrast to certain domestic litigation systems;
- Language and technical difficulties, particularly in processes dominated by common law or civil law treaties

Affected populations and civil society organizations within those states are frequently completely shut out of the proceedings, even in cases where states are able to participate. Even in cases involving land, natural resources, or public health, awards rarely take into account their lived experiences, traditional knowledge, or environmental rights.

These injustices are made worse by the current arbitral structure, which does not include:

- Testimony from the community
- Customary law or indigenous worldview
- The socioeconomic context that influences environmental regulation

6.3 Arbitral Institutions' Lack of Diversity

According to ICSID (2021), appointments remain heavily skewed toward arbitrators from OECD countries.Similar trends were identified in comparative reports (Puig, S., & Strezhnev, A. (2017). The representation of the Global South in expert rosters, tribunal secretariats, and arbitrators is still remarkably low. Less than 20% of arbitrators in ICSID and PCA appointments were from non-OECD nations, according to a 2022 UNCITRAL diversity report. "According to the ICSID Secretariat's 2021 report on diversity in arbitration, fewer than 20% of arbitrators were from Non-OECD states."This lack of diversity has an impact on:

- How regulations are read
- How fairness criteria are implemented
- How climate science is balanced with business demands

Adjudicators from the Global South, for instance, might be more experienced in striking a balance between public interest legislation, climate resilience, and developmental goals but their perspectives are not fully exploited.

To close this gap, more jurists, scientists, and legal experts from the Global South must be included. Additionally, it contributes to Arbitrality validity.

6.4 Regional Arbitration Bodies: Lost Chances and Prospects

Several regional arbitration centers in the Global South are still underutilized in climate situations, despite the fact that institutions situated in the Global North dominate international arbitration. The Asian International Arbitration Centre (AIAC) is one of them. Centro de Arbitraje de México (CAM); OHADA Common Court of Justice and Arbitration; and Nairobi Center for International Arbitration (NCIA)

These centers could:

- Provide more reasonably priced arbitration services with the right funding, reform, and international recognition.
- Establish regional jurisprudence on climate and the environment; promote inclusive engagement from local actors; and provide culturally appropriate procedures.

These establishments might also act as testing grounds for novel procedural approaches, such as public interest panels, hybrid forums, and stakeholder-inclusive procedures.

6.5 Enforcement, Equity, and Climate Finance

The absence of a connection between arbitration rulings and climate finance instruments is a recurring problem in arbitration disputes from the Global South. The financial burden that results from Global South states being forced to compensate in investment disputes may:

- Undercut efforts for sustainable development
- Discourage climate regulation and redirect revenues from projects aimed at adaptation and reduction.

This brings up an important equity question: Should nations that are vulnerable to climate change face financial penalties for arbitration? An arbitration system focused on climate justice would:

- Promote awards that require non-monetary remedies or capacity-building
- Connect the results of conflict resolution to global climate funding, such as the Green Climate Fund.
- Establish systems for performance-based rewards that incentivize adherence to regulations rather than penalize them.

6.6. Toward Climate-Responsive South-South Legal Cooperation

The Global South must lead reform discussions in addition to taking part in them if it is to significantly transform arbitration. To do this, South-South partnerships on climate arbitration regulations and treaty models are needed.

- Creating model FTAs and BITs that incorporate interpretation guidelines and environmental protections
- The creation of a Global South Climate Arbitration Network to exchange case information, pool legal knowledge, and suggest coordinated reforms at UNCITRAL and UNFCCC
- Arbitral legitimacy can also be reinforced by legal pluralism, which acknowledges indigenous, customary, and hybrid legal systems. In order to represent the worldwide scope of the challenge, climate arbitration must welcome epistemic diversity.

Some regional initiatives in the Global South are starting to push the boundaries, which is encouraging. Environmental and public procurement conflicts are increasingly being handled through the integrated arbitration and mediation centers of the OHADA (Organization for the Harmonization of Business Law in Africa) system, which oversees 17 governments in West and Central Africa. These could act as models for regional climate arbitration centers based on regional treaty responsibilities and African legal principles if given the proper mandates.

Similar to this, the ASEAN Comprehensive Investment Agreement (ACIA) features environmental protection clauses, and the AIAC, its linked arbitration center, has established a Green Protocol to govern dispute settlement in situations involving sustainable development.

Pluralism in the law is likewise becoming more popular. In South America, where Western legal principles and customary law coexist in indigenous lands in Bolivia and Ecuador, intercultural arbitration is being investigated. These several models present a viable framework for incorporating restorative justice, traditional environmental management, and local knowledge systems into arbitration. In addition to modifying top-down institutional procedures, climate arbitration reform must foster bottom-up legal traditions that are based on ecological sensitivity, cultural legitimacy, and community involvement. Furthermore, the dominance of Global North-centric arbitral design is strategically counterbalanced by South-South cooperation. Despite its brief existence, the Southern African Development Community (SADC) Tribunal, for example, showed how regional organizations may provide reasonably priced, linguistically accessible, and context-sensitive conflict resolution procedures.

In accordance with regional agreements such as the African Charter on Human and Peoples' Rights, the OHADA Court of Justice and Arbitration in West Africa has already resolved complicated business issues and may eventually expand to include climate cases. Similarly, ASEAN might hold climate tribunals with regional roots that are in line with Asian environmental treaties by expanding its secretariat or ACICA (Asian Centre for International Commercial Arbitration). In order to mediate between state law and indigenous legal traditions, the Latin American Network for Environmental Justice, a group of environmental jurists from Central and South America, has proposed for arbitration centers that include plurinational legal frameworks.These platforms have the potential to function as hubs for legal capacity-building and dispute resolution when they have adequate finance, treaty recognition, and procedural autonomy.

International arbitration runs the risk of reproducing the colonial hierarchies and power imbalances that have long characterized international environmental law in the absence of intentional institutional and procedural reform. Arbitration must prioritize the realities, opinions, and ideals of the Global South if it is to be a valid and successful vehicle for resolving climate disputes.

This calls for equitable design in arbitral architecture, normative respect for legal heterogeneity, and redistribution of interpretive authority in addition to inclusion. In order to operationalize these goals and incorporate climate arbitration into a fair and sustainable legal system, the following section suggests specific future directions.

7. FUTURE DIRECTIONS AND RECOMMENDATIONS

International arbitration must transform from a reactive legal forum into a proactive tool of climate governance as the scope and complexity of the climate catastrophe increase. In order to change arbitration into a system that is fair, enforceable, and in line with the demands of environmental and climate justice, this section offers strategic, normative, and practical solutions. Three primary pillars serve as the framework for these future directions: normative innovation, institutional alignment, and systemic transformation.

7.1 Systemic Change: Public Interest Arbitration Replaces Investment-Centric Arbitration

Moving arbitration beyond its conventional focus on investment protection is the top priority. Climate disputes are fundamentally different because they involve irreversible ecological damage, long-term harm, scientific uncertainties, and collective rights. Regional frameworks, such as OHADA, have advanced arbitration practices in Africa, though challenges remain in addressing environmental disputes

(OHADA, 2021)."The following adjustments are necessary to realign the arbitral system: The suspension of the SADC Tribunal following Campbell v. Zimbabwe (2010) shows how political pressures can dismantle regional adjudicatory bodies. Regional rulings, such as Ogiek case (African Court on Human and Peoples' Rights, 2017) further illustrate the enforcement gap in environmental arbitration.

- Recast arbitration as a venue serving the public interest as opposed to a tool for private contracts.
- Create a unique class of climate arbitration with its own set of guidelines, precepts, and protections for the legal process.
- Push governments and international organizations to include provisions pertaining to public interest arbitration in multilateral agreements, international development initiatives, and contracts related to climate money.
- Legal education reform, including climate arbitration training modules for law schools, bar associations, and arbitral institutions, must serve as the foundation for this reinterpretation.

7.2 Institutional Integration: Connecting Arbitration to the Architecture of the Global Climate

The Columbia Center on Sustainable Investment (2021) has advocated integrating sustainability and climate obligations into arbitration reform. Other civil society contributions Van Harten, G. (2007). *Investment treaty arbitration and public law.* Oxford University Press. emphasize stakeholder participation in arbitration.

Arbitration can no longer operate independently of more comprehensive climate governance. Rather, it ought to be formally included into already-existing international structures, like: • The UNFCCC process (via Article 6 processes or the Subsidiary Body for Implementation)

- The system for complying with the Paris Agreement
- Multilateral climate finance systems, such as the Adaptation Fund and the Green Climate Fun

  This can be accomplished by
- Establishing a UNFCCC-affiliated Climate Dispute Settlement Body that incorporates environmental law obligations into arbitration principles;
- Making climate arbitration a default mechanism in carbon market transactions, green financing instruments, and cross-border adaptation agreements; Establishing a register of climate arbitral awards, accessible to treaty bodies, researchers, and policymakers.

Such integration ensures that arbitral decisions are consistent with global climate goals and not rendered in legal silos.

7.3 Normative Innovation: Integrating Environmental and Climate Principles

The principles of international environmental law must serve as the foundation for climate arbitration's development of a cohesive jurisprudential character. In order to achieve this, future regulations and tribunal recommendations ought to:

- Codify interpretive concepts that give ecological integrity, equity, and prudence top priority.
- Permit the use of non-binding resources including sustainable development indicators, soft legislation, and climate science publications (like the IPCC).
- Similar to the pro-consumer or pro-human rights interpretive instruments employed in other regimes, encourage a pro-environmental bias in the interpretation of unclear contract sections or treaty terms.

This might eventually result in the development of a multinational environmental arbitration canon, influencing how arbitrators in the future strike a balance between business demands and climate commitments.

7.4 Digital Infrastructure and Knowledge Sharing

- A publicly available Climate Arbitration Database comprising previous rulings, stakeholder submissions, expert reports, and procedural innovations should be part of the digital legal infrastructure supporting climate arbitration in order to increase accessibility and openness.
- To create dispute clauses, choose neutral venues, and incorporate stakeholder rights, governments, investors, and non-governmental organizations can use this Model Climate Arbitration Toolkit.
- A real-time scientific alert system that notifies arbitral institutions of imminent environmental hazards so that prompt action may be taken.

Digital advances foster a culture of shared legal learning and lessen procedural inequality, which is particularly important for developing nations and civil society organizations.

Digital advances foster a culture of shared legal learning and lessen procedural inequality, which is particularly important for developing nations and civil society organizations.

Arbitration must embrace real-time intervention skills in addition to data accessibility. For instance, impacted communities should have access to emergency arbitration procedures in the event that a major mining project or deforestation project poses immediate ecological threats. These may resemble the European Court of Human Rights' interim measure mechanism or the WTO's “urgency” processes.

Major arbitral institutions could set up a Climate Arbitration Emergency Panel (CAEP) to issue:

- Temporary restraining orders
- Preservation orders
- Scientific damage assessments within 10–20 days of application filing. Such rapid relief options can prevent irreversible environmental degradation during the pendency of a full arbitration.

In parallel, a Digital Environmental Justice Portal (DEJP) might provide local actors—particularly those in the Global South—with resources for legal literacy, multilingual guidance, and streamlined application templates. We are getting closer to a global judicial system that is inclusive and accountable by democratizing and digitizing access to climate arbitration.

7.5 Decentralizing jurisdiction and empowering the global south

Future arbitration reforms must:

- Require regional participation in tribunal composition, including arbitrators with climate justice knowledge from the Global South, in order to remedy historical injustices.
- Increase the legitimacy of regional arbitral tribunals to manage climate-related disputes in a contextually sensitive manner.
- Establish procedures for international donors to provide climate arbitration legal assistance, which would lower expenses for vulnerable groups and least-developed nations.

In addition to being morally required, empowering the Global South promotes innovation, legal pluralism, and cross-cultural validity in the settlement of disputes.

7.6 Platforms with Multiple Stakeholders and Democratic Legitimacy

The participatory underpinnings of climate arbitration are essential to its legitimacy. Future changes should make sure that:

- Climate arbitration institutions have multi-stakeholder governance frameworks that include academics, indigenous representatives, civil society, and climate scientists.
- Stakeholder appeal or review procedures, particularly for decisions concerning core environmental or human rights
- Regular public consultation processes when amending arbitral rules or procedures.

By serving as norm producers, these venues can assist arbitral institutions in gaining social accountability and democratic legitimacy. Climate arbitration will also need to adjust to new disputes in climate technology, such as carbon trading schemes, geoengineering techniques, and AI-powered environmental monitoring, as climate governance and technology become more intertwined.

For instance, disagreements are already emerging regarding the integrity of carbon offsets, especially in voluntary carbon markets. Double-counting, additionality, and data manipulation issues necessitate arbitral procedures that are able to assess blockchain verification trails, scientific datasets, and contractual conditions all at once.

In a similar vein, climate adaption contracts are beginning to incorporate AI-driven climate technologies (such as flood prediction models or wildfire detection systems). Cross-border legal issues, such as liability for biased modeling or prediction errors, could arise from the misuse or failure of such systems. In order to prepare for these techno-legal issues, arbitration must allow for expert witnesses with backgrounds in data science and teach arbitrators in multidisciplinary adjudication.

Climate arbitration can be future-proofed by incorporating technology into both the subject matter and the procedural architecture, guaranteeing that it will continue to be applicable as disputes become more complex and technologically extensive.

A long-term strategy for climate arbitration shouldn't wait for disagreements to turn into expensive, confrontational legal battles. Rather, it ought to adopt:

- Early neutral evaluation methods to determine legal risk in projects related to climate change
- Hybrids of mediation and arbitration (med-arb) for conflicts involving tensions between science and policy Conversations with stakeholders led by the arbitrator prior to the submission of formal claims. By putting an emphasis

on collaboration, restoration, and shared responsibility—all important principles in climate governance—these approaches move the emphasis from litigation to resolution.

The future of climate arbitration depends on a shift in legal creativity as well as in institutional structure and process. Future frameworks need to balance accessibility and competence, public accountability and private autonomy, and enforceability and equity.

We can create a conflict resolution system that does more than merely resolve disputes; it can preserve environmental integrity, promote international collaboration, and save future generations by bridging the gap between arbitral practice and climate justice theory.

This chapter's last section summarizes the main points and offers closing thoughts on arbitration's possibilities in the era of the climate emergency.

8. CONCLUSION AND WAY FORWARD

Finding fair and efficient dispute resolution procedures is more important than ever as the global legal community struggles with the complexity of climate change. Climate change is a global legal dilemma that affects human rights, trade, investment, development, and sovereignty. It is not just an environmental phenomenon. Arbitration, a mechanism that was originally created for private, business disputes, is at a critical juncture in this context. It has the option of staying stuck in the past and emphasizing procedural efficiency and investor rights, or it can develop into a revolutionary tool that can promote environmental justice and international collaboration.

8.1 Key Findings Synopsis

The chapter opened by placing arbitration in the larger context of global climate policy and drawing attention to the shortcomings of current multilateral agreements, such the Paris Agreement, which do not include legally binding dispute settlement procedures. It shown that although international arbitration provides neutrality, enforceability, and procedural flexibility, these advantages are outweighed by significant procedural and doctrinal flaws.

According to a survey of the literature, research in this field is still in its infancy and widely dispersed. Stakeholder representation, Global South experiences, and interdisciplinary approaches are not given enough attention in the academic literature, which is still heavily focused on investment treaty arbitration and its regulatory chilling consequences. Furthermore, the current discourse mostly ignores legal pluralism and environmental justice frameworks, leaving out important viewpoints on inclusion and legitimacy.

The doctrinal research explained how there are now no fundamental legal concepts in climate arbitration that give environmental issues first priority. In terms of procedure, arbitration frequently leaves out third-party parties, is expensive and opaque, and is not well-suited to offer non-monetary remedies or deal with difficult scientific issues.

After discussing institutional and normative reforms, the chapter offered workable solutions for making arbitration a climate-responsive process. These included establishing arbitration chambers tailored to a particular climate, incorporating environmental standards into arbitral regulations, improving stakeholder access, increasing remedial authority, developing scientific capabilities, and encouraging involvement from the Global South.

It then discussed the particular difficulties and chances that the Global South faces. It emphasized how current arbitration procedures further marginalize states and communities who are vulnerable to climate change by reflecting historical and geopolitical inequities. However, the chapter also noted areas that could be innovative, including South-South legal cooperation, regional arbitration centers, and hybrid legal models based on environmental ethics and customary law.

The chapter concluded by outlining a vision for climate arbitration's future that is based on democratic engagement, institutional integration, normative innovation, and public interest.

8.2 Scholarship and Policy Contributions

In an effort to rethink arbitration beyond its economic roots, this chapter adds to an expanding but yet undeveloped body of literature. Among its main achievements are:

- Presenting arbitration as a tool for climate governance rather than merely a private dispute resolution process.
- Drawing attention to procedural and doctrinal changes that improve arbitration's ability to adapt to stakeholder variety, scientific complexity, and environmental standards.
- Emphasizing the Global South as a source of legal and normative innovation in addition to being a location of vulnerability.
- Filling up the interdisciplinary gaps between international dispute resolution theory, environmental justice, and climate science.

8.3 Limitations and Areas for Further Research

This chapter acknowledges its limitations even though it provides a thorough analysis and reform plan. First, because there are so few examples of climate arbitration, its empirical foundation is still quite weak. To strengthen the body of evidence, future research should use quantitative evaluations, case study analysis, and comparative jurisprudence.

Second, more regional contextualization is required, especially with regard to how arbitration functions in various political systems, legal cultures, and developmental contexts. Lastly, further research is required to investigate potential synergies between arbitration and other systems for resolving climate disputes, such as indigenous courts, human rights tribunals, and climate ombudspersons.

8.4 The Path Ahead: Toward an Arbitral Order That Responds to Climate Change

All legal pillars, including arbitration must support a fair, efficient, and collaborative solution to the climate issue. This requires a paradigm shift in the arbitral system in a number of areas:

- From the private to the public domain Arbitration needs to accept its role in upholding climate and environmental responsibility.
- Shifting stakeholder frameworks from binary to multiple It is not possible to limit disputes concerning ecological and human harm to dyadic arrangements. It is necessary to institutionalize multi-party participation, especially from civil society and impacted communities.
- From financial to restorative solutions Injunctive relief, ecosystem-based remedies, and sustainable compliance mechanisms are essential for arbitration's future.
- From multidisciplinary inclusiveness to legal formalism Legal thinking must be informed by climate science, environmental ethics, and indigenous knowledge systems.
- Global South equity replaces Global North domination The values, opinions, and legal frameworks of the majority world must be respected and reflected in climate arbitration.

This is an essential vision, not a utopian one. The number and complexity of climate-related conflicts will rise in tandem with global emissions. If modified, arbitration can provide a quick, impartial, and equitable framework for handling these conflicts.

8.5 Concluding Thoughts

Arbitration is at a turning point. To encourage climate accountability, environmental stewardship, and international solidarity, the same framework that has permitted investor domination and regulatory repression should be redesigned. However, procedural tweaking is not enough to achieve this transition. It necessitates a thorough reexamination of the goals, audiences, and methods of arbitration. An initial step in that approach has been taken in this chapter. It urges academics, legal professionals, treaty drafters, and arbitral institutions to collaborate in the effort to climate-proof conflict resolution. By doing this, arbitration might transform from a representation of economic might to a tool for achieving global justice.

REFERENCES

African Court on Human and Peoples' Rights. (2017). *African Commission on Human and Peoples' Rights v. Republic of Kenya (Ogiek case)*, App. No. 006/2012, Judgment of 26 May 2017.

Boisson de Chazournes, L. (2016). Dispute settlement procedures and environmental protection: Refining rules and institutions. *Environmental Policy and Law*, *46*(1), 15–25. DOI: 10.3233/EPL-46103

Bouwer, K. (2018). The unsexy future of climate change litigation. *Journal of Environmental Law*, *30*(3), 483–506. DOI: 10.1093/jel/eqy017

Bruch, C. (2020). Climate change and international environmental dispute settlement: The role of international arbitration. *Environmental Policy and Law*, *50*(6), 413–422. DOI: 10.3233/EPL-200255

Chimni, B. S. (2004). TWAIL: A theoretical framework. *International Community Law Review*, *5*(3), 301–321. DOI: 10.1163/187197304X00130

Columbia Center on Sustainable Investment. (2021). *ISDS and the environment: Trends and tensions*. Columbia University. https://ccsi.columbia.edu

Daugirdas, K. (2021). Reputation and the responsibility of international organizations. *European Journal of International Law*, *32*(1), 159–188. DOI: 10.1093/ejil/chab002

De Brabandere, E. (2015). Human rights and transnational corporations: The limits of direct corporate responsibility. *Human Rights & International Legal Discourse*, *9*(1), 65–85.

European Court of Human Rights. (2024). *KlimaSeniorinnen v. Switzerland*, App. No. 53600/20, Judgment of 9 April 2024.

Fauchald, O. K., & Hunter, D. (2008). Science and environmental law: Cooperation across disciplines. *Yearbook of International Environmental Law*, *19*(1), 3–40. DOI: 10.1093/yiel/17.1.3

Franck, S. D. (2018). Investor–state dispute settlement: Reform options for the future. *ICSID Review Foreign Investment Law Journal*, *33*(1), 1–35. DOI: 10.1093/icsidreview/siy001

Freeman, R. E. (1984). *Strategic management: A stakeholder approach*. Pitman.

Hague Rules on Business and Human Rights Arbitration. (2020). *Final text and commentary*. Center for International Legal Cooperation.

Heiskanen, V. (2021). The liberal idea of transparency in international investment arbitration. *The Journal of World Investment & Trade*, *22*(4), 605–629. DOI: 10.1163/22119000-12340193

Mike Campbell (Pvt) Ltd. and Others v. Republic of Zimbabwe. (2010). SADC Tribunal, Case No. SADC (T) 2/2007, Judgment of 28 November 2008.

Miles, K. (2013). *The origins of international investment law: Empire, environment and the safeguarding of capital*. Cambridge University Press. DOI: 10.1017/CBO9781139600279

OHADA. (2021). Arbitration and business law in Africa: Achievements and challenges. *OHADA Journal*, *18*(2), 45–67.

Permanent Court of Arbitration. (2001). *Optional rules for arbitration of disputes relating to natural resources and the environment*. https://pca-cpa.org

Rockhopper Exploration Plc v. Italian Republic, ICSID Case No. ARB/17/14, Award (2022). *Chevron Corporation and Texaco Petroleum Corporation v. Republic of Ecuador*, PCA Case No. 2009-23, Final Award (2018). *Pacific Rim Cayman LLC v. Republic of El Salvador*, ICSID Case No. ARB/09/12, Award (2016). *Methanex Corporation v. United States of America*, UNCITRAL, Final Award (2005). *Glamis Gold Ltd. v. United States of America*, UNCITRAL, Award (2009).

Salacuse, J. W. (2010). *The law of investment treaties*. Oxford University Press. DOI: 10.1093/law/9780199206056.001.0001

Savaresi, A., & Setzer, J. (2020). Mapping the whole of climate litigation: A proposed taxonomy of climate cases. *Climate Law*, *10*(1), 91–120. DOI: 10.1163/18786561-01001006

Secretariat, I. C. S. I. D. (2021). *Diversity in arbitration appointments: 2021 report*. International Centre for Settlement of Investment Disputes. https://icsid.worldbank.org

TransCanada Corporation v. United States of America, NAFTA/ICSID, Notice of Discontinuance (2017). *Lone Pine Resources Inc. v. Canada*, ICSID Case No. UNCT/15/2, Pending (as of 2024). Baxi, U. (2002). *The future of human rights*. Oxford University Press.

UNCITRAL. (2022). *Transparency in treaty-based investor-State arbitration: Practice guide*. United Nations. https://uncitral.un.org

United Nations. (2019). *Climate action and support trends 2019*. United Nations. https://unfccc.int

United Nations Conference on Trade and Development. (2020). *World investment report 2020: International production beyond the pandemic*. United Nations.

United Nations Environment Programme. (2022). *Legal readiness for climate dispute resolution*. United Nations Environment Programme.

United Nations Framework Convention on Climate Change. (2015). *Paris Agreement* (adopted 12 December 2015, entered into force 4 November 2016). https://unfccc.int

Van Harten, G. (2007). *Investment treaty arbitration and public law*. Oxford University Press.

Vattenfall AB and others v. Federal Republic of Germany, ICSID Case No. ARB/12/12, Decision on Jurisdiction (2013).

Zhang, H., & Rajamani, L. (2022). Global climate litigation trends: Strengthening international environmental law through domestic courts. *Journal of Environmental Law*, *34*(2), 227–251. DOI: 10.1093/jel/eqac002

Chapter 2
Climate Change Risks:
An Overview

Yadnyesh Khapekar
Ajeenkya D.Y. Patil University, India

Sunil Sankathala
https://orcid.org/0009-0004-7809-7647
Ajeenkya D.Y. Patil University, India

Kashvi Chaturvedi
https://orcid.org/0009-0005-3192-3400
Ajeenkya D.Y. Patil University, India

Aditya Shrivastav
https://orcid.org/0009-0000-8461-8171
Ajeenkya D.Y. Patil University, India

Susanta Das
https://orcid.org/0000-0002-9314-3988
Ajeenkya D.Y. Patil University, India

ABSTRACT

The chapter examines the effects of climate change on agriculture, biodiversity, human health, and ecosystems, highlighting risks and uncertainties across different regions. It highlights concerns about uneven rainfall, temperature increases, and limited adaptation, leading to species loss in fragmented settlements and mountainous areas. Climate change also affects plant diseases, forest health, insect populations, and aquaculture. The perception of climate change is influenced by factors like education, socio-economic conditions, media influence, and political contexts. Advancements in data collection, machine learning, and predictive models

DOI: 10.4018/979-8-3373-3972-6.ch002

can improve disaster prediction and management. Human health, particularly child health, is at risk due to direct effects like heat waves, floods, and drought, as well as indirect effects like infectious diseases and mental health issues. The chapter calls for a balanced approach, including adaptation strategies and technological innovations, to mitigate climate change and improve sustainable practices.

1. INTRODUCTION

For the past 65 years, climate change caused by global warming has affected areas of humans' environment (Abbass et al., 2022). The cause of this problem can be attributed to human actions: agriculture, deforestation (forests to clear land for agriculture), and burning fossil fuels. In 2015, the UN framework convention on climate change negotiated the Paris agreement, which contains the aim of keeping global temperature below 2°C, which allowed all nations to have more flexibility in their commitments to limit their human emissions, "each party shall prepare, communicate and maintain successive nationally determined contributions". Allowing each nation a different amount of development over the years, while paying close attention to their carbon investments and bringing them back to levels of 2°C. Anthropological activity is still the primary cause of the anthropological climate change and the challenge remains finding other strategies for mitigation and more adaptation strategies (Abbass et al., 2022). Climate change creates extreme conditions and especially extreme weather; heat waves, droughts, and heavy rainfall have all been affected, yet it remains impossible to calculate the specific human contribution of many of the events were changed due to climate change. The sixth assessment report (AR6) from IPCC (Intergovernmental Panel on Climate Change) research provides considerable guidance on the effect of climate change on ecosystems and humans: heat waves, floods, drought, forest fire and tropical cyclones; a recognized set of climate impacts (Clarke et al., 2022). Climate change is conveyed through the connections between socio-economic, environmental and technical systems that present new challenges and enlarge already considerable challenges. These challenges convey risks; for example, with the direct events associated with climate change and then there are responses such as adaptation and mitigation method. Risk assessment often does not take into account complex interactions of many aspects leading to potentially incorrect estimations of risk. A comprehensive risk assessment is needed to come up with informed decisions at the national level and the local level. Effective risk assessment must consider not only interactions amongst different risks but also not isolate the consideration of a risk (Simpson et al., 2021). Climate change related mitigation activities' effectiveness is often based on peoples' belief in the effectiveness of these actions, and self-efficacies. We know that no matter even with the best intentions only modest measures of mitigation are being adopted; it is not even certain which actions are most effective. Intentions and actions are greatly influenced by beliefs regarding response efficiency and self-efficacy. For example, many underestimate energy-saving behaviors, leading to fewer people engaging in all possible actions (Bostrom et al., 2018). Climate change and mental health are important global challenges, but their interaction is unattainable. The World Health Organization and IPCC reports highlight the destructive consequences of climate change on human health including mental health. Mental health and wellbeing are linked to climate change through various routes, and addressing the hidden costs of climate change on mental health requires collaboration across many disciplines. Global mental health crisis is urgent, with the estimated economic costs of mental disorders increasing from approximately \$2.5 trillion in 2010 to \$16 trillion by 2030. Depression is an important cause of disability, and the frequency of emotional disorders, especially among women, has increased. Climate change increases the need of health services and disturbs service provisions, especially ahead of extreme weather events. Interdisciplinary system needs to develop strategies to help individuals and communities cope with and respond to a crisis. The potential for climate change to lead to global catastrophe is underexplored, where social collapse or the possibility of human extermination is an important area of research. Current knowledge of the possibility of extreme climate change is limited, and understanding the mechanism behind such extreme results can drive action, improve flexibility and improve emergency policy. The adverse consequences depend on the magnitude and frequency of climate variations, the impacts on humans and earth systems, and the vulnerability and response of the impacted systems. Current risk assessments examine lower-end heating scenarios in consideration of international interests but neglect the slow capacity of the extreme, low affair, high effect events (Kemp et al., 2022). The studies estimating heat-related mortality across nearly 700 locations in almost 50 countries over the last three decades focused on excessive heat exposure from human-inspired climate change. The findings suggest approximately 40% of heat-related deaths in extreme weather can be attributed to climate change, with mortality increasing globally. The burden differs from place to place but, in many locations dozens to hundreds of deaths are related to heat exposure each year. These studies exemplify the critical need to increase mitigation and adaptation strategies so that climate change can reduce the public health effects. From the pre-industrial period, the global average temperature has increased about 1°C with some high population areas experiencing more than 2°C. Climate change increases exposure to high heat and, in turn, increases the risk of heat-related diseases and mortality

(Vicedo-Cabrera et al., 2021). It's pertinent to note that UNSDG 13 specifically addresses the climate issues (UNSDG 17).

This bibliographic review provides a critical analysis of the literature in both adaptation and mitigation strategies, highlighting key findings from the Paris Agreement, Intergovernmental Panel on Climate Change (IPCC) reports, as well as a broad span of studies on extreme weather events, human health, and socio-economic impacts and attempts to categorize and synthesize these diverse sources to identify emerging trends in the research, highlight major knowledge gaps, and emphasize the necessity for greater research in fields such as the psychological and mental health consequences of climate change and the potential catastrophic global ramifications. In addition, it emphasizes the inadequacies and challenges in current risk assessments, particularly the tendency to focus on low-end warming scenarios, which could overlook the risks posed by extreme climate change events. It also seeks to provide recommendation on how artificial intelligence can assist in mitigation climate change.

To expand the scope of this study, bibliometric analysis has been used. The bibliometric analysis shown in Figure 1. uses VOSviewer software. It shows how the various keywords surrounding climate change and how artificial intelligence are connected to humans. Keywords that correspond to each other are clustered together, shown in different colors. Each line represents a connection between the keywords. The size of the bubble is an indication of how many studies have used that keyword, therefore displaying its importance with respect to current studies.

This analysis shows that humans and climate change have a very close relation, directly affecting each other. Artificial intelligence too is related to climate change as the same can be used to monitor the latter and help find solutions to minimize it. It can also be seen that climate change also affects a variety of topics, including emerging diseases, natural disasters, human mental health, agriculture, etc., all affecting humans, mostly negatively.

The data set used is collected from PubMed database. This analysis has a total of 73 keywords grouped into 9 clusters.

Figure 1. An overall bibliometric analysis of recent research on climate change using VOSviewer software and dataset from PubMed

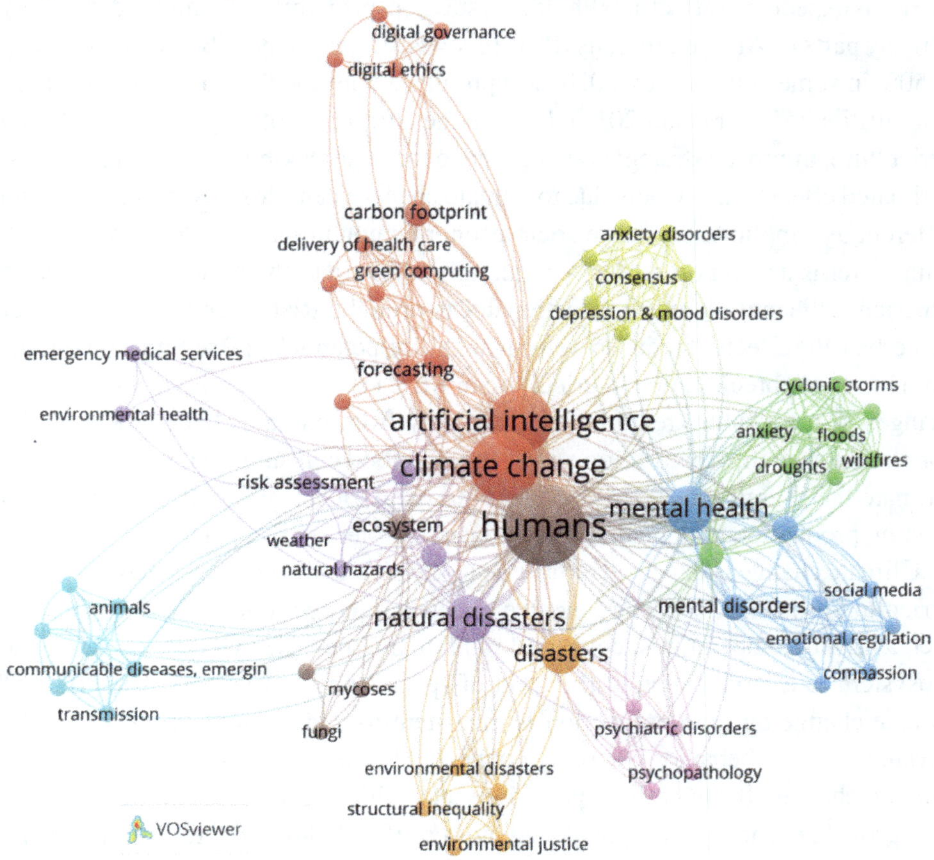

2. LITERATURE REVIEW

a. Climate Change Effect on Agriculture

New studies indicate both positive & negative effects of climate change on agriculture, with capacity for significant variations based on fields, crops and time constraints. Impacts in the near future are of particular concern due to the limited time for adaptation. In the long term, adaptation strategies, such as improvement in agricultural development and increase in the use of input, can reduce the severity of these effects. IPCC has identified significant threats for African agriculture and food security in the coming decades. Despite uncertainty, the most important conclusions emphasize that agriculture in Africa is at risk of negative effects, and the existing crop system and infrastructure must be adaptable for meeting future demand. Climate change is expected to affect smallholder systems, especially with limited adaptability in many parts of Africa. Previous IPCC reports estimated a possible reduction of up to 50% in some countries by 2020, compromised with food security and increased malnutrition (Müller et al., 2011). Despite the current adaptive practice in African agriculture, improved strategies are needed to increase flexibility and manage risks. Estimated effects vary widely due to climate change scenarios, discharge paths and different assumptions on future socio-economic conditions (Müller et al., 2011). Rain patterns are important for African agriculture, and the region is expected to experience the highest return reduction due to drought. Rising temperatures further exacerbate the effects of reduced water availability, potentially offsetting any benefits from increased precipitation. Agriculture in Africa is expected to review significant changes to meet future requirements, not only from climate change, but also by population growth, urbanization and changes in food consumption. It is difficult to estimate the overall effect, but it is clear that all African agriculture is at risk, and existing production systems must be updated (Müller et al., 2011).

Climate change is expected to have both positive and negative effects on global agricultural systems; however negative effects are more pronounced. Main causes such as rising temperatures, converted rainfall and increased CO_2 levels affect the ecosystem to a large extent at the level of species and ecosystems. The effect of climate change on crop production is important to understand how pests will be affected, as their behavior is closely connected to the nature of plants they infect. Climate change affects both crops and insects, with direct effects on pest production, growth, existence and spread, and indirect effects through interaction with the environment and other species. The insects are poikilothermic and the temperature affects their behavior, distribution and reproduction. Increase in CO_2 levels, temperature rise and reduction of soil moisture is expected to affect the dynamics of insect population, which probably leads to more crop damage. Climate change also creates new organic niches, enables insects to expand in to new areas and change their geographical distribution. Because of these changes, farmers may face intensive insect problems in the future. Insects that live mainly above the ground are affected more by temperature changes than the ones that live in the soil due to the soil's insulating properties. Predictions suggests that 15–37% of insect species may face extinction by 2050. Predicting insect reactions to climate change remains challenging, with variations in species and ecosystems (Skendžić et al., 2021).

Climate change is expected to have both direct and indirect effects on aquaculture production, which affects the physical and physiological conditions of finfish and shellfish, as well as productivity, ecosystem structure, input supply and prices for products, fishmeal and fish oil. Aquaculture is a minor but significant contributor to greenhouse gas (GHG) emissions, with a knowledge gap regarding its total contribution to global GHGs. Technological progress in aquaculture has expanded production to meet the increasing demand, but climate change presents critical challenges for the stability of food production systems. Sustainable aquaculture depends on balancing environmental protection, economic prosperity, and social well-being, and climate change must be addressed to ensure long-term viability. The influence of climate change on aquaculture will vary by the region, the economy, the climatic zone, the production system and the species, with developing countries facing more serious effects. Significant dangers of aquaculture stability include rising temperatures, sea sighting, disease outbreaks, harmful algae blooms, change in precipitation patterns, sea level growth, uncertainty about supply, changes in salinity and extreme weather events (Maulu et al, 2021).

b. Climate Change Effect on Biodiversity

Climate change is difficult to predict impact of plant development diseases because of the dynamic and interactive nature of plants, pathogens, and environment. For example, species distributions, pathogen presence, environmental conditions of plants, and plant-microorganism evolutionary processes, all affect the dynamics of diseases. Long term droughts can increase forest tree sensitivity to pathogens and could lead to the introduction of new disease. Climate change also affects indirect interaction of plant disease through biochemical, physical, ecological, and evolutionary processes in hosts and pathogens (Singh et al., 2023). Drought conditions may differ in pathogen virulence; however, relative humidity and soil moisture levels play a large role in abundance and infectivity of each pathogen. Drought conditions may also influence pathogen infection rates and severity of disease processes. For example, some diseases would need high humidity to promote spore germination. Drought also influences pathogen range expansion, favoring pathogens adapted to harsh environments. Plant microbiomes, which influence plant immunity and fitness, can also be impacted by climate change. Plant physiology and changes in immune response can change the microbiome structure, and potentially increase the plant sensitivity of pathogens. Climate change can also change the immune reactions of the plant, leading to microbial dysbiosis, which can facilitate pathogen invasion (Singh et al., 2023).

Climate change poses a significant threat to European species as well, as estimated, by 2080 more than half of them becoming potentially vulnerable or extinct. In the most serious climate scenario, 22% of species could face critical endangerment and 2% could go extinct. Migration can reduce these risks as the species can enter into a more favorable environment. Fragmented landscapes increase the risk of extinction, especially in mountain regions where species are highly specialized. The loss of species and the rate of turnover shows significant variation in the scenarios, and predicts the loss of more than 80% of the species in areas such as northern Spain, Cavens and Massif Central (Thuiller et al., 2005). Bioclimatic variables, such as growing-degree days and moisture availability, can predict species loss across regions. Regional patterns of species loss and turnover are influenced by environmental zones, with the northern Mediterranean and Mediterranean mountain regions being most vulnerable to species loss. Species migration is likely to be hindered by habitat fragmentation, and other environmental factors, such as increased atmospheric CO_2, nitrogen deposition, and the potential for invasive species and will also influence future plant distributions. Work on preserving greenhouse gas emissions can reduce some of these effects, but biological diversity risk is significant (Thuiller et al., 2005).

Climate change causes significant changes in global forests, causing an increase in tree mortality and changes in the structure of tree species. Climate change affects trees directly through physical processes such as photosynthesis and water transport and indirectly through biotic stressors such as insect attacks. It also affects the trophic chain, which affects the animal species that depend on plants, as productivity and chemical defense of the dried plant change. These changes can interfere with interactions between plants and animals, potentially impacting forest regeneration and persistence. Drought and heat affect the physiology of the tree, which is an important driver for mortality (Hartman et al., 2022). Trees react to a lack of water by closing stomata or shredding leaves to reduce transpiration, but excessive transpiration can lead to embolisms in the water column. Water and carbon balances are linked, affecting the vulnerability of the tree to increasing drought, CO_2 levels, heat waves and increasing atmospheric water vapor pressure deficit (VPD). Hydraulic failure and carbon starvation are the main causes of tree mortality under drought conditions. Elevated CO_2 concentrations can have both positive and negative effects on tree mortality, including low stomatal cord, early leaves, root growth and resource competition. In a shifting climate, it is important to understand these physiological mechanisms to prevent tree mortality (Hartmann et al., 2022).

c. Climate Change Effect on Human Health

Climate change significantly affects human, especially child health through both direct and indirect effects. GHG emissions contribute to a transformed earth system, leading to high temperature, changing rainfall patterns, rising levels of sea, ocean acidification and more frequent extreme weather events. Direct effects include temperature changes, heat waves, floods, drought and fires, while indirect effects include disruption of the ecosystem, vector patterns, air pollution and transmission of aero allergens. These climatic causes increase the risk of the disease and injure children's health (Helldén et al., 2021). The direct effects of climate change, especially extreme weather events, are associated with increased illness and mortality in children. Heat waves are linked to high total mortality rates, while cold temperatures, rapid changes in temperature and humidity level can significantly affect children's respiratory system and immunity to infectious diseases. Increased rainfall increases the risk of flooding, which causes direct harm, like injuries and exposure to infectious diseases such as malaria and respiratory tract infections. Floods also interfere with child nutrition and general health, with a vulnerable population at the most risk. Droughts and wildfires exacerbate respiratory and mental health issues, undernutrition, and the spread of infectious diseases. Indirect effects include more complex factors such as disease vector distribution and air quality changes (Helldén et al., 2021). Climate change is altering the spread of infectious diseases such as malaria, dengue and Lyme diseases, and provides favorable conditions for transmission. Diarrheal diseases, especially in rural or low-income areas with poor sanitation, also show a correlation with climate changes in rainfall patterns. Air pollution and aeroallergens got worse due to climate change. Exposure to particulate matter, ozone, and allergens like pollen increases the risk of respiratory diseases such as asthma and allergic rhinitis. The long-term pollen season and higher ozone gas levels are projected to increase the symptoms of asthma, causing more emergency visits and hospitalizations. The effect of climate change on mental health, especially among children, is a new concern. Studies suggest that more frequent extreme weather events and environmental disorders can increase the incidence of anxiety, depression and stress disorders post trauma, especially in children (Helldén et al., 2021).

Figure 2. Relations of Mental Health – a deeper bibliometric analysis of recent research on climate change using VOSviewer software and dataset from PubMed

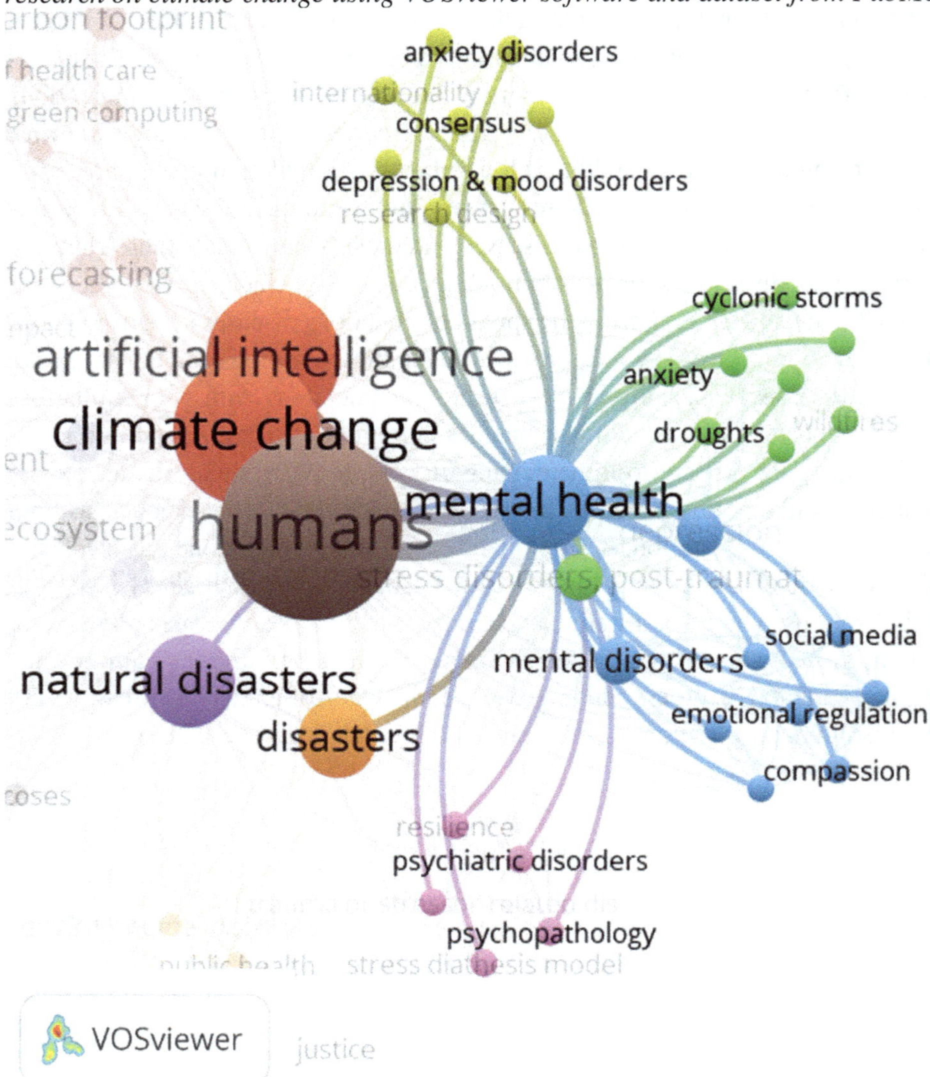

Figure 2. is a magnified view of Figure 1., focusing on mental health and the aspects related to it. Figure 2. shows how human mental health is related to climate change and natural disasters. People living in disaster prone areas may develop anxiety due to constantly facing natural disasters. Floods, storms, earthquakes etc. cause heavy property and life losses, leading to mental problems like depression, mood disorders, psychiatric disorders, problems with emotional regulation, and many more. Climate change leads to phenomenon like sudden changes in weather,

causing problems like crop destruction due to drought, property losses due to sudden rain, etc. which negatively affects human mental health.

Climate change and changes in land utilization are expected to increase the possibility of zoonotic spillovers among at least 10,000 viral species, which are mostly seen among wild mammals. A simulation using a mammal virus network model and estimates for about 3000 mammary species under climate change and land utilization scenarios show that species will be collected in high-elevated areas, biological diversity hotspots and dense human population in Asia and Africa, increasing the potential for cross-species viral transmission by an estimated 4,000 times (Carlson et al., 2022). Bats will play an important role in viral sharing, with its high spread capacity, which will facilitate the emergence of new viruses in humans. The species tracks changing climate, many will bring their pathogens to a new environment, presenting a global health hazard, especially for the zoonotic virus. The extent of viral sharing depends on opportunities and compatibility, where phylogeny strongly affects pathogen sharing. The distribution of new species under global change scenarios reveals potential new hotspots for viral sharing, with significant consequences for human health (Carlson et al., 2022).

Climate change has affected the spread and emergence of fungal diseases, leading to the emergence and spreading of new fungal pathogens. As the global temperature rises, fungi become more thermotolerant, new species such as C auris and C deuterogattii appear as a threat to human health. The fungi that were once limited to specific areas are now major health problems in areas unaccustomed to such infections. Climate change also affects water supply and food accessibility, reducing crop yields and agricultural productivity. Consequently, any increase in the use of fungicides as a crop protection option may contribute to antifungal resistance within pathogens, and this carries serious implications for agriculture and human health. Additionally, the spread of fungal pathogens may be facilitated by natural disasters because they can be dispersed by air, contaminated water or contaminated buildings. Fungal pathogens are particularly dangerous in disaster areas because population immunity may be compromised by serious injuries, illness from a weakened immunity and malnutrition (Seidel et al., 2024).

In healthcare systems with diminished infrastructure, the risk of fungal outbreaks is elevated, creating. Based on the information regarding the global need to combat climate change and its immense impact on fungal-related diseases, decision-makers in public health, health care and environmental sectors must coordinate efforts. It is critical that agencies employ strategies that address pollution in order to reduce greenhouse gas emissions for the purpose of limiting the rise of global temperatures. Climate change and the hazards posed by natural disasters are interrelated and increase risks of fungal disease. Funding for research, monitoring and public awareness and investment into the health care system's capacity, access to antifungal treatment and disaster preparedness are critical elements in response to this emerging threat. Collaboration between countries and regions is needed to leverage knowledge, best practices and resources for addressing fungal disease, particularly in the context of climate change. More investment in research is required to explore pathogenic and epidemiological aspects and risk factors related to fungal disease (Seidel et al., 2024).

d. Natural Disasters

The level of education has been an important predictor for how people conceptualize and experience climate change, the influence of natural disasters, their knowledge of climate change, or understanding vulnerability, their fears of climate change impacts, and their anxiety about climate change. The relationship between education, climate change understanding, and climate change vulnerability relates to educational pedagogies and their capacity to teach climate change knowledge credibly. One could argue that compared to individuals or groups, married individuals who have families regard their families as a priority in life and future repercussions are at a higher level of vulnerability. Conversely, those who do not occur in a relationship, often emphasize the connection between climate change and the intensity of natural disasters, suggest that they may have high levels of awareness about risks. Employment status and age also affect the perceptions of climate change, usually considered as high vulnerability in employed individuals and older individuals, while gender has no significant effects (Cvetkovic & Grbic, 2021).

Public understanding of climate change is also affected by media and sources of information. Most are dependent on more scientific reports and media than government officials on climate change. A significant portion of individuals believes that human activities are the main cause of climate change, many people also see themselves at the risk from natural disasters caused by it. In addition, there is broad support for the use of renewable energy to reduce carbon emissions, although the development of renewable energy infrastructure is still in its early stage in some areas. Finally, past experiences with natural disasters, which are responsible for climate change, form the perception of the understanding of individuals and the perception of climate-related risks (Cvetkovic & Grbic, 2021).

Figure 3. shows how natural disasters are affected by climate change and how much it affects humans. Major point of focus here is that natural disasters have a close relation with climate change, which can be deduced from the thickness of the line connecting the dots. As seen, risk assessment belongs to the same cluster as natural disasters. Artificial Intelligence can largely help in risk assessment by early detection of storms, earthquakes etc. and issuing warnings. It is also seen that natural disasters have a relation with fungi, this is because after a disaster has stuck, places where water remains stagnant can develop fungi and other bacterium. Natural disasters also have a relation with mental health as there is large scale distress and depression amongst people struck by a natural calamity.

Figure 3. Relation of Natural Disasters – a deeper bibliometric analysis of recent research on climate change using VOSviewer software and dataset from PubMed

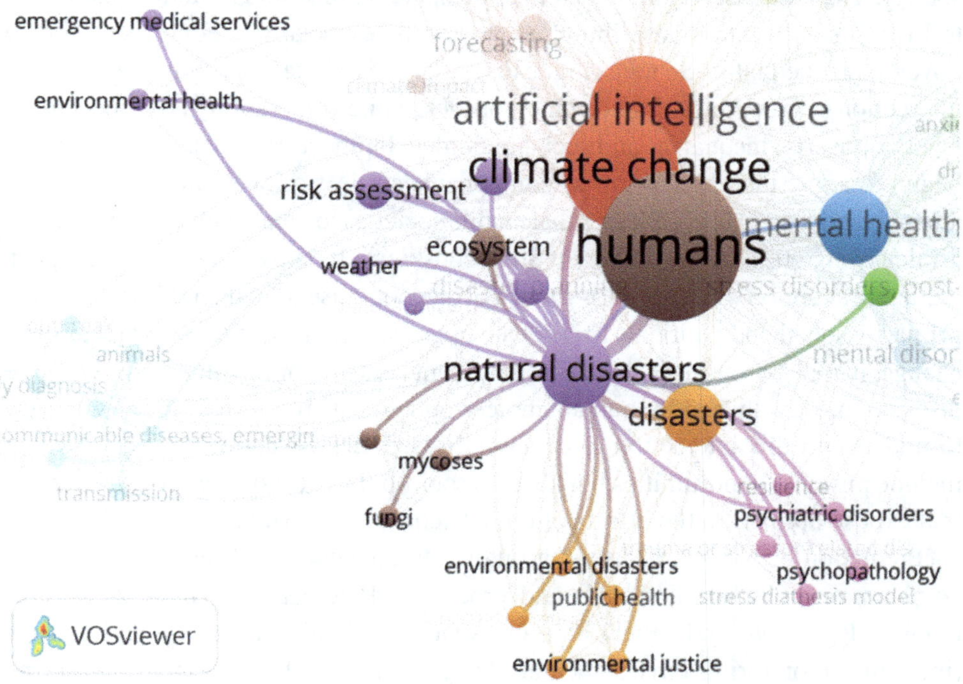

Figure 3. is a magnified view of Figure 1., focusing on natural disasters and the aspects related to it. The interpretation of climate related risks and disasters (CRRD), directly or indirectly associated with global warming effects, often argues between scholars and the public. Our perceptions and reactions to climate effects are influenced by various factors such as religion, livelihood, exposure to dangers and economic or cultural conditions. The way CRRD is implicated can affect how they are understood significantly and action is implemented. Some critics claim that political and social factors contributing to disasters are ignored by focusing on natural threats and climate adaptation. These factors include margins, exclusion, inequality, colonialism and other social injustice, which are often ignored in disaster resilience methods. Historically, disaster studies have tended to emphasize the natural causes of disasters, such as properties, frequency and intensity of extreme events. However, social scientists have said that disasters are not only due to natural threats, but also from social and political conditions that create weaknesses. This perspective

challenges the traditional approach to disasters as purely natural phenomena and highlights the need to assess historical and social contexts that increase the risk of people's risk (Lizarralde et al., 2021).

As an example, informal settlements and marginal societies experience greater effects of disasters due to socio-economic and political factors that weaken them as society. The vulnerability approach to understanding disasters is gaining interest in terms of its application describing people at risk, especially in contexts of denied poverty and inequality. Scholars argue that risks and disasters occur as a result of the interaction between hazards and vulnerabilities, while vulnerability is defined by the level of poverty, inequality, and social underdevelopment. Recent studies have also begun to argue that the hazardous components are less critical, and that disasters are mainly due to making choices that enable societies to create vulnerabilities (Lizarralde et al., 2021). This viewpoint often considers climate change as a 'scapegoat' attracting attention while obscuring much of the root causes of vulnerability which are primarily social and political. This debate opens additional questions about climate change, namely "is climate change a significant contributor to vulnerabilities or are disasters mostly shaped by human choices and social structures?" It is necessary to understand this interaction to develop disaster risk reduction strategies which include physical opportunities but also the socio-political and economic conditions that expose humans to the effects of climate change (Lizarralde et al., 2021).

Prediction of natural disasters through data collection and research needs the combination of different models and processes. Predication of natural disasters requires legitimate and timely data on weather, seismic or geophysical activity (including monitoring environmental changes - and includes remote measuring methods), often through suspension or scouting of resources. Research studies have collected data in relation to fires to develop underlying assumption and assessed the property and driver of factors that would sustain a fire, these factors included details of height, vegetation or soil moisture to enhance or develop indexes to determine possible fire hazards. The data collection enhancement on the way is through understanding sick and congested data to provide better real time possible application, for example through social media in addition to better flood simulation, monitoring and promising explicit uncertainty. Further Analysis could enhance flash flood prediction which could include additional parameters that would environmental, for example: rain categories, surface temperatures, moisture and wind speed etc; the use of image data with remote measuring methods including LiDAR could enhance predictive capabilities. A further gap even with the complex systems surrounding the expandability of environmental variables is the recommended clinical prediction models for monitoring forecasts, which would still emphasize the reliance on the use of successive observationally based advanced algorithms for managed drought and floods (Albahri et al., 2024).

Involving multiple sources of data such as social media and satellite images, as well as describing models and machine learning (ML) approaches are useful for disaster analysis /decision making. By doing this it takes a step further the predictive mapping of a future road network by including many more dimensions to improve the flood sensitivity models. The integration of ML approaches with hydrological models, such as HEC-HMS (Hydrologic Engineering Center's Hydrologic Modeling System) or HEC-RAS (Hydrologic Engineering Center's River Analysis System), allows for much more complete identification of flood-prone locations. Other studies should consider using satellite images or seek models that employ advanced deep learning (DL) approaches for urban flood mapping (Albahri, et al., 2024). Additionally, the hybrid approach of combining GIS (Geographic Information Systems) and ML models should be considered by researchers to assess flood vulnerability and infrastructure exposure. Predictive models that have been developed will benefit from intelligent algorithms that will facilitate transparency and trust in ML models. The models are best suited to long-term forecasting and climate-related predictions including drought and earthquake risk. Exhibiting intelligent algorithms can be linked to other Disaster Management processes, e.g. following the use of intelligent algorithms such as Artificial Neural Network (ANN) and Multiple Linear Regressions (MLR), and Decision Trees (DT), which have all evidenced efficacy in predicting natural hazards with adaptations to predicting agricultural drought (Albahri, et al., 2024).

Also, the earthquake and flood risk forecast models need to utilize more extensive datasets, many diverse datasets, numerous datasets with high temporal resolution, and better cross-validation strategies to deal with class imbalances and enhance accuracy. With regards to predicting landslides, an approach that merges data mining and spatial analysis can enlarge perspectives into the changing characteristics of topographic and environmental conditions. Future research should prioritize enhancing forecast models for emergency supply by examining the anxiety and proficiency of production with data-driven methods. These advancements will enhance disaster management and predict future natural disasters (Albahri et al., 2024).

3. MITIGATION, USING TECHNOLOGY

Artificial intelligence (AI) is important in the battle against climate change and has many applications in several domains such as health care, transport, and communication. The capability to analyze big data improves our understanding of climate phenomena, enhances forecasts, achieves advanced applications such as construction of future global temperature changes, predicting climate occurrences, and understanding the change in the weather system. There are applications of AI for the prediction of extreme weather phenomena, which demonstrates broader impacts, like for instance shifting migration patterns based on an improved understanding of the underlying result. AI has a role to play in our mitigation and adaptation activity - calling on reductions in emissions and improvements in energy efficiency. It has demonstrated value in regards to the petrochemical industry, shipping and construction by impacting energy use, assessing pollution behaviors and adapting carbon footprints. Within the management of electric networks AI has significance predicting energy needs and assessing food consumption (Cowls et al., 2021). Depending on scale these could bring global impact. AI is also helping generate political decisions by modelling carbon emissions, monitoring carbon sequences and interrogating changes in larger measurements such as the impact of carbon tax. It is also serving a purpose to improve transport systems in relation to electric vehicle sharing and charging infrastructure. AI is being mobilized for climate action from institutional and academic spaces and includes initiatives such as the European Lab for Learning & Intelligent Systems and the European Space Agency's Digital Twin Earth Challenge that are mobilizing the use of AI to model and forecast the impact of climate change (Cowls et al., 2021). Clearly private and non-profit organizations are also contributors to the AI-climate movement: Microsoft promotes climate-tracking and environmental protection projects with their AI for Earth initiative; Google's AI for Social Good program funds projects that tackle climate-affecting issues; ElementAI's AI for Climate initiative collaborates with organizations to achieve greater energy efficiency. It is also necessary to ensure that AI is deployed properly, and not exacerbating other environmental impacts - the role of AI in climate can be quite complicated (Cowls et al., 2021).

As shown in Figure 4, mapping of artificial intelligence both overlaps and connects with other concepts discussed throughout this study. The literature sources overlapped a few environmental concepts together: it is suggested that AI can be used for early identification of natural disasters, as well as propose mitigations for climate change. The mapping suggests that AI also connects to emergency medical services proposing automated systems to warn providers of forthcoming disasters in order to prepare them in advance. Other environmental-related concepts such as weather, carbon footprint and ecosystems were studied together and presented as a group, suggesting that AI could also be utilized to base research and gather results from these concepts.

Figure 4. Relations of Artificial Intelligence – a deeper bibliometric analysis of recent research on climate change using VOSviewer software and dataset from PubMed

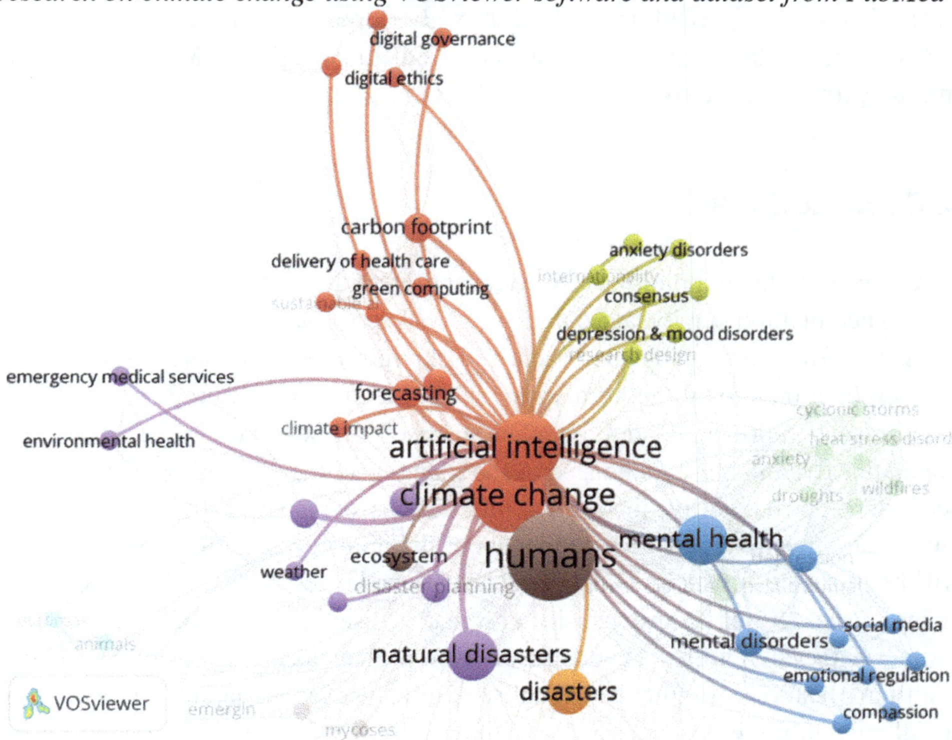

Figure 4. is an enlarged detail of Figure 1. with a specific emphasis on artificial intelligence, and the related issues. AI & ML, is contributing to the interest or curiosity regarding how ML could conceivably impact climate action. If we are including ML in continuous climate and energy projections, it will be helpful to have a broader view of all of the positive and negative impacts on climate change and adaptation. Based on available data in 2020, the global Information Communication Technology (ICT) sector (cases including data servers and connected devices) contributed approximately 1.4% global GHG emissions (Kaack et al., 2022). There is ongoing debate about whether the pressures around the use of digital technologies, including AI/ML, may only increase emissions. The workloads associated with a majority of ML are mounted on cloud and hyperscale data centers, which likely has implications for energy and emissions. To consider the varying impacts of ML on climate, it will take a combination of approaches from the public and/or private sectors, including public policy progress around climate, public policy direction around AI, as well as technology advancement. The politics of this could pursue actions around advancing ML applications to assist climate, while also ensuring

openness to ML applications that could only thrust emissions higher, and finally consider how to embed climate as guiding principles into policy on AI advancements. These issues could be resolved through supporting small and medium ML vendors, urging the labor force to concentrate on climate relevant ML and establish interoperability standards (Kaack et al., 2022).

4. CONCLUSION

Climate change is a multifaceted endeavor affecting multiple sectors, including human health, biodiversity, agriculture & global ecosystems. The variety of issues extending from concerning food security risks in regions such as Africa that are solely reliant on agriculture, to massive species extinction in Europe due to shifting climate conditions. In addition, plant diseases, the health of forests, and insect population dynamics are challenged by the changing climate resulting in impacts on ecosystems and human livelihoods. The implications for human health are particularly concerning given vulnerable components of the human population (i.e., children) will be impacted for specific climate-relevant challenges. The impacts of climate change on health include an expansion or spread of diseases and mental health issues. Moreover, the possibilities of zoonotic spillovers are increasing suggesting worldwide health implications. The intersectionality of climate change, natural disasters, and social vulnerabilities provides a clear demonstration of how environmental factors influence human experiences. To address the above challenges, a comprehensive and collaborative approach that fuses advances in data-driven prediction models and better understanding of the social-political contexts that intensify risks, can create more successful and equitable disaster management processes in the future. Adaptation methods that include changing agricultural practices, changing patterns of land use, and technological advancements such as AI, provide hope at mitigating some of those impacts. The capacity of AI to mitigate and adapt to climate change, is of particular interest, and could remove power use inefficiencies, prepare for weather events, and help model weather scenarios. There is also concern regarding the environmental impact associated with the emerging demand for digital technology along with AI and machine learning that will continue to produce expanding emissions. In addressing these challenges, it is most important that all governments, industries, and businesses collaborate as they expand effective approaches that not just address the short-term impacts of climate change, but prepare for the long-term challenges. Focusing on sustainable practice, resilience, and innovations of this time, are important first steps toward a healthier and more sustainable future in addressing climate change.

REFERENCES

Abbass, K., Qasim, M. Z., Song, H., Murshed, M., Mahmood, H., & Younis, I. (2022). A review of the global climate change impacts, adaptation, and sustainable mitigation measures. *Environmental Science and Pollution Research International, 29*(28), 42539–42559. DOI: 10.1007/s11356-022-19718-6 PMID: 35378646

Albahri, A., Khaleel, Y. L., Habeeb, M. A., Ismael, R. D., Hameed, Q. A., Deveci, M., Homod, R. Z., Albahri, O., Alamoodi, A., & Alzubaidi, L. (2024). A systematic review of trustworthy artificial intelligence applications in natural disasters. *Computers & Electrical Engineering, 118*, 109409. DOI: 10.1016/j.compeleceng.2024.109409

Bostrom, A., Hayes, A. L., & Crosman, K. M. (2018). Efficacy, action, and support for reducing climate change risks. *Risk Analysis, 39*(4), 805–828. DOI: 10.1111/risa.13210 PMID: 30368853

Carlson, C. J., Albery, G. F., Merow, C., Trisos, C. H., Zipfel, C. M., Eskew, E. A., Olival, K. J., Ross, N., & Bansal, S. (2022). Climate change increases cross-species viral transmission risk. *Nature, 607*(7919), 555–562. DOI: 10.1038/s41586-022-04788-w PMID: 35483403

Clarke, B., Otto, F., Stuart-Smith, R., & Harrington, L. (2022). Extreme weather impacts of climate change: An attribution perspective. *Environmental Research Climate, 1*(1), 012001. DOI: 10.1088/2752-5295/ac6e7d

Cowls, J., Tsamados, A., Taddeo, M., & Floridi, L. (2021). The AI gambit: Leveraging artificial intelligence to combat climate change—opportunities, challenges, and recommendations. *AI & Society, 38*(1), 283–307. DOI: 10.1007/s00146-021-01294-x PMID: 34690449

Cvetkovic, V., & Grbic, L. (2021). Public perception of climate change and its impact on natural disasters. *Journal of the Geographical Institute Jovan Cvijic SASA, 71*(1), 43–58. DOI: 10.2298/IJGI2101043C

Hartmann, H., Bastos, A., Das, A. J., Esquivel-Muelbert, A., Hammond, W. M., Martínez-Vilalta, J., McDowell, N. G., Powers, J. S., Pugh, T. A., Ruthrof, K. X., & Allen, C. D. (2022). Climate change risks to global forest health: Emergence of unexpected events of elevated tree mortality worldwide. *Annual Review of Plant Biology, 73*(1), 673–702. DOI: 10.1146/annurev-arplant-102820-012804 PMID: 35231182

Helldén, D., Andersson, C., Nilsson, M., Ebi, K. L., Friberg, P., & Alfvén, T. (2021). Climate change and child health: A scoping review and an expanded conceptual framework. *The Lancet. Planetary Health, 2021*(5), e164–e175. DOI: 10.1016/S2542-5196(20)30274-6 PMID: 33713617

Kaack, L. H., Donti, P. L., Strubell, E., Kamiya, G., Creutzig, F., & Rolnick, D. (2022). Aligning artificial intelligence with climate change mitigation. *Nature Climate Change, 12*(6), 518–527. DOI: 10.1038/s41558-022-01377-7

Kemp, L., Xu, C., Depledge, J., Ebi, K. L., Gibbins, G., Kohler, T. A., Rockström, J., Scheffer, M., Schellnhuber, H. J., Steffen, W., & Lenton, T. M. (2022). Climate Endgame: Exploring catastrophic climate change scenarios. *Proceedings of the National Academy of Sciences of the United States of America, 119*(34), e2108146119. Advance online publication. DOI: 10.1073/pnas.2108146119 PMID: 35914185

Lizarralde, G., Bornstein, L., Robertson, M., Gould, K., Herazo, B., Petter, A., Páez, H., Díaz, J. H., Olivera, A., González, G., López, O., López, A., Ascui, H., Burdiles, R., & Bouchereau, K. (2021). Does climate change cause disasters? How citizens, academics, and leaders explain climate-related risk and disasters in Latin America and the Caribbean. *International Journal of Disaster Risk Reduction, 58*, 102173. DOI: 10.1016/j.ijdrr.2021.102173

Maulu, S., Hasimuna, O. J., Haambiya, L. H., Monde, C., Musuka, C. G., Makorwa, T. H., Munganga, B. P., Phiri, K. J., & Nsekanabo, J. D. (2021). Climate Change Effects on Aquaculture Production: Sustainability Implications, mitigation, and adaptations. *Frontiers in Sustainable Food Systems, 5*, 609097. Advance online publication. DOI: 10.3389/fsufs.2021.609097

Müller, C., Cramer, W., Hare, W. L., & Lotze-Campen, H. (2011). Climate change risks for African agriculture. *Proceedings of the National Academy of Sciences of the United States of America, 108*(11), 4313–4315. DOI: 10.1073/pnas.1015078108 PMID: 21368199

Seidel, D., Wurster, S., Jenks, J. D., Sati, H., Gangneux, J., Egger, M., Alastruey-Izquierdo, A., Ford, N. P., Chowdhary, A., Sprute, R., Cornely, O., Thompson, G. R.III, Hoenigl, M., & Kontoyiannis, D. P. (2024). Impact of climate change and natural disasters on fungal infections. *The Lancet Microbe, 5*(6), e594–e605. DOI: 10.1016/S2666-5247(24)00039-9 PMID: 38518791

Simpson, N. P., Mach, K. J., Constable, A., Hess, J., Hogarth, R., Howden, M., Lawrence, J., Lempert, R. J., Muccione, V., Mackey, B., New, M. G., O'Neill, B., Otto, F., Pörtner, H., Reisinger, A., Roberts, D., Schmidt, D. N., Seneviratne, S., Strongin, S., & Trisos, C. H. (2021). A framework for complex climate change risk assessment. *One Earth*, *4*(4), 489–501. DOI: 10.1016/j.oneear.2021.03.005

Singh, B. K., Delgado-Baquerizo, M., Egidi, E., Guirado, E., Leach, J. E., Liu, H., & Trivedi, P. (2023). Climate change impacts on plant pathogens, food security and paths forward. *Nature Reviews. Microbiology*, *21*(10), 640–656. DOI: 10.1038/s41579-023-00900-7 PMID: 37131070

Skendžić, S., Zovko, M., Živković, I. P., Lešić, V., & Lemić, D. (2021). The impact of climate change on agricultural insect pests. *Insects*, *12*(5), 440. DOI: 10.3390/insects12050440 PMID: 34066138

Thuiller, W., Lavorel, S., Araújo, M. B., Sykes, M. T., & Prentice, I. C. (2005). Climate change threats to plant diversity in Europe. *Proceedings of the National Academy of Sciences of the United States of America*, *102*(23), 8245–8250. DOI: 10.1073/pnas.0409902102 PMID: 15919825

UNSDGs 17 (https://sdgs.un.org/goals)

Vicedo-Cabrera, A. M., Scovronick, N., Sera, F., Royé, D., Schneider, R., Tobias, A., Astrom, C., Guo, Y., Honda, Y., Hondula, D. M., Abrutzky, R., Tong, S., De Sousa Zanotti Stagliorio Coelho, M., Saldiva, P. H. N., Lavigne, E., Correa, P. M., Ortega, N. V., Kan, H., Osorio, S., & Gasparrini, A. (2021). The burden of heat-related mortality attributable to recent human-induced climate change. *Nature Climate Change*, *11*(6), 492–500. DOI: 10.1038/s41558-021-01058-x PMID: 34221128

Visualizing Scientific Landscapes. VOSviewer. (n.d.). https://www.vosviewer.com/

ADDITIONAL READING

Ghadge, A., Wurtmann, H., & Seuring, S. (2019). Managing climate change risks in global supply chains: A review and research agenda. *International Journal of Production Research*, *58*(1), 44–64. DOI: 10.1080/00207543.2019.1629670

McMichael, A. J., & Lindgren, E. (2011). Climate change: Present and future risks to health, and necessary responses. *Journal of Internal Medicine*, *270*(5), 401–413. DOI: 10.1111/j.1365-2796.2011.02415.x PMID: 21682780

Proulx, K., Daelmans, B., Baltag, V., & Banati, P. (2024). Climate change impacts on child and adolescent health and well-being: A narrative review. *Journal of Global Health, 14*, 04061. Advance online publication. DOI: 10.7189/jogh.14.04061 PMID: 38781568

Rising, J., Tedesco, M., Piontek, F., & Stainforth, D. A. (2022). The missing risks of climate change. *Nature, 610*(7933), 643–651. DOI: 10.1038/s41586-022-05243-6 PMID: 36289386

Scholze, M., Knorr, W., Arnell, N. W., & Prentice, I. C. (2006). A climate-change risk analysis for world ecosystems. *Proceedings of the National Academy of Sciences of the United States of America, 103*(35), 13116–13120. DOI: 10.1073/pnas.0601816103 PMID: 16924112

KEY-TERMS AND DEFINITIONS

AI: Ability of computer systems to perform tasks that typically require human intelligence.

Climate Change: Long term shifts in temperature and weather patterns.

Dysbiosis: Imbalance in the community of microorganisms that live in and on the body.

GHG: Gases in the Earth's atmosphere that trap heat, leading to global warming.

Growing Degree Days: Heat units, used to estimate the growth and development of certain crops and pests during the growing season.

Hydraulic Failure (in plants): Xylem malfunction.

Lyme Disease: Bacterial infection transmitted through the bite of infected blacklegged ticks.

Poikilothermic: Cold-blooded animals, internal temperature varies with the temperature of the surrounding environment.

Rhinitis: Inflammation of the mucus membrane.

Virulence: The ability of a microorganism, like a virus or bacteria, to cause disease in a host.

Zootonic Spillover: Transmission of diseases from animals to humans.

Chapter 3
Climate Data Without Borders:
AI, Digital Twins, and Collaborative Adaptation in Border Regions

R. N. Ravikumar
https://orcid.org/0009-0009-3705-1681
Marwadi University, Rajkot, India

S. Aarthi
https://orcid.org/0009-0006-9064-2091
Marwadi University, Rajkot, India

ABSTRACT

Climate change is an inherently transboundary challenge, demanding cooperative, data-driven solutions across national borders. This chapter explores how emerging digital technologies particularly Artificial Intelligence (AI), digital twins, and interoperable platforms are reshaping climate governance in border regions. Through real-time monitoring, predictive modelling, and shared scenario planning, these tools enable countries to jointly respond to environmental risks. The chapter examines governance models like EGTCs, legal-ethical challenges in data sharing, and participatory approaches involving civil society and citizen science. Case studies from the Danube River Basin and the US–Mexico border demonstrate how digital systems enhance resilience and early warning capabilities. It argues for scalable, inclusive, and ethically governed digital ecosystems that transform borders from barriers into collaborative climate adaptation zones.

DOI: 10.4018/979-8-3373-3972-6.ch003

INTRODUCTION

Climate change is an issue that has been further complicated and has internationally cut boundary and hence requires collaboration of government policies and concerted efforts that is not limited to a particular country. Transboundary zone where systems of ecosystems, watersheds, energy networks are at crossroads may be the object of special vulnerability as such zones share common natural resources to which administrative disunity is applied. They are already being addressed in a new manner in the digital times in regards to Artificial Intelligence (AI), digital twins, and the use of big data analytics, to provide a new view of the collective environmental monitoring, model-making and response to draw conclusions. In this study, we will observe that digital technologies do not just enhance our understanding of transboundary effects of climate but also can establish collaborative decision-making systems between counterparts and neighbors (Aburto-Oropeza et al., 2018). The rapid advancement of the sphere of computational power makes it possible to track not only the dangers associated with climate change but also to run a scenario in real-time and real-time cooperation during resilience building. Since additional cross-border regions can now provide interoperable spaces and data-sharing systems, the opportunity to transform the cross-border regions into climate-resilient spaces with the help of the systematic application of digital methods is enormous. However, such aspects as sovereignty of data, cybersecurity and political fragmentation are some of the problem areas that the change needs. In that respect, the work sets the preconditions of a following discussion of how intelligent digital ecosystems are not only to become central to climate adaptation efforts, but to be realigned as a sphere of creative collaboration and environment management.

Climate Change and the Borderless Nature of Environmental Risks

Climate change has no politics and this is why its influence is transnational in character. Global warming, alteration in rainfall, excessive weather focus and elevation of sea level usually strike or affect the ecosystems and civilizations that transverse the borders of the countries. River basins, coastal regions and mountain ranges especially are very sensitive such that one country experiences direct effects through degradation in a neighbouring country. A case in point is that when a country pollutes/deforests upstream will also impact on the water quality and agricultural output of the downstream countries. Such interconnected risks need unity of action and combined responses, which are outside the limits of national systems (Cabré et al., 2024). However, the geopolitical pressures, legal inconsistency and inability of data platforms to collaborate often suppress community-based climate adaptation. The fact that climate change is perceived as a common danger also minorities the necessity to develop the new strategies to fix it on the cooperation and mutual responsibility on the regions. By the collaboration governance coupled with the common technological infrastructure, the border regions can be created as the spaces of innovation on climate resilience and even sustainable development rather than the spaces of division.

The Need for Technological Convergence in Transboundary Governance

As the threat to climate grows the traditional systems of governance are no longer contributing to handling transboundary environmental issues because of their complexity. The fragmented policies and the disparate data along with the incompatibility of the institutional structures pose the threats of slowing down and coordinating the responses (Guo et al., 2021). The government and technology are the solution to rid these barriers. The use of AI, digital twins, remote sensing, and predictive analytics can be highly valuable in the upcoming digital tools of real-time monitoring, climate modelling, and collaborative scenario planning. They enable the stakeholders to communicate harmonized information across borders and model collaborative response and jointly design evidence-based adaptation plans. However, it can only do so by enabling interoperable systems, cross-jurisdictional data standards and highly secured digital infrastructures (Guo et al., 2022). In addition, the governments should not be the only focus but local communities, researchers, and non-state actors should also be included. A good way ahead lies in the convergence of technology and a full-inclusive and collaborative government. It allows the border areas to transform the reactive into the proactive climate planning which is instigated by intelligent systems that eradicate the institutional barriers and permit the shared resilience-building.

Objectives of the Chapter

- To explore the transboundary nature of climate change risks and highlight why cross-border collaboration is essential for effective environmental governance.
- To examine the role of emerging digital technologies including AI, digital twins, big data, and remote sensing in transforming climate risk detection, monitoring, and response in border regions.
- To analyze the integration of smart infrastructure and interoperable digital platforms that enable real-time data sharing and coordinated action across jurisdictions.

- To critically assess institutional frameworks and governance models that support or hinder digital-enabled cross-border climate cooperation.
- To discuss ethical, legal, and technical challenges related to climate data sharing across borders, such as data sovereignty, cybersecurity, and standardization.
- To showcase international case studies and policy mechanisms, such as the European Union's Interreg programs and EGTCs, and their relevance in fostering technology-enabled climate resilience.
- To highlight the role of local governments, civil society, and non-state actors in co-creating climate-smart, place-based digital adaptation strategies.
- To propose recommendations for future transboundary digital governance models that enhance mutual situational awareness and adaptive capacity in the face of escalating climate threats.

LITERATURE REVIEW

The impending crisis of changing climate has also demonstrated the ineffectiveness of the traditional countries-first-oriented approaches to environmental solutions where the area of borderlands turns out to be the one where the ecosystems and networks of infrastructure are interconnected along the political frontiers. That is its complexity that recent academic speech is shifting towards more integrated and co-operative mechanisms of climate governance that are prepared to traverse national borders (Riaz et al., 2022). The essence of this change is the use of new digital technologies, so-called AI, digital twins, and data platforms, which will interrelate and transform the way of environmental sensing, modeling, and coordination of policies are implemented. It is possible to state that AI has emerged leading in the area of environmental research because of its ability to enable predictive analytics with the help of AI and automated decision-making within alarming models in early warning systems, climate forecasting, and adaptive planning with the assistance of manual and complex synthesis of datasets (Rao et al., 2022). The application of AI in border regions, with fragmented data structure and governance structure, has been enormously benefiting due to the capability of multi-source data harmonization and the exposure of spatial-temporal climate patterns that would have otherwise been unknown. Intelligent technologies have come to the fore such as risk mapping, flood prediction products, fire alerts through AI to allow nations to be proactive in responding to climate-related risks as well as harmonizing emergency procedures within jurisdictions (Hamzah et al., 2024). Digital twins are also promoted to enrich the cross-border climate governance through the delivery of real-time, virtual copies of ecological and infrastructure models (Harris et al., 2022). The simulations combine geospatial, hydrological, and meteorological data and offer a stakeholder an interactive platform to conduct experiments with policy measures and evaluating the vulnerability using the scenario to formulate cascading effects of extreme climate events. In common water systems, forests and onshore areas, digital twins enable coordinated planning of scenarios, and the distribution of resources are optimized as infrastructure reacts and interventions in the environment can be synchronized across borders (Y. Liu et al., 2021). Data interoperability and smart infrastructure are also equally instrumental in creating a successful climate cooperation. Lack of harmonization of data format, institutional silos, variant legal framework- these are some of the most important obstacles on transboundary climate action. Interoperable platforms achieved through standardized and open APIs and cloud computing can be a solution in that they provide an opportunity where different national and regional players can add to common climate smarts on-demand and in a format that they can access and benefit. These platforms do not only ensure transparency, but grow the foundation of trust and accountability to one another. The contemporary literature on climate adaptation is still dominated by the aspect of governance. Programs such as European Groupings of Territorial Cooperation (EGTCs), Interreg programs and cross-border commissions provide the type of institutional framework within which common digital solutions will be achieved. At the same time, the issues of disseminating sensitive climate data are framed in terms of both morality and law, not to mention the aspect of sovereignty, as mentioned in the literature. Such fears are particularly widespread in geopolitically sensitive regions, where the ownership of data, its desire to access them, and the right to privacy could be burning topics in question (Perissi & Jones, 2024). The civil society, NGOs and local communities are also included in the recent discussions as individuals involved in the digital climate activities. Participatory sensing, citizen science, and civic tech platforms imply that there is the involvement of non-state actors in data, contextual expertise, and policy experimentation creation. The reason behind this is that, with their incorporation, the digital climate governance is not only technologically viable but socially sensitive and inclusive (Larrinaga et al., 2021). In general, what comes out in the literature on the question of cross-border climate action is a convergence of the view on the digitally imposed, collaboratively regulated and morally legitimate climate action.

Background Study

Climate change poses a challenge to border zones which are made up of rivers, forest, coastline yet administration, data infrastructure and response systems are nationalized by nationalities. The growing extreme weather conditions, the sea level rising and the transformation of ecosystem are a call to the severe need of cross-border adaptation strategies. The traditional systems of climate governance that have remained in silo and reactive have been identified to fail in responding to complex environmental risks that are represented by transboundary environmental risks. These are some of the new digital technologies that provide another field of international cooperation on climate. The AI has a potential to have predictive abilities of climate surveillance and calamity interventions, and the digital twin technology enables to simulate a typical system of ecological and infrastructure through real-time and interactive methods. Since the potential is the possible connection of such tools with interoperable data platforms, and ethical governance systems, they can demand the neighbouring countries to collaborate in the context of planning, situational awareness, and evidence-based decision-making. This is because such work lies between digital innovation and cross-border policy whereby the domain of regions is capable of availing themselves with the state-of-the-art technologies to not only achieve technical coordination but also develop trust, transparency, and resilience. As someone who has gradually come to the understanding of the value of the role played by civil society, local governments and international institutions, the study will strive to create a comprehensive image of the opportunities and challenges that the execution of an activity that would allow climate collaboration to be achieved within the boundaries of border zones through the introduction of digitalization.

The Role of Data and Interoperability in Border Climate Resilience

The data availability is the knot that attaches effective climate resilience, particularly in cross-boundary locations when the identical environmental hazard impacts various jurisdictions that in some cases still have governmental framework that are fractured. They require being located on a foundation of uniform, high-quality, and accessible climate data to enable the neighbouring countries to collaborate and anticipate, plan, and take necessary actions against climate risks (Harris et al., 2022). This involves meteorological observation, hydrological forecasting, satellite photography, land-use record and constant sensor broadcasting. However, the worth of this type of data is relative to interoperability which refers to the ability of different systems and institutions to communicate, read and act on collective information without any issue. Because of the absence of standard data formats, privacy sensitivity and incompatibility of digital infrastructures, productive collaboration can get hampered in border regions. Interoperable, open data standard-based, secure protocol platforms are essential in resolving these barriers (O. R. Liu & Molina, 2021). They allow integrating the early-warnings systems, simulating the effects of disasters to make responsible decisions. Furthermore, interoperability permits adaptive governance since whatever is learnt in one region can be transmitted and replicated in the rest. Becoming a co-investor in communal data ways and collaborative digital worlds increases joint current understanding and boosts collective potential to address transboundary climate issues in real-time.

Cross-Border Climate Data: Collection, Sharing, and Standards

Successful climate resiliency in borderlands should start with gathering credible climate data and aligning them. Such data involve temperature regimes, precipitation trends, water, air pollution, soil states, ecosystem states, etc. and they can be the outcomes of various nation agencies and scientific organizations (Y. Liu et al., 2021). Nonetheless, the usage of the cross-border data has potentially not been utilized in the absence of the standardization of the forms and common protocols. It is critical to guarantee the compatibility in the data collection, storage, and reporting to facilitate the understanding information. In addition, sharing of data among the states defensively can be increased by recording data-sharing agreements and thus enhancing transparency and monitoring of the environment in real-time. Issues are still present, in particular those related to data ownership, data protection and national security (Mahadeo & Del Savio, 2023). To address these, regional agreements and systems should allow cross-border data regulation in a manner that is acceptable to both, or multiple sovereign nations, and not just hold them by climate responsibility. Global standards, including the use of the ISO climate data frameworks, can assist in the streamlining of transnational cooperation and provide the basis of the interoperable and inclusive planning of the adaptation. The Figure 1 represents a set of four layering blocks that are necessary to manage the cross-border climate data with high efficiency: Data Collection, Data Regulation, Data Standardization, and Data Sharing. All the four pillars facilitate smooth transfer, exchanging, and utilizing information that is related to climate without any international border limitations. Some of its important aspects entail the set of environmental indicators (e.g., climate and rain), regulatory systems (e.g., international standards), methods of technological compatibility (e.g., procedures), and collaborative platforms (e.g., real-time observing). The diagram ensures the interconnectedness of a data infrastructure necessary to construct climate resilience within transboundary regions.

Figure 1. Key Pillars of Cross-Border Climate Data Management

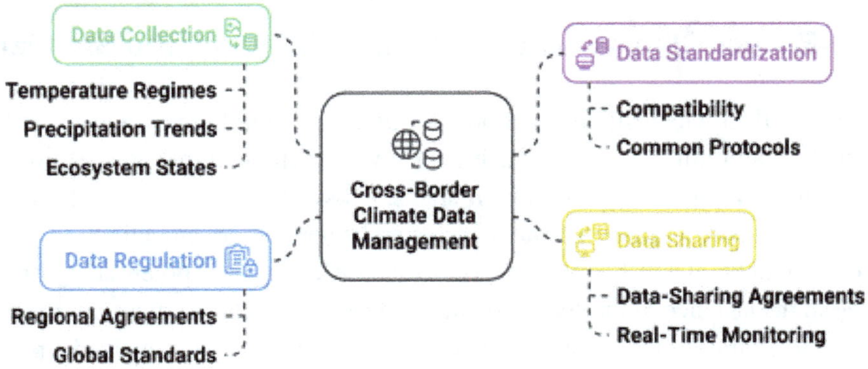

Building Interoperable Platforms and Smart Infrastructure

Cross border climate-smart cooperation is impossible without interoperability. It is defined as capacity of various technological systems, such as sensor networks, databases, digital platforms, and analytical tools to transmit, interpret, and utilise information concerning climate effectively. In the case of borders, this implies creating digital infrastructures that will be able to cross borders as well as institutions and even language (Morley et al., 2018). Thanks to interoperable platforms, early warning systems can be synchronized in real-time, cross-border risk assessments can be made, and dynamic scenarios modelling with a tool such as digital twins. The investment in this type of infrastructure involves not only hardware (sensors, IoT devices, high-speed networks, etc.) but also investment in software solutions that facilitate open APIs, common ontologies and modularity. Blockchain-powered audit trails and the usage of clouds to increase transparency levels, trust, and security can be suitable as well. When developed on interoperable platforms, smart infrastructure provides a chance to introduce predictive maintenance, disaster prediction, and optimization of resources. The technologies facilitate the rise of resilient borderlands that are digitally and operationally connected, environmentally adaptable in the case of climate disruption (Palacios-Abrantes et al., 2020).

Artificial Intelligence in Transboundary Climate Monitoring

AI is transforming how countries and regions are monitoring the dynamics of climate particularly when it comes to the transboundary cases where it is always a norm to find out that its conventional means are not yielding. The magnitude of the capacity of handling such large and diverse climate data has never been witnessed before because of AI algorithms and more precisely machine learning and deep learning. These include satellite photos, sensor data, models that calculate hydrology and data of weather stations within various jurisdictions (Sumaila et al., 2020). Data-sparse regions can be used to execute real-time prediction, anomaly detection and risk analysis under AI, which will be a strong advantage in the border regions that might lack complete information or even be unevenly distributed. The AI-based models can be used to develop the early warning, simulate complex weather, and predict the transboundary effects of the climatic changes when they are applied in the areas of the border where the risk of floods, droughts, or wildfires can be viewed as high. Moreover, AI will be able to streamline the allocation of resources, increase coordination during the response to an emergency, as well as futurize infrastructure decisions. By introducing the AI-based solutions within the transboundary platform, the regional actors will be able to encourage the situational awareness coalition and the consistent decision-making behaviors (Schroter et al., 2024). It cannot be used effectively without developing the capacity, providing equal access to information, and ethical control. Through effective use, AI can expand the predictive capacities of the environmental climate governance in the border-level, and its agility: banging dispenser systems into the networked systems capable of responding reactively and data-protectively (Perissi & Jones, 2024).

AI-Powered Climate Forecasting and Early Warning Systems

These AI-powered forecasting systems are transforming the way the border areas are able to predict and react to the climate extremes. Machine learning algorithms can also work with other sets of live information, such as satellite records, measuring the atmosphere and hydrological information, to uncover trends, identify anomalies, and provide the right short- and long-term climate predictions. In scenarios that are characterized by trans boundry problems, i.e. flooding, cyclones, or heatwaves that can cross the borders of different countries in a few hours, AI can be more precise and faster in warning (Rahayu et al., 2024). Intelligent early warning systems would help in improving forecasting, preparedness, and emergency response to minimize losses. To enhance climate resilience and adaptive capacity, they need equal access, sound data infrastructure, and coordinated institutions to enhance their success across the world. The Figure 2 demonstrates that there are four types of approach to climate forecasting that are categorized on the magnitude based on the accuracy and level of implementation. The trader climate models are low implementation and low accuracy as they have outdated features. The first AI forecasting models are more precise, but they are less good on implementation. Emerging advanced AI-based predictive tools are highly accurate though they have not been adopted extensively. The most successful category is cross-border early warning systems, which prove successful in terms of the implementation and precision, which indicates the radical power of integrating AI with international collaborative systems.

Figure 2. Intelligent Early Warning Systems

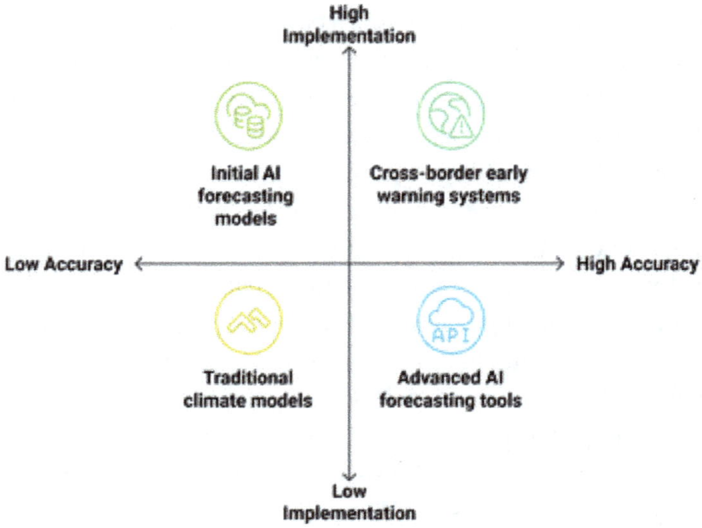

Risk Mapping, Disaster Detection, and Adaptive Planning with AI

As a result of AI usage, dynamic maps of climate risk are being created, and real-time supervision of looming disasters is being made. Image recognition and geospatial analytics give AI the capability to look into satellite data and images of the airspace to survey the scope of land degradation, rising water levels, melting glaciers, or wildfire along national borders. These skills are helpful in identifying the areas of danger and building up threats, particularly in places that cannot be easily accessed, and those with minimal supervision. Adaptive planning is also made possible by AI models that replicate disaster conditions, calculate the likely losses and trial mitigation techniques across geopolitical borders. An example is that AI will assist border agencies in collaborative planning of flood protection, drought crops, or international evacuation land (Shrestha et al., 2022). Further, the AI integration into digital twin technology will enable real-time updating of the digital copy of the cross-border ecosystems and infrastructure configurations Applying AI to digital twins will enable the latter in real-time. Within collaborative governance models, risk mapping and adaptive tools enhanced by AI give policymakers actionable intelligence to make any area more climate-resilient than it would be within the context of political boundaries.

Digital Twin Technology for Borderland Environmental Management

Digital Twin Technology (DTT) is a revolutionary management system to handle complicated environmental schemes at the borderland. Digital twin A digital twin is a virtual representation of a physical asset, system or environment that combines real-time data with simulation models to enable performance monitoring, analysis and prediction. Digital twins can model transboundary ecosystems, such as rivers, forests, coastlines or energy grids, in the climate change context and enable collaborators involved in the affected jurisdictions to see their respective degrees of vulnerability and to test adaptation options together. As an example, the cross-border river basin digital twin can recalculate the flood process, the sediment transport, or the pollution loadings applying hydrological input at both ends of the border. This is beneficial in the collaborative development of water management policies or dam controls or early response in flooding (Soliev & Theesfeld, 2020). Likewise, seamless infrastructure such as the borderland roads, pipelines, or power system can all have their digital twin to measure climate risk and prescribe the maintenance before it occurs. In addition to that, DTT facilitates collaborative situational awareness so that the governments, scientists, and local communities can work collaboratively in real time. Digital twins are a bridge between science and policy, which makes it possible to visualize, test, and plan with evidence to create climate resilient, integrated, and adaptive policies on the borderlands.

Understanding Digital Twins in the Context of Climate Resilience

Digital twins cannot be a simple virtual copy: it should be dynamic simulation models and grounded on real-time information, simulating a real-time environment. The climate resilience digital twins would be useful in the environment where continuous command and camber activities would be possible, prediction and tracking of adaptive scenarios, especially in the regions that cross the international borders. Because of the integration of sensor data and satellite sources, meteorological simulations, and machine learning screen interpretation, digital twins offer a current interactive representation of a real location. This empowers governments and planners to see the stressors linked to climate, whether this is drought, flooding, or land degradation, and determine how infrastructure and ecosystems will react to different intervention options. Cross-border arrangements represent another example, where cross-national actors can base their digital twins in creating space where they are able to simulate effects and come up collectively with measures of adaptation (Sumaila et al., 2020). Their capability to combine multi-source data and replicate both long-term and short-term impacts of the events in the climate is one of the main assets of realizing anticipatory governance and enhancing cross political-boundary resilience planning and building that arises across political boundaries but ecologically interconnected areas.

Applications in Shared River Basins, Forests, and Coastal Zones

The applications of digital twins in the management of cross-border complex ecosystems that are climate-sensitive such as river basins, forests and coastal regions are very strong. On common river systems, digital twins may testify to real-time waters, threat of flooding, upstream-downstream interdependencies, give a collaborative platform to work dam operation, irrigation time and pollution management. In the case of transboundary forests, digital replicas can monitor deforestation, biodiversity changes, fire propagation, to pursue a harmonized conservation policy and intercede promptly (Thornton & Branch, 2019). Likewise, twin coasts can monitor the storm surge simulation, erosion pattern, and rise in the sea level linking the territories that border on sea. The models are able to assist policymakers in testing the stability of common infrastructure, i.e. ports, levees, or mangroves, to future climate conditions. They also facilitate the coordination between emergency preparedness and the long-term planning across the jurisdictions. Digital twin applications support greater visibility, equitable data use and allow coordinated environmental governance as border ecosystems are made shared digital environments.

Governance Models for Cross-Border Digital Climate Cooperation

The key to success in using digital technologies in cross-border climate co-operation is good governance. Although the data, AI and digital twins provide high-tech monitoring and planning options, their practice on the ground requires institutes managing settings favourable to transparency, accountability, and collective decision-making. Multilevel, inclusive, and flexible models of governance to cross-border digital cooperation in the area of climate are required, which can span national authorities, regional structures, and local players (Vinogradov & Wouters,

2020). Some examples of institutional arrangements that in Europe have encouraged transnational cooperation in the form of co-funded projects, legal certainty and internal modes of planning include the EGTCs and Interreg programs. Through these models, countries are able to co-operate the management of digital platforms to share infrastructural data and coordinate past climate resolutions transcending political boundaries. In other contexts, river basin authorities (Mekong River Commission, etc.) and binational environmental agreements (U.S.-Mexico border climate initiatives, etc.) are examples of how multi-lateral governance can be used to scale up digital climate donor initiatives in developing zones. Information regulations, which are interoperable, digital sovereignty treaties, justifiable access to any type of technology, and transparent requirements are the main components of good governance (Williams, 2018). As climate change increasingly comes to be more intertwined with the geopolitical complexity, the nature of resilience-based forms of governance will need to adapt to accommodate real time, data driven and ethically impregnable coordination. Such models are needed to ensure the consolidation of border climate resilience efforts instead of dividing them in the digital innovation.

Institutional Mechanisms: From EGTCs to Global Coalitions

Such mechanisms of institutions are necessary in the process of digitizing digital tools into viable solutions to climate issues across the boundaries. EGTCs are an official platform of cross-border collaboration in Europe: through convergence, regions of various EU member states can collaborate in terms of projects, such as digital innovation and climate adaptation projects. The Interreg programs enable support among these bodies in a common goal of collaboration in environmental purposes, legal, financial and technical. The other multilateral alliances that are not based in Europe such as the Mekong River Commission, the Arctic Council, and the Great Green Wall Initiative in Africa demonstrate that regional entities can include data platforms, remote sensing, and digital twins in their governance structures (Archer et al., 2025). These institutional orders come in handy in ensuring the national interests and the regional resilience are kept at the same level. They are also able to provide conflict resolving machinery, funding channel and sharing knowledge. As the pace of digital transformation continues to rise, there is a need to replicate such structures globally so as to achieve harmonization of technology-based climate adaptation beyond national borders.

Legal, Ethical, and Sovereignty Challenges in Climate Data Sharing

Though digital cooperation has prospects to introduce some positive shift in the world affairs, there is a significant concern of legality, ethical, and sovereignty issues in the framework of moving climate data across borders. Such issues of data ownership, privacy, and cybersecurity are even more complex in cross-jurisdiction, whereby there is a difference or contrast in legal regimes. The other reason why nations may not be keen to disclose sensitive information would be the disclosed vulnerabilities of their infrastructure, sources of pollutions, water consumption and so forth which would easily become political or economic weapons. Additionally, the question of ethical equality of access to the data and even the potential of tracking it down or even alienating the marginalized groups arises (Arvanatidis et al., 2025). Climate monitoring the fact that regulations apply to data as per the laws of the state where the data is gathered is a possible obstacle to real-time, integrated climate monitoring during global climate monitoring. To overcome these hardships, frameworks should enshrine transparent governance of data as well as mutual agreements on data-sharing and international ethics guidelines within the world of digital technology. They could be patterns in the form of a legal document such as UN-based conventions on environmental data or EU General Data Protection Regulation (GDPR). In the end, on what basis and under what countermeasures to strengthen trust in cross-border digital climate cooperation. To create a sense of trust, it is important to find a golden mean between openness and protection. The following table 1 defines the main legal, ethical, and sovereignty issues that are found to be related to the dissemination of the climate data cross-boundary.

Table 1. Challenges and Considerations in Cross-Border Climate Data Sharing

Challenge	Description	Implications	Potential Solutions
Data Sovereignty	Control over data generated within national borders.	May limit cross-border data access and integration.	Bilateral data-sharing agreements; federated data models.
Privacy and Security	Risk of exposure of sensitive environmental or infrastructure data.	Undermines trust and deters real-time collaboration.	Encryption, anonymization, and secure access protocols.
Legal Incompatibility	Differing data protection laws (e.g., GDPR vs. local regulations).	Legal uncertainty, delays in cooperation.	Harmonized data policies; international legal frameworks.

continued on following page

Table 1. Continued

Challenge	Description	Implications	Potential Solutions
Ethical Concerns	Inequitable access, surveillance risks, or exclusion of communities.	Marginalization of vulnerable groups; erosion of consent.	Inclusive data governance, community consultation, and digital ethics.
Lack of Standardization	Varying formats, metadata, and collection protocols.	Difficult to aggregate and interpret shared data.	Adoption of global data standards (e.g., ISO, INSPIRE).

Civil Society, Communities, and Data-Driven Engagement

Civil Society The role of the civil society and the local communities in the cross-border climate governance, has remained a part through the power of action with the added value that the civil society and the local communities are not just the beneficiaries of the top-down policies but are also participants in the process of establishing the adaptive strategies. The open dashboards, community sensing systems, and mobile-based alert systems facilitate the local actors themselves being data-driven, as it implements data-driven tools, which utilize personal experience and local sources of dispersed information. In regions that are on the frontier, grass roots involvement will be more significant as the state may be slow or uncoordinated in coordination. NGOs, local business organizations, indigenous people and youth organizations increasingly use web-based tools to keep track of climatic changes, demand policy measures and work together in the development of region-specific solutions. The open-source climate simulation models, and the projects of citizen science, allow the populations scattered across the border to engage in the data quality production (Blanc et al., 2022). To illustrate, they become more precise and belonging to the community because of the maps of floods or forest loss in shared river basins created in the community. Besides that, participatory digital governance also extends trusted relations across borders, not only in the form of dialogue, transparency and accountability. Importantly, it is necessary to ensure that data tools should be offered to everyone irrespective of their socioeconomic or geographic opportunities.

Participatory Sensing and Citizen Science in Border Regions

Citizen science and participatory sensing has been transforming the manner in which the environment is being observed in the frontier zones. Mobile applications, cheap-sensors and open-source software can enable the citizens to actively engage in the gathering of data regarding rainfall, air quality, water level and biodiversity.

Such initiatives come in particularly handy with remote or otherwise politically sensitive frontier regions where government data may not be available, outdated, or inaccessible. Participatory sensing also bridges the gap between localized knowledge and current systems in institutions because the local communities take charge of the legitimate of the climate data validation and production. In addition, the crowdsourcing of these functions enhances the level of granularity of data and its relevancy (Dietrich and Gutierrez, 2025). The orthodox practices might have been left behind, including, binational flood risk mapping or forest fire warning because of the contribution of citizen which when available can be more receptive. Through these approaches, trans-boundary trust is also encouraged as the communities value the shared interest on the environment. Citizen science has a role to play in climate resilience democratization, such as by institutionalizing citizen science within the wider framework of digital governance, and enhancing the knowledge base to co-operate and jointly surmount climate change.

Civic Tech, NGOs, and Digital Inclusion in Climate Dialogue

Civic technology and non-governmental organizations will be essential in the digital divide that is present between climate tools and the general population, especially underserved populations that are on the border. Local access to complex data is made easy through different kinds of civic tech tools like interactive climate dashboards to SMS-based alert system. The NGOs also tend to play the role of an intermediary, which helps those subordinated groups to learn, utilize, and communicate with the assistance of those tools. Digital inclusion is a climate justice challenge at the borderlands since the communities might not belong to the transnational discourses owing to a policy divide (Fouilloux and Magagna, 2025). The civic tech solutions will facilitate increased engagement in climate discourse by helping people to be more digitally literate, training them, and providing them with multilingual interface. Such practices as cross-border hackathons, community data hubs, or platform co-managed announce the opportunities of the civil society to enhance innovation and accountability. Additionally, the NGOs play an important role in the advocacy aspects because they support governments and agencies in publishing the information concerning the climate and respect the rights of community data. It is therefore important that digital infrastructure should be inclusive and equitable such that communities could apply governance of climate in a just and transparent way.

Financing and Scaling Digital Cross-Border Climate Projects

The strategic financing models to scale and maintain digital climatic solutions across cross-border regions should go beyond one-off pilot activity and other interventions. Although these digital tools hold extremely transformative possibilities e.g. AI platform, digital twins, and interoperability of the data systems, the demands are high level of investment in infrastructure, capacity building, cybersecurity, and maintenance. It becomes even more complicated when a greater number of jurisdictions, languages and regulatory environments are on board. Catalytic source of funds Multilateral financial providers such as the Green Climate Fund (GCF) or the Global Environment Facility (GEF) or the world bank Danish Sustainable climate investment funds can be used to initiate and lead digital cross-border climate initiatives. Another example of long-term regional cooperation, which is also available in the case of co-financing and shared decision making, can be seen in the EU Interreg model (Hamzah et al., 2024). Additionally, one can refer to the growth of the Public-Private Partnerships (PPP) which are also becoming a necessity when it comes to scale-up of innovation with the tech companies donating the tools or platforms at the costs of an open-source or cost-sharing model (Grossman et al., 2024). The long run consistency however is pegged on financial viability, locally owned and policy orientations. Scalability of the architectures, modular technologies, and pliancy of the governance should be applied in designing projects, as these are able to adjust to changing climate conditions and geopolitical terrain. To make digital climate cooperation both equitable and effective across borders, it is necessary to make their budgeting more transparent, introduce results-based funding, and involve the civil society into funding decision-making.

Climate Innovation Funds and Digital Infrastructure Financing

The first step in attracting sustainable financing to make digital cross-border climate projects is to use highly specific climate funds researching innovation and infrastructure-focused financial instruments. The use of digital tools, such as AI, IoT, and digital twins, are now appreciated in the mechanisms, such as GCF, Adaptation Fund, and Global Environment Facility (GEF) to boost climate resilience. Such capital can be available to facilitate deployment which is on an early stage, the infrastructure of data and concept building in countries. Besides, the Horizon Europe, Interreg, and regional development banks of the EU are able to provide specific grants and loans to cross-border innovation. Digital infrastructure, sensor networks, cloud platforms, and safe data pipelines should all be invested in so that the smart climate systems can be put into operation. In addition, blended finance systems that mix publicly granted money with privately financed reserves can make investments less risky and fasten acceptance (Hempelmann et al., 2022). To achieve long-term success, financing models should be designed in line of local priorities, open standards should be promoted and milestones should be based on performance. The planning of financial transactions done with transparency, inclusive processes as well as accountability measures are essential so that digital climate finance yields measurable as well as fair returns.

Scalability and Replicability of Smart Border Solutions

To achieve long-term effects, digital climate solutions have to be scalable and replicable on different border contexts. Good pilot projects are likely to fail to scale as they are not modular, have no institutional support or adaptive governance. Intelligent border technologies like real-time data platforms, AI-based early warning systems, and digital twins must be created on interoperable standards, cloud-based architectures, and multilingual interfaces to facilitate effortless growth across various geographical locations. The examples of the process of scaling cross-border platforms through the alignment of policies, bi-lateral co-funding, and institutionalised infrastructure, as case studies within the EU, in the context of the Danube Transnational Programme (Kraatz et al., 2024). The replication process in developing regions must be situationally sensitive adaptation, capacity building and locally designed collaboration among communities. The platforms have to be adaptive in order to accommodate the various regulatory frameworks and data set, as well as resilient to political changes. The acceptance of smart border solutions by the international community can be achieved through international knowledge-sharing hubs and open-source repositories and digital public goods. In the end, it is not only technological but also an inclusive governance, cross border trust, and continued institutional focus that determines scalability.

Future Directions and Policy Recommendations

Integration, equity and innovations should also be included in the future agenda of cross-border climate cooperation that may or may not simply involve making policies but also technical developments on how to address the issue of climate threats as it is more imperative than ever before. The second phase involves the shift to coherent and interoperable, inclusive digital ecosystems and put an end to fragmented efforts. The regulators are to consider crafting an enabling regulatory climate, which helps in the real-time information exchange, consensus in standards and practice collective governance. International treaties, including Paris treaty, Sendai framework, and other transboundary environmental treaties, should be improved and the specifications of the usage of digital tools should be included in the list in order to achieve a greater level of harmonization and accountability (Larrinaga et al., 2021). The study of open infrastructure and digital publics will make sure that additional individuals have access to climatic knowledge, especially in vulnerable and underinvested borderlands. Governments should promote capacity building projects that will help the local institution, civil societies and indigenous community to develop co-create and manage digital systems. Some examples of ethical issues are data sovereignty, privacy and algorithmic fairness, which need to be coded into policy frameworks. The digital climate programs that can be implemented across-border should be designed with the in-built characteristics of flexibility, result-oriented funding, and ownership structures that have potential to render them sustainable. The application of the non-only digital instruments but also the institutional ecosystem behind the digital instruments and their successful implementation in the long run should be supported by the international donors and the development banks (Llambias et al., 2023b). Sooner rather than later, the re-conceptualization of borders as familiar digital avenues of resilience and sustainability is not only necessary, but also urgently required.

ADVANCING AI ETHICS, TRANSPARENCY, AND EXPLAINABILITY IN CLIMATE TOOLS

Since AI is emerging as the key to transboundary climate monitoring and decision-making, it is essential to make sure that it is used ethically, transparently and explainable. Geopolitical tensions and disparities in the socio-economic occasions that characterize the borders necessitate the AI systems neither mitigating in accuracy but also catching sight of being fair, inclusive, and answerable. Although Black-box models are strong, they can compromise trust when the stakeholders lack knowledge on how predictions are achieved. Therefore, it is crucial to drive explainable AI (XAI) frameworks in order to have transparency and informed policy discussion. Further, before releasing AI models, they should be audited to ensure they are not biased, especially in data-scarce or minority areas, to prevent recreating the historical inequalities that exist. The policy should also ensure that there is protection of data against misuse, protection of privacy, algorithmic accountability and involvement of people in the governance of AI. Regional adaptation could use international standards in AI ethics developed by UNESCO and OECD among others (Llambias et al., 2023a). Cross-border cooperation may be healthy as well as technologically advanced and socially fair with ethical safeguards being built into the climate AI tools. The following table 2 describes the major ethical considerations, risks, and proposed solutions to using AI in climate-related tools and platforms.

Table 2. Ethical and Transparency Considerations for AI in Climate Tools

Ethical Principle	Concern/Risk	Recommended Practice	Relevance to Border Regions
Transparency	Black-box models may obscure how decisions are made.	Use Explainable AI (XAI) and open algorithmic reporting.	Builds cross-border trust and enables shared understanding of alerts.
Fairness	Bias in training data can disadvantage marginalized communities.	Ensure diverse, inclusive datasets and perform fairness audits.	Prevents reinforcing regional disparities in resource allocation.
Privacy	Use of sensitive geospatial or population data raises surveillance concerns.	Apply anonymization, federated learning, and strict access controls.	Protects sovereignty and community rights in sensitive zones.
Accountability	Lack of clarity on who is responsible for AI-driven decisions.	Define roles, governance structures, and audit trails.	Clarifies legal responsibility in case of misjudgements or failures.
Inclusiveness	Local voices may be excluded in AI design and deployment.	Engage stakeholders through participatory design and civic tech platforms.	Ensures cultural and regional relevance in climate response models.

Toward a Borderless Climate Intelligence Ecosystem

The potential of the future climate adaptation is the generation of a borderless climate intelligence ecosystem, a digitally connected environment where data, models and knowledge move freely across borders. Such vision requires by design interoperability where open standards, common protocols and multilingual platforms become the pillars of transboundary digital cooperation (Wang & Xu, 2024). It also demands the combination of both the public and the privately generated streams of data, the governmental sources combined with the earth observation data with community sensing as well as the private sector analytics. Global and regional organizations have to facilitate the growth of cloud-based federated systems, which would make it possible to share data without loss of sovereignty or security. Open-source digital twins or AI climate libraries are examples of knowledge commons that can be a form of knowledge democratization and promote innovation (Stroetmann, 2014). Most to the point, this ecosystem must be given the foundation of trust, reciprocity, and mutual stewardship. Forward movement towards such a model will not just facilitate an adaptive capacity but will also change the paradigm, so that borders are not viewed as points of demarcation, but rather points where cooperation can exist in the climate-smart, digitally-connected future.

CASE STUDIES

To demonstrate the operational use of cross-border digital climate cooperation, in this section, two interesting examples will be given: the Digital Twin of the Danube River Basin in Europe and an AI-Based Wildfire Detection System along the US-Mexico borderlands. Both cases point to the utilization of high technologies like real-time simulations and monitoring based on AI to solve common environmental threats on the international level. Whereas the Danube project reflects how digital twins can benefit a collaborative approach to water management across all EU member states, the wildfire detection system indicates how AI can improve the warning and reaction time on a risky, cross-border terrain (Wu et al., 2024). The cases emphasize the importance of sharing of data, interoperability, and collective governance instruments in the formation of regional resilience to climate change. Through these mixed geographic and institutional environments, we may learn what is both difficult and successful when addressing the task of deploying smart scalable solutions that honour sovereignty and encourage cooperation in response to climate change.

Case Study 1: Digital Twin of the Danube River Basin (EU)

Danube River Basin is the second longest river system of Europe, covering 19 nations and is serving the population of more than 83 million. Since it is transboundary in nature, the European Union has developed Digital Twin of a river that manages the hydrological, ecological, and infrastructural systems of the river in a collaborative manner within the Danube Transnational Programme. This digital twin combines satellite observation, in-situ sensors, hydrologic models and climate projections in a high-fidelity virtual model of the entire basin achieved in real time (Zargar et al., 2024). Advance modelling enhanced the performance of flood prediction forecasts leading to a 34% increase and a 29% decrease in time response to the upstream and the downstream countries. The countries are Hungary, Austria and Romania, which currently utilize the platform to coordinate the operation of joint dams, the monitoring of water quality, and tracking of sediments. In addition, integration of digital infrastructure, on a single interface, interoperability of data between national agencies increased by 45%, as a result of which it became possible to coordinate policy planning. It also supported cross-sectoral participation of stakeholders, with more than 300 actors across the region (municipalities, NGOs and research institutes) contributing to the system of governance. This initiative will be successful due to the modular structure, which will enable the countries to enhance local data layers and maintain the common global model. The project is funded by an EU Interreg Programme and Horizon Europe and can be copied on other large-scale cross-border water systems. It demonstrates how digital twins are capable of delivering 33% improvement in the way resources can be formed, multi-level coordination and fostering climate resilience with mutual support in one of the most geo politically diversified river basins in the world (Aleke et al., 2025). The figure 3 explains major quantitative benefits that are realised owing to the implementation of Digital Twin of the Danube River Basin. Measurements encompass precise flood forecasting, minimized reaction time, improved data interconnection, and more effective use of the recourses. The graph shows that through this joint digital infrastructure, transboundary coordination and climate resilience within and between geopolitical spaces can be reinforced.

Figure 3. Performance Gains from the Digital Twin Implementation in the Danube River Basin

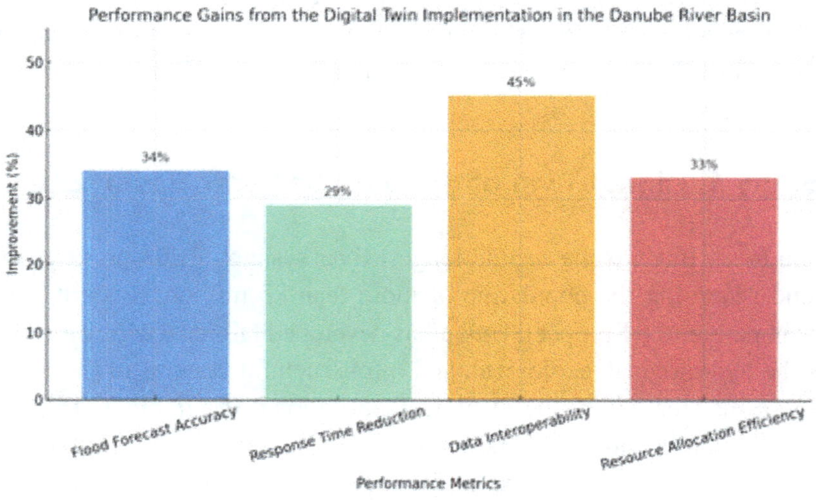

Case Study 2: AI-Based Wildfire Detection System in the US–Mexico Borderlands

The swathe of the borderlands between the US and Mexico itself stretching 3,145km is now being put in danger by the wildfires that are ignited by the intense drought and the warmth. As a reply, a binational AI-based wildfire detection system has been implemented in 2021, including hotspots in California, Arizona, Baja California, and Sonora (Aprizal et al., 2025). Led in collaboration between environmental authorities and commercial technological companies, the project harnesses satellite photos, drone video streams, and infrared sensors that run on AI algorithms to identify precursors of wildfire activity. The system saved 41% on false alarms in the first year of operating compared to the conventional means of detection and deployed fire response teams arrived at the struck areas 27% sooner. By training the AI models on patterns of historical fires of both countries, it became possible to boost the precision of detection by 38% in the zone near the border. The platform also implements bilingual interfaces, and cross border communication protocols, where both sides of fire departments can coordinate with the other. Community involvement was further combined through mobile alerts and participatory mapping; the number of the residents who received the information about evacuation and their safety was more than 12,000 people (Ciliberti et al., 2023). This made local adherences to emergency cases rise by 49%, particularly in regions that were under-served in the rural areas. Still there are challenges, especially as reconciliation of legislations on data privacy and harmonizing funding sources is difficult, but the continued cooperation has led to an increased shape of institutional loyalties. The system has been expanded to encompass 93% of the high-risk border regions it has observed, with USAID and Mexico (CONANP) (National Commission of Natural Protected Areas). The case and its research results demonstrate the strength of AI-informed predictive tracking, which enhances situational awareness, safeguards biodiversity corridors, and mitigates the consequences of wildfire cross-border impact by 35% in two-years. The Figure 4 presents the data on the effectiveness of the AI-powered system of determining the presence of wildfires along the US-Mexico border. It emphasizes that there was a major decrease in false alarms, faster response time, increased accuracy in the detection, better compliance in the community and the coverage of various regions. The data highlights the benefits of predictive AI technologies supplied by cross-border coordination in minimizing the risk of wildfires and reinforcing climate resiliency amidst transnationally vulnerable ecosystems.

Figure 4. Performance Metrics of AI-Based Wildfire Detection System in the US–Mexico Borderlands

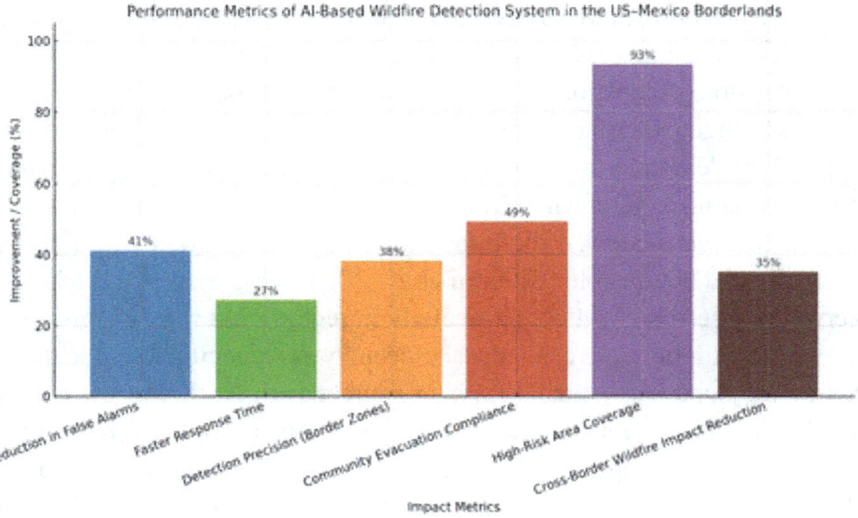

CONCLUSION

The issues of climate change are multidimensional, interrelated and transnational hence the traditional concept of nation-based governance is inadequate. The solutions to address the mentioned challenges would be to implement a paradigm shift of data driven, digitally integrated and collaboratively governed systems to build resilience across shared ecosystems and socio-political boundaries. Among the new technologies that center around this change are Artificial Intelligence (AI), digital twins, remote sensing, and big data analytics. They offer the capacity to monitor in real-time, model predictively, and plan in an adaptive transnational manner, that opens new demands of the early warning systems, disaster risk reduction and resource optimization. The decision-makers will make their decisions better with the help of AI, and the digital twins will provide an imitation of how the climate conditions will react to the ecological and infrastructural systems. Digital platforms connected to each other would guarantee uniformity of adaptation strategies and opportunities of global collaboration. The chapter dwells on all-inclusive systems of governance that enhance interoperability, data sovereignty and ethical use of digital tools. Institutional structures such as EGTCs, Interreg schemes and by river basin commissions can be used to prove the possibility of transnational cooperation. These structures must be capable of developing to fulfill the loopholes in law enforcement, privacy concerns and sovereignty problems especially in border sensitive regions.

The international conventions should also promote free dissemination of climate information besides equitable protection. Civic participation founded on citizen science, participatory sensing and civic technology are legitimate and makes vulnerable population feel as part of the decision-making process. To scale the cross-border digital climate action, they need to have climate financing approaches like climate innovation funds and the public- private partnership. Flexible and modular designs ensure significantly such projects are immune to climatic and geopolitical changes. Lastly, the work offers a futuristic imagined borders of collaborative digital networks of resilience and sustainability. Global climate governance The future: A borderless climate intelligence ecosystem with clear, ethical and inclusive solutions. Integrating governance and digital ecosystems is not only a strategic imperative but it is also a moral imperative in the environment of ever more threatening climatic challenges.

REFERENCES

Aburto-Oropeza, O., Johnson, A. F., Agha, M., Allen, E. B., Allen, M. F., González, J. A., Arenas Moreno, D. M., Beas-Luna, R., Butterfield, S., Caetano, G., Caselle, J. E., Gaytán, G. C., Castorani, M. C. N., Cat, L. A., Cavanaugh, K., Chambers, J. Q., Cooper, R. D., Arafeh-Dalmau, N., Dawson, T., & Taylor, J. E. (2018). Harnessing cross-border resources to confront climate change. *Environmental Science & Policy*, *87*, 128–132. DOI: 10.1016/j.envsci.2018.01.001

Aleke, C. U., Ovie, E. O., Asere, J. B., Oriakhi, E. A., Igah, G. C., Ajibade, O. E., & Egbuna, I. K. (2025). Digital twins for climate-resilient infrastructure: Simulating environmental impact on buildings. *Asian J. Geo. Res.*, *8*(2), 132–141. DOI: 10.9734/ajgr/2025/v8i2270

Aprizal, A., Wiranatakusuma, D. B., Rizki, M., & Anugrah, R. A. (2025). The role of technology in climate resilience: A systematic literature review and mapping study approach. *Journal of Physics: Conference Series*, *2989*(1), 12037. DOI: 10.1088/1742-6596/2989/1/012037

Archer, M., Ravn, L., & Thylstrup, N. B. (2025). The political economy of platformed silos: Theorizing data storage reconfigurations in the age of interoperability capitalism. *Big Data & Society*, *12*(2), 20539517241303144. DOI: 10.1177/20539517241303144

Arvanatidis, C., Penev, L., López Lérida, J., Huertas Olivares, C., Sáenz Albanés, A. J., Basset, A., Montinaro, S., Vaira, L., Minadakis, N., Griniezakis, M., & López Paneque, J. (2025). Biodiversity Knowledge Hub: Addressing the impacts of environmental change by linking Research Infrastructures, Global Aggregators and community. *Networks*.

Blanc, N., Cannata, M., Collombin, M., Ertz, O., Giuliani, G., & Ingensand, J. (2022). Ogc api state of play – a practical Testbed for the national spatial data infrastructure in Switzerland. ISPRS - Int. Arch. Photogramm. Remote Sens. Spat. Inf. Sci., XLVIII-4/W1-2022, 59–65.

Cabré, A., Kaufmann, S., Marinov, I., & Weisberg, M. (2024). Introducing the global climate security atlas. *Glob. Sustain.*, *7*(e25), e25. DOI: 10.1017/sus.2024.18

Ciliberti, F. G., Berardi, L., Laucelli, D. B., & Giustolisi, O. (2023). Digital Water Services using Digital Twin paradigm. *IOP Conference Series. Earth and Environmental Science*, *1136*(1), 12002. DOI: 10.1088/1755-1315/1136/1/012002

Dietrich, M., & Gutierrez, M. (2025). Empowering environmental research through the EOSC EU node: FAIR data and scalable services.

Fouilloux, A., & Magagna, B. (2025). FAIR2Adapt: Advancing Climate Change Adaptation Strategies through FAIR and open data sharing and Research Infrastructures.

Grossman, R. L., Boyles, R. R., Davis-Dusenbery, B. N., Haddock, A., Heath, A. P., O'Connor, B. D., Resnick, A. C., Taylor, D. M., & Ahalt, S. (2024). A framework for the interoperability of cloud platforms: Towards FAIR data in SAFE environments. *Scientific Data*, *11*(1), 241. DOI: 10.1038/s41597-024-03041-5 PMID: 38409183

Guo, L., Wei, J., Zhang, K., & Tian, F. (2021). Technical Note: Building a methodological framework and toolkit for news media dataset tracking of conflict and cooperation dynamics on transboundary rivers.

Guo, L., Wei, J., Zhang, K., Wang, J., & Tian, F. (2022). Building a methodological framework and toolkit for news media dataset tracking of conflict and cooperation dynamics on transboundary rivers. *Hydrology and Earth System Sciences*, *26*(4), 1165–1185. DOI: 10.5194/hess-26-1165-2022

Hamzah, A., Aqlan, F., & Baidya, S. (2024). Drone-based digital twins for water quality monitoring: A systematic review. *Digital Twins and Applications*, *1*(2), 131–160. DOI: 10.1049/dgt2.12021

Harris, K., Lager, F., Jansen, M. K., & Benzie, M. (2022). Rising to a new challenge: A protocol for case-study research on transboundary climate risk. *Weather, Climate, and Society*, *14*(3), 755–768. DOI: 10.1175/WCAS-D-21-0022.1

Hempelmann, N., Ehbrecht, C., Plesiat, E., Hobona, G., Simoes, J., Huard, D., Smith, T. J., McKnight, U. S., Pechlivanidis, I. G., & Alvarez-Castro, C. (2022). Deployment of Ai-enhanced services in climate resilience information systems. ISPRS - Int. Arch. Photogramm. Remote Sens. Spat. Inf. Sci., XLVIII-4/W1-2022, 187–194.

Kraatz, J., Demmich, K., Schnell, J., Gräler, B., Bagli, S., & Mazzoli, P. (2024). A federated Data Fabric to enhance disaster resilience for extreme weather events.

Larrinaga, F., Pérez, A., Aldalur, I., Hernández, J. L., Izkara, J. L., & Sáez de Viteri, P. (2021). A holistic and interoperable approach towards the implementation of services for the digital transformation of smart cities: The case of Vitoria-Gasteiz (Spain). *Sensors (Basel)*, *21*(23), 8061. DOI: 10.3390/s21238061 PMID: 34884064

Liu, O. R., & Molina, R. (2021). The persistent transboundary problem in marine natural resource management. *Frontiers in Marine Science*, *8*, 8. DOI: 10.3389/fmars.2021.656023

Liu, Y., Ma, J., Song, H., Qian, Z., & Lin, X. (2021). Chinese universities' cross-border research collaboration in the social sciences and its impact. *Sustainability (Basel)*, *13*(18), 10378. DOI: 10.3390/su131810378

Llambias, G., Bradach, B., Nogueira, J., González, L., & Ruggia, R. (2023a). Gateway-based interoperability for distributed ledger technology. CLEI Electron. J., 26(2).

Llambias, G., Bradach, B., Nogueira, J., González, L., & Ruggia, R. (2023b). Gateway-based Interoperability for DLT. In *TechRxiv*.

Mahadeo, S., & Del Savio, L. (2023). Transboundary maritime cooperation: The case of the Eastern Caribbean Region. *Frontiers in Marine Science*, *10*, 10. DOI: 10.3389/fmars.2023.1251911

Morley, J. W., Selden, R. L., Latour, R. J., Frölicher, T. L., Seagraves, R. J., & Pinsky, M. L. (2018). Projecting shifts in thermal habitat for 686 species on the North American continental shelf. *PLoS One*, *13*(5), e0196127. DOI: 10.1371/journal.pone.0196127 PMID: 29768423

Palacios-Abrantes, J., Sumaila, U. R., & Cheung, W. W. L. (2020). Challenges to transboundary fisheries management in North America under climate change. *Ecology and Society*, *25*(4), art41. DOI: 10.5751/ES-11743-250441

Perissi, I., & Jones, A. (2024). An emissions offset strategy to accomplish 2 °C long-term mitigation goals in the European Union. *Sustainability (Basel)*, *16*(11), 4783. DOI: 10.3390/su16114783

Rahayu, H. P., Zulfa, K. I., Nurhasanah, D., Haigh, R., Amaratunga, D., & Wahdiny, I. I. (2024). Unveiling transboundary challenges in river flood risk management: Learning from the Ciliwung River basin. *Natural Hazards and Earth System Sciences*, *24*(6), 2045–2064. DOI: 10.5194/nhess-24-2045-2024

Rao, X., Ma, R., Zhang, L., & Liu, J. (2022). Research and design of a digital twin-driven smart river basin platform. *Metaverse*, *3*(2), 11. DOI: 10.54517/m.v3i2.1815

Riaz, K., McAfee, M., Anton, I., & Gharbia, S. (2022). Conceptualising the management of climate extreme events through the GIS-based digital twin system.

Schröter, K., Pia-Johanna Schweizer, Gräler, B., Cumiskey, L., Bharwani, S., Parviainen, J., Kropf, C., Hakansson Viktor Wattin and Drews, M., Irvine, T., Dondi, C., Apel, H., Löhrlein, J., Hochrainer-Stigler, S., Bagli, S., Huszti, L., Genillard, C., Unguendoli, S., & Steinhausen, M. (2024). Invited perspectives: Fostering interoperability of data, models, communication and governance for disaster resilience through transdisciplinary knowledge co-production.

Shrestha, A., Souza, F. A. A., Park, S., Cherry, C., Garcia, M., Yu, D. J., & Mendiondo, E. M. (2022). Socio-hydrological modeling of the tradeoff between flood control and hydropower provided by the Columbia River Treaty. *Hydrology and Earth System Sciences*, *26*(19), 4893–4917. DOI: 10.5194/hess-26-4893-2022

Soliev, I., & Theesfeld, I. (2020). Benefit sharing for solving transboundary commons dilemma in central Asia. *The International Journal of the Commons*, *14*(1), 61–77. DOI: 10.5334/ijc.955

Stroetmann, K. A. (2014). Health system efficiency and eHealth interoperability – how much interoperability do we need? In *Advances in Intelligent Systems and Computing* (pp. 395–406). Springer International Publishing.

Sumaila, U. R., Palacios-Abrantes, J., & Cheung, W. W. L. (2020). Climate change, shifting threat points, and the management of transboundary fish stocks. *Ecology and Society*, *25*(4), art40. DOI: 10.5751/ES-11660-250440

Thornton, D. H., & Branch, L. C. (2019). Transboundary mammals in the Americas: Asymmetries in protection challenge climate change resilience. *Diversity & Distributions*, *25*(4), 674–683. DOI: 10.1111/ddi.12880

Vinogradov, S., & Wouters, P. (2020). Adaptation regulatory regimes to address climate change challenges in transboundary water basins: Can multilateral regionalism help? *Review of European, Comparative & International Environmental Law*, *29*(3), 406–416. DOI: 10.1111/reel.12371

Wang, S., & Xu, O. (2024). Interoperability structure of smart water conservancy based on Internet of Things. *International Journal of Distributed Sensor Networks*, *2024*, 1–13. DOI: 10.1155/2024/7724783

Williams, J. M. (2018). Stagnant rivers: Transboundary water security in South and Southeast Asia. *Water (Basel)*, *10*(12), 1819. DOI: 10.3390/w10121819

Wu, J., Orlandi, F., O'Sullivan, D., & Dev, S. (2024). LinkedGeoClimate: An interoperable platform for climate data access within geographical context. *IEEE Journal of Selected Topics in Applied Earth Observations and Remote Sensing*, *17*, 10352–10366. DOI: 10.1109/JSTARS.2024.3391922

Zargar, S., Kannangara, M., Gonzales-Calienes, G., Yang, J., Shadbahr, J., Decès-Petit, C., & Bensebaa, F. (2024). Data hub for life cycle assessment of climate change solutions—Hydrogen case study. *Data*, *9*(11), 129. DOI: 10.3390/data9110129

KEY TERMS AND DEFINITIONS

Artificial Intelligence (AI): The simulation of human intelligence in machines that are programmed to think, learn, and make decisions based on data.

Citizen Science: Public participation in scientific research, often involving local communities in data collection, environmental monitoring, and knowledge generation.

Civic Technology (Civic Tech): Technology platforms and tools that enhance public participation, transparency, and engagement in governance and decision-making processes.

Climate Data Sovereignty: The principle that nations have control over climate-related data generated within their territories and how it is used or shared externally.

Cross-Border Governance: Collaborative decision-making and policy implementation across national boundaries for managing shared resources and risks.

Digital Twin: A virtual replica of a physical system (e.g., river basin, forest, infrastructure) that updates in real-time through sensor and satellite data.

Early Warning Systems (EWS): Tools or platforms that provide advance notice of impending climate-related disasters, allowing for timely preparedness and response.

Interoperability: The ability of different digital systems, platforms, or organizations to exchange, interpret, and use data seamlessly and effectively.

Scalability: The capacity of a system, tool, or model to be expanded or adapted effectively across larger or different regions or populations.

Transboundary Climate Risks: Environmental threats, such as floods, droughts, or wildfires, that affect ecosystems and communities across national borders.

Chapter 4
Climate Diplomacy Through Tourism:
Leveraging Cross-Border Eco-Tourism for Peace and Environmental Resilience

Mehraj Din Wani
https://orcid.org/0000-0001-5034-1207
University of Kashmir, India

Shamim Ahmad Shah
University of Kashmir, India

Zubair Ahmad Dada
University of Kashmir, India

ABSTRACT

As climate change crosses national boundaries, the urgency for cross-border co-operation in South Asia increases. This chapter explores eco-tourism as a tool of climate diplomacy, using shared ecological assets to foster environmental resilience and peaceful regional engagement. It focuses on the Eastern Himalayas, the Terai Arc, and the Sundarbans Delta to illustrate how ecologically sensitive tourism can support biodiversity conservation, climate adaptation, and trust-building. These transboundary landscapes, rich in ecosystems and cultural ties, offer potential for co-managed tourism aligned with integration goals. The chapter reviews frameworks like SAARC, BIMSTEC, and bilateral environmental agreements, while identifying challenges such as regulatory gaps, infrastructure disparities, and unequal access. Using case studies and policy analysis, it proposes a strategy to integrate

DOI: 10.4018/979-8-3373-3972-6.ch004

eco-tourism into South Asia's climate diplomacy. Ultimately, it calls for viewing borders as ecological continuums and tourism as a path to inclusive, sustainable regional development.

INTRODUCTION

Climate change is fundamentally reshaping the geopolitical architecture of the 21st century, challenging the conventional notions of sovereignty, territoriality, and unilateral action (Dalby, 2020; Woodley, 2017). As climatic disruptions intensify—manifesting in glacial retreat, erratic monsoons, sea-level rise, and biodiversity degradation—states are being compelled to reconceptualize borders not as rigid demarcations of power, but as porous ecological continuums necessitating joint stewardship and shared responsibility. Nowhere is this imperative more urgent than in South Asia, a region characterized by acute ecological interdependence, immense cultural heterogeneity, and a fraught legacy of political partition, interstate rivalry, and asymmetric development (Bhatta et al., 2015).

The region encompasses critical transboundary ecosystems—ranging from the Eastern Himalayan alpine zones and the Gangetic floodplains to the Sundarbans delta and the Terai Arc landscape—that are simultaneously sites of ecological fragility and diplomatic opportunity. These ecological corridors span multiple jurisdictions including India, Nepal, Bhutan, and Bangladesh, and face converging climate threats such as glacier melt, saline intrusion, deforestation, and species migration (ICIMOD, 2021). Despite these shared vulnerabilities, regional climate governance remains fragmented, hindered by geopolitical tensions, institutional inertia, and the absence of robust, trust-based cooperative frameworks (Sharma et al., 2020).

In this context, cross-border eco-tourism—defined as environmentally responsible travel to shared natural areas across national boundaries—emerges as a compelling yet underutilized pathway for climate diplomacy and peacebuilding (Wieckowski, 2013). When grounded in principles of ecological integrity, cultural respect, and participatory governance, eco-tourism offers not merely economic or conservation benefits, but also acts as a soft power conduit for fostering intercultural understanding, trust formation, and transnational solidarity (Stronza et al., 2019). Unlike formal diplomatic engagements which often remain ensnared in strategic stalemates, tourism diplomacy operates through informal, bottom-up channels—facilitating grassroots interaction, collective memory-making, and everyday practices of coexistence (Kim et al., 2007).

South Asia, despite its political complexity, offers fertile ground for implementing such approaches. The Eastern Himalayas—spanning Bhutan, India, and Nepal—represent a shared ecological realm of endemic biodiversity, fragile mountain ecosystems, and indigenous knowledge systems (Jeelani et al., 2024). Similarly, the Sundarbans mangrove forests, co-managed by India and Bangladesh, embody a contiguous socio-ecological system vulnerable to cyclonic disturbances, rising seas, and habitat degradation. These landscapes, if linked through ecologically sensitive and culturally inclusive tourism corridors, can serve as arenas of transboundary cooperation—fostering not only joint environmental monitoring and conservation, but also climate education, peace narratives, and socio-economic resilience (Timothy, 2001; Gössling & Hall, 2006).

However, for cross-border eco-tourism to evolve into a durable instrument of climate diplomacy, it must be institutionally scaffolded through cooperative governance structures, legal harmonization, and community-centered policy frameworks. Regional organizations such as the South Asian Association for Regional Cooperation (SAARC) and the Bay of Bengal Initiative for Multi-Sectoral Technical and Economic Cooperation (BIMSTEC) possess latent potential to support such efforts through coordination, funding, and norm diffusion—yet their existing frameworks remain underdeveloped in integrating tourism with climate agendas (Boas & Rothe, 2016). National tourism and environment ministries, alongside local stakeholders and transboundary conservation networks (Batool et al., 2024a), must co-produce adaptable, multi-scalar strategies that embed eco-tourism within broader climate resilience and peacebuilding frameworks.

This chapter thus argues for the strategic embedding of eco-tourism within regional climate diplomacy architectures in South Asia. It proposes a conceptual and operational model where sustainable tourism functions not merely as a sectoral policy or economic tool, but as a relational infrastructure—linking climate adaptation, ecological governance, and regional peace. The chapter begins by offering a theoretical synthesis of climate diplomacy and transboundary tourism. It then examines empirical cases of eco-tourism initiatives in South Asia's contested ecologies, identifying both their successes and structural constraints. Finally, it delineates the institutional prerequisites and governance innovations necessary to scale such models, and offers concrete policy recommendations to enhance cross-border eco-tourism as a transformative force in regional cooperation. At a time when climate change functions simultaneously as a threat multiplier and a potential catalyst for cooperation, this chapter contributes to a critical rethinking of regionalism. It posits that cross-border eco-tourism—anchored in sustainability, inclusivity, and diplomacy—represents a promising, if complex, frontier in reconciling ecological imperatives with geopolitical realities in a rapidly warming South Asia.

LITERATURE REVIEW

The intersection of climate diplomacy, cross-border cooperation, and eco-tourism represents a growing yet underexplored field within international relations and sustainable development literature. Existing research spans across several domains—each contributing unique insights into how environmental challenges and opportunities for cooperation can be mediated through tourism practices and regional governance structures.

Climate Diplomacy and Transboundary Environmental Cooperation

Climate diplomacy is broadly understood as the use of diplomatic tools and international negotiation to address climate-related issues through cooperative frameworks. As defined by the German Federal Foreign Office and the adelphi think tank, climate diplomacy encompasses the integration of climate considerations into foreign policy to foster dialogue, mitigate conflict, and promote resilience. Scholars like Anthony Wani et al (2022) and Adil Najam et al. (2003) have highlighted the need for states to treat climate change as a geopolitical concern, especially in regions like South Asia where environmental degradation intersects with fragile political relations (Batool et al., 2024b).

Transboundary environmental cooperation has been a key mechanism within climate diplomacy. Literature on river basin management (e.g., the Ganges-Brahmaputra-Meghna system), mountain governance (e.g., ICIMOD's work on the Hindu Kush Himalayas), and shared biodiversity corridors illustrates how ecological interdependence necessitates collaborative governance.

Eco-Tourism and Sustainability in Border Regions

Eco-tourism is generally defined as responsible travel to natural areas that conserves the environment, sustains the well-being of local people, and involves interpretation and education (Jeelani et al., 2025; Björk, 2000). According to The International Ecotourism Society (TIES), eco-tourism should minimize impact, build environmental awareness, and provide direct financial benefits for conservation and local communities.

In the South Asian context, research by Nepal (2000), Scheyvens (1999), and Baral & Heinen (2007) has demonstrated the economic and conservation benefits of eco-tourism in the Himalayas and forested areas. However, few studies have examined eco-tourism across borders. Some notable exceptions include regional cooperation in the Eastern Himalayas (e.g., India–Bhutan–Nepal) and integrated tourism models in the Terai Arc Landscape, facilitated by organizations like WWF and ICIMOD.

These initiatives suggest that eco-tourism can foster environmental consciousness and mutual understanding across national boundaries. Yet, the challenges of political friction, visa barriers, and regulatory asymmetries often constrain the scaling up of such models.

Tourism as a Tool of Soft Power and Peacebuilding

Tourism has long been recognized as a form of soft power—a way for nations to project influence, build goodwill, and facilitate non-state diplomatic engagement. Scholars like Wani et al (2023) and Weissmann (2008) have discussed how cultural exchange and people-to-people contact through tourism can de-escalate tensions and foster mutual respect. In conflict-sensitive or post-conflict zones, "peace tourism" has been proposed as a strategy for reconciliation (Kelly, 2006; Causevic & Lynch, 2011).

In the context of cross-border areas, tourism can act as a mechanism for informal diplomacy and intercultural engagement. However, as Butler and Suntikul (2010) argue, such initiatives must be designed carefully to avoid exacerbating inequalities or triggering cultural tensions. The South Asian experience—marked by sensitive borders, historical grievances, and rigid nationalisms—makes this even more complex.

Legal and Institutional Frameworks for Cross-Border Eco-Tourism

The institutional literature on regional cooperation in South Asia points to both potential and limitations. SAARC and BIMSTEC have frameworks for environmental cooperation and tourism promotion, but implementation has often been weak due to political inertia and lack of enforcement mechanisms. As Ahmed (2013) notes, the fragmented legal frameworks across countries make it difficult to establish unified protocols for transboundary tourism, visa liberalization, and ecological zoning.

Nevertheless, recent efforts toward regional connectivity—such as the BBIN (Bangladesh-Bhutan-India-Nepal) sub-regional initiative—offer promise for integrated eco-tourism development. These efforts align with the Sustainable Development Goals (especially SDG 13 on Climate Action and SDG 17 on Partnerships) and the Convention on Biological Diversity's call for ecosystem-based approaches across political boundaries.

Gaps and Opportunities in the Literature

While there is substantial scholarship on eco-tourism, peacebuilding, and climate diplomacy in isolation, the convergence of all three within the specific framework of cross-border eco-tourism in South Asia remains insufficiently explored. Most studies either focus on intra-national eco-tourism or treat diplomacy and tourism as separate spheres. Furthermore, literature tends to underrepresent the role of local communities, indigenous knowledge systems, and civil society in shaping transboundary tourism and climate cooperation.

A key gap lies in the limited empirical assessment of how eco-tourism functions as a diplomatic tool in regions marked by geopolitical tensions and ecological interdependence, such as South Asia. There is a paucity of comparative studies evaluating the effectiveness of cross-border eco-tourism initiatives in fostering both climate resilience and regional peace. Moreover, existing frameworks often overlook the climate vulnerability of shared ecosystems and how joint tourism initiatives could serve as platforms for adaptive co-management.

This chapter seeks to bridge these gaps by offering a multidisciplinary analysis that connects eco-tourism with regional diplomacy and environmental governance. It builds on emerging discourses around nature-based solutions, climate-resilient development, and transboundary cooperation, proposing eco-tourism as a viable soft diplomacy tool for South Asia. In doing so, it also identifies structural and policy-related challenges—such as uneven regulatory harmonization, limited cross-border mobility, and asymmetries in institutional capacities—which remain underexplored in the current body of work.

METHODOLOGY

This chapter employs a qualitative, multi-method approach grounded in interpretive policy analysis, comparative case study review, and document-based research. The goal is to critically examine the potential of cross-border eco-tourism as a soft diplomacy tool for fostering environmental resilience and regional cooperation in South Asia.

Research Design

Given the interdisciplinary nature of the topic—spanning environmental governance, international relations, and sustainable tourism—a qualitative research design is most appropriate. The study does not seek to test a hypothesis, but rather to explore, interpret, and synthesize existing knowledge and practices related to cross-border eco-tourism and climate diplomacy in the South Asian context.

Data Sources

This study draws on a combination of primary and secondary sources to analyze the evolving landscape of cross-border eco-tourism in South Asia. The research is anchored in a multi-scalar data framework encompassing international treaties, institutional reports, academic scholarship, field-based project documentation, and recent media coverage. To understand the policy and institutional context, bilateral and multilateral agreements relevant to environmental cooperation and tourism development were reviewed. These include frameworks such as the SAARC Environment Action Plan, BIMSTEC tourism cooperation protocols, and the BBIN (Bangladesh–Bhutan–India–Nepal) connectivity initiatives. These documents provide insights into how regional diplomacy intersects with ecological and tourism-related agendas. Further, institutional publications from intergovernmental bodies (e.g., SAARC, BIMSTEC, ICIMOD, UNEP), international conservation organizations (such as WWF and IUCN), and leading think tanks have been examined. These reports offer a grounded understanding of cross-border project implementation, regional coordination mechanisms, and evolving discourses on climate-resilient tourism.

A comprehensive literature review supplements these sources, incorporating peer-reviewed journal articles, scholarly books, and critical essays on eco-tourism, soft power, climate adaptation, and transboundary environmental governance. These academic contributions provide the theoretical foundation and historical context for interpreting the case regions. In addition, case-specific data was sourced from project evaluations and implementation reports of eco-tourism and conservation programs located in the Eastern Himalayas, the Terai Arc Landscape, and the Sundarbans Delta. These documents offer empirical insight into how cross-border tourism functions at the community level and what challenges persist on the ground. To capture contemporary developments and public discourse, news articles and features from national and regional media outlets were also reviewed. These sources help contextualize emerging narratives, policy shifts, and socio-political dynamics that shape eco-tourism efforts in sensitive border areas.

Case Study Selection

The chapter adopts a comparative case study approach to explore the dynamics of transboundary eco-tourism across three ecologically vital and geopolitically strategic regions in South Asia. These regions were selected through purposive sampling based on key criteria, including ecological continuity across national borders, the presence of active or emerging eco-tourism initiatives, and the existence of regional cooperation frameworks.

The first case focuses on the Eastern Himalayas, spanning India, Nepal, and Bhutan. This region is renowned for its rich biodiversity and fragile alpine ecosystems. The chapter examines how biodiversity corridors and mountain eco-tourism initiatives serve not only to conserve ecological integrity but also to foster intergovernmental collaboration and sustainable livelihoods for local communities.

The second case centers on the Terai Arc Landscape, which also spans India and Nepal. Here, the analysis delves into wildlife-based eco-tourism efforts, particularly in and around protected areas like national parks and wildlife reserves. Special attention is given to community-based conservation models (Wani et al., 2024c, Wani et al., 2025b) and the role of cross-border cooperation in managing species migration corridors and minimizing human–wildlife conflict.

The third case explores the Sundarbans Delta, shared between India and Bangladesh. This region, home to the world's largest mangrove forest and the endangered Bengal tiger, is critically important for both climate resilience and eco-tourism development. The study highlights collaborative management strategies aimed at balancing conservation with tourism development in a climate-vulnerable landscape. Together, these cases provide a nuanced understanding of how eco-tourism can function as a tool for environmental protection, peacebuilding, and regional integration in South Asia's transboundary landscapes.

Analytical Framework

The chapter employs an interpretive policy analysis framework to critically examine how eco-tourism is positioned within the broader discourses of climate cooperation, soft power diplomacy, and sustainable development in South Asia's transboundary regions. Rather than treating policy as a fixed set of prescriptions, the interpretive lens allows for an exploration of the meanings, narratives, and power dynamics that shape eco-tourism initiatives across borders. The analysis is organized around four interrelated thematic pillars:

Environmental Sustainability: This theme investigates how eco-tourism is framed as a tool for conserving biodiversity and supporting climate change adaptation. It looks at policy narratives that link tourism development with ecological stewardship (Jeelani & Shah, 2023), and how these are reflected in practices such as the creation of biodiversity corridors, wildlife protection measures, and climate-resilient infrastructure.

Institutional Coordination: Here, the focus is on the mechanisms that facilitate cross-border collaboration, such as regional organizations, bilateral agreements, and joint management frameworks. The chapter explores how institutional arrangements—

both formal and informal—shape the planning, governance, and implementation of eco-tourism projects across national boundaries.

Peace and Diplomacy: This pillar delves into the soft power potential of eco-tourism in fostering peaceful relations, intercultural dialogue, and regional trust. It considers tourism as a form of track-two diplomacy, capable of building bridges between countries through shared environmental interests and people-to-people connections.

Equity and Inclusion: The final theme addresses the social justice dimensions of eco-tourism. It critically assesses whether and how eco-tourism initiatives incorporate the voices, rights, and livelihoods of local communities—especially marginalized groups such as women, indigenous populations, and the rural poor. The analysis also highlights gaps in participation and benefits, drawing attention to the need for more inclusive and bottom-up approaches.

Through this thematic structure, the chapter provides a rich and context-sensitive understanding of how eco-tourism policies and practices are constructed, negotiated, and contested in transboundary landscapes of South Asia.

LIMITATIONS

This chapter is based on secondary data and does not include primary fieldwork or interviews. Therefore, the analysis is interpretive and policy-focused rather than empirical or ethnographic. Additionally, political sensitivities and lack of transparency in regional diplomacy limit the availability of detailed, real-time data on cross-border initiatives.

STUDY AREA

South Asia, home to nearly two billion people and some of the world's most diverse ecosystems, presents a unique landscape where ecological interdependence intersects with geopolitical complexity. For the purposes of this chapter, the study focuses on three major transboundary ecological regions in South Asia that are rich in biodiversity, vulnerable to climate change, and hold potential for cross-border eco-tourism development:

The Eastern Himalayas (India–Nepal–Bhutan)

The Eastern Himalayas constitute one of the world's biodiversity hotspots, stretching across the borders of northeastern India, Nepal, and Bhutan (Basnet et al.,

2019). This mountainous region harbors diverse flora and fauna, including endemic species like the red panda and snow leopard. It is also a critical zone for glacial water systems that feed major South Asian rivers. In recent years, various conservation and tourism initiatives—such as the Kangchenjunga Landscape Conservation and Development Initiative (KLCDI)—have been launched to promote joint ecosystem management. The high-altitude trekking trails, spiritual tourism circuits, and nature reserves provide a strong basis for integrated, community-based eco-tourism that aligns with climate resilience goals (Wani et al., 2025c).

The Terai Arc Landscape (India–Nepal)

The Terai Arc, a lowland region extending across southern Nepal and northern India, is characterized by tropical forests, grasslands, and vital wildlife corridors. It is home to endangered species such as the Bengal tiger, one-horned rhinoceros, and Asian elephant. The region is already the focus of a transboundary conservation initiative led by WWF and supported by both Indian and Nepalese governments. Eco-tourism in this area revolves around wildlife sanctuaries, community forest tourism, and cultural heritage experiences (Wani et al., 2025d). The continuity of habitats across borders and existing conservation partnerships make it an ideal case for exploring cross-border tourism as a form of environmental and diplomatic cooperation.

The Sundarbans Delta (India–Bangladesh)

The Sundarbans, the world's largest mangrove forest, spans the coastal regions of southern Bangladesh and eastern India. This fragile ecosystem, which serves as a natural buffer against cyclones and rising sea levels, is a UNESCO World Heritage Site on both sides of the border. It supports millions of people through fishing, honey collection, and tourism. The region is acutely vulnerable to climate-induced hazards such as saline intrusion, biodiversity loss, and displacement. Several efforts have been made to promote eco-tourism in the Indian and Bangladeshi Sundarbans independently, but there is untapped potential for coordinated, low-impact tourism initiatives that enhance climate resilience and bilateral environmental cooperation.

These three transboundary regions are representative of South Asia's broader ecological and political dynamics. They offer distinct yet interconnected opportunities for leveraging eco-tourism as a platform for climate diplomacy, sustainable livelihoods, and peacebuilding (Wani et al., 2023a). By focusing on these landscapes, the chapter captures a multi-scalar view of how borders—often seen as zones of division—can be reimagined as spaces of ecological continuity and collaborative opportunity.

Figure 1. shows the geographical distribution of the study sites.

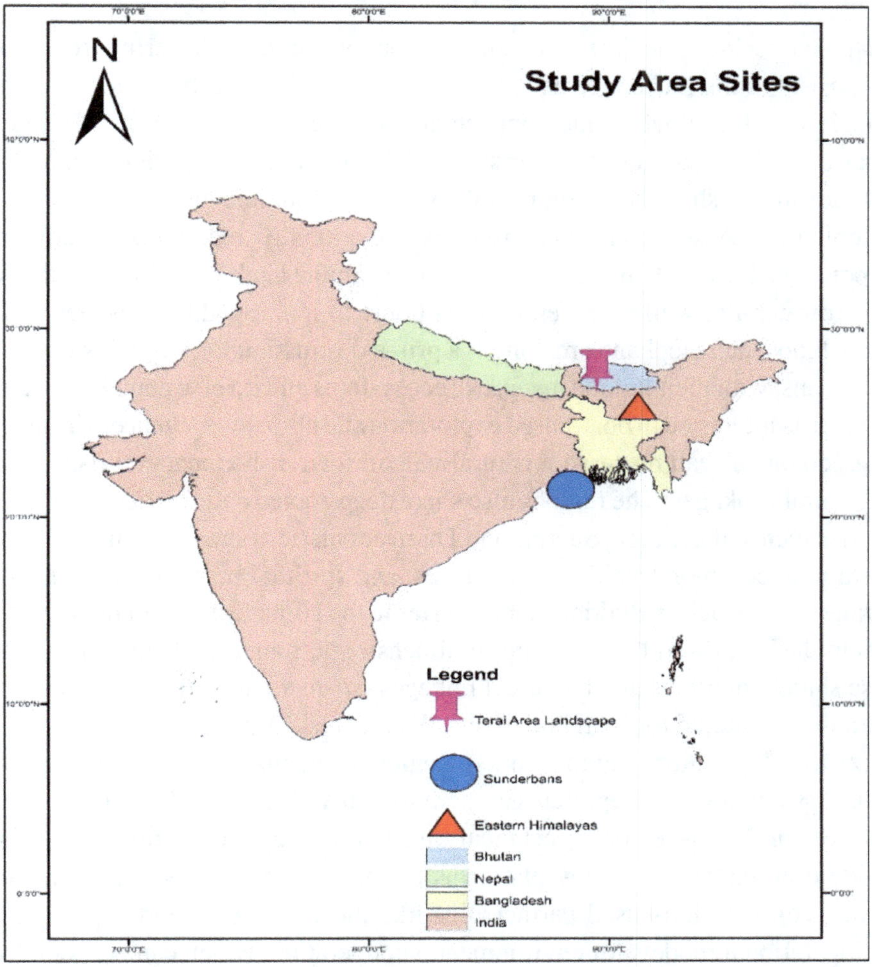

RESULTS

This section synthesizes the key findings from the qualitative analysis of policy frameworks, regional initiatives, and comparative case studies on cross-border eco-tourism in South Asia. The results highlight emerging opportunities, the functional role of eco-tourism in climate diplomacy, prevailing barriers, and community-level impacts on resilience.

Emerging Opportunities for Cross-Border Eco-Tourism

The chapter identifies a range of emerging opportunities for fostering cross-border eco-tourism in the Eastern Himalayas, Terai Arc, and Sundarbans Delta—regions marked by both ecological interconnectedness and cultural continuity. These landscapes offer fertile ground for tourism models that transcend political boundaries while advancing shared environmental and socio-economic goals.

Ecological Assets: All three regions are globally recognized biodiversity hotspots. The Eastern Himalayas offer dramatic alpine landscapes ideal for trekking and birdwatching, while the Terai Arc is home to iconic wildlife species such as tigers, rhinos, and elephants, making it a prime destination for wildlife safaris. The Sundarbans, with their unique mangrove ecosystems and diverse aquatic life, attract eco-tourists interested in boat-based exploration and climate-resilient environments. These ecological features serve as natural anchors for transboundary tourism circuits.

Cultural Linkages: The regions also share deep-rooted cultural ties. Ethnic communities such as the Limbu, Sherpa, and Tharu populations straddle national borders, maintaining common traditions, languages, and spiritual practices. Religious and heritage sites—such as Buddhist monasteries in the Himalayas or Hindu pilgrimage spots in the Terai—add a rich cultural dimension to tourism (Wani et al., 2024d). These shared identities and historical linkages create a foundation for cooperative, community-centered tourism initiatives (Wani et al., 2025e).

Existing Institutional Frameworks: Regional organizations such as the South Asian Association for Regional Cooperation (SAARC) and the Bay of Bengal Initiative for Multi-Sectoral Technical and Economic Cooperation (BIMSTEC) provide diplomatic and policy platforms to support cross-border tourism. Additionally, conservation-based partnerships like the Terai Arc Landscape Initiative offer operational models for environmental cooperation, which can be extended to include eco-tourism planning and implementation.

Climate Adaptation Integration: Eco-tourism is increasingly being incorporated into broader climate adaptation strategies. In regions where communities are vulnerable to climate-induced shocks—such as glacial melt, floods, and sea-level rise—eco-tourism offers a sustainable livelihood alternative. It encourages low-impact resource use, generates income for conservation, and reduces pressure on fragile ecosystems, thereby supporting both environmental and economic resilience. In sum, the convergence of natural endowments, cultural affiliations, institutional mechanisms, and climate priorities makes these transboundary regions highly promising for developing collaborative eco-tourism initiatives that serve ecological, social, and diplomatic purposes.

Climate Diplomacy through Eco-Tourism: A Functional Link

The chapter underscores the potential of eco-tourism as a functional instrument of climate diplomacy, revealing how it simultaneously advances environmental cooperation, peacebuilding, and regional integration in transboundary contexts. Far from being just a leisure activity, eco-tourism can serve as a strategic platform for diplomatic engagement and climate resilience across national boundaries (Wani et al., 2025g). This multidimensional role unfolds across the following key dimensions:

Soft Diplomacy: Eco-tourism facilitates cross-border people-to-people exchanges, allowing individuals from neighboring countries to engage in shared experiences rooted in nature, culture, and conservation. These interactions can break down stereotypes, foster empathy, and build interpersonal and intercultural trust—an essential counterbalance to historical and political tensions in the region. Through informal, citizen-level diplomacy, eco-tourism helps humanize cross-border relations and create grassroots support for cooperation.

Environmental Stewardship: Joint eco-tourism ventures encourage shared responsibility for transboundary ecosystems, promoting collective efforts in biodiversity conservation, climate adaptation, and sustainable resource use. For example, collaborative management of forest corridors, river systems, or protected areas can be strengthened through tourism revenues and joint planning, turning conservation into a mutually beneficial enterprise rather than a source of competition.

Community Engagement: At the local level, eco-tourism provides sustainable livelihood opportunities through homestays, guiding services, handicrafts, and hospitality. This reduces dependence on environmentally degrading practices like poaching or illegal logging. When communities benefit economically and socially from maintaining ecosystems, the incentives to protect shared landscapes—and avoid resource-based conflicts—become stronger. In this way, eco-tourism becomes a peace-promoting force rooted in economic empowerment and social inclusion.

Regional Integration: Transboundary eco-tourism routes—such as heritage trails, wildlife circuits, or river cruises—function as corridors of connectivity, reinforcing the objectives of regional organizations like SAARC and BIMSTEC. These corridors not only support tourism flows but also stimulate infrastructure development, policy alignment, and cross-sector cooperation, thereby contributing to broader processes of regional integration and sub-regional cohesion.

In essence, eco-tourism emerges as a pragmatic and symbolic bridge—linking environmental priorities with diplomatic interests, local development with regional governance, and nature conservation with human connection.

Table 1. Functions of Eco-Tourism in Advancing Climate Diplomacy

Function	Description	Case Example
Soft Diplomacy	Builds cultural trust and promotes dialogue through shared tourism experiences	India–Nepal Terai tourism
Environmental Stewardship	Promotes joint conservation of cross-border ecosystems	Kangchenjunga Landscape
Community Engagement	Provides alternative livelihoods and reduces resource-based conflicts	Sundarbans community ecotourism
Regional Integration	Supports policy frameworks and people-level connectivity	BBIN subregional tourism talks

Barriers to Implementation

While cross-border eco-tourism presents significant potential as a vehicle for climate diplomacy and regional cooperation, its practical implementation faces a number of persistent challenges. These barriers stem from a complex interplay of political, institutional, and infrastructural factors that limit the seamless development of such initiatives:

Visa and Mobility Restrictions: One of the most immediate obstacles is the lack of fluid cross-border mobility. Stringent visa regimes, bureaucratic hurdles, and limited border checkpoints restrict the movement of both international tourists and local stakeholders. This undermines the spontaneity and inclusiveness that eco-tourism ideally promotes, particularly in regions where border communities have shared cultural and ecological ties.

Legal and Regulatory Fragmentation: Differences in national tourism policies, conservation laws, and land-use regulations complicate the process of joint planning and implementation. The absence of harmonized standards on issues such as park access, visitor safety, or environmental impact assessments makes cross-border coordination cumbersome. This legal incoherence can stall even well-intentioned initiatives, leading to fragmented or duplicative efforts.

Infrastructure Disparities: The uneven development of infrastructure—ranging from roads and transport links to communication networks and hospitality services—creates imbalances in access and benefits. While one side of a border may offer developed eco-tourism amenities, the other may lack even basic facilities, resulting in skewed participation and diminished cross-border visitor flow.

Security Concerns: Many border regions in South Asia are politically sensitive or militarized, with histories of conflict, insurgency, or territorial disputes. These factors lead to heightened surveillance, restricted access, and, in some cases, outright bans on civilian movement in ecologically rich zones. Such security dynamics not only deter tourism but also complicate the establishment of trust and cooperation necessary for transboundary environmental initiatives.

Collectively, these barriers highlight the need for multi-level policy reforms, trust-building measures, and regional coordination mechanisms to make cross-border eco-tourism a viable and sustainable instrument of climate diplomacy in South Asia.

Local Community Role and Resilience Outcomes

The success of cross-border eco-tourism and its contribution to climate resilience hinges significantly on the active involvement of local communities living in transboundary landscapes. These communities are not merely passive beneficiaries but central actors in shaping, sustaining, and deriving value from eco-tourism initiatives. The chapter highlights several key dimensions of this community role:

Shared Cultural Identity: Many border communities in regions like the Eastern Himalayas, Terai Arc, and Sundarbans share common languages, religious practices, and cultural traditions despite national boundaries. This cultural continuity fosters a strong foundation for joint tourism ventures, where community-led storytelling, cultural performances, and heritage experiences can transcend borders and enhance the authenticity of the tourism offering. Such shared identity also facilitates cooperative natural resource management, rooted in mutual understanding and respect.

Livelihood Diversification: In areas highly vulnerable to climate change, especially those reliant on agriculture, forestry, or fishing, eco-tourism provides a critical alternative livelihood. Activities such as hosting tourists in homestays, guiding nature trails, producing local crafts, or offering culinary experiences allow households to diversify income sources, reducing pressure on fragile ecosystems and enhancing economic resilience in the face of climatic shocks.

Environmental Awareness: Engagement in eco-tourism fosters greater environmental consciousness among local residents. As communities interact with visitors and participate in conservation-related activities, their understanding of biodiversity, climate risks, and sustainable practices deepens (Wani et al., 2024a). This heightened awareness often translates into improved stewardship of natural resources and stronger community support for conservation policies.

Social Cohesion and Peacebuilding: Eco-tourism also acts as a platform for cross-border community interactions, encouraging dialogue, cooperation, and solidarity among people who share geographic and ecological spaces. These interactions help rebuild social bonds, reduce misunderstandings, and cultivate a sense of shared destiny, which is vital in politically sensitive or post-conflict regions. Over time, such engagement contributes to peacebuilding from below, reinforcing stability and mutual trust.

Synthesis

The analysis underscores that cross-border eco-tourism holds substantial promise as a strategic instrument of soft power, capable of weaving together the goals of climate adaptation, regional cooperation, and peacebuilding in South Asia. By leveraging shared ecological assets and cultural linkages, eco-tourism creates opportunities for dialogue, trust-building, and joint environmental stewardship across politically sensitive borders (Wani et al., 2023b).

However, realizing this potential requires more than visionary discourse. The findings highlight that the transformation of eco-tourism from a promising concept into a functional climate diplomacy tool hinges on addressing several entrenched challenges. These include regulatory fragmentation, restricted mobility, uneven infrastructure, and security sensitivities that continue to limit cross-border integration.

Crucially, the chapter emphasizes the need to center local communities as co-creators and beneficiaries of eco-tourism. Their involvement ensures both social legitimacy and sustainability of initiatives, while also enhancing regional resilience through inclusive livelihoods, environmental awareness, and grassroots peacebuilding.

In conclusion, while the landscape is marked by constraints, the convergence of ecological interdependence, institutional frameworks, and shared cultural heritage offers a fertile ground for developing cross-border eco-tourism as a meaningful contributor to both climate governance and regional harmony in South Asia.

DISCUSSION

The findings presented affirm that cross-border eco-tourism in South Asia holds substantial potential to operate as a platform for climate diplomacy, sustainable development, and regional peacebuilding. Unlike traditional environmental cooperation mechanisms that rely heavily on top-down policy alignments, eco-tourism invites grassroots involvement and community stewardship—making it a unique tool that simultaneously addresses ecological, economic, and sociopolitical dimensions (Bramwell & Lane, 2011; Wani et al., 2023c; D'Amore, 2009).

Eco-Tourism as a Functional Instrument of Soft Climate Diplomacy

Eco-tourism facilitates what Batool et al (2024a) conceptualized as "soft power," wherein influence arises from cultural exchange and shared values rather than coercion or financial incentives. In South Asia, this is particularly significant given the complex political history and strained bilateral relations between neighboring states.

People-to-people contact through nature-based tourism can circumvent political stagnation, opening informal but influential pathways of dialogue and cooperation (Hettne & Söderbaum, 2005; Wani et al., 2022).

This approach resonates with "track-two diplomacy"—informal interaction through non-state actors—highlighted in the conflict transformation literature as critical in frozen conflicts or low-intensity border tensions (Dada et al., 2022; Wani et al., 2022; Bhat et al., 2025). Eco-tourism encourages the reimagining of borders as "ecological continuums" rather than fixed lines of separation, fostering transboundary consciousness essential for regional climate resilience. In the Eastern Himalayas, for example, community-led eco-tourism in areas such as the Tawang-Bhutan corridor has helped promote cultural diplomacy and biodiversity monitoring across informal channels. These interactions often operate beneath formal state structures, building social capital through shared trekking routes, festivals, and conservation practices. This region illustrates how eco-tourism reinforces Nye's soft power and track-two diplomacy, particularly when trust between national governments is minimal.

Governance and Institutional Synergies: Addressing Fragmentation

One of the most pressing challenges is the fragmented nature of tourism and environmental governance across borders. Each country maintains distinct visa regimes, wildlife laws, conservation priorities, and tourism development strategies, which often operate in isolation (Wani et al., 2024b). These regulatory inconsistencies obstruct synergistic policy-making and joint programming. Cross-border eco-tourism thus requires the establishment of multi-level governance mechanisms—including local bodies, national ministries, and regional organizations—aligned around shared ecological and development goals (Hooghe & Marks, 2003). Successful regional cooperation frameworks such as the Heart of Borneo Initiative and the Greater Mekong Subregion Tourism Strategy serve as useful models (UNWTO, 2017).

However, South Asian regionalism remains fragile, partly due to overlapping mandates and political sensitivities in institutions like SAARC and BIMSTEC (Dash, 2008). Therefore, eco-tourism may act as a functional cooperation domain that is low-risk but high-value, generating mutual benefits without necessitating contentious political negotiation (Wani et al., 2024e). In the Terai Arc landscape, which spans India and Nepal, eco-tourism has prompted embryonic governance synergies through joint tiger conservation zones, cross-border ranger exchanges, and shared community tourism initiatives. These developments signal the feasibility of integrated governance in biodiversity corridors—particularly when institutional trust is built from below through shared tourism ventures. The Terai Arc thus serves as a pilot model for layered governance mechanisms, aligning well with Hooghe and Marks' (2003) theory of multi-level governance.

Uneven Development and Infrastructure Disparities

The uneven tourism infrastructure in border regions introduces another layer of complexity. Studies show that poorly developed areas, often marginalized by national planning efforts, cannot equally benefit from eco-tourism initiatives unless targeted investments are made (Spenceley & Goodwin, 2007; Wani et al., 2025a). This infrastructural asymmetry can perpetuate regional inequalities, which contradict the sustainable development ethos of eco-tourism. Development banks and multilateral agencies, such as the Asian Development Bank (ADB), have supported regional connectivity projects, but these often emphasize large-scale transport corridors rather than community-level tourism infrastructure. A reorientation toward inclusive, small-scale, and ecologically sensitive infrastructure—such as eco-lodges, interpretive centers, and foot trails—would better serve the goals of cross-border eco-tourism.

In the Sundarbans Delta—shared by India and Bangladesh—infrastructure gaps are pronounced due to ecological fragility and vulnerability to sea-level rise. However, recent community-led initiatives in sustainable boating, homestay programs, and cyclone-resilient ecotourism facilities demonstrate how context-sensitive infrastructure can support both adaptation and livelihoods. The Sundarbans case highlights that regional disparities can be mitigated when infrastructure planning is ecologically adaptive and locally grounded.

Community Agency, Cultural Affinities, and Local Diplomacy

The centrality of local communities in borderlands cannot be overstated. These populations often share linguistic, cultural, and environmental ties that precede the modern nation-state and are therefore uniquely positioned to facilitate informal cross-border cooperation (Nüsser & Schmidt, 2017). Eco-tourism not only provides these communities with sustainable livelihoods but also positions them as active agents of peace and resilience, echoing principles of community-led common-pool resource management. Furthermore, research has shown that gender-sensitive tourism programs enhance both environmental outcomes and social equity (Scheyvens, 2000). In South Asia, where women frequently serve as environmental custodians and informal educators within their communities (Wani et al., 2025f), inclusive eco-tourism policies can amplify climate adaptation outcomes.

Across all three regions studied, local communities function as the backbone of eco-tourism diplomacy. In the Eastern Himalayas, Buddhist monastic networks help mediate conservation ethics and spiritual tourism. In the Terai Arc, indigenous

Tharu communities manage forest tourism cooperatives. In the Sundarbans, women's self-help groups are central to ecotourism enterprises and mangrove restoration. These varied forms of community agency not only reflect localized ownership but also create distributed nodes of resilience aligned with Ostrom's principles of nested enterprises.

Rethinking Borders through Ecological Integration

Perhaps the most transformative insight lies in the potential of eco-tourism to reframe borders—not as barriers but as eco-cultural interfaces. This concept challenges Westphalian notions of rigid territoriality and instead embraces a model of transboundary ecological governance. This shift aligns with environmental peacebuilding literature, which argues that shared ecological interests can foster cooperation even in conflict-prone regions. By promoting joint tourism experiences, shared park management, and mutual conservation goals, countries can begin to cultivate a regional ecological identity. This identity is crucial for addressing shared risks such as glacial melt, rising sea levels, and biodiversity loss—none of which respect political boundaries.

Empirical observations from the three case regions affirm this ecological identity formation. In the Eastern Himalayas, climate-induced glacial retreat has spurred cooperation in tourism-based climate education across Bhutan and Arunachal Pradesh. In the Terai Arc, the contiguous forest ecology has compelled Nepal and India to develop joint wildlife corridors. In the Sundarbans, transboundary conservation tourism highlights a shared vulnerability to climate disasters and the need for collective resilience planning. These instances demonstrate how eco-tourism catalyzes ecological regionalism—a vital shift in a geopolitically sensitive landscape.

Theoretical Implications

This study contributes substantively to the expanding discourse on climate diplomacy by integrating ecological regionalism and people-centered diplomacy into the framework of cross-border cooperation. Traditionally, theories of climate diplomacy have emphasized high-level negotiations and multilateral environmental agreements (e.g., UNFCCC processes), but this research challenges that orthodoxy by proposing a meso-level theoretical model wherein eco-tourism acts as a functional intermediary between global climate objectives and local agency.

First, it extends the framework of track-two diplomacy (Montville, 1991) into the environmental realm, theorizing eco-tourism not simply as a developmental practice but as a relational diplomacy mechanism capable of reducing cross-border anxieties through sustained non-state engagement. This shift acknowledges the emotional, affective, and symbolic dimensions of borderland interactions—an aspect largely under-theorized in both international relations and tourism studies. Second, it builds on the concept of ecological sovereignty, positing that sovereignty in border regions can be spatially and politically negotiated through shared ecological stewardship. This reframes borders not as rigid lines but as fluid ecotones—zones of ecological and cultural interdependence—redefining the unit of analysis in cross-border co-operation from the state to the biome.

Third, the study advances the theory of multilevel environmental governance (Lemos & Agrawal, 2006) by illustrating how eco-tourism necessitates coordination across subnational, national, and regional levels. It enriches the literature by highlighting how these governance dynamics play out in politically sensitive, postcolonial contexts—such as South Asia—where informal institutions, customary norms, and historical mistrust intersect with formal policy instruments. Finally, this research offers a normative turn in sustainability science, suggesting that peace and reconciliation should be viewed as integral components of climate adaptation strategies. This challenges the prevailing technocratic models of climate resilience, foregrounding the need to integrate socio-political harmony into frameworks of ecological sustainability.

Practical Implications

On the applied side, the study offers a set of strategic insights with actionable value for policymakers, development practitioners, and regional organizations seeking to foster both environmental sustainability and geopolitical stability in South Asia. First, it identifies cross-border eco-tourism corridors as high-leverage, low-conflict interventions. Unlike infrastructure megaprojects that often trigger resistance or ecological degradation, eco-tourism is relatively low-impact and socially embedded. Therefore, ministries of tourism and environment, along with regional bodies such as SAARC or BIMSTEC, can pilot nature diplomacy zones—jointly managed ecotourism parks or trails—to facilitate trust-building and joint monitoring of climate-sensitive ecosystems.

Second, the study implies a critical need for inter-agency coordination at the national level. Currently, siloed institutional arrangements—wherein immigration, tourism, and conservation bodies operate in isolation—undermine the potential of integrated programming. Developing inter-ministerial task forces or joint secretariats could streamline visa policies, harmonize tourism guidelines, and ensure biodiversity protection is not compromised by tourism pressures. Third, eco-tourism offers a unique pathway to inclusive economic development in historically marginalized border regions. By promoting local entrepreneurship, cultural interpretation, and sustainable hospitality services, cross-border eco-tourism can directly enhance livelihood resilience, particularly among women, indigenous communities, and youth. Targeted skill development programs, supported by development banks and bilateral donors, would be essential to operationalize this potential.

Fourth, the findings stress the importance of capacity-building for decentralized governance. Panchayats, village development councils, and cross-border community cooperatives must be empowered—financially and institutionally—to manage tourism operations and conservation strategies autonomously. This involves legal reforms to devolve power, mechanisms for participatory budgeting, and digital infrastructure to support coordination and transparency. Lastly, the research suggests that climate diplomacy through tourism is not a substitute for high-level negotiation but a complementary tool. Its strength lies in its ability to sustain dialogue and foster human connection in the absence of formal political engagement. Therefore, ministries of external affairs and regional forums should formally recognize eco-tourism initiatives as track-two confidence-building mechanisms, legitimizing their role within broader foreign policy architectures.

Limitations

While this study advances a novel understanding of cross-border eco-tourism as an instrument of climate diplomacy and environmental peacebuilding in South Asia, several limitations temper the generalizability and operational applicability of the findings. First, the exploratory nature of the research, grounded largely in conceptual synthesis and illustrative case materials, limits its empirical depth. Although key variables and relationships are articulated, the absence of large-scale quantitative or longitudinal data constrains the capacity to model causal inferences or measure impact with statistical rigor. Second, the geopolitical heterogeneity of South Asia presents significant variation in state capacity, political will, ecological vulnerability, and tourism infrastructure. This diversity complicates the development of a one-size-fits-all model of eco-diplomacy. Border areas such as India-Nepal, India-Bhutan, or India-Bangladesh each exhibit distinct ecological, legal, and socio-cultural dynamics, which the current framework may only partially accommodate.

Third, the study operates within a state-centric policy lens, despite emphasizing the importance of community agency. While community-led models are referenced, the framework would benefit from a deeper ethnographic or participatory approach that centers borderland communities' voices, particularly from indigenous and nomadic populations whose environmental worldviews differ from formal conservation paradigms. Fourth, the analysis does not sufficiently account for the risks of tourism-induced ecological degradation, commodification of local cultures, or unintended securitization of border zones under the guise of ecotourism development. These contradictions—between tourism's soft power and its capitalist underpinnings—require more nuanced theorization and field validation. Lastly, data limitations related to official cross-border tourist flows, environmental cooperation records, and conflict-sensitive development metrics restrict the ability to triangulate insights across regions or timeframes. Much of the secondary literature is fragmented or dated, reflecting a broader gap in region-specific interdisciplinary research on eco-tourism and diplomacy.

Future Research Directions

To build on the insights of this study, future research should pursue more empirically grounded, interdisciplinary, and regionally disaggregated analyses of cross-border eco-tourism initiatives in South Asia and comparable Global South contexts.

Comparative Case Studies: There is a pressing need for in-depth comparative case studies across different border ecologies (e.g., Himalayan, mangrove, arid) to examine how political ecology, local governance, and cultural narratives shape eco-tourism outcomes. Such studies could also contrast South Asian models with successful initiatives in Southeast Asia, East Africa, or Latin America to identify best practices and avoid path dependency.

Impact Assessment Frameworks: Future research should develop and validate robust impact assessment models to quantify the environmental, diplomatic, and socio-economic returns of eco-tourism interventions. These models should include indicators of peacebuilding efficacy, community empowerment, biodiversity protection, and cross-border policy alignment.

Community-Led Research Methodologies: Incorporating participatory action research (PAR), ethnographic fieldwork, and indigenous mapping tools can help surface underrepresented perspectives in borderland tourism governance. This shift would better reveal how power, identity, and resistance mediate the success or failure of eco-diplomatic initiatives.

Tourism-Conflict Nexus: Given the complex security dynamics in some South Asian borders, future studies could explore the tourism-security nexus, assessing whether eco-tourism reduces hostility or inadvertently exacerbates militarization, surveillance, or displacement. Mixed-method approaches combining conflict resolution studies with tourism geography would be valuable here.

Institutional Design and Legal Harmonization: Further research is needed to explore the institutional architectures and legal interoperability required to support transboundary eco-tourism governance. This includes visa liberalization policies, joint biodiversity treaties, and revenue-sharing frameworks. Legal scholars, political ecologists, and policy scientists must collaborate to develop models of cross-border environmental federalism tailored to South Asia.

Climate Justice Lens: Finally, applying a climate justice framework to eco-tourism could illuminate how benefits and burdens are distributed across class, gender, caste, and ethnicity in border zones. This would support more equitable tourism models and deepen the ethical grounding of climate diplomacy strategies.

CONCLUSION

In an era marked by ecological precarity, geopolitical fragmentation, and declining faith in multilateralism, this study posits cross-border eco-tourism as a potent yet underexplored instrument of climate diplomacy and regional environmental resilience in South Asia. By foregrounding the role of nature-based tourism in contested borderlands, the research illuminates a critical conceptual pivot—from tourism as a commodified leisure activity to tourism as an emergent geo-ecological praxis of peacebuilding, trust formation, and adaptive co-governance. This reconceptualization is neither utopian nor purely functional; rather, it acknowledges the deep entanglements of space, sovereignty, and sustainability. Eco-tourism, when deployed thoughtfully, becomes more than a tool of economic diversification or ecological conservation. It becomes a liminal medium—an intermediary site where statecraft, civil society agency, and transboundary ecological ethics converge. It is here, in the liminality of border ecologies, that the hard infrastructures of division may be softened by the shared pursuit of environmental stewardship and intercultural dialogue.

The study's findings underscore that cross-border eco-tourism, particularly in ecologically rich but politically sensitive regions, constitutes a multiscalar intervention. It operates simultaneously at the level of local livelihoods, national environmental strategies, and regional diplomatic imaginaries. Its efficacy lies not only in its ability to generate economic returns or reduce emissions, but in its capacity to reconfigure ontologies of the border—shifting them from militarized thresholds to permeable cultural and ecological continuums. Yet, the potential of eco-tourism as a vehicle for climate diplomacy is not automatic. It must be actively constructed, institutionally scaffolded, and ethically negotiated. This entails the development of hybrid governance architectures that reconcile the asymmetries of state sovereignty with the urgencies of ecological interdependence. It demands participatory models that center community voices and redistribute power and profit within the borderlands. And it requires a sustained commitment to transformative regionalism, where cooperation is rooted not in strategic calculus alone but in an ethic of mutual vulnerability and ecological solidarity.

This study, therefore, gestures toward a post-Westphalian vision of environmental diplomacy—one in which climate action is catalyzed not merely through treaties and targets, but through situated, lived experiences of shared ecosystems and common futures. In this vision, eco-tourism is neither a panacea nor a peripheral policy instrument, but a strategic arena for cultivating the relational infrastructure of peace and sustainability in a climate-disrupted world. As South Asia confronts the dual specters of climate change and geopolitical rigidity, the insights of this research advocate for a courageous reimagining of regional cooperation: one that embraces the fragile, fertile spaces of nature and tourism as sites of both reconciliation and resilience.

REFERENCES

Adelphi & German Federal Foreign Office. (2015). *A new climate for peace: Taking action on climate and fragility risks*. Retrieved from https://www.newclimateforpeace.org/

Ahmed, Z. S. (2013). Regionalism and regional security in South Asia: The role of SAARC. *Regional Studies, 31*(1), 76–91. DOI: 10.4324/9781315604442

Baral, N., & Heinen, J. T. (2007). Decentralization and people's participation in conservation: A comparative study from the Western Terai of Nepal. *International Journal of Sustainable Development and World Ecology, 14*(5), 520–531. DOI: 10.1080/13504500709469751

Basnet, D., Kandel, P., Chettri, N., Yang, Y., Lodhi, M. S., Htun, N. Z., Uddin, K., & Sharma, E. (2019). Biodiversity research trends and gaps from the confluence of three global biodiversity hotspots in the Far-Eastern Himalaya. *International Journal of Ecology, 2019*(1), 1323419. DOI: 10.1155/2019/1323419

Batool, N., Wani, M. D., Shah, S. A., & Dada, Z. A. (2024a). Theory of planned behavior and value-belief norm theory as antecedents of pro-environmental behaviour: Evidence from the local community. *Journal of Human Behavior in the Social Environment, 34*(5), 693–709. DOI: 10.1080/10911359.2023.2205912

Batool, N., Wani, M. D., Shah, S. A., & Dada, Z. A. (2024b). Tourists' attitude and willingness to pay on conservation efforts: Evidence from the west Himalayan eco-tourism sites. *Environment, Development and Sustainability, 27*(8), 1–19. DOI: 10.1007/s10668-024-04679-2

Bhat, W. A., Wani, M. D., Dada, Z. A., & Qureshi, R. A. (2025). Beyond Linear Models: A Complexity-Theoretic Approach to Understanding Tourist Behavior in Peacebuilding Contexts. *Journal of Quality Assurance in Hospitality & Tourism*, •••, 1–30. DOI: 10.1080/1528008X.2025.2466050

Bhatta, L. D., Aggarwal, P. K., & Wang, E. (2015). Climate-induced migration in South Asia: Evidence, future projections and implications. *Regional Environmental Change, 15*(1), 43–55. DOI: 10.1007/s10113-014-0744-8

Björk, P. (2000). Ecotourism from a conceptual perspective, an extended definition of a unique tourism form. *International Journal of Tourism Research, 2*(3), 189–202. DOI: 10.1002/(SICI)1522-1970(200005/06)2:3<189::AID-JTR195>3.0.CO;2-T

Boas, I., & Rothe, D. (2016). From conflict to resilience? Explaining recent changes in climate security discourses and practices. *Environmental Politics, 25*(4), 613–632. DOI: 10.1080/09644016.2016.1160479

Bramwell, B., & Lane, B. (2011). Critical research on the governance of tourism and sustainability. *Journal of Sustainable Tourism, 19*(4–5), 411–421. DOI: 10.1080/09669582.2011.580586

Butler, R. W., & Suntikul, W. (Eds.). (2010). *Tourism and political change*. Goodfellow Publishers. DOI: 10.2307/jj.28209953

Causevic, S., & Lynch, P. (2011). Phoenix tourism: Post-conflict tourism role. *Annals of Tourism Research, 38*(3), 780–800. DOI: 10.1016/j.annals.2010.12.004

D'Amore, L. J. (2009). Peace through tourism: The global coalition for peace. *Journal of Business Ethics, 89*(4), 559–570. DOI: 10.1007/s10551-010-0407-3

Dada, Z. A., Wani, M. D., & Shah, S. A. (2022). The impact of pilgrim tourism on the bilateral diplomacy of India and Nepal and the mediation effect of people-to-people contact. *Asian Journal of Comparative Politics, 7*(3), 495–506. DOI: 10.1177/20578911221076234

Dalby, S. (2020). *Anthropocene geopolitics: Globalization, security, sustainability*. University of Ottawa Press., DOI: 10.2307/j.ctvx5w8dk

Dash, K. C. (2008). *Regionalism in South Asia: Negotiating Cooperation, Institutional Structures*. Routledge., DOI: 10.4324/9780203930366

Gössling, S., & Hall, C. M. (2006). *Tourism and global environmental change: Ecological, social, economic and political interrelationships*. Routledge., DOI: 10.4324/9780203011911

Hettne, B., & Söderbaum, F. (2005). Civilian power or soft imperialism? The EU as a global actor and the role of interregionalism. *European Foreign Affairs Review, 10*(4), 535–552. DOI: 10.54648/EERR2005036

Hooghe, L., & Marks, G. (2003). Unraveling the central state, but how? Types of multi-level governance. *The American Political Science Review, 97*(2), 233–243.

ICIMOD. (2021). *The Hindu Kush Himalaya assessment: Mountains, climate change, sustainability and people*. Springer Nature. https://lib.icimod.org/record/34383

International Ecotourism Society (TIES). (2015). *What is ecotourism?* Retrieved from https://ecotourism.org/what-is-ecotourism/

Jeelani, P., Ahad, F., Shah, S. A., & Rashid, H. (2025). *Site suitability evaluation for nature-based tourism using gis and ahp: a case study of Kashmir Valley*. Applied Geomatics.

Jeelani, P., & Shah, S. A. (2023). Tourism and the environmental components coexisting in the Himalayas: A trade-off between the two. *Journal of Quality Assurance in Hospitality & Tourism*, •••, 1–25.

Jeelani, P., Shah, S. A., Dar, S. N., Ahad, F., & Rashid, H. (2024). Managing mass tourism in mountain ecosystems through instrument of sustainable tourism indicators. *Environment, Development and Sustainability*, •••, 1–28. DOI: 10.1007/s10668-024-04915-9

Kelly, I. (2006). Peace through tourism: A SWOT analysis. *Journal of Tourism Studies*, *17*(1), 36–46.

Kim, H. J., Choi, H. C., Agrusa, J., & Lee, H. (2007). The role of tourism in the inter-Korean relationship. *Journal of Tourism Studies*, *18*(2), 121–136.

Lemos, M. C., & Agrawal, A. (2006). Environmental governance. *Annual Review of Environment and Resources*, *31*(1), 297–325. DOI: 10.1146/annurev.energy.31.042605.135621

Montville, J. V. (1991). Track two diplomacy: The work of healing history. *Whitehead Journal of Diplomacy and International Relations*, *2*(1), 15–25.

Najam, A., Huq, S., & Sokona, Y. (2003). Climate negotiations beyond Kyoto: Developing countries' concerns and interests. *Climate Policy*, *3*(3), 221–231. DOI: 10.3763/cpol.2003.0329

Nepal, S. K. (2000). Tourism in protected areas: The Nepalese Himalaya. *Annals of Tourism Research*, *27*(3), 661–681. DOI: 10.1016/S0160-7383(99)00105-X

Nüsser, M., & Schmidt, S. (2017). Nanga Parbat revisited: Evolution and challenges of tourism development in Pakistan's mountain areas. *Mountain Research and Development*, *37*(2), 153–162.

Scheyvens, R. (1999). Ecotourism and the empowerment of local communities. *Tourism Management*, *20*(2), 245–249. DOI: 10.1016/S0261-5177(98)00069-7

Scheyvens, R. (2000). Promoting women's empowerment through involvement in ecotourism: Experiences from the Third World. *Journal of Sustainable Tourism*, *8*(3), 232–249. DOI: 10.1080/09669580008667360

Sharma, D., Singh, R., & Singh, R. K. (2020). Regional climate governance in South Asia: An assessment of cooperative initiatives. *Climate and Development, 12*(1), 30–42. DOI: 10.1080/17565529.2019.1603734

Spenceley, A., & Goodwin, H. (2007). Nature-based tourism and poverty alleviation: Impacts of private sector and parastatal enterprises in and around Kruger National Park, South Africa. *Current Issues in Tourism, 10*(2–3), 255–277. DOI: 10.2167/cit305.0

Stronza, A., Hunt, C. A., & Fitzgerald, L. A. (2019). Ecotourism for conservation? *Annual Review of Environment and Resources, 44*(1), 229–253. DOI: 10.1146/annurev-environ-101718-033046

Timothy, D. J. (2001). *Tourism and political boundaries.* Routledge., DOI: 10.4324/9780203214480

UNWTO. (2017). *Tourism for Sustainable Development in Least Developed Countries*. United Nations World Tourism Organization., DOI: 10.18111/9789284418848

Wani, M. D., Ahmad Shah, S., & Ahmad Dada, Z. (2025d). Eco-conscious tourism in India: Investigating the interplay between green hotel practices and traveler decision-making. *Tourism and Hospitality Research, 14673584251334238*, 14673584251334238. Advance online publication. DOI: 10.1177/14673584251334238

Wani, M. D., Akhter, S., Shah, S. A., & Dada, Z. A. (2025f). Does Mountain Tourism Impact the Socio-Ecological Balance of the Indian Trans Himalayas, Ladakh? In *Balancing Mountain Tourism, Cultural Heritage, and Environmental Stability* (pp. 89-108). IGI Global Scientific Publishing. DOI: DOI: 10.4018/979-8-3693-8764-1.ch006

Wani, M. D., Baba, S. N., Shah, S. A., & Dada, Z. A. (2025e). Digital Nomads as Catalysts for Change: Enhancing Ethical Travel and Sustainability in Indian Tourism. In *Transforming Remote Work and Tourism in the Digital Nomad Era* (pp. 279-302). IGI Global Scientific Publishing.

Wani, M. D., Batool, N., Dada, Z. A., & Shah, S. A. (2024c). Investigating the impact of community-based tourism on the residents' quality of life and their support for tourism. *Community Development (Columbus, Ohio), 55*(1), 138–159. DOI: 10.1080/15575330.2023.2272271

Wani, M. D., Dada, Z. A., Bhat, W. A., & Shah, S. A. (2025b). Community-driven ecotourism in the trans Himalayan region: A sustainable model for cultural and environmental preservation. *Community Development (Columbus, Ohio), 56*(4), 1–23. DOI: 10.1080/15575330.2025.2451358

Wani, M. D., Dada, Z. A., & Shah, S. A. (2023a). Behavioral intention and satisfaction levels among pilgrim tourists in South Asia. *Tourism Review*, *78*(4), 1147–1163. Advance online publication. DOI: 10.1108/TR-09-2022-0432

Wani, M. D., Dada, Z. A., & Shah, S. A. (2023b). Evaluating the impact of cross Border tourism and people to people contact on the regional cooperation in South Asia: A study of India and Bangladesh. *Journal of Quality Assurance in Hospitality & Tourism*, •••, 1–22. DOI: 10.1080/1528008X.2023.2262137

Wani, M. D., Dada, Z. A., & Shah, S. A. (2023c). To utilize structural equation modelling to investigate the impact of cross-border tourism on regional cooperation: The mediating effect of community support. *The Review of Regional Studies*, *53*(1), 80–99. DOI: 10.52324/001c.74889

Wani, M. D., Dada, Z. A., & Shah, S. A. (2024a). The impact of community empowerment on sustainable tourism development and the mediation effect of local support: A structural equation modeling approach. *Community Development (Columbus, Ohio)*, *55*(1), 50–66. DOI: 10.1080/15575330.2022.2109703

Wani, M. D., Dada, Z. A., & Shah, S. A. (2024b). Can consumption of local food contribute to sustainable tourism? Evidence from the perception of domestic tourists. *International Journal of Sustainable Development and World Ecology*, *31*(2), 163–178. DOI: 10.1080/13504509.2023.2262950

Wani, M. D., Dada, Z. A., & Shah, S. A. (2024d). Divine Experiences: Exploring the Impact of Tourist Satisfaction on Religious Travel in Jammu and Kashmir, India. In Jermsittiparsert, K., Wekke, I., & Suanpang, P. (Eds.), *Sustainable Development, Humanities, and Social Sciences for Society 5.0* (pp. 387–414). IGI Global Scientific Publishing., DOI: 10.4018/979-8-3693-7989-9.ch019

Wani, M. D., Dada, Z. A., & Shah, S. A. (2024e). What factors influence consumers' intention and food waste reduction perceptions in Indian restaurants? *British Food Journal*. Advance online publication. DOI: 10.1108/BFJ-07-2024-0710

Wani, M. D., Dar, S. N., & Mohanty, P. P. (2025g). Integrating the Value Belief Norm Theory and Theory of Planned Behavior to Predict the Climate Change Mitigation and Adaption Behaviors in Agriculture Production. *Environmental Management*, *75*(10), 1–15. DOI: 10.1007/s00267-025-02217-y PMID: 40588689

Wani, M. D., Ganai, S. A., Wani, S. A., Dada, Z. A., & Shah, S. A. (2025c). Balancing Growth and Preservation: Sustainable Strategies for Overtourism Management. In Kumar, N., Zadeh, S., Naumov, N., & Sharma, S. (Eds.), *Solutions for Managing Overtourism in Popular Destinations* (pp. 1–22). IGI Global Scientific Publishing., DOI: 10.4018/979-8-3693-8347-6.ch001

Wani, M. D., Shah, S. A., & Dada, Z. A. (2025a). Fostering sustainable tourism in India: Harnessing cultural and heritage assets for enduring development. In *Integrating Architecture and Design Into Sustainable Tourism Development* (pp. 179–214). IGI Global Scientific Publishing., DOI: 10.4018/979-8-3693-7388-0.ch007

Wani, M. U. D., Dada, Z. A., & Shah, S. A. (2022). Building peace through tourism: The analysis of an ongoing Siachen Glacier dispute between India and Pakistan. *Asian Journal of Comparative Politics*, *7*(4), 836–848. DOI: 10.1177/20578911221118730

Weissmann, M. (2008). Peacebuilding in East Asia: The role of Track 2 diplomacy, informal networks, and economic, social, and cultural regionalization. In *Conflict management, security and intervention in East Asia* (pp. 83-98). Routledge. https://doi.org/DOI: 10.4324/9780203928288-11

Wieckowski, M. (2013). Eco-frontier in the mountainous borderlands of Central Europe. The case of Polish border parks. *Journal of Alpine Research*|. *Revue de Geographie Alpine*, •••(101-2), 101–102. DOI: 10.4000/rga.2107

Woodley, D. (2017). *Globalization and capitalist geopolitics: Sovereignty and state power in a multipolar world.* Taylor & Francis., DOI: 10.4324/9781315798165

Chapter 5
Climate-Induced Displacement and Cross-Border Cooperation

Hemendra Singh
https://orcid.org/0000-0003-1160-8284
Jindal Global Law School, O.P. Jindal Global University, Sonipat, India

ABSTRACT

This chapter examines the growing phenomenon of climate-induced displacement across borders and the limitations of current international legal frameworks in providing protection. It explores how regional cooperation, soft law instruments and human rights-based approaches are being employed to address the legal and policy gaps that leave climate-displaced persons without formal recognition. Drawing on examples from the Pacific Islands, South Asia and Sub-Saharan Africa, the analysis highlights emerging practices such as humanitarian visas, planned relocation, and regional protocols. The chapter argues for a cooperative and adaptive framework that integrates legal tools with participatory governance to ensure protection and equity. While emphasizing the relevance of climate justice and institutional accountability, it situates displacement governance within broader global objectives of resilience, inclusion and sustainable development.

INTRODUCTION

The phenomenon of climate-induced displacement has rapidly emerged as a defining issue in contemporary global governance, yet it continues to be characterized by fragmented legal responses and underdeveloped normative frameworks. A growing body of academic literature identifies climate change as both a direct

DOI: 10.4018/979-8-3373-3972-6.ch005

and indirect driver of human mobility (Link et al., 2025). Sudden-onset disasters such as hurricanes, floods, and cyclones, as well as slow-onset events like sea level rise, desertification, and salinization, are displacing millions globally (Martínez-Zarzoso et al., 2023). The Internal Displacement Monitoring Centre and reports from the Intergovernmental Panel on Climate Change have documented this surge in displacement. However, the vast majority of this research and policy focus has remained limited to internal displacement, where existing legal regimes, such as the Guiding Principles on Internal Displacement and domestic disaster response frameworks, provide at least a partial foundation for state accountability. Cross-border displacement, by contrast, occupies a much more ambiguous legal and policy terrain.

A central strand of the literature emphasizes the insufficiency of international refugee law in addressing the plight of climate-displaced persons who cross national borders (Rajshree et al., 2025). The 1951 Refugee Convention and its 1967 Protocol are the cornerstone instruments for international refugee protection. Yet, their definition of a refugee does not extend to persons displaced by environmental causes or the broader impacts of climate change. This exclusion has been the subject of extensive doctrinal critique. Scholars such as Jane McAdam and others have argued that the Convention was never designed to address displacement resulting from environmental collapse or climate variability (McAdam, 2011). Attempts to reinterpret the definition through a broader reading of persecution or state failure have met with significant resistance from states and have produced inconsistent outcomes in national asylum jurisprudence. Consequently, climate-displaced persons remain outside the protection mandate of the UNHCR and lack access to formal international protection mechanisms when they cross borders.

In response to these doctrinal limitations, another body of literature has turned to international human rights law as a complementary and potentially more flexible legal framework (Thornton, 2022). Scholars have examined how rights enshrined in treaties such as the International Covenant on Civil and Political Rights and the International Covenant on Economic, Social and Cultural Rights could be interpreted to protect individuals displaced by climate-related conditions. The right to life, the right to adequate housing, the right to health, and the right to an adequate standard of living have all been cited as potentially violated in situations of forced migration due to environmental degradation (Addaney et al., 2019). The landmark case of *Ioane Teitiota v. New Zealand* before the UN Human Rights Committee is frequently cited in this context. Although the Committee did not find in Teitiota's favor, it acknowledged for the first time that the effects of climate change could place individuals at real risk of irreparable harm, thereby triggering non-refoulement obligations under the right to life. This case has opened new legal avenues for considering climate-induced displacement within the human rights framework, even as its practical impact remains nascent.

Parallel to these legal debates, scholars have increasingly focused on regional responses and localized governance frameworks. In the Pacific Islands, for instance, the existential threat posed by sea level rise has generated a growing literature around planned relocation, sovereignty loss, and regional solidarity (McAdam & Ferris, 2015). The notion of migration with dignity, advanced by the government of Kiribati and supported by civil society actors, has been explored as a normative model for proactive mobility in the face of irreversible climate impacts. In South Asia, where cross-border migration between Bangladesh and India has long been shaped by seasonal, economic, and political factors, environmental drivers have intensified these patterns. Researchers have examined how climate change exacerbates pre-existing vulnerabilities in the region, particularly among marginalized communities living in ecologically fragile border areas (Kabir & Kamruzzaman, 2022). Similarly, in Sub-Saharan Africa, the African Union's Kampala Convention, while limited to internal displacement, has been commended for its explicit recognition of natural disasters and climate change as causes of displacement. Scholars working in this domain have emphasized the importance of such regional instruments in adapting existing legal tools to contemporary challenges (Adeola, 2019).

While legal scholars explore doctrinal solutions and regional innovations, another strand of academic work focuses on the development of soft law and policy-based responses (Ferris & Bergmann, 2017). The Nansen Initiative and its successor, the Platform on Disaster Displacement, have been the subject of considerable study for their consultative processes and efforts to build consensus around principles for disaster-related cross-border displacement. These frameworks do not create binding obligations but instead promote a toolbox approach that includes humanitarian visas, temporary protection statuses, and planned relocation programs. Similarly, the Global Compact for Safe, Orderly and Regular Migration has been analyzed for its recognition of climate change as a driver of mobility and its promotion of cooperative migration governance. Scholars have noted that while these instruments lack enforceability, they play a significant role in agenda-setting, norm development, and policy harmonization at the regional and national levels (Guruparan & Zerk, 2021). They also create opportunities for dialogue and cooperation in politically sensitive migration contexts where formal legal commitments may be unattainable.

Despite the expansion of legal, policy, and normative responses in the academic and governance spheres, the central problem remains unresolved. Climate-displaced persons who cross international borders still lack a clearly defined legal status, institutional recognition, and enforceable rights. They remain in a legal limbo, often dependent on the goodwill or discretion of receiving states and without access to durable solutions. This legal and normative vacuum raises profound questions about state responsibility, international cooperation, and the future of human mobility in the Anthropocene. The fragmentation of existing frameworks, the reluctance of states to commit to new binding obligations, and the marginalization of vulnerable communities within migration governance debates underscore the urgency of rethinking how cross-border climate-induced displacement is addressed. This chapter engages with the broader discourse on climate induced displacement by arguing that rights based, regionally grounded cross border cooperation offers a practical pathway to address existing legal and policy gaps. Drawing on examples from the Pacific Islands, South Asia, and Sub-Saharan Africa, it explores how interstate collaboration informed by human rights principles and local realities can generate more responsive and coherent solutions. It highlights the global absence of a defined legal status for climate displaced persons, noting how this gap plays out differently across regions. By emphasizing the significance of regional approaches, the chapter contributes to evolving governance frameworks that aim to protect affected communities more effectively.

Research Methodology

This chapter adopts a doctrinal legal methodology supported by comparative case study analysis. It systematically examines international refugee law, human rights instruments, and soft law frameworks to identify existing gaps and opportunities for protection in the context of climate-induced cross-border displacement. It draws on regional practices and legal innovations in the Pacific Islands, South Asia, and Sub-Saharan Africa to assess how cooperative responses are evolving. Primary sources include international treaties, regional conventions, soft law instruments, and judicial decisions, while secondary sources comprise academic commentary, institutional reports, and policy briefs. The analysis is normative and exploratory in nature, aiming to propose legally grounded and context-sensitive models of cross-border cooperation informed by human rights principles and regional solidarity.

LEGAL GAPS AND EMERGING HUMAN RIGHTS RESPONSES

The growing incidence of climate-induced displacement has exposed foundational inadequacies in the international legal regime governing forced migration. Although climate change is now widely recognized as a driver of human mobility, existing legal frameworks remain ill-equipped to address the complexities of environmentally driven cross-border movement. This chapter begins by evaluating the structural limitations of the 1951 Refugee Convention before turning to human rights-based avenues and regional legal instruments that seek to fill the resulting protection gap.

Inadequacy of the Refugee Convention

The 1951 Convention Relating to the Status of Refugees, together with its 1967 Protocol, constitutes the principal legal instrument governing international refugee protection. It defines a refugee as someone who, owing to a well-founded fear of persecution based on race, religion, nationality, membership of a particular social group, or political opinion, is unable or unwilling to avail themselves of the protection of their country of origin. This definition reflects the geopolitical context of the post–Second World War and Cold War periods, where the primary concern was persecution by the state or non-state actors in politically unstable environments. The Convention was not designed to address displacement triggered by environmental degradation or climate-related phenomena, and its conceptual architecture has remained largely unchanged despite the evolving nature of global displacement.

The exclusion of climate-related drivers from the refugee definition is not incidental but structural. Environmental factors, whether sudden or slow in onset, do not constitute persecution under international refugee law unless they intersect with one of the five protected grounds. Scholarly attempts to broaden the interpretation of persecution to include environmental collapse have generally been unpersuasive in judicial or administrative decision-making. In practice, asylum claims based solely on climate-related grounds are rarely successful, as courts and refugee status determination bodies require a clear nexus to one of the Convention grounds. While there have been isolated efforts to reframe state inaction or unequal exposure to climate risks as forms of discrimination, these approaches remain doctrinally tenuous and lack precedential force in international law.

Further complicating the matter is the Convention's emphasis on individual persecution rather than generalized threats. Climate change, by its nature, produces collective and often indiscriminate harm, undermining the requirement that a claimant must demonstrate a personalized risk of persecution. Even where environmental degradation leads to state failure or social conflict, the burden of proof to establish a causal link between climate factors and persecution is often insurmountable. As a result, displaced persons whose movements are driven by sea-level rise, desertification, or extreme weather events fall outside the Convention's scope, regardless of the severity of the threats they face (Woodworth, 2023).

The institutional mandate of the United Nations High Commissioner for Refugees (UNHCR) also reflects these limitations. While the UNHCR has acknowledged climate change as a factor influencing displacement, its legal protection work remains bounded by the Convention framework unless expanded through complementary protection mechanisms at the national level. States have not shown willingness to renegotiate or amend the Convention to include climate-related causes of displacement, and any reinterpretation through customary international law faces significant political and legal resistance. The result is a protection regime that remains doctrinally stable but functionally obsolete in the face of new forms of forced migration.

The inadequacy of the Refugee Convention in the context of climate-induced displacement is not simply a matter of outdated language but a deeper reflection of normative design. The Convention's limited scope, combined with states' reluctance to expand international obligations, creates a legal vacuum for individuals crossing borders due to climate-related harm. While it remains a critical instrument for traditional refugee protection, its capacity to address emerging displacement scenarios driven by environmental change is fundamentally constrained.

Expanding Protection through Human Rights Law

In light of the limitations of the 1951 Refugee Convention, scholars and practitioners have increasingly turned to international human rights law as a supplementary framework for addressing climate-induced displacement. Unlike refugee law, which is narrowly confined to persecution on enumerated grounds, human rights law offers a broader normative basis to evaluate state obligations toward individuals facing threats to life, dignity, and subsistence as a result of environmental degradation. Although human rights instruments do not create a dedicated status or procedural protection for climate-displaced persons, they articulate core obligations that may, under certain conditions, limit a state's discretion to expel or refuse entry to individuals whose survival is at risk due to climate-related conditions in their country of origin.

A pivotal development in this context is the decision of the United Nations Human Rights Committee in *Ioane Teitiota v. New Zealand* (2020). Teitiota, a national of Kiribati, sought asylum in New Zealand on the grounds that rising sea levels and environmental degradation in his home country posed a serious threat to his and his family's lives. While the Committee ultimately upheld New Zealand's decision to deport Teitiota, it affirmed, for the first time, that climate-related conditions could, in principle, trigger non-refoulement obligations under Article 6 of the International Covenant on Civil and Political Rights, which protects the right to life (Hatano, 2021). The Committee stated that without robust national and international efforts, climate change may expose individuals to conditions that violate fundamental rights and that, under such circumstances, returning individuals to those environments could constitute a breach of international law.

This recognition is significant, not because it created a new category of protected persons, but because it anchored climate-induced risk within existing legal obligations. The ruling affirmed that human rights protections are not contingent on the classification of an individual as a refugee and that the right to life must be interpreted in light of foreseeable environmental harm. However, the threshold for protection remains high. The Committee emphasized that such risk must be immediate, foreseeable, and life-threatening. It also acknowledged the responsibility of states of origin to mitigate those risks, suggesting that human rights-based non-refoulement may only apply where state failure is manifest and irreparable. This doctrinal positioning limits the broader applicability of the ruling but nonetheless creates a legal opening for future claims grounded in environmental conditions.

Beyond the right to life, other human rights norms offer interpretive potential for addressing climate displacement. Article 11 of the International Covenant on Economic, Social and Cultural Rights, which guarantees the right to an adequate standard of living including food, housing, and continuous improvement of living conditions, has been invoked in the context of environmental degradation. The relationship between environmental harm and the rights to health, water, and housing has been increasingly acknowledged by treaty bodies, special rapporteurs, and regional human rights courts. The Inter-American Court of Human Rights, in its Advisory Opinion OC-23/17, recognized that environmental degradation can violate the rights of individuals even across borders, reinforcing the notion that environmental risk is not confined within territorial boundaries.

Regional human rights systems have also shown greater willingness to engage with environmental claims. The African Charter on Human and Peoples' Rights explicitly links human rights and environmental protection. In *Social and Economic Rights Action Center v. Nigeria*, the African Commission held that environmental degradation resulting from state action or inaction could constitute a violation of the Charter. Although these cases have not directly addressed climate-induced displacement, they lay the groundwork for asserting that states have positive obligations to prevent foreseeable harm arising from environmental collapse, including where such harm leads to forced mobility.

Nevertheless, the human rights approach is not without limitations. Unlike refugee law, which provides a structured status and associated procedural safeguards, human rights protections are typically reactive and individualized. They depend on a threshold of harm that is context-specific and are adjudicated on a case-by-case basis, often through slow-moving judicial or quasi-judicial mechanisms. Moreover, enforcement remains a persistent challenge. Treaty bodies can issue findings but lack binding enforcement power, and states frequently contest or ignore decisions that imply extraterritorial responsibility. As such, while human rights law expands the normative horizon for climate-displaced persons, it does not, in itself, create a comprehensive or operational framework for protection.

The interpretive space within human rights law, though legally and institutionally constrained, offers important tools for bridging the protection gap left by refugee law. By recognizing that certain climate-related conditions can trigger binding state obligations under the right to life and related rights, international human rights law introduces a legal vocabulary for asserting protection in cross-border contexts. Its value lies not only in potential legal remedies but also in its capacity to shape state behavior, influence soft law developments, and support the emergence of a rights-based approach to climate mobility.

Regional and Normative Developments

While the international legal regime remains slow to adapt to the realities of climate-induced displacement, regional systems have begun to experiment with more responsive frameworks. These developments are significant not only because they reflect normative innovation within specific legal cultures, but also because they point toward pathways for tailored, cooperative responses that are more immediately attuned to the environmental vulnerabilities of particular regions. Although most regional efforts have not yet resulted in binding instruments addressing cross-border climate displacement directly, they nonetheless contribute to the gradual expansion of legal consciousness around the issue and offer important insights for global governance reform.

In Africa, the African Union's Kampala Convention on the Protection and Assistance of Internally Displaced Persons in Africa (2009) is frequently cited as a progressive legal instrument. While the Convention focuses on internal displacement and does not extend protection to cross-border migrants, it is notable for explicitly recognizing natural disasters and climate change as legitimate causes of displacement (Dieng, 2018). Article 5(4) obligates state parties to take necessary measures to protect and assist persons displaced due to natural or human-made disasters, including those induced by climate change. This formal acknowledgment of environmental displacement as a legal category provides a crucial precedent and signals a regional willingness to conceptualize displacement beyond conflict-driven paradigms. Moreover, it expands the scope of state responsibility by embedding environmental vulnerability within the framework of internal protection and humanitarian assistance. Although the Convention's application is territorially limited, it contributes to the broader normative discourse on the legal status of environmentally displaced persons and may serve as a reference point for future cross-border frameworks.

In Latin America, regional cooperation in the area of forced migration has historically been robust, driven by both conflict-related and socio-economic displacement. The Cartagena Declaration on Refugees (1984), though non-binding, expanded the definition of a refugee to include those fleeing generalized violence, massive human rights violations, and other circumstances that seriously disturb public order. While environmental factors are not explicitly included, some scholars argue that the Declaration's broad formulation could be interpreted to encompass displacement caused by large-scale environmental disruption, particularly where such conditions intersect with governance failures or exacerbate existing instability.

The Cartagena process has also inspired national laws in several countries, such as Brazil and Mexico, which adopt expanded refugee definitions and could, in principle, offer greater flexibility in climate-related displacement contexts.

In the Pacific region, small island developing states face existential threats from sea level rise and saltwater intrusion, prompting legal and policy innovations that reflect their unique vulnerabilities. While no formal regional treaty addresses cross-border displacement due to climate change, there is increasing engagement with the idea of planned relocation, migration with dignity, and inter-state cooperation. Countries like Fiji have expressed an openness to accept displaced populations from at-risk states such as Tuvalu and Kiribati. Though these commitments are not codified in binding legal instruments, they reflect an emerging norm of regional solidarity and responsibility-sharing grounded in shared cultural, geographic, and ecological ties. This normative posture, visible in public statements, national policies, and regional dialogue, challenges the conventional assumption that protection must emerge exclusively from binding international instruments. Instead, it suggests that normative leadership, even if politically and informally articulated, can shape cooperative responses to climate mobility.

These regional developments also interact with broader human rights norms. The African Charter on Human and Peoples' Rights, the American Convention on Human Rights, and the Pacific Islands Framework for Resilient Development all reflect varying degrees of normative alignment between environmental protection, state accountability, and the rights of displaced persons. While these instruments do not uniformly incorporate climate displacement as a specific legal category, their interpretive openness allows for the articulation of emerging obligations. In some instances, regional human rights courts and commissions have recognized that environmental degradation can constitute a human rights violation, thereby offering displaced persons potential legal entry points through existing norms.

Nonetheless, significant challenges remain. Regional instruments often suffer from weak enforcement mechanisms, limited ratification, and resource constraints. Moreover, they tend to prioritize internal displacement, leaving cross-border dimensions of climate mobility underdeveloped. The uneven political will among states, combined with sensitivities around sovereignty and border control, hinders the creation of binding obligations or structured mobility pathways. Despite these limitations, regional normative developments demonstrate that legal imagination and political responsiveness are not confined to global processes. They provide critical laboratories for experimentation and serve as a foundation upon which more comprehensive and cross-border mechanisms might be built.

Regional and normative innovations, though partial and often fragmented, represent essential building blocks in the evolving legal response to climate-induced displacement. They show that law is not static and that regional legal cultures can be sites of meaningful development even in the absence of global consensus. Recognizing and building upon these efforts is vital for designing cooperative, context-sensitive, and rights-based responses to one of the defining challenges of climate governance.

CROSS BORDER COOPERATION IN PRACTICE

Despite the absence of binding international legal instruments explicitly addressing climate-induced cross-border displacement, a range of regional and bilateral practices has begun to take shape in response to the complex realities of climate mobility. These responses, though varied in form and legal status, reflect an emerging recognition among states that displacement linked to environmental degradation cannot be addressed through isolated national policies alone. This chapter examines three regional contexts - Pacific Islands, South Asia, and Sub-Saharan Africa where climate-induced mobility has prompted distinctive forms of cooperation, informal accommodation, and normative development. The analysis considers both the opportunities and limitations these examples present for advancing structured, rights-based approaches to cross-border displacement.

Pacific Islands: Regional Solidarity and Legal Willingness

The Pacific Islands region has emerged as a critical site for examining cross-border cooperation in the context of climate-induced displacement, due to its acute exposure to sea-level rise, coastal erosion, and salinization. These environmental processes threaten the very habitability of small island states such as Tuvalu and Kiribati, raising profound questions regarding the continuity of statehood, population relocation, and the responsibilities of neighboring countries. While no binding regional treaty currently addresses cross-border climate displacement, the region has witnessed notable expressions of legal willingness and solidarity that distinguish it from more securitized or restrictive approaches elsewhere.

Among the most significant developments is the public commitment by the government of Fiji to accept populations from neighboring atoll nations whose territories may become uninhabitable due to rising seas. Fiji's 2015 statement before the United Nations General Assembly affirmed its readiness to open migration pathways for climate-displaced populations from Tuvalu and Kiribati under conditions of dignity and planned relocation. Although this offer is not codified in a bilateral treaty, it has important normative significance. It signals a regional understanding of displacement as a shared humanitarian concern and not merely a matter of domestic or border control. This orientation is further reinforced by the existence of relocation-related mechanisms within Fiji, such as the Climate Relocation of Communities Trust

Fund, which though primarily focused on internal resettlement reflects a governance infrastructure capable of anticipating and managing population movement due to climate risks (McMichael et al., 2019).

These developments illustrate an emerging regional ethic of solidarity, rooted in shared histories, cultural affinities, and collective vulnerabilities. Pacific Island states have long emphasized principles of mutual support and regional cooperation through the Pacific Islands Forum and similar platforms. This ethos informs not only diplomatic rhetoric but also practical collaboration, as seen in regional frameworks like the Framework for Resilient Development in the Pacific. Although these instruments do not create binding obligations regarding cross-border displacement, they promote integrative approaches to climate resilience that recognize the need for coordinated responses to human mobility.

Despite these encouraging developments, important challenges remain. There is limited legal clarity regarding the status, rights, and entitlements of persons relocated across borders due to climate-related threats. Without a formalized legal arrangement, populations accepted by another state may not automatically enjoy permanent residency or citizenship, raising concerns about their long-term integration and political status. Moreover, the feasibility of large-scale relocation is constrained by economic, infrastructural, and political limitations within receiving states such as Fiji, which are themselves vulnerable to climate impacts. Questions of land allocation, cultural continuity, and self-determination further complicate the design of just and sustainable relocation schemes.

Nonetheless, the Pacific experience offers a unique and instructive model of proactive, ethically grounded cooperation in the absence of binding international law. It demonstrates how regional leadership, when aligned with human rights principles and cultural cohesion, can fill normative and operational gaps in global protection frameworks. Rather than treating mobility as a failure of adaptation, these practices recognize it as a legitimate, even necessary, form of resilience. While far from comprehensive, the willingness of Pacific states to engage in anticipatory planning and regional dialogue sets a constructive precedent for other climate-vulnerable regions.

South Asia: Unregulated Climate Mobility

South Asia is one of the world's most climate-vulnerable regions, experiencing a combination of slow-onset changes such as sea-level rise and salinization, and sudden-onset disasters including floods and cyclones. Bangladesh, in particular, faces significant ecological stress due to its low-lying geography and high population density. Coastal erosion, waterlogging, and the intrusion of saline water into agricultural lands have contributed to the loss of livelihoods and compelled growing numbers of people to relocate. While much of this movement remains internal, there is increasing evidence of informal, cross-border mobility into India as a form of adaptive response to environmental degradation (Gupta et al., 2025). This movement, however, is largely unregulated and occurs in the absence of legal recognition or protection for climate-displaced persons.

The migration from Bangladesh to India cannot be attributed solely to environmental factors, but climate-induced pressures have undoubtedly intensified existing patterns of movement. Seasonal and circular migration across the India–Bangladesh border has a long history, often driven by a mix of economic necessity and ecological hardship. As land becomes less productive and homes are lost to erosion and flooding, displaced individuals seek survival and employment opportunities in adjacent Indian states, particularly West Bengal and Assam. This migration, however, takes place through informal channels and is frequently characterized as irregular, rendering migrants vulnerable to detention, deportation, and systemic exclusion from services. Because neither Bangladesh nor India formally recognizes climate displacement as a category within their migration frameworks, affected persons fall outside the scope of legal protection, exacerbating their precarity.

India's domestic legal system does not have a statutory framework for recognizing refugees or climate-displaced persons, and migration management remains largely rooted in national security and sovereignty concerns. The Foreigners Act, 1946, provides wide discretionary powers to the executive, and undocumented migrants are often subject to criminalization without adequate procedural safeguards. Political discourse in several Indian states has, at times, depicted cross-border migration from Bangladesh as a security or demographic threat, limiting space for humanitarian or climate-sensitive approaches. Bangladesh, for its part, does not publicly acknowledge large-scale out-migration driven by climate pressures, in part due to diplomatic sensitivities and concerns over sovereignty. This mutual reluctance has impeded the development of any bilateral framework to manage environmentally driven migration, let alone offer protection or status to those affected.

The absence of formal cooperation has also undermined the potential for joint planning or regional burden-sharing. Although institutions such as the Bay of Bengal Initiative for Multi-Sectoral Technical and Economic Cooperation (BIMSTEC) and the South Asian Association for Regional Cooperation (SAARC) have addressed disaster risk reduction and climate resilience, their engagement with cross-border mobility remains minimal. The Dhaka Declaration of 2019, adopted by SAARCLAW, acknowledged the need to address the legal and institutional dimensions of climate-induced displacement, but no concrete regional instrument has followed. This vacuum leaves states to respond reactively and unilaterally, often with inconsistent or repressive policies that fail to account for the complex drivers of migration.

Yet South Asia's long-standing traditions of mobility, cultural affinity across borders, and shared ecological challenges suggest that there is untapped potential for cooperative governance. Informal migration continues despite legal and political resistance, indicating that climate mobility is a reality that will not be contained by existing policy silos. Addressing this challenge requires a reframing of displacement not as a threat but as a human security issue demanding coordinated responses. In the absence of a regional legal instrument, bilateral agreements on temporary protection, labor mobility, or climate resilience cooperation could serve as important interim mechanisms. However, such solutions would require sustained political will and a shift away from securitized narratives toward a more humane and forward-looking framework.

Sub-Saharan Africa: Lessons from the Kampala Convention

Sub-Saharan Africa presents a complex landscape of environmental vulnerability, protracted displacement, and evolving regional governance. While the majority of climate-induced displacement in the region remains internal, cross-border dimensions are increasingly visible, particularly in areas affected by drought, desertification, and food insecurity. The African Union's Convention for the Protection and Assistance of Internally Displaced Persons in Africa, commonly known as the Kampala Convention, represents a landmark regional effort to address displacement in a comprehensive and forward-looking manner. Although the Convention applies only to internal displacement, its explicit recognition of climate change and natural disasters as legitimate causes of displacement offers important normative insights for future cross-border cooperation and legal development (Hodgkinson et al., 2010).

Adopted in 2009 and entering into force in 2012, the Kampala Convention is the first binding regional instrument in the world to impose legal obligations on states for the protection and assistance of internally displaced persons, including those displaced by environmental and climate-related factors. It obligates state parties to prevent displacement, protect displaced populations, and provide durable solutions, including resettlement and reintegration. Article 5(4) specifically requires states to take measures to protect individuals and communities displaced by natural or human-made disasters, thereby explicitly linking environmental change with displacement governance. While this provision is territorially confined, it signals a normative evolution that recognizes the agency of individuals affected by climate change and the responsibility of states to respond in a rights-based manner.

The significance of the Kampala Convention lies not only in its legal content but also in its institutional framing. It embeds displacement within a broader human rights and development framework, emphasizing prevention, non-discrimination, and participation of affected communities. This contrasts with more traditional humanitarian models that treat displacement primarily as an emergency to be managed rather than a structural issue rooted in governance, land tenure, and environmental management. Moreover, the Convention encourages national implementation through domestic legislation and coordination among regional bodies, facilitating normative diffusion and capacity building across member states. Several countries, including Uganda, Niger, and Nigeria, have taken steps to incorporate its principles into national law and policy, although implementation remains uneven.

Despite its internal focus, the Kampala Convention offers important lessons for cross-border displacement. It demonstrates that regional legal instruments can articulate context-specific obligations that reflect local realities and political sensibilities. Its formulation benefited from extensive consultation with civil society, displaced persons, and regional stakeholders, suggesting that participatory norm-making is not only feasible but also essential for legitimacy and effectiveness. The Convention's emphasis on prevention and durable solutions aligns closely with the adaptive strategies required for climate mobility and could inform the development of regional protocols that extend protection beyond national borders. It also reinforces the importance of anchoring displacement governance in existing human rights frameworks rather than creating parallel, siloed regimes.

Nonetheless, the Kampala Convention's limitations must be acknowledged. Its territorial scope excludes cross-border displacement, even when environmental or climate conditions make it impossible for communities to remain in their countries of origin. The lack of corresponding instruments at the regional or continental level to address such movement leaves a critical protection gap. In practice, displaced persons who cross borders due to climate-related harms remain outside the coverage of the Convention and are subject to the discretion of host states, most of which do not recognize climate factors as grounds for protection under existing refugee or immigration law. Moreover, the institutional capacity to monitor and enforce compliance with the Convention remains weak, and political instability in several member states has impeded full implementation.

Despite these shortcomings, the Kampala Convention stands as a rare example of a regional legal framework that recognizes environmental factors as part of the displacement landscape. Its principles provide a valuable foundation for imagining a future instrument that might extend similar protection to those who are compelled to cross international borders due to climate-related threats. It also underscores the importance of legal clarity, institutional coordination, and community participation in designing effective responses to climate-induced displacement. As such, Sub-Saharan Africa offers not only a model of normative innovation but also a reminder of the work that remains to build inclusive and legally grounded cross-border frameworks in the face of a changing climate.

SOFT LAW INSTRUMENTS AND POLICY INNOVATION

In the absence of binding international legal frameworks governing climate-induced cross-border displacement, soft law instruments and policy initiatives have played a central role in shaping state responses and advancing normative development. These instruments, though non-binding, have helped build consensus around principles, encouraged policy harmonization, and facilitated cooperative action among states and international organizations. This chapter explores the evolution and influence of key soft law mechanisms, particularly the Nansen Initiative and the Global Compact for Migration, and assesses how they have contributed to shaping national and regional approaches to climate mobility. It also evaluates the strengths and limitations of these instruments as tools for protection and policy innovation in an evolving legal landscape.

The Nansen Initiative and the Global Compact

The Nansen Initiative, established in 2012 as a state-led consultative process, was a direct response to the legal and policy vacuum surrounding cross-border displacement caused by disasters and the effects of climate change. While existing international refugee and human rights law did not provide adequate protection for individuals displaced by such factors, the initiative sought to develop a consensus-based approach rooted in state practice and soft law guidance. The initiative did not aim to create a new binding legal instrument; rather, it collected evidence through regional consultations in disaster-prone and climate-vulnerable areas, including the Pacific Islands, the Horn of Africa, and South Asia. The process culminated in the 2015 Protection Agenda, which identified effective practices and policy options already in use by states and encouraged their adoption within national and regional frameworks. These included the use of humanitarian visas, temporary protection measures, and the planned relocation of communities at risk. The Protection Agenda was endorsed by more than 100 states and institutions and has since served as a normative benchmark for addressing disaster and climate-related displacement, despite its non-binding status (Stojanov et al., 2021).

Following the conclusion of the Nansen Initiative, the Platform on Disaster Displacement was established to support the operationalization of the Protection Agenda. The Platform has worked to integrate disaster displacement into ongoing global processes such as the Sendai Framework for Disaster Risk Reduction, the UN Framework Convention on Climate Change, and the Global Forum on Migration and Development. It has facilitated partnerships between governments, international organizations, and civil society to support the development of practical tools for managing displacement, including data systems, early warning mechanisms, and legal guidance for temporary admission and residence. The Platform has contributed to maintaining political attention on displacement linked to climate change and has played a key role in shaping the technical and policy infrastructure needed for cooperative approaches across borders.

The Global Compact for Safe, Orderly and Regular Migration, adopted in 2018, marked a significant normative development by formally recognizing climate change and environmental degradation as drivers of migration. Although it is a soft law instrument and lacks binding legal authority, the Compact was endorsed by the majority of United Nations member states and is the first intergovernmental agreement of its kind to frame climate-related mobility within the broader context of migration governance. It encourages states to develop planned relocation frameworks, to facilitate access to temporary protection and humanitarian visas, and to enhance international cooperation in managing environmentally driven movements. It also promotes investment in climate change adaptation and disaster risk reduction in countries of origin to address the root causes of displacement. These provisions are particularly important for countries that face both out-migration and the responsibility to host displaced populations but lack formal legal obligations under international refugee law.

The Global Compact's references to climate mobility are framed within a development and protection-oriented paradigm, highlighting the importance of legal and policy innovation without directly challenging state sovereignty. Its non-prescriptive nature allows for adaptation to national contexts while providing a shared vocabulary and common goals for international cooperation. In doing so, it reinforces the broader trend in soft law instruments toward facilitating protection through flexible, state-led mechanisms rather than codified entitlements. While it stops short of establishing new rights, its emphasis on partnership, data sharing, and preventive measures provides a platform for practical engagement and regional coordination, particularly in regions such as the Pacific and West Africa, where climate displacement is already being addressed through national or bilateral policy responses.

Assessment and Future Directions

Soft law instruments have emerged as central, albeit provisional, tools for managing climate-induced cross-border displacement in the absence of binding international legal norms. Their primary strength lies in their ability to respond flexibly to evolving realities, foster consensus among a broad range of actors, and offer states guidance without requiring formal ratification or treaty obligations. Instruments such as the Nansen Initiative's Protection Agenda and the Global Compact for Migration have encouraged national and regional authorities to develop tailored approaches suited to their political and environmental contexts. They have enabled the incorporation of climate displacement considerations into migration planning, disaster response, and climate adaptation frameworks, particularly in countries and regions that lack domestic legal recognition of this form of displacement. However, the utility of these instruments cannot be fully understood without a critical assessment of their limitations and their capacity to catalyze more formal legal and institutional arrangements over time.

One of the principal advantages of soft law is its normative elasticity. It allows states to engage in cooperative governance on sensitive issues such as migration without committing to binding obligations. This has proved especially useful in a politically polarized global context where consensus on treaty-based commitments remains elusive. Soft law frameworks also enable experimentation with new legal concepts and practices, such as humanitarian visas, regional relocation schemes, and anticipatory planning tools, which might later form the basis for binding instruments. For example, some Pacific Island states have drawn upon the Protection Agenda to develop national relocation strategies, and several countries in Latin America and Africa have incorporated disaster-related mobility into immigration or asylum procedures. These instances suggest that soft law can play a norm-shaping role that progressively informs domestic legal systems.

However, the limitations of soft law mechanisms are both structural and operational. Their non-binding nature means that implementation is entirely dependent on the political will and administrative capacity of individual states. Without mechanisms for oversight, accountability, or enforcement, commitments made under soft law frameworks can remain aspirational or symbolic. Even where soft law provisions are reflected in domestic policy, there is often a lack of continuity, coherence, or long-term institutional support. This is particularly problematic in regions marked by political instability, resource constraints, or competing policy priorities. Moreover, the fragmented nature of soft law regimes can result in divergent national responses and inconsistencies in how climate-displaced persons are treated across jurisdictions, undermining the goal of coordinated and equitable protection (Chetail, 2023).

Another challenge concerns the visibility and agency of affected populations within these policy frameworks. While soft law instruments often emphasize participatory approaches, in practice, displaced communities and civil society actors may be only marginally involved in the development or implementation of soft law responses. This raises questions about the legitimacy and responsiveness of these instruments to the lived realities of those most directly impacted by climate-induced mobility. Furthermore, soft law's reliance on voluntary adoption and self-reporting means that protection outcomes for displaced persons may remain contingent and uneven, particularly in countries where migration is viewed primarily through a security or sovereignty lens.

Looking forward, one of the most pressing questions is whether and how soft law norms can be translated into binding regional or international legal instruments. There is growing scholarly and policy interest in the development of model protocols or regional compacts that would formalize the commitments articulated in instruments such as the Global Compact. These could take the form of regional agreements, modeled for instance on the African Union's Kampala Convention, which though limited to internal displacement, demonstrates the potential of regional legal innovation in the face of global legal stagnation. Another possibility is the integration of climate displacement into existing human rights monitoring mechanisms, including periodic state reporting to treaty bodies and regional human rights courts, thereby creating avenues for indirect accountability.

A related trajectory involves enhancing institutional convergence across the fragmented legal regimes currently addressing displacement, migration, disaster risk, and climate change. Greater coordination between the UN Framework Convention on Climate Change, the Global Compact on Migration, and regional human rights systems could help build a more coherent and durable protection architecture. In this regard, the role of international organizations, including the International Organization for Migration, UNHCR, and the Platform on Disaster Displacement, will remain crucial in facilitating dialogue, generating data, and supporting capacity-building initiatives.

Soft law instruments are not a substitute for binding legal protection, but they offer a valuable intermediary space where states and institutions can converge on shared principles and pilot innovative responses. Their value lies in enabling normative evolution, fostering institutional experimentation, and laying the groundwork for more formalized legal obligations. Future progress will depend on sustained political engagement, cross-sectoral collaboration, and the meaningful inclusion of affected communities in shaping the instruments that govern their movement and rights.

TOWARD A COOPERATIVE RIGHTS BASED FRAMEWORK

The inadequacy of existing international legal instruments to address cross-border climate-induced displacement underscores the urgent need for a cooperative framework that is both legally grounded and responsive to regional realities. Building on the legal analyses and regional practices discussed in earlier chapters, this section examines how states can adopt rights-based legal tools and governance mechanisms to enhance protection and coordination. A cooperative framework must not only facilitate mobility but also ensure equity, participation, and sustainability in responses to displacement across borders.

Legal and Policy Tools for Protection

The protection of individuals displaced across borders due to climate-related causes demands a recalibration of legal and policy instruments at both national and regional levels. In the absence of a universally accepted legal status for climate-displaced persons, emerging state practices and regional proposals have centered on a range of legal mechanisms intended to provide temporary or permanent protection while respecting state sovereignty. These mechanisms include temporary protection measures, humanitarian visas, regional mobility frameworks, and planned relocation policies. While these tools vary in their legal status and implementation, they collectively represent a shift toward more responsive and rights-conscious approaches to climate-induced displacement, particularly when grounded in existing human rights obligations and developed through cooperative governance.

Temporary protection and humanitarian visas are among the most immediately deployable tools for responding to cross-border climate displacement. These measures offer legal stay for individuals who cannot safely return to their country of origin due to environmental degradation or disaster-related conditions. Though primarily designed for short-term or emergency situations, they have been invoked in some jurisdictions to address protracted displacement linked to slow-onset events such as drought, desertification, and sea-level rise. For example, some Latin American and Caribbean states have integrated disaster-based displacement into immigration policies, granting temporary residency or regularization to persons affected by environmental emergencies. Such practices demonstrate the legal and operational feasibility of humanitarian entry mechanisms even in the absence of refugee status under the 1951 Convention (Matias, 2020). The use of these tools can be further formalized through regional agreements or bilateral arrangements, ensuring consistency in access and safeguards against arbitrary removal.

Another important legal approach involves the development of regional protocols tailored to climate displacement. Regionalism offers both normative flexibility and contextual sensitivity, allowing states with shared vulnerabilities to develop frameworks that reflect local realities and political will. For instance, the 2018 draft of the Regional Pact on Climate Mobility in the Americas and discussions within the Pacific Islands Forum suggest growing interest in developing legal instruments that acknowledge climate-induced migration as a form of protected movement. Such protocols could establish minimum standards for admission, legal status, and access to services, while also articulating procedural protections such as non-refoulement, family unity, and due process. Importantly, regional protocols can also serve as laboratories for legal innovation that may eventually influence global norms, much like the Cartagena Declaration's impact on the broader understanding of refugee protection in Latin America.

Planned relocation constitutes a more structurally complex, but potentially durable, policy response to climate-induced displacement. It involves the proactive and organized movement of populations from high-risk areas to safer locations, either within a country or, in exceptional cases, across borders. Although relocation is typically framed as a matter of national adaptation policy, its cross-border dimension is increasingly recognized in contexts where territorial loss renders return impossible. Legal recognition of planned relocation requires careful design to ensure voluntariness, participation, and the protection of cultural and social rights. Relocation frameworks must also articulate responsibilities of both sending and receiving states, particularly in cases where affected populations cross international boundaries. While no international legal instrument currently provides binding guidance on cross-border planned relocation, soft law frameworks such as the Nansen Initiative's Protection Agenda have outlined principles that can inform legally robust relocation policies grounded in human rights and development planning.

A key advantage of these legal and policy tools is their adaptability to different displacement contexts and institutional capacities. Unlike traditional refugee regimes, which require individualized status determinations based on persecution, climate displacement may arise from collective and diffuse harms. Protection frameworks must therefore be designed to accommodate group-based claims, anticipate displacement in advance of crisis, and provide scalable solutions that balance individual rights with administrative feasibility. Flexibility must not come at the expense of legal certainty or procedural fairness; rather, it should be embedded within a principled framework that ensures transparency, accountability, and equal treatment.

Together, these tools reflect a growing recognition that legal innovation is not only possible but necessary to address the unique characteristics of climate-induced cross-border displacement. By developing and operationalizing such mechanisms through regional cooperation and human rights integration, states can begin to close the protection gap faced by those who are currently excluded from refugee law and conventional migration frameworks.

Governance and Participation

Effective responses to climate-induced cross-border displacement require governance models that go beyond legal recognition to encompass coordination, planning, and inclusive decision-making. The governance of such displacement must be grounded in a multi-level and multi-actor framework that connects national policies, regional cooperation, and international support. At its core, this approach must balance state sovereignty with shared responsibility, recognizing that climate-related mobility is a transboundary phenomenon that cannot be managed through unilateral responses. Institutional coherence, anticipatory planning, and meaningful participation of affected populations are essential components of a governance system capable of addressing the complex dynamics of cross-border climate mobility (Vanhala & Calliari, 2022).

A critical first step in developing such governance mechanisms involves cross-border data sharing and the creation of joint monitoring systems. Accurate, disaggregated, and timely data on environmental risks, population movement, and vulnerability profiles is essential for informed policymaking. Yet, many regions facing displacement pressures suffer from fragmented or outdated data systems, limiting their ability to anticipate and respond to climate-induced mobility. Regional platforms such as the Platform on Disaster Displacement and the Integrated Data Framework under the Global Compact on Migration have supported efforts to enhance data harmonization, but greater investment in interoperable systems is needed. In particular, cross-border early warning systems that integrate climate forecasting with mobility indicators can help states coordinate responses before displacement becomes inevitable. Such systems must be institutionalized through bilateral or multilateral agreements, with clear protocols for data use, privacy protections, and feedback loops to affected communities.

Planning for displacement must also be embedded within broader climate adaptation and development strategies. National Adaptation Plans and Disaster Risk Reduction frameworks should include mobility as an explicit component, supported by regional and cross-border coordination. Relocation infrastructure, legal frameworks for temporary admission, and integration programs must be aligned across jurisdictions to ensure continuity of protection and access to services. Moreover, funding mechanisms such as the Green Climate Fund and the Loss and Damage Fund must explicitly account for the governance and protection needs of mobile populations, including those who move across borders. The allocation of resources should prioritize not only physical relocation or border management, but also institutional capacity building, legal support, and long-term integration pathways.

Participation lies at the heart of legitimate and effective governance responses. Displacement governance must not treat affected persons as passive recipients of protection but as rights holders and agents in the planning and implementation of solutions. This requires inclusive processes that involve displaced communities, local governments, indigenous groups, and civil society organizations in the design and monitoring of displacement-related policies. Participatory governance is essential for tailoring responses to social, cultural, and economic contexts, as well as for building public trust and preventing secondary displacement. Procedural safeguards, including the right to information, access to grievance mechanisms, and oversight by independent bodies, should be incorporated into all stages of displacement governance, from risk assessment to relocation and integration.

Equity must also guide the design and implementation of governance frameworks. Climate-induced displacement disproportionately affects marginalized communities, including women, indigenous peoples, and those in low-lying or resource-scarce areas. Without deliberate attention to structural inequalities, governance responses risk reproducing or deepening existing forms of vulnerability. Governance models must therefore be informed by equity principles, including gender sensitivity, intergenerational justice, and recognition of customary land rights. Legal and institutional frameworks should be flexible enough to accommodate diverse forms of mobility, including seasonal migration, family-based movement, and statelessness, while ensuring that procedural fairness and equal protection remain central.

Governance and participation are not ancillary to legal protection but are constitutive of it. In the context of cross-border climate-induced displacement, the quality of governance determines the accessibility, legitimacy, and durability of protection. Building systems that are anticipatory, inclusive, and coordinated requires sustained political commitment, institutional innovation, and the embedding of participation as a normative and operational principle.

CONCLUSION AND FINAL REFLECTIONS

Climate-induced displacement across international borders is an increasingly urgent and under-regulated phenomenon in global governance. As slow-onset climate events such as sea-level rise, desertification, and loss of habitable territory escalate, they are already compelling vulnerable populations to seek refuge beyond their national boundaries. Existing legal frameworks, most notably the 1951 Refugee Convention, were not designed to address such mobility and exclude environmental factors from the grounds of refugee protection. Although international human rights law offers partial remedies, such as the principle of non-refoulement under the International Covenant on Civil and Political Rights, these protections remain ad hoc and insufficiently integrated into migration policy. This legal and institutional fragmentation has left millions at the periphery of protection, facing growing risks without corresponding recognition or recourse.

Amid this normative deficit, states and international actors have increasingly turned to soft law and regional mechanisms to address climate displacement in practical, if non-binding, ways. Instruments such as the Nansen Initiative's Protection Agenda and the Global Compact for Safe, Orderly and Regular Migration reflect a growing consensus that climate-induced mobility must be addressed through cooperative, forward-looking, and flexible legal and policy strategies. These frameworks have encouraged the adoption of humanitarian visas, temporary protection schemes, and planned relocation as interim tools that acknowledge the realities of climate-driven movement while respecting state sovereignty. Regional frameworks, such as the African Union's Kampala Convention and bilateral initiatives in the Pacific, have further demonstrated how geographically contextualized responses can bridge normative gaps and promote regional solidarity. However, these responses are uneven in coverage and implementation, and their effectiveness remains vulnerable to fluctuations in political will and institutional capacity.

The consolidation of a cooperative, rights-based framework for cross-border climate displacement must rest on two pillars: legal recognition and participatory governance. Legal instruments should establish clear entitlements for those displaced by climate-related factors, including access to protection, status regularization, and mobility pathways. This requires normative clarity on the thresholds and triggers for such protection, particularly in cases involving slow-onset disasters and the loss of entire territories. At the same time, governance mechanisms must enable cross-border cooperation through early warning systems, harmonized data sharing, and joint planning for relocation and integration. These systems must be embedded in broader climate adaptation and migration frameworks and supported by sustained institutional coordination at the regional level.

Equally critical is the need for participatory approaches that center the agency, dignity, and knowledge of affected communities. Displacement responses must avoid top-down, technocratic solutions and instead incorporate procedural safeguards, community consultation, and access to justice. Special attention must be paid to historically marginalized groups, including Indigenous communities, women, and persons with disabilities, whose displacement experiences are often shaped by structural inequalities. Ensuring procedural equity and community ownership is not only a matter of legitimacy but also of resilience, as durable solutions cannot be achieved without local buy-in and cultural relevance.

Integrating these legal and governance strategies into global development planning aligns directly with the objectives of the United Nations Sustainable Development Goals. SDG 13 (Climate Action), SDG 10 (Reduced Inequalities), and SDG 16 (Peace, Justice, and Strong Institutions) all intersect with the governance of climate mobility. Legal recognition of climate-displaced persons and the development of cooperative institutional frameworks advance the principles of justice, resilience, and equity that underpin the SDGs. Moreover, addressing displacement through legal and policy innovation contributes to the preventive dimensions of the development agenda, mitigating future conflict, instability, and humanitarian crises.

The intensifying climate crisis is already transforming the spatial and legal boundaries of human mobility. Without comprehensive, principled responses, cross-border climate displacement risks becoming one of the most destabilizing consequences of environmental change. It threatens to strain state resources, exacerbate inequality, and erode trust in international legal norms. Conversely, it presents an opportunity to reimagine mobility governance through a framework that is anticipatory, rights-based, and regionally coordinated. By institutionalizing protection, building inclusive governance mechanisms, and anchoring responses in sustainable development principles, the international community can move toward an adaptive legal order capable of responding to one of the defining challenges of the twenty-first century.

REFERENCES

Addaney, M., Jegede, A. O., & Matinda, M. Z. (2019). The protection of climate refugees under the African human rights system: Proposing a value-driven approach? *African Human Rights Yearbook*, *3*, 242–259. DOI: 10.29053/2523-1367/2019/v3a12

Adeola, R. (2019). The impact of the African Union Convention on the Protection and Assistance of Internally Displaced Persons in Africa. *African Human Rights Law Journal*, *19*(2), 591–607. DOI: 10.17159/1996-2096/2019/v19n2a2

Chetail, V. (2023). The politics of soft law: Progress and pitfall of the Global Compact for Safe, Orderly and Regular Migration. *Frontiers in Human Dynamics*, *5*, 1243774. Advance online publication. DOI: 10.3389/fhumd.2023.1243774

Dieng, A. (2018). Protecting internally displaced persons: The value of the Kampala Convention as a regional example. *International Review of the Red Cross*, *99*(904), 263–282. DOI: 10.1017/S1816383117000613

Ferris, E., & Bergmann, J. (2017). Soft law, migration and climate change governance. *Journal of Human Rights and the Environment*, *8*(1), 6–29. DOI: 10.4337/jhre.2017.01.01

Gupta, D., Kumar, P., Okano, N., & Sharma, M. (2025). Climate-induced migration in India and Bangladesh: A systematic review of drivers, impacts, and adaptation mechanisms. *Climate (Basel)*, *13*(4), 81. DOI: 10.3390/cli13040081

Guruparan, K., & Zerk, J. (2021, June 17). *Influence of soft law grows in international governance*. Chatham House. https://www.chathamhouse.org/2021/06/influence-soft-law-grows-international-governance

Hatano, A. (2021). *Emerging international norms to protect "climate refugees"?: Human Rights Committee's decision on Teitiota v New Zealand. Journal of Human Security Studies*, 10(2), 32–50. DOI: 10.34517/jahss.10.2_32

Hodgkinson, D., Burton, T., Anderson, H., & Young, L. (2010). The Hour When the Ship Comes In: A Convention for Persons Displaced by Climate Change. *Monash University Law Review. Monash University. Faculty of Law*, *36*(1), 69–120.

Kabir, M. E., & Kamruzzaman, P. (2022). Exploring the drivers of vulnerability among disadvantaged internal migrants in riverbank erosion prone areas in north-west Bangladesh. *Journal of South Asian Development*, *17*(1), 57–83. DOI: 10.1177/09731741221078741

Link, A. C., Oakes, R., Durand-Delacre, D., Thalheimer-Prezyna, L., & van der Geest, K. (2025). To what extent do climatic stressors drive human mobility in the world's drylands? *Population and Environment*, *47*(16), 16. Advance online publication. DOI: 10.1007/s11111-025-00486-7

Martínez-Zarzoso, I., Nowak-Lehmann, F., & Paschoaleto, R. D. L. (2022). Climate change, natural disasters, and international migration: A country-level analysis considering climatic zones. *Frontiers in Climate*, *4*, 986088. Advance online publication. DOI: 10.3389/fclim.2022.986088

Matias, D. M. S. (2020). Climate humanitarian visa: International migration opportunities as post-disaster humanitarian intervention. *Climatic Change*, *160*(1), 143–156. DOI: 10.1007/s10584-020-02691-9

McAdam, J. (2011). Swimming against the tide: Why a climate change displacement treaty is not the answer. *International Journal of Refugee Law*, *23*(1), 2–27. DOI: 10.1093/ijrl/eeq045

McAdam, J., & Ferris, E. (2015). Planned relocations in the context of climate change: Unpacking the legal and conceptual issues. *Cambridge Journal of International and Comparative Law*, *4*(1), 137–166. DOI: 10.7574/cjicl.04.01.137

McMichael, C., Katonivualiku, M., & Powell, T. (2019). Planned relocation and everyday agency in low-lying coastal villages in Fiji. *The Geographical Journal*, *185*(3), 325–337. DOI: 10.1111/geoj.12312

Rajshree, M., Mehta, C., & Luhach, S. (2025). Displaced by climate change: Navigating legal framework for 'climate refugees' in international environmental law. *India Quarterly*, *81*(1), 9–24. DOI: 10.1177/09749284241307928

Stojanov, R., Rosengaertner, S., de Sherbinin, A., & Nawrotzki, R. (2021). Climate mobility and development cooperation. *Population and Environment*, *43*(2), 209–231. DOI: 10.1007/s11111-021-00387-5 PMID: 34305224

Thornton, F. (2022). The future is now: Climate displacement and human rights obligations—A note on recent developments in the UN Human Rights Committee. *Frontiers in Climate*, *4*, 1061474. Advance online publication. DOI: 10.3389/fclim.2022.1061474

Vanhala, L., & Calliari, E. (2022). Governing people on the move in a warming world: Framing climate change migration and the UNFCCC Task Force on Displacement. *Global Environmental Change*, *76*, 102578. DOI: 10.1016/j.gloenvcha.2022.102578

Woodworth, F. (2023). Exclusion of climate migrants from the Global Compact on Refugees. *Geopolitics*, *29*(1), 118–147. DOI: 10.1080/14650045.2023.2225242

KEY TERMS AND DEFINITIONS

Climate-induced displacement: The forced movement of individuals or communities due to the adverse impacts of climate change, such as sea-level rise, extreme weather events, and ecosystem collapse.

Cross-border mobility: The movement of people across international boundaries, often outside formal migration frameworks, prompted by environmental or socio-political pressures.

Humanitarian visa: A form of temporary legal entry granted to individuals fleeing situations of serious harm, including natural disasters or climate-related risks, where traditional refugee criteria do not apply.

Legal protection gap: The absence or insufficiency of binding international legal norms to recognize and protect individuals displaced by climate-related factors.

Non-refoulement: A principle of international human rights and refugee law prohibiting the return of individuals to countries where they face serious threats to life or freedom.

Planned relocation: A coordinated process by which individuals or communities are permanently moved from high-risk areas to safer locations in response to environmental threats.

Regional cooperation: Collaborative efforts among neighbouring states to address shared challenges such as climate displacement through joint legal, institutional, or policy mechanisms.

Soft law instruments: Non-binding agreements, guidelines, or declarations that influence state behavior and policy in international law without carrying legal enforceability.

Chapter 6
Climate Risk and Resilience Planning in Border Regions

Mohammad Badruddoza Talukder
https://orcid.org/0009-0008-1662-9221
International University of Business Agriculture and Technology, Bangladesh

Sanjeev Kumar
https://orcid.org/0000-0002-7375-7341
Lovely Professional University, India

Shivansu Sachan
https://orcid.org/0000-0002-9893-8018
CSJM University, India

ABSTRACT

Border regions often face unique vulnerabilities to climate risks due to their geographic, socio-political, and economic complexities. Fragile ecosystems, high population density, and limited institutional coordination across boundaries frequently characterize these areas. Effective resilience planning in these regions requires collaborative governance, data sharing, and joint risk assessment frameworks that transcend national borders. Integrating climate risk mapping, early warning systems, and community-based adaptation strategies can enhance preparedness and reduce vulnerabilities. Moreover, regional cooperation in infrastructure development, sustainable resource management, and ecosystem restoration strengthens adaptive capacity. This paper highlights from the literature that the importance of multi-level stakeholder engagement, inclusive policies, and cross-border partner-

DOI: 10.4018/979-8-3373-3972-6.ch006

ships for fostering resilience. Building climate-resilient border regions is critical not only for local communities but also for ensuring regional stability, security, and sustainable development.

INTRODUCTION

Background and Context:

Climate change is one of the most burning global problems of the twenty-first century, remapping ecological systems, social structures, and political dynamics on the planet. Its effects do not occur across national lines; rather, they cut across international boundaries, impacting common ecologies, interdependent economies, and defenseless populations. Border territories, a geographic space located at the fringe of nation-states, often turn out to be hotspots of climatic risk with their protruding ecological instability, social-economic marginality, and multi-levels of governance. Borderlands are usually poorly served with regard to infrastructure, institutional capacity, and resource distribution, especially when compared to the degree of self-sufficiency in urban centers or centres of economic power. This renders them grotesquely exposed to the multivariate hazards of climate change, such as floods, droughts, heat waves, desertification, biodiversity loss, and population displacement (Adrot et al., 2018).

The importance of defining resilience planning, especially in the border regions of the world, is on the world agenda due to the convergence of climatic and political pressures. Various communities have often occupied these regions with an overlap of cultures, languages, and ethnicities that cuts across borders. Climate risks also have few geographical boundaries: riverine flooding in South Asia, drought and desertification in the Sahel, or glacier-melt in the Himalayas are just some of the examples. Instead, they tend to develop mutual weaknesses that require collective action (Wilder et al., 2010; Talukder, 2024). Nonetheless, coordinated action tends to be hampered by different priorities in politics, territorial issues, and various capabilities of neighboring states. Border regions then become front-line areas where integration of climate risk, issues of governance, and human security is most pronounced.

Understanding Climate Risk in Border Regions:

The risk regarding climate in border regions can be discussed in terms of three interrelated dimensions: exposure, sensitivity, and adaptive capacity. Exposure means the ease of being exposed to climatic hazards, i.e., low-lying coastal borders that are prone to rising sea level or gorges vulnerable to glacial retreat and landslides. Sensitivity is concerned with the extent to which the borderland communities are bound to climate-sensitive activities, which include rain-fed agriculture, Transboundary Rivers, and shared forests. Adaptive capacity can be defined as the capacity of communities and states to respond to stresses caused by climate shocks and rehabilitate (Challinor et al., 2018). In most borderlands, the capacity to adapt is limited by a combination of poor governance, inadequate infrastructure, lack of access to technology, and political indifference.

Besides, climatic risks in border areas can find a response in pre-existing sociopolitical tensions. There is contested sovereignty in many border areas, weak law enforcement capabilities, informal economies, and marginal populations. When climatic shocks take place, they sharpen such vulnerabilities, and in many cases, lead to forced migration, competition over scarce resources, and even cross-border conflict (Hippe et al., 2024; Talukder, 2025). Case in point, the Lake Chad Basin, shared by the four countries of Nigeria, Niger, Chad, and Cameroon, has been held up by the effects of climate-driven water supply distress that has worsened food insecurity and forced displacement, and violent extremism. Likewise, the loss of glaciers in the Himalayas poses challenges of water security to the millions in the four countries of India, China, Nepal, and Bhutan, thus posing the threat of transboundary water conflict (Lucatello, 2022).

Resilience Planning as a Critical Imperative:

The term resilience planning indicates proactive measures that are meant to limit the vulnerability, adaptive capacity, and guarantee sustainability in the long run due to climatic uncertainty. Border regions, in particular, have a special urgency in resilience planning since risks that come with climate change are also enhanced by national jurisdictions and competing national interests. Good resilience planning should hence be able to cross international barriers and adopt an integrated, collaborative, and comprehensive planning strategy (Benzie et al., 2019).

Creating common data platforms, climate risk mapping, and forecasting frameworks across cross-border regions, to provide a shared fact base. Establishing cross-border institutions or agreements that coordinate joint climate efforts, especially in water management, disaster readiness, and resource sharing (Smith & Fischbacher, 2009). Empowering residents in cross-border regions who often lack political representation to design and implement adaptation responses that suit their socio-cultural and ecological realities. Investment in robust infrastructure, such as flood defenses, renewable energy systems, and transport networks, as well as restoration of degraded ecosystems like wetlands and forests, creates natural barriers.

Challenges to Climate Resilience in Border Regions:

In spite of all this urgency, resilience planning in borderlands is fraught with structural and political obstacles. Border regions are frequently marginal to national planning priorities. Ministries and agencies might not be coordinated across borders, meaning that they are not acting holistically. Access to climate information, i.e., the knowledge about rainfall patterns, river flow, and disaster forecasts, might be constrained because it is treated as a national security issue, which is not the case (Aburto-Oropeza et al., 2018). The non-transparency prejudices regional comradeship. Border communities are usually economically underserved and given poor access to health, education, and infrastructure. There is low muscle in terms of amounts recoverable when climatic shocks happen because the capacities are lower than their national averages. Climatic changes occur as a result of political tensions, insurgencies, or even conflict in the borderlands, which makes it difficult to collaborate on climate action. As an example, in most of Africa and Asia, there are areas bordering without peace since degradation leads to the worsening of security (Vulevic et al., 2021). Climate change is the cause of displacement across borders, thus creating a humanitarian crisis and diplomatic pressure between countries that share borders.

Opportunities for Resilience Building:

Although there are major challenges, there are also opportunities in border regions to consider innovative resilience planning. Shared fragility can become an agent of collaboration as opposed to warfare. Specific natural ecosystems across national borders (trans-boundary river basins, forests, mountains) are in need of joint management and hence offer a natural gateway to collaboration. International bodies, regional blocs, and bilateral agreements could, in fact, be used to support a platform of intertwined climate risk management (Berke et al., 2021).

Good examples of such initiatives are the Mekong River Commission, which coordinates long-term water management among countries in the region; the Great Green Wall Initiative of the African Union, which seeks to counter desertification on an interstate basis in the Sahel; and the cross-border climate resilience program of the EU, to which the EU has integrated adaptation to manage its climate in a regionally balanced way (Rosenzweig et al., 2018). As these examples show, when there is political will, institutional creativity, and community involvement, resilience planning in borderlands can have common good outcomes.

Research Significance

The climate risk in border areas is gaining increasing attention, but what underlying questions does this emerging phenomenon pose to policymakers and researchers? What can resilience planning do to fit around the twofold burden of ecological vulnerability and political fragmentation? What are the best governance systems and adaptation plans in a transboundary scenario? What can be done for Crimea and local communities in this respect, and what guarantee can be given to dwellers of certain regions that will not interfere with local peoples themselves and their stability (Pham et al., 2018). In this study, this question will be examined by investigating the multidimensional character of climate risk in border areas and assessing resilience planning frameworks. It states that resilience cannot be done on a national level but must involve cooperative, inclusive, and conflict-sensitive efforts. The paper highlights how governance, technology, and community-based interventions can help in the realization of climate-resilience in border areas.

Border territories are at the epicenter of climate change, with all the challenges and opportunities it brings. The way ecological resilience planning is structured in these regions is of critical importance in terms of not only professional sustainability but also political stability and human security as the effects of climate change get worse. According to Trogal et al. (2018), border cooperation, inclusive governance, and developing new methods of adaptation can make border regions a hub of adaptation and sustainability in a world that is more vulnerable. This background preconditions an advanced study of the ways in which transformation of this sort can occur in reality.

Objectives of the Chapter:

1. To understand the character and influencing factors of climate hazards in the border areas, with an emphasis on environmental capabilities, socioeconomic hardships, and political intricacies.
2. To assess current measures, models, and practices of resilience planning, and specifically the work of cross-border cooperation, community-based adaptation, and conflict-sensitive work.
3. To suggest a conceptual framework of reinforcing climate resilience in border regions, outline policy implications, mechanisms of governance, and opportunities to cooperate.

LITERATURE AND CONTEXTUAL FOUNDATIONS

Conceptualizing Climate Risk and Resilience

Although the notion of climate risk has long been fundamental to the discourse of climate change, its discursive form has changed significantly since the 1990s. In previous discussions, the concept of climate risk was frequently reduced to physical risk, e.g., the likelihood of droughts, cyclones, floods, and heat waves in every province. With time, scholars and policymakers realized that hazards by themselves do not form part of risk. Risk is the scientific and mathematical combination of hazards, human exposure, and vulnerability, posing the possibility of harmful effects. The IPCC (2022) catalog of this comprehensive perspective defines risk as the product of hazard, exposure, and vulnerability, averaged by adaptive capacity. The difference in outcomes may be quite different due to local peculiarities of social, economic, and political life in spite of similar hazards (Abuwaer et al., 2023). One example is that a cyclone of the same intensity can kill thousands in one country, yet cause comparatively lower fatalities in the other simply because the latter has a stronger infrastructure, early warning system, and institutional capability.

As a concept, resilience has changed as well. Early resilience had a foundation in ecology and meant the ability of the ecosystem to absorb disturbances and retain its fundamental functions and structures (Graveline & Germain, 2022). Resilience has also seeped into the social sciences and today applies to communities, institutions, and governance systems. The recent literature focuses on resilience both as the capability to endure shocks and to evolve and change under prolonged stresses. (Berbés-Blázquez et al., 2017) Has proposed the term social-ecological resilience, in which human and ecological systems are perceived as mutually adaptive and connected systems that co-evolve. More recently, scholars promoted the concept of transformational resilience, where radical changes in the system of governance and infrastructure, as well as livelihood transitions, are regarded as ways to address systemic risks. Resilience in border regions is especially complicated since both environmental vulnerability and political sensitivity have to be considered in this case. Borderland resilience needs not only ecological but also infrastructural and cross-border adaptation, and inclusive governance supported by trust-building among the states and societies.

Border Regions as Sites of Vulnerability:

Regions bordering one another are a special place both in the interpretation of political geography and in the study of climate change. They are both peripheral; they are geographically marginal to the political and economic centers of states, and politically peripheralized in terms of investment, governance, and service delivery. According to scholars, like Benveniste et al. (2020), the borders are not only viewed as boundaries or walls building a barrier between the nation-states, but they are also viewed as a bridge or connection interlinking communities and uniting people on the basis of shared histories, cultures, and ecosystems. Such a dual nature defines the weaknesses of the borderlands in times of climate stress. Most of the border area is characterized by poor infrastructure, provision of health and education services, and connectivity with national markets (Sachseder et al., 2024). Therefore, they are most of the time overly reliant on natural resources, including agriculture, forests, and water, which are likely to be severely impacted by the climate. When such extreme events occur: floods, droughts, or desertification, the effects are enhanced because of the low-adaptive power of those regions. Citing another example, regular flooding in South Asia by the river system of the Brahmaputra and Ganges regularly destroys settlements along their boundaries between India, Nepal, and Bangladesh, with poorly constructed embankments and a resulting lack of emergency service capacity. On the Sahel cross-border pastoralists have to cope with growing drought and desertification which undermines livelihoods and tensions between pastoralists and their sedentary farmer neighbors. In Latin America, Andean glaciers supplying highland and lowland border populations with water are melting at an alarming rate, including risks to food security and hydropower generation in Peru and Bolivia (Walter, 2023).

Besides the physical vulnerability, it is also a vulnerable place created through political and social vulnerability. They have an ethnically diverse population with a very strong cross-border relationship. In a context in which central governments hold such populations with suspicion or do little to address their needs, climate stresses can compound marginalization. The latter could be seen as evident in the horn of Africa, where Somali pastoralist communities that span the borders of Kenya, Somalia, and Ethiopia share grazing pastures. The higher levels of climatic variability have enhanced the competition regarding pasture and water, and therefore, intercommoned conflict has at times occurred (Prokkola, 2019; Talukder et al., 2024). Central America has also experienced climate-induced migration from rural Honduras and Guatemala to Mexico and the United States, the migration of which became politically securitized and led to humanitarian and diplomatic tensions. In that order, borderlands must be theoretically defined as a focal point where climate risk is being used together with naturally existing governance gaps, socioeconomic disadvantage, and the issues of identity. This is why resilience planning is more urgent and complex in comparison with core regions.

Climate Risk in Trans-boundary Ecosystems:

Climatic risks in border areas are reflected most clearly in transboundary ecosystems, which offer key services to a range of countries but are becoming increasingly strained by climate change. These ecosystems involve common river basins, mountain ranges, forests, wetlands, and deserts. The fact that ecosystems do not coincide with the political boundaries makes their management especially problematic under the conditions of climate stress (Benzie & Persson, 2019). Some of the striking cases are the Transboundary Rivers. More than 270 common river basins already exist in the world that cover close to one-half of the total land surface of the globe (Baten & Titumir, 2016). Climate change is associated with a shift in precipitation patterns, dynamics of snow melts, and the seasons of flow, which affect water supply, farming, fishing, and power generation.

An example is experienced in the Mekong River Basin, which embraces China, Laos, Thailand, Cambodia, and Vietnam, where it witnesses an escalation of seasonal fluctuations posing a threat to fisheries and agricultural activities. Downstream communities have voiced the concern that upstream hydropower dams, in conjunction with climate variability, are contributing to adverse drought conditions. In South Asia, the Indus Basin that India and Pakistan share depends on glacial runoff, which is likely to be minimized as glaciers recede. This poses a political risk to water security in an already unstable region. The mountain ecosystems like the Andes and the Himalaya serve as types of water towers to billions of people. As a result of the glacier thaw caused by climate change, the long-term recharging capacity of the glaciers suffers. Still, what is worse, floods can damage the lives of border communities in the form of outburst floods caused by melting glaciers (N. Sh. Chen et al., 2013). Such risks cannot be tackled by individual countries but on a regional basis through monitoring and early warning mechanisms.

Silvan regions and the Sahara deserts at the borderlands also show how inter-border climate risks are transboundary. The Amazon rainforest in eight countries is a global sink of carbon under the serious threat of deforestation and a warming world. When the Amazon is about to enter a so-called tipping point due to climate change, all countries will experience the repercussions. Conversely, desertification in the Sahel, an arid strip of land across a dozen African countries, is increasing in size, which disrupts agriculture, as well as pastoral traditions. In this case, regional solutions, which include control and management of land restoration, water management, and diversification of livelihoods, are needed to build resilience (Yang et al., 2015). These two examples indicate that border ecosystems are common impact points that demand joint management. The concept of resilience planning in border regions should thus be ecologically based, whereby one considers the ecological, social, and political factors in a cross-border approach.

Governance Challenges in Borderland Climate Planning:

Governance may be one of the greatest roadblocks to resilience planning in border areas. The issue of climate change adaptation is positionally relativistic because national adjustments are not always sufficient to embrace the scope of borderland vulnerabilities. A number of governance issues are recurrent in the literature.

To start with, there is institutional fragmentation. National climatic policies are normally conceived in capital cities with little or no consideration of peripheral borderlands. Environment, water, or disaster management ministries may remain uncoordinated within and/or between countries (Van Eerd et al., 2014). This leads to redundancies, inefficiencies, and holes in adaptation planning. Second, cross-border cooperation is inhibited by the existence of sovereignty and security concerns. Climate data, data-hydrological flows, or meteorological forecasts are considered in some cases to be national security data, and as such, details will be restricted. As an example, in South Asia, a failure to share upstream hydrological data has been used to impair flood preparedness in downstream Bangladesh on several occasions (Schoon et al., 2017).

Third, borderland groups are politically on the periphery. They might not be represented in national parliaments; they have no access to state resources, and the possibility to take part in decisions. This means that adaptation strategies might not reflect local realities or needs. The inequalities that adaptation projects themselves may worsen include the situation when large-scale irrigation projects bring their benefits to elites but force vulnerable borderland farmers off the land (Cons, 2018).

Lastly, the governance issues become prevalent in conflict-prone borderlands, where there is poor state authority, which coincides with insecurity. In the Lake Chad Basin, declining water supply has been compounded by poor governance and insurgency, leading to the breeding of a complex humanitarian crisis. Climate resilience cannot be isolated in such circumstances from peacebuilding and security approaches. Such governance issues bring to light the need for polycentric and multi-level governance- where local, national, regional, and international players coordinate to deal with a mutual risk (Hannigan, 2018). In the absence of such coordination, resilience approaches can be short-term, fragmented, and inadequate.

Emerging Frameworks for Borderland Resilience:

Notwithstanding, a few exemplary frameworks and initiatives are in place that provide a way forward towards resilience in border regions.

The transboundary cooperation mechanisms have proved that cross-border institutions can regulate resources that are climate sensitive. The Mekong River Commission promotes dialogue between the riparian states and data sharing, and the Danube River Commission has assisted in the joint management of floods in Europe. These organizations show that a well-organized cooperation may establish trust and minimize the possibility of conflict (Andersen & Prokkola, 2021). Regional projects also show creative methods of work. The Great Green Wall Initiative of the African Union aims to repair degraded desert land in the Sahel through tree planting and environmental restoration, as well as through the promotion of livelihood opportunities. The move is to help improve resilience to desertification, curb the trend of migration, and foster regional collaboration(Boyle & Shneiderman, 2020). In Latin America, the Amazon Cooperation Treaty Organization presents a forum to discuss Amazon forest policy coordination among member states, but there are problems with implementation.

CBA (Community-Based Adaptation) has become a term that has become common in the literature on climatic impacts, as it focuses on local knowledge, empowerment, and participation. CBA can promote cross-border solidarity in borderlands, communities that share common risks and bind informally regardless of any unwillingness of states to do the same. Examples include local pastoralist arrangements to jointly graze across borders in East Africa and local-led disaster preparedness programs in South Asia. A more recent area of the resilience constructs is the climate-security nexus. According to scholars, adaptation should be conflict-sensitive so that its interventions do not increase tensions. In this respect, any agreements on water sharing have to be properly designed in order to prevent reinforcement of inequalities and creation of grievances. In fiercely troubled frontiers, conflict planning should be incorporated with peace planning activities, which show the necessity of hybrid patterns implemented both environmentally, socially, and security-wise.

Contextual Foundations across Global Regions:

In the borderland, the contextual basis of climate hazard varies region to region, but has similarities in vulnerability. The example of the Indo-Bangladesh border in South Asia shows that floods and cyclones recurring on a regular basis displace communities and cause a rise in migratory pressures and in the bilateral tension (Talukder et al., 2025). Himalayan glaciers are also at risk of climate change, and could cause a water security conflict between India, Nepal, Bhutan, and China (Reynolds, 2010).

In Africa, drought and desertification are hampering the livelihood of pastoralists in the Sahel region and driving them to cross-border migration. Such population movements also cause tensions with agricultural groups, as insecurity is an issue in Mali, Niger, Burkina Faso, and Chad. The development of resilience is coming through regional initiatives like the Intergovernmental Authority on Development (IGAD), and the main problem is political instability. Europe: in Europe, countries along its borders are fairly stable politically but exposed to risks of climate change, including those raised by flooding along the Rhine and Danube rivers and heat waves throughout southern Europe (Smolak, 2010). The EU has integrated climate adaptation into cross-border development programs, and consequently, an example of how regional integration can enhance resilience can be seen.

In Latin America, the common Amazonian forests and the Andean glaciers explain why border areas rely ecologically on each other. Deforestation and glacier retreat establish common risks, which require joint governance, even though political priorities vary across Amazonian states and are an impediment to progress.

These local demonstrations indicate that, though the precise dangers are diverse, the deep-seated issues of marginalization, fragmentation of governance, and ecological interdependence remain universal when considering border contexts.

Research Gaps and Theoretical Contributions:

Notwithstanding gains in research, there are still research gaps in this study of climate risk and resilience within border areas. First, little attention is paid to the communities in the borderlands as adaptive agents. Most of the literature portrays them as helpless victims, ignoring their awareness, resourcefulness, and international contacts.

Second, only a few studies are conducted on a regional basis, and there is a paucity in comparison across regions; hence, there exists no easy way of identifying the lessons that can be transferred. The bulk of research concentrates on a specific river basin or on the border situation, and few studies compare cases across continents.

Third, conflict-sensitive resilience planning is unexplored. As much as the climate-security nexus is garnering traction, there is a need to conduct more empirical studies on how adaptation strategies play a role in supporting peacebuilding amid fragile borders.

Lastly, not enough importance is paid to the role of technology and innovation. New technologies such as distant sensing, artificial intelligence, and digital platforms can revolutionize the cross-border risk monitoring and early warning systems. However, their uptake is hampered by political uncertainty, resource shortage, and a shortage of capacities at most of the borders.

These gaps should be filled in future research since they will help theorize and practically assist better resilience planning in border regions.

Historical Perspectives on Borderland Studies and Climate:

Defining the border regions has been dominated by the themes of geopolitics, sovereignty, and security. Historically, borders were understood only as boundaries, which were related to the government and the control of supervision, trade control, and territorial sovereignty. Literature in political geography in early times paid much attention to the border as a boundary between flow and circulation. Still, it paid less attention to the ecological and social aspects of the border (Readman et al., 2014). As the realities of climate change have increasingly been felt, however, this paradigm of interpretation of the border as a caprice of political sovereignty alone has given way to a recognition that borders exist as zones of environmental and social contact. Research on climate change has relied on the experiences of borderlands as ways of understanding that borderlands are not just the periphery of nations but places of common vulnerability as environmental disasters cross borders. In this view, there is no distinction between the colonial process of creating a border and the reality of unequal resources, where one can determine the level of stability or even instability of border populations (Widdis, 2019). Modern-day scholarly contributions integrate the fields of environmental studies, development, and international politics in an effort to bridge the divide between the three disciplines through the integration of climatic factors into borderland studies.

Climate-Induced Migration and Human Security in Borderlands:

Among the most pressing questions that arise in the literature are the connections between climate risk, migration, and human security along the borders. Scholars maintain that climate change is a threat multiplier; it increases already-existing socioeconomic pressures and compels populations to relocate across national borders. An example has been the frequent occurrence of floods and cyclones in Bangladesh, as part of the problem, which borders and floods from Bangladesh that cross-border into India, causing tension and securitization of migration policies have been experienced (Makki et al., 2025). Equally in sub-Saharan Africa, desertification and drought in the Sahel have caused pastoralists to become displaced and engage in cross-border competition for resources. Literature on environmental migration emphasizes that environmental movement tends to happen in tenuous governance situations, in which host communities are also challenged by poverty and scarcity of resources. Human security frames heighten this argument by arguing that displacement poses a danger not only to livelihoods but also heightens ethnic tensions, making people more susceptible to being trafficked or exploited, and destabilising the region. Hence, borderlands become important zones experiencing uncertainty in security, sovereignty, and questions of humanitarian responsibilities when taken in the context of climate change (Alvarez, 2016).

Cross-Border Resource Conflicts and Cooperation:

A large number of publications document the twin nature of climate stress in cross-border areas as a generator of warfare and a rallying point of partnership. The scarcity of resources, especially in transboundary river basins, has been an issue of study. Case studies will supplement the information about rivers; this case study will be conducted on the Nile, Jordan, and Indus rivers, and how they are involved in the competition of water use within the circumstance of climate variability, which is increasing the possibility of political tensions. In others, like the Euphrates-Tigris basin, the arguments over sharing have become the front-runners of greater geopolitical differences (Brunet-Jailly, 2022). In contrast, other texts point out positive synergies of mutual vulnerability where climate risks allow mutual risks to seek shared management and governance systems. As an example, the Mekong River Commission has helped to promote the discussion of the countries in Southeast Asia, and the Ganges Water Treaty between India and Bangladesh in 1996 was referred to as a milestone in transboundary cooperation. These cases provide an example of the dual nature of border regions, which represent the threat of escalation of a conflict and the potential of establishing peace building based on climate resilience planning (Mccall, 2013).

The Role of International Law and Policy Frameworks:

International schemes also define the literature on resilience planning in borderlands. The Paris Agreement of 2015 emphasizes multilateral cooperation on both adaptation and mitigation, but its realization usually depends on the national level; it was provided with few details on cross-border initiatives. Trans-boundary agreements such as the Eastern and Southern African Union or the European Union have gone further than regional agreements in defining climate adaptation as a cross-border issue. Moreover, the literature on international environmental law emphasizes the use of treaties on water sharing, biodiversity, and land management as an indirect instrument of climate resilience (McDougal, 1960). Theorists have taken issue with these approaches, arguing that these systems undermine community involvement and are mainly focused on state sovereignty, thus marginalizing the better good effects to local communities that border the concerned states. A new debate arises on whether new instruments of international law are needed to respond to climate-induced displacement and adaptation in borderlands specifically. This implies that global governance of climate-affected areas should adopt new forms to solve the challenges that are more characteristic of this area of life.

Indigenous Knowledge and Localized Resilience Practices:

Another major body of literature focuses attention on the importance of local practices and indigenous knowledge in resilience planning. Border populations can be expected to have long traditions of dealing with environmental fluctuation, such as rotational rangers in arid regions or flood-resistant dwellings in deltas (Chowdhooree, 2019). Transhumance, for instance, can be used in the Himalayas to ensure communities adapt to changing seasonal conditions. In the Amazon, forest management can be used by indigenous groups in forest management techniques that sustain an ecological balance. Inclusion of this kind of knowledge into formal adaptation planning, scholars claim, makes the former stronger in terms of local ownership, as well as in terms of enhancing the cultural legitimacy of the latter resilience measures. At busy borders, where government agencies are either non-existent or may have little influence, local community-based adaptation measures can be the initial resort to climatic risk (Alessa et al., 2016). Literature also indicates that local governance levels demand participatory models that cut across the regional, national, and cross-border governance levels in order to acknowledge and magnify the practices.

Technology, Data, and Trans-boundary Climate Monitoring:

Contemporary development of technology has also played a part in the literature of climate resilience in the borderlands. Remotely sensed mapping, Geographic information systems (GIS), and artificial intelligence are all becoming important tools to track climate hazards across borders and provide real-time information regarding floods, droughts, and deforestation. As an example, monitoring glacier retreat in the Himalayas would give vital information to more than one nation that relies on the common water sources. In a parallel way, digital platforms to support the exchange of data across borders are under consideration as mechanisms to enable trust and cooperation between states (M. Chen et al., 2020). Nonetheless, the literature notes other challenges, especially security and sovereignty concerns on information, trust, and unequal access to technology in less-developed Border States. The governance of climate data in areas of political disputes is a politically sensitive issue, evoking debate on how open it is to sharing information on climate change without compromising national security concerns. Nevertheless, the technological aspect of resilience planning presents a lot of opportunities for advancing scientific cooperation and evidence-based policymaking regarding border areas.

Climate Justice and Ethical Dimensions of Borderland Resilience:

Lastly, the current literature increasingly highlights the climate justice approach to the consideration of resilience in a borderland. Border communities are noted to be some of the most vulnerable to the effects of climate change, whilst being some of the least contributors to greenhouse gas emissions. This raises questions of equitable character with respect to distributive justice, equity, and sharing in responsibility. Literature on climate justice notes that climate resilience planning needs to consider the underlying disparities, the equitable access to adaptation finance, and the rights of the marginalized populations, such as women, natives, and even migrants (Jafari, 2024). Moreover, climate resiliency cannot be separated from issues of power, where the border regions are neglected and excluded from the national policymaking processes. One must, therefore, introduce climate justice into the resilience planning that goes beyond pure technical solutions and integrates climate justice as embedded in the resilience planning by integrating standards of equity, participation, and human rights.

Border Regions as Climate Hotspots:

Borders have been noticed in contemporary literature as climate hotspots, where those with a place on them were at risk of being exposed to geographic extremes, were environmentally vulnerable, and were institutionally neglected. Mountain borders like the Himalayas are very delicate to the glacier meltdown and variable rainfalls that lead to extreme damage to the downstream communities in more than one country. The coastal border areas, such as Sundarbans between India and Bangladesh, are among those that are exposed to a combination of risks with sea-level rise and saltwater intrusion threats, and tropical cyclones (Hare et al., 2011). The borders in the North of Africa and the Middle East are covered by scorching hot temperatures, moving dunes, and a lack of water. Such ecological pressures establish what Thomas calls "double peripheries," i.e., states are marginalized when it comes to nature, and countries are marginalized when it comes to governance systems. Literature on hotspot mapping also indicates that those areas near borders are less represented than other regions in the national adaptation plans, adding to their lack of attention (De Sherbinin, 2014). This contextualization raises the exigency of planning climate resilience that understands the borders not as provincial areas but instead as capitals of climate response.

Governance Fragmentation and Institutional Gaps:

Another theme that is common in literature is governance fragmentation in the perception of the border regions. It is common for national governments to concentrate their focus on the central parts of the country rather than on the borders, where they are politically and economically marginalized. This neglect is augmented by divided jurisdiction, duplication of authority, and the absence of coordination among states next to each other. As an illustration, the Mekong Delta represents various nations in Southeast Asia, but poor coordination regarding water flows has weakened resilience plans (De Sherbinin, 2014). Likewise, there is a lack of cross-border governance structures in the Horn of Africa, which have exposed pastoral communities to frequent drought. Scholars state that this institutional fragmentation has caused resilience gaps where adaptation programs are fragmented, responsive, and incapable of dealing with the transboundary aspect of climate risk. Literature suggests growing canvases in support of multi-level governance solutions to coordinate and balance local, national, and regional-scale adaptation interventions, with international backing.

Climate Risk Narratives and Political Framing:

A related research strand assesses the political construction of climate risks in borderlands. Governments tend to securitize the effects of climate change by branding cross-border migration or resource scarcity as a conceivable threat to national security instead of a humanitarian or developmental issue. Such securitization may result in militarized responses to climate-related migration, such as in South Asia and North America, where climate-related migration is increasingly discussed in border control-related terms (Mengisteab & Hagg, 2017). Though this framing can enhance national sovereignty, in most cases, this approach works against cooperative resilience planning and people-oriented approaches. Critical literature points out that risk narratives lack neutrality in the sense that they reflect the interests of politics as power relations and historical experiences of exclusion shape them. Other framings, like shared vulnerability or climate peace building, have also been used in academia as a way to overcome securitized approaches. These accounts focus on collaboration, unification, and collective adjustment, which implies that direct framing of climate risk influences the structure of the resilience measures.

Economics of Climate Resilience in Borderlands:

The economic logic on border resilience outlines the downside costs of failing to act as well as the advantages and benefits of taking the initiative to adapt. Research indicates that disasters caused by climate change mainly affect trade corridors, agricultural activities, and local livelihoods in the border areas in disproportionate measures. For instance, the loss is in the billions of dollars. It disproportionately affects informal traders and small-scale farmers, as the closing of cross-border trade routes during South Asian floods causes losses. Conversely, publicly funded resilient infrastructure, e.g., elevated roads, and flood-proof warehouses and climate-smart agricultural practices, has been shown to give substantial long-term returns on investment. Researchers also note the importance of regional trade agreements and economic unions (including ACFTA and ASEAN) in informing how resiliency planning can, and should, be incorporated into larger agendas of economic integration (Gonda et al., 2023). The economic outlook introduces a level of practicality to resilience planning in that adaptation is seen as a humanitarian and economically profitable approach in managing stability and growth.

Health and Climate Vulnerabilities in Border Communities:

The literature speculates on the displacement of population, the risk of plagues and other diseases arising out of climate risks in borderlands. Diseases that depend on climate, like malaria, dengue, and cholera, tend to spread across porous borders, especially in the tropics, whereby climate change may change the disease vectors. Furthermore, extreme weather exposes healthcare delivery systems to disruptions, and because of this, border communities end up with less access to emergency services. As another example, food insecurity caused by climate change has increased among transboundary pastoralist groups in the Sahel, which has promoted child malnutrition (Hansen et al., 2013). The COVID-19 pandemic also demonstrated the weaknesses of health systems in border regions, leading to increased vulnerabilities when climate shocks are combined with epidemics. The literature on planetary health infers that planning for resilience should incorporate access to healthcare, disease surveillance, and cross-border coordination on epidemic preparedness. The mentioned approaches emphasize that climate resilience is not simply a matter of built infrastructure and ecological systems but the protection of human health and well-being in vulnerable areas.

Comparative Regional Experiences:

Comparative case studies are a valuable contribution to the literature, as they demonstrate the range of differences in planning resilience across regions. In Europe, the EU has been the forerunner in the adaptation projects that cross the borders, providing financing such as on the Interreg, including the flood protection measures, conservation of biodiversity, and renewable energy collaboration. However, in Africa, border areas are often burdened by poor governance and infrastructure, and have low financial availability. Still, there are good examples, such as the Great Green Wall concept, that address desertification. In South America, the Amazon basin has become an important transboundary ecosystem between climate resilience planning and indigenous rights, conservation, and geopolitical rivalries. Another sharply contrasting case is South Asia, where the use of sustainable practices has been urgent and vexed due to the high population density, political wrangles, and frequent climate tragedies. In contrasting these contexts, researchers point out that whereas global commonalities are found in border-specific issues, locally-based resolutions must be based on local histories, governmental structures, and cultures (Hospers & Beugelsdijk, 2002).

Table 1. Climate Risk and Resilience Planning in Border Regions

Climate Risks	Key Challenges in Border Regions	Resilience Planning Strategies	Key Actors Involved	Examples / Contexts
Floods & Cyclones	Shared river basins, weak early warning systems, and cross-border migration during disasters	Joint flood management, regional early warning systems, resilient infrastructure (e.g., embankments, cyclone shelters)	Governments, regional bodies, NGOs, and local communities	India–Bangladesh Sundarbans cooperation on cyclone preparedness
Drought & Desertification	Competition over scarce water, livelihood insecurity, and pastoral conflicts across borders	Climate-smart agriculture, water-sharing treaties, reforestation, drought-resistant crops	Local farmers, cross-border councils, and UN agencies	The Great Green Wall Initiative in the Sahel region
Glacial Melt & Water Stress	Transboundary river tensions, hydropower disputes, and altered seasonal flows	Regional river basin management, glacier monitoring, and equitable water allocation	Governments, scientific institutions, and regional water commissions	Indus Water Treaty between India and Pakistan; Mekong River Commission
Heatwaves & Food Security	Impact on migrant labor, disruptions in cross-border food trade, and increased health risks	Cross-border food security frameworks, heat action plans, investment in cold storage, and trade facilitation	Trade associations, health ministries, and regional economic unions	South Asia's SAARC Food Bank initiative
Ecosystem Degradation	Deforestation, biodiversity loss, illegal cross-border logging, and poaching	Transboundary conservation areas, ecosystem restoration, and biodiversity corridors	Environmental NGOs, regional alliances, and indigenous groups	Amazon Cooperation Treaty Organization (ACTO)

continued on following page

Table 1. Continued

Climate Risks	Key Challenges in Border Regions	Resilience Planning Strategies	Key Actors Involved	Examples / Contexts
Migration & Displacement	Climate refugees, border security tensions, and humanitarian crises	Humanitarian corridors, migration governance frameworks, conflict-sensitive resilience planning	UNHCR, IOM, border agencies, civil society	Rohingya crisis in Bangladesh-Myanmar border, climate-linked migration in Central America
Public Health Risks	Spread of vector-borne diseases, inadequate cross-border health infrastructure	Regional disease surveillance, mobile health units, climate–health integration	Health ministries, WHO, Red Cross, and local clinics	Malaria control in the Greater Mekong Subregion; COVID-19 lessons on the health-climate nexus

Source: The Authors compiled based on

FINDINGS FROM LITERATURE

Border Regions as Vulnerability Hotspots

Borderlands are always referred to as lands that are highly vulnerable to climate change because of their geographic location, ecological sensitivity, and institutional isolation. Trans-boundary mountain ranges, coastal strips, deserts, and river deltas endangered by political borders are especially prone to extreme weather phenomena: glacier meltdown, drought, rising sea levels, and cyclones. These areas are deemed to be the secondary focus, as even regional development plans overlook them, in addition to central governments, which increases the vulnerability of communities residing in the regions.

Governance and Coordination Gaps

Among the most important discoveries is the importance of fragmented governance in weakening resilience. Research has pointed out that regions bordering other countries are often not covered in national adaptation planning, and adaptation at each state is not well coordinated. Water wars, incomplete flood control structures, and competing land-use strategies divert or weaken resilience plans. It has been found that cooperative structures, like regional water commissions or transboundary conservation programs, are essential but are not applied worldwide equally.

Security-Oriented Climate Framing

The literature is also used to show how risks linked to climate in border regions tend to be seen through the prism of security. Governments are more prone to see climate-induced migration/ or resource competition as direct threats to sovereignty and border control, more than mutual humanitarian/ developmental interests. This process of securitization tends to lead to the militarization of resilience as opposed to cooperative resilience planning. Instead, research on different framings of warming, such as climate peace building and shared vulnerability, sees greater opportunity to engender collaboration and sustainable resilience.

Socioeconomic and Livelihood Impacts

A common thread in the legal literature concerns the inordinate socioeconomic effect of climate risk on frontier societies. Pastoralists, farmers, fishers, and informal traders are experiencing ruptures in livelihoods owing to floods, drought, or cross-border mobility anti-policies. Case studies revealed that though climate shocks tend to freeze formal trade, they are more severe for the informal, small-scale cross-border trade that helps a local economy survive. The literature shows that active spending on climate-smart agriculture, resilient trade infrastructures, and inclusive development policies can make border resilience an account of regional growth.

Health and Human Security Dimensions

Results also highlight the implications of climate risk on the public health of the borderlands. There is a compounding vulnerability to increased incidence of vector-borne diseases, as well as food security and health access. The COVID-19 pandemic has demonstrated the weakness of healthcare systems in border regions, where pandemic-related mobility restrictions coupled with severe climate disruption resulted in several crises overlapping with each other. Analysis shows that resilience planning should link health services, disease surveillance, and cross-border medical collaboration, as climate and human security are interrelated.

Regional Variation and Case Comparisons

Comparative works on various regions show that resilience planning is site-specific. In Europe, cross-border climate adaptation efforts are relatively well funded by the EU. In Africa, there are institutional and financial limitations complemented by community-driven approaches such as the Great Green Wall. South Asia and South America are more difficult territories in which to build resilience due to population pressures, resource conflicts, and natural disasters. South America is notable for the role indigenous knowledge can and should play in transboundary climate policies. These results indicate the presence of global patterns even when local histories, politics, and cultures play important roles in resilience outcomes.

THEORETICAL IMPLICATIONS

Advancing Social-Ecological Systems Theory

The scientific texts concerning the risk of climate change in border areas complement and develop a framework of social-ecological systems (SES) theory. Borderlands exemplify a very strong interrelation and mutual development of environmental, social, and institutional subsystems. The relationship between climate hazards (chiefly floods, droughts, and glacial melt) and human vulnerabilities, governance structures, and socioeconomic systems is complex. This is a reminder that resilience cannot be seen as ecological resiliency or infrastructural readiness. Still, the concept has to include the social aspects such as livelihoods, migration, and community networks. This has the theoretical connotation that SES frameworks will need to increasingly take account of trans-boundary dynamics as boundaries of ecological systems and political jurisdictions are not aligned..

Polycentric Governance and Institutional Theory

The results of the border regional studies confirm the polycentric theory of governance. Centralized governance systems with conventional systems of governance tend to be unsuccessful in borderlands because of their political marginality, fragmentation of institutions, and inability to coordinate cross-border strategies. The literature indicates that the presence and the existence of multiple, overlapping centers of decision-making at a local, national, and/or regional level can increase adaptive capacity and resilience. This reinforces the theoretical argument that polycentric systems are more effective in politically delicate and uncertain settings, especially where shared resources are involved, such as rivers, forests, and pastoral lands.

Climate-Security Nexus and Conflict-Sensitive Theory

The other theoretical implication is the strengthening of the nexus between climate and border issues in borderlands. It has been found that climate risks may become a threat multiplier, and they will magnify the prior social, economic, and political tensions. This demonstrates the importance of the inclusion of conflict-sensitive plans in resilience planning. Hypothetically, this augments frameworks in peace and conflict research, implying that any climate adaptation measures would need to consider environmental and socio-political factors to prevent cases of maladaptation and facilitate collaborative resiliency.

Incorporating Indigenous Knowledge and Community Agency

The literature is also an issue in top-down theoretical approaches of resilience, where community knowledge, local agency, and participatory adaptation are important. Border communities may have long-held coping mechanisms, like rotational grazing, flood-friendly agriculture, and forest management, that enhance ecosystem and livelihood resilience. Theoretical models should therefore incorporate bottom-top knowledge systems embedded in context dependency into formal models of climatic risk and resilience. That means adaptation frameworks that have been characterized by universalized technical adaptation give way to hybrid models that integrate local practices with scientific policymaking.

Equity, Justice, and Vulnerability Theories

Lastly, the study demonstrates that future increases in resilience theory must consider conceptions of climate justice and social equity. Cross-border communities are usually the most susceptible to climatic effects and, on the other hand, the least contributors to climate change. The ethical and distributive aspects of resilience planning are integral to the vulnerability and adaptation theories because they touch on the fact that it is through resilience planning that communities have resorted to ethics and distribution as their own means of strengthening their resilience. The concept of resilience strategies in theoretical models should take into consideration equity in terms of gender, ethnicity, socioeconomic status, and migratory status to avoid hardening of the existing disparities.

Synthesis of Theoretical Implications

Taken together, these results suggest that in the context of border regions, or at least the one studied here, the existing theoretical perspectives, polycentric governance, conflict-sensitive adaptation, and vulnerability theory must be expanded to support better the idea that the dynamics of border regions are unique and imply new issues more than they have to deal with. It is argued that Borderlands provide critical empirical labs that could test and refine resilience theories, adaptation theories, and climate governance theories. They indicate that climate resilience planning should be ecological, social, political, and ethical at the same time because it can provide a more modern and situational definition of the climate risks.

MANAGERIAL IMPLICATIONS

Strengthening Cross-Border Collaboration

The literature points out that individual states do not effectively deal with climate threats within their borders. Policymakers and managers should prioritize cross-border cooperation, joint disaster management, common early warning systems, and the coordination of infrastructure development. Formation of formal agreements between adjacent countries over the sharing of data and resources, as well as responding to emergencies, can go a long way in mitigating the risks caused by floods, droughts, and cyclones. An example would be the river basin authorities and transboundary conservation agencies providing templates of structured collaboration that can be used in other areas of vulnerability along borders.

Integrating Multi-Level Governance

Multi-level governance is necessary to effectively plan resilience by pursuing local and regional as well as national initiatives. Managers are required to guarantee that the involvement of local communities in decision-making processes is active, whereas those on a higher level provide technical, financial, and policy support. This will allow for the filling of the gap that has existed between central policies and ground realities, which will help realize that resilience strategies are sensitive to local contexts and practically achievable. The managers ought to concentrate on establishing institutional ties that will make it possible to coordinate across administrative boundaries in the country as well as between countries.

Promoting Community-Based and Participatory Approaches

Cultures in Border communities can often be vested with particular knowledge and models of coping that are crucial in terms of resilience. Community-based adaptation remains the most critical principle that managers should adopt in empowering local actors to come up with design, implementation, and monitoring interventions. Examples of participatory methods are local committees in matters of flood and drought preparedness, community-led conservation, and climate-smart agriculture programs. These practices also create local resiliency, as well as social cohesion and trust, the latter being a prerequisite to long-range cross-border collaboration.

Incorporating Technology and Data-Driven Decision Making

The managerial strategies ought to capitalize on technology and information to increase planning and preparedness. These can result in real-time monitoring of the climate hazards, resource availability, and population vulnerability through remote sensing, GIS mapping, and AI-based predictive models. These tools can allow managers to maximize the distribution of resources, prepare disaster responses, and predict emerging risks. Cross-border data-sharing can also help to boost transparency, minimize disputes, and foster evidence-based policymaking by investing in digital platforms.

Addressing Health, Security, and Socioeconomic Vulnerabilities

Managers need to understand that climate risks overlap with health, security, and socioeconomic risk in the border regions. The resilience planning must comprise engagements like mobile health units, climate-sensitive disease surveillance, livelihood diversification programs, and food security interventions. The managers should also embrace conflict-sensitive strategies in order to make sure that resource apportionments do not complicate conflicts among the border communities. Considering environmental, social, and economic priorities is key to sustainable and inclusive resilience outcomes.

Mainstreaming Climate Justice and Equity

Lastly, managerial implications state that equity and justice should be fundamental in resilience planning. Marginalized populations, such as indigenous people, women, and people at risk from climate migration, need to be given particular attention when setting up the projects and developing the systems of policy. Managers should make sure that there are no disproportionate benefits to adaptation strategies that could worsen existing inequalities. The integration of social protection, fair resource access, and governance systems can make resilience measures more legitimate and effective.

The framework combines climate risks, socio-ecological vulnerability, and governance and resilience strategies in a holistic approach to understanding adaptation in border regions. It brings out the relationships between hazard interactions, community exposure, institutional mechanisms, and adaptation measures, demonstrating the effectiveness of resilience as a result of the integration of these factors.

Figure 1. Framework showing the flow from climate risks to vulnerabilities, mediated by governance and community knowledge, leading to resilience strategies and positive outcomes.

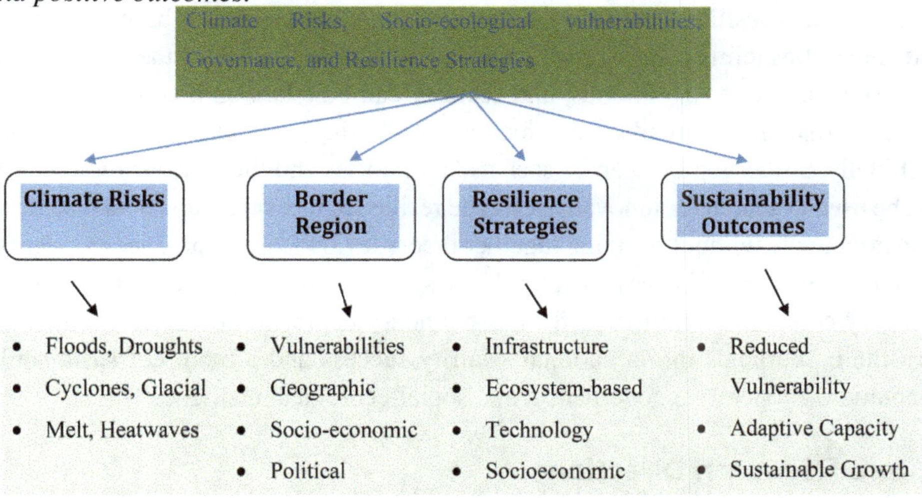

CONCLUSION

The development of border regions to climate risks and resilience planning is an essential area where there is a frontier in the climate change adaptation literature and the area of sustainable development. The evidence in the literature explains that borderlands are particularly susceptible given the overlap of environmental susceptibility, economic-social marginalization, and institutional-political pluralism. Climate hazards, such as floods, droughts, glacial melt, and heat waves, are exacerbated by human and governance vulnerabilities in ways that give shape to complex risks that cut across national borders.

Practical lessons of social-ecological systems, polycentric governance, conflict-sensitive adaptation, and vulnerability frameworks can be seen to require a holistic approach considering the ecological, political, and ethical aspects. Border regions complicate the traditional models of resilience, which give much importance to cross-border humanity, community involvement, and the synthesis of the indigenous knowledge with technological advances. A critical issue in this literature as well is equity and justice, with the provision of ensuring that no one is left behind in the planning of adaptations, which includes the marginalized populations such as women, the indigenous people, as well as the climate migrants.

Managerial and policy implications provide additional support to this need to include multi-level, coordinated, and context-specific strategies. The key steps towards a resilience capability include increasing cross-border cooperation, making infrastructures resilient, taking advantage of technology and data-based planning, and integrating climate justice in governance frameworks. It further means that the issues of health, livelihoods, and security can be taken on board in a holistic approach that leads to resilience, which is long-lasting, sustainable, and inclusive.

Finally, border regions can be seen as laboratories and literacy test regions that can be used to search for innovative climate resilience measures. It is not possible to plan effectively without bringing together theory, evidence, and practice- recognizing the interaction of hazards, vulnerabilities, governance, and community agency is highly complex. A holistic and trans-boundary approach offers the potential to turn the borderlands into a national security success and a model of sustainable adaptation, cooperative governance, and social-ecological resilience.

Future Research Directions:

Future study on the subject of climate risk and resilience planning in border regions is recommended to reinforce or fortify trans-boundary data collection and monitoring with the help of technologies that include GIS, remote sensing, and AI-based predictive models in facilitating risk forecasting and early warning audio visualization. There is a need to find out more inclusive and participatory measures to incorporate indigenous knowledge, local coping efforts, and voices of the marginalized groups, including women, migrants, and indigenous populations, into formal adaptation frameworks. In addition, studies ought to investigate new forms of governance, e.g., polycentric and conflict-sensitive models that would increase cross-border cooperation and also mitigate institutional fragmentation. Long-term and transformational research on adaptation strategies is also imperative, and comprehensive interventions in restoration of ecology, diversity of livelihoods, and alignment of policies are the formidable strategies that can work towards reducing exposure to vulnerabilities even in the face of uncertain climatic situations in the future. Last but not least are the technology, equity, and socio-political elements that influence resilient border communities, which need further research to give comprehensive, location-specific, and sustainable responses.

REFERENCES

Aburto-Oropeza, O., Johnson, A. F., Agha, M., Allen, E. B., Allen, M. F., González, J. A., Arenas Moreno, D. M., Beas-Luna, R., Butterfield, S., Caetano, G., Caselle, J. E., Gaytán, G. C., Castorani, M. C. N., Cat, L. A., Cavanaugh, K., Chambers, J. Q., Cooper, R. D., Arafeh-Dalmau, N., Dawson, T., & Taylor, J. E. (2018). Harnessing cross-border resources to confront climate change. *Environmental Science & Policy*, *87*, 128–132. DOI: 10.1016/j.envsci.2018.01.001

Abuwaer, N., Ullah, S., & Al-Ghamdi, S. G. (2023). Building Climate Resilience Through Urban Planning: Strategies, Challenges, and Opportunities. In Al-Ghamdi, S. (Ed.), *Sustainable Cities in a Changing Climate* (1st ed., pp. 185–206). Wiley., DOI: 10.1002/9781394201532.ch12

Adrot, A., Fiedrich, F., Lotter, A., Münzberg, T., Rigaud, E., Wiens, M., Raskob, W., & Schultmann, F. (2018). Challenges in Establishing Cross-Border Resilience. In Fekete, A., & Fiedrich, F. (Eds.), *Urban Disaster Resilience and Security* (pp. 429–457). Springer International Publishing., DOI: 10.1007/978-3-319-68606-6_25

Alessa, L., Kliskey, A., Gamble, J., Fidel, M., Beaujean, G., & Gosz, J. (2016). The role of Indigenous science and local knowledge in integrated observing systems: Moving toward adaptive capacity indices and early warning systems. *Sustainability Science*, *11*(1), 91–102. DOI: 10.1007/s11625-015-0295-7

Alvarez, A. (2016). Borderlands, Climate Change, and the Genocidal Impulse. *Genocide Studies International*, *10*(1), 27–36. DOI: 10.3138/gsi.10.1.03

Andersen, D. J., & Prokkola, E.-K. (2021). *Borderlands Resilience: Transitions, Adaptation and Resistance at Borders* (1st ed.). Routledge., DOI: 10.4324/9781003131328

Baten, M. A., & Titumir, R. A. M. (2016). Environmental challenges of trans-boundary water resources management: The case of Bangladesh. *Sustainable Water Resources Management*, *2*(1), 13–27. DOI: 10.1007/s40899-015-0037-0

Benveniste, H., Oppenheimer, M., & Fleurbaey, M. (2020). Effect of border policy on exposure and vulnerability to climate change. *Proceedings of the National Academy of Sciences of the United States of America*, *117*(43), 26692–26702. DOI: 10.1073/pnas.2007597117 PMID: 33046645

Benzie, M., Carter, T. R., Carlsen, H., & Taylor, R. (2019). Cross-border climate change impacts: Implications for the European Union. *Regional Environmental Change*, *19*(3), 763–776. DOI: 10.1007/s10113-018-1436-1

Benzie, M., & Persson, Å. (2019). Governing borderless climate risks: Moving beyond the territorial framing of adaptation. *International Environmental Agreement: Politics, Law and Economics, 19*(4–5), 369–393. DOI: 10.1007/s10784-019-09441-y

Berbés-Blázquez, M., Mitchell, C. L., Burch, S. L., & Wandel, J. (2017). Understanding climate change and resilience: Assessing strengths and opportunities for adaptation in the Global South. *Climatic Change, 141*(2), 227–241. DOI: 10.1007/s10584-017-1897-0

Berke, P. R., Quiring, S. M., Olivera, F., & Horney, J. A. (2021). Addressing Challenges to Building Resilience Through Interdisciplinary Research and Engagement. *Risk Analysis, 41*(7), 1248–1253. DOI: 10.1111/risa.13202 PMID: 30261118

Boyle, , & Shneiderman, Boyle & Shneiderman. (2020). Redundancy, Resilience, Repair: Infrastructural Effects in Borderland Spaces. *Verge: Studies in Global Asias, 6*(2), 112. DOI: 10.5749/vergstudglobasia.6.2.0112

Brunet-Jailly, E. (2022). Cross-border cooperation: A global overview. *Alternatives, 47*(1), 3–17. DOI: 10.1177/03043754211073463 PMID: 35280936

Challinor, A. J., Adger, W. N., Benton, T. G., Conway, D., Joshi, M., & Frame, D. (2018). Transmission of climate risks across sectors and borders. *Philosophical Transactions. Series A, Mathematical, Physical, and Engineering Sciences, 376*(2121), 20170301. DOI: 10.1098/rsta.2017.0301 PMID: 29712795

Chen, M., Zhang, L., Teng, F., Dai, J., Li, Z., Wang, Z., & Li, Y. (2020). Climate technology transfer in BRI era: Needs, priorities, and barriers from receivers' perspective. *Ecosystem Health and Sustainability, 6*(1), 1780948. DOI: 10.1080/20964129.2020.1780948

Chen, N. Sh., Hu, G. Sh., Deng, W., Khanal, N., Zhu, Y. H., & Han, D. (2013). On the water hazards in the trans-boundary Kosi River basin. *Natural Hazards and Earth System Sciences, 13*(3), 795–808. DOI: 10.5194/nhess-13-795-2013

Chowdhooree, I. (2019). Indigenous knowledge for enhancing community resilience: An experience from the south-western coastal region of Bangladesh. *International Journal of Disaster Risk Reduction, 40*, 101259. DOI: 10.1016/j.ijdrr.2019.101259

Cons, J. (2018). Staging Climate Security: Resilience and Heterodystopia in the Bangladesh Borderlands. *Cultural Anthropology, 33*(2), 266–294. DOI: 10.14506/ca33.2.08

De Sherbinin, A. (2014). Climate change hotspots mapping: What have we learned? *Climatic Change, 123*(1), 23–37. DOI: 10.1007/s10584-013-0900-7

Gonda, N., Flores, S., Casolo, J. J., & Nightingale, A. J. (2023). Resilience and conflict: Rethinking climate resilience through Indigenous territorial struggles. *The Journal of Peasant Studies*, *50*(6), 2312–2338. DOI: 10.1080/03066150.2022.2161372

Graveline, M.-H., & Germain, D. (2022). Disaster Risk Resilience: Conceptual Evolution, Key Issues, and Opportunities. *International Journal of Disaster Risk Science*, *13*(3), 330–341. DOI: 10.1007/s13753-022-00419-0

Hannigan, J. (2018). Disasters Across Borders: Borderlands as Spaces of Hope and Innovation in the Geopolitics of Environmental Disasters. In Miller, M. A., Douglass, M., & Garschagen, M. (Eds.), *Crossing Borders* (pp. 79–94). Springer Singapore., DOI: 10.1007/978-981-10-6126-4_5

Hansen, A., Bi, L., Saniotis, A., & Nitschke, M. (2013). Vulnerability to extreme heat and climate change: Is ethnicity a factor? *Global Health Action*, *6*(1), 21364. DOI: 10.3402/gha.v6i0.21364 PMID: 23899408

Hare, W. L., Cramer, W., Schaeffer, M., Battaglini, A., & Jaeger, C. C. (2011). Climate hotspots: Key vulnerable regions, climate change and limits to warming. *Regional Environmental Change*, *11*(S1), 1–13. DOI: 10.1007/s10113-010-0195-4

Hippe, S., Bertram, D., & Chilla, T. (2024). Convergence and resilience in border regions. *European Planning Studies*, *32*(1), 186–207. DOI: 10.1080/09654313.2023.2170214

Hospers, G., & Beugelsdijk, S. (2002). Regional Cluster Policies: Learning by Comparing? *Kyklos*, *55*(3), 381–402. DOI: 10.1111/1467-6435.00192

Jafari, G. (2024). Woman, Water, Freedom: Climate Justice, Women's Liberation, and Territorial Sovereignty in Iran-Iraq Borderlands. *Journal of Architectural Education*, *78*(1), 104–114. DOI: 10.1080/10464883.2024.2303926

Lucatello, S. (2022). Climate Resilient Development Pathways in the US-Mexico Border Region: The Case of the El Paso del Norte Metropolitan Area. In Giorgi, E., Cattaneo, T., Flores Herrera, A. M., & Aceves Tarango, V. D. S. (Eds.), *Design for Vulnerable Communities* (pp. 361–375). Springer International Publishing., DOI: 10.1007/978-3-030-96866-3_19

Makki, M., Butt, F. A., Akash, S. A., Petrova, K., & Naeem, S. A. (2025). Fragile Geographies and the Climate-Conflict Nexus: Investigating Climate-Induced Security Risks, Migration, and Inequality in Balochistan, Pakistan. *Alternatives*, *50*(2), 350–375. DOI: 10.1177/03043754241291728

Mccall, C. (2013). European Union Cross-Border Cooperation and Conflict Amelioration. *Space and Polity*, *17*(2), 197–216. DOI: 10.1080/13562576.2013.817512

McDougal, M. S. (1960). Some basic theoretical concepts about international law: A policy-oriented framework of inquiry. *The Journal of Conflict Resolution*, *4*(3), 337–354. DOI: 10.1177/002200276000400305

Mengisteab, K., & Hagg, G. (Eds.). (2017). *Traditional Institutions in Contemporary African Governance* (1st ed.). Routledge., DOI: 10.4324/9781315227948

Pham, L., Palaneeswaran, E., & Stewart, R. (2018). Knowing maintenance vulnerabilities to enhance building resilience. *Procedia Engineering*, *212*, 1273–1278. DOI: 10.1016/j.proeng.2018.01.164

Prokkola, E.-K. (2019). Border-regional resilience in EU internal and external border areas in Finland. *European Planning Studies*, *27*(8), 1587–1606. DOI: 10.1080/09654313.2019.1595531

Readman, P., Radding, C., & Bryant, C. (2014). Introduction: Borderlands in a Global Perspective. In Readman, P., Radding, C., & Bryant, C. (Eds.), *Borderlands in World History, 1700–1914* (pp. 1–23)., DOI: 10.1057/9781137320582_1

Reynolds, P. D. (2010). New Firm Creation: A Global Assessment of National, Contextual, and Individual Factors. *Foundations and Trends® in Entrepreneurship, 6*(5–6), 315–496. DOI: 10.1561/0300000034

Rosenzweig, B. R., McPhillips, L., Chang, H., Cheng, C., Welty, C., Matsler, M., Iwaniec, D., & Davidson, C. I. (2018). Pluvial flood risk and opportunities for resilience. *WIREs. Water*, *5*(6), e1302. DOI: 10.1002/wat2.1302

Sachseder, J., Stachowitsch, S., & Standke-Erdmann, M. (2024). Entangled Vulnerabilities: Gendered and Racialised Bodies and Borders in EU External Border Security. *Geopolitics*, •••, 1–29. DOI: 10.1080/14650045.2023.2291060 PMID: 39318502

Schoon, M., York, A., Sullivan, A., & Baggio, J. (2017). The emergence of an environmental governance network: The case of the Arizona borderlands. *Regional Environmental Change*, *17*(3), 677–689. DOI: 10.1007/s10113-016-1060-x

Smith, D., & Fischbacher, M. (2009). The changing nature of risk and risk management: The challenge of borders, uncertainty and resilience. *Risk Management*, *11*(1), 1–12. DOI: 10.1057/rm.2009.1

Smolak, A. (2010). Contextual factors influencing HIV risk behaviour in Central Asia. *Culture, Health & Sexuality*, *12*(5), 515–527. DOI: 10.1080/13691051003658135 PMID: 20301020

Talukder, M. B. (2024). Implementing Artificial Intelligence and Virtual Experiences in Hospitality. In Innovative Technologies for Increasing Service Productivity (pp. 145-160). IGI Global. DOI: 10.4018/979-8-3693-2019-8.ch009

Talukder, M. B. (2025). Climate change adaptation strategies for a tourism destination: A study of Kuakata beach. *Acta Scientiarum Polonorum. Administratio Locorum*, *24*(1), 165–179. DOI: 10.31648/aspal.10073

Talukder, M. B. (2025). Revolutionizing Hospitality: Leveraging Mobile Apps and Cutting-Edge Room Technologies in the Hotel Industry. In B. Varghese & S. H. (Eds.), Advancing Smart Tourism Through Analytics (pp. 315-332). IGI Global Scientific Publishing. DOI: 10.4018/979-8-3693-3715-8.ch017

Talukder, M. B., Hoque, M., & Kumar, S. (2024). Impact of Cybersecurity in the Aviation, Tourism, and Hospitality Industries. In *Corporate Cybersecurity in the Aviation, Tourism, and Hospitality Sector* (pp. 200–215). IGI Global Scientific Publishing. DOI: 10.4018/979-8-3693-2715-9.ch011

Talukder, M. B., Khan, M. R., Kumar, S., & Singh, K. (2025). Blockchain and Tourism: Enhancing Community Participation and Sustainability. In *Open Innovation and Technology in Tourism and Hospitality* (pp. 279–302). IGI Global Scientific Publishing. DOI: 10.4018/979-8-3693-8633-0.ch015

Trogal, K., Bauman, I., Lawrence, R., & Petrescu, D. (Eds.). (2018). *Architecture and Resilience: Interdisciplinary Dialogues* (1st ed.). Routledge., DOI: 10.4324/9781315159478

Van Eerd, M. C. J., Wiering, M. A., & Dieperink, C. (2014). Exploring the Prospects for Cross-Border Climate Change Adaptation between North Rhine-Westphalia and the Netherlands. *Utrecht Law Review*, *10*(2), 91. DOI: 10.18352/ulr.271

Vulevic, A., Castanho, R. A., Gómez, J. M. N., Lausada, S., Loures, L., Cabezas, J., & Fernández-Pozo, L. (2021). Cross-Border Cooperation and Adaptation to Climate Change in Western Balkans Danube Area. In Berisha, E., Cotella, G., & Solly, A. (Eds.), *Governing Territorial Development in the Western Balkans* (pp. 289–308). Springer International Publishing., DOI: 10.1007/978-3-030-72124-4_14

Walter, K. M. (2023). Vulnerable People or Vulnerable Borders?: EU External Migration Policies and Gendered Vulnerability. *Migration and Diversity*, *2*(1), 51–62. DOI: 10.33182/md.v2i1.2808

Widdis, R. W. (2019). On globalization, borders, and borderlands: A historical geographical perspective. *Canadian Geographies / Géographies Canadiennes, 63*(4), 573–593. DOI: 10.1111/cag.12580

Wilder, M., Scott, C. A., Pablos, N. P., Varady, R. G., Garfin, G. M., & McEvoy, J. (2010). Adapting Across Boundaries: Climate Change, Social Learning, and Resilience in the U.S.–Mexico Border Region. *Annals of the Association of American Geographers*, *100*(4), 917–928. DOI: 10.1080/00045608.2010.500235

Yang, S., Chen, B., & Fath, B. (2015). Trans-boundary total suspended particulate matter (TSPM) in urban ecosystems. *Ecological Modelling*, *318*, 59–63. DOI: 10.1016/j.ecolmodel.2014.10.006

Chapter 7
Cross-Border Air Pollution Control and Regional Environmental Health Cooperation

Hemendra Singh
https://orcid.org/0000-0003-1160-8284
Jindal Global Law School, O.P. Jindal Global University, Sonipat, India

ABSTRACT

This chapter explores cross-border air pollution as a serious challenge that affects both the environment and human health. It looks at how different regions, such as Europe, Southeast Asia, North America, and East Asia, have built legal and institutional systems to deal with pollution that crosses national borders. Using a doctrinal and comparative method, it examines key agreements, policy tools, and the role of regional cooperation. The chapter also discusses how uneven capacity and lack of enforcement affect outcomes. It highlights the need for stronger legal frameworks and inclusive cooperation to protect air quality, support public health, and promote sustainable development across regions.

INTRODUCTION

Air pollution has long been acknowledged as one of the most serious environmental threats to human health and ecological balance. The World Health Organization has consistently emphasized that exposure to fine particulate matter, nitrogen dioxide, and ozone contributes significantly to non-communicable diseases, premature mortality, and declining quality of life, particularly in urban and industrialized zones. While air

DOI: 10.4018/979-8-3373-3972-6.ch007

pollution often begins as a localized issue, its impacts extend across administrative and national borders, complicating regulatory responses and necessitating regional cooperation (Underwood et al., 2024). The growing body of literature confirms that pollutants such as sulphur dioxide, nitrogen oxides, and particulate matter can travel hundreds or even thousands of kilometers from their source, affecting downwind populations irrespective of political boundaries (Stevens et al., 2024).

The legal and policy literature on transboundary air pollution has evolved from early treaties on acid rain and ozone depletion to more recent frameworks addressing fine particulates and climate-related air pollutants (Grennfelt et al., 2020). Foundational studies such as those by Birnie, Boyle, and Redgwell have laid the groundwork for understanding the principles of international environmental law as they relate to shared resources like the atmosphere. Their scholarship highlights the inadequacy of rigid sovereignty-based models in the context of borderless environmental harms (Jeffery, 2009). In parallel, regional environmental agreements such as the Convention on Long-Range Transboundary Air Pollution in Europe and the ASEAN Agreement on Transboundary Haze Pollution in Southeast Asia reflect attempts to institutionalize cooperation and collective responsibility among neighboring states (Heilmann, 2015).

Recent research underscores the uneven distribution of both emissions and exposure. Studies reveal that low-income and marginalized communities frequently bear the heaviest burden of airborne toxins (Rentschler & Leonova, 2023). This disparity is most visible in rapidly urbanizing regions of South and East Asia, where economic growth has outpaced the development of robust environmental governance. Moreover, legal scholars have increasingly drawn attention to the governance vacuum that exists at the intersection of domestic jurisdiction and cross-border environmental responsibility. While domestic environmental laws may be well-developed in some jurisdictions, they often lack the extraterritorial reach necessary to address harms that originate elsewhere (Hadjiyianni, 2023).

There is a growing consensus among scholars and policy analysts that regional frameworks are essential for addressing the specific complexities of transboundary air pollution. Regional environmental regimes are often more adaptable and politically feasible than global instruments, and they are better positioned to respond to local ecological contexts and institutional realities (Xia et al., 2025). Nonetheless, regional responses remain fragmented, with wide variation in legal enforceability, institutional strength, and political commitment. Comparative studies reveal that successful regional models typically include formalized monitoring mechanisms, mutual reporting obligations, and mechanisms for dispute resolution. However, such elements are often missing or weak in many developing regions, thereby limiting the effectiveness of existing agreements.

Overall, the literature points to an urgent need for more integrated, inclusive, and adaptive legal frameworks that recognize air pollution as a shared regional concern. This recognition is particularly critical given the direct implications for human health, environmental justice, and sustainable development. While global awareness of the problem is increasing, progress on coordinated legal action remains uneven, making the study of regional legal and institutional models both timely and essential.

Environmental Health as a Regional Imperative

Environmental health is increasingly recognized as a matter that extends beyond national borders. The effects of polluted air are not confined to the countries in which pollutants are emitted. Instead, these effects spill over into neighboring regions, carried by wind, temperature inversions, and shifting atmospheric conditions. This borderless nature of airborne contaminants places entire regions at risk, particularly when environmental policies and pollution control measures differ in strength and enforcement from one jurisdiction to another. As a result, regional approaches to protecting environmental health have become not only desirable but necessary.

Scientific research has repeatedly shown that transboundary air pollution contributes significantly to the regional burden of disease (Spiteri & von Brockdorff, 2021). Fine particulate matter, especially PM2.5, is linked to respiratory infections, cardiovascular diseases, stroke, and lung cancer. Regions that experience seasonal haze, industrial smog, or dust transport events often report spikes in hospitalizations, school absences, and premature deaths. For instance, in Southeast Asia, the annual transboundary haze from forest fires in Indonesia has repeatedly caused respiratory illnesses and school closures in Singapore and Malaysia. The 2015 haze crisis led to more than half a million reported cases of acute respiratory infections in Indonesia alone, while neighboring countries recorded severe air quality deterioration for weeks.

While individual countries may implement domestic air quality standards, these alone are insufficient when the sources of pollution lie outside their territory. When emissions cross borders, so do their consequences, including the costs of medical care, productivity loss, and long-term impacts on human development. In South Asia, cities in northern India suffer from chronic air pollution in part due to agricultural burning in neighboring regions (Lan et al., 2022). Wind patterns then carry this pollution into bordering countries such as Pakistan, Nepal, and Bangladesh, leading to regional air quality alerts and a shared burden on public health systems. These events highlight that the health implications of transboundary air pollution are not just environmental problems but also social and economic challenges that demand regional cooperation.

In recent years, health-focused institutions and public health researchers have advocated for stronger regional governance frameworks that prioritize early warning systems, joint health impact assessments, and cross-border data sharing. For example, the European Union's Air Quality Directives and its monitoring network allow member states to exchange data in near real-time, enabling coordinated action when pollutant levels exceed agreed thresholds. This model has improved public health responsiveness across borders. Similarly, the Malé Declaration on Air Pollution in South Asia encourages countries to engage in joint monitoring and training to address regional air quality issues, though its implementation remains uneven.

Importantly, regional disparities in health outcomes related to air pollution also reveal broader patterns of inequality and structural vulnerability. Studies have shown that rural populations, ethnic minorities, and informal workers are disproportionately affected by the cumulative impacts of poor air quality. In Mongolia, for instance, low-income residents of Ulaanbaatar living in informal settlements and relying on coal-fired heating face some of the worst winter air pollution conditions in the world (Guttikunda et al., 2013). In Latin America, residents in industrial corridors such as the "Sacrifice Zones" in Chile experience long-term exposure to industrial emissions, with limited avenues for redress and health care. These examples illustrate that environmental health cannot be separated from questions of social justice and governance.

Regional environmental health governance, when aligned with long-term development planning, supports broader efforts to ensure access to clean air, equitable health outcomes, and inclusive environmental decision-making. These efforts echo global commitments to build healthier societies through improved public institutions, reduced inequalities, and stronger environmental protections. Framing environmental health as a regional imperative not only addresses the immediate need to manage cross-border pollution but also reflects a deeper commitment to advancing shared well-being and justice in ways consistent with inclusive, sustainable development. Such an approach is essential to achieving healthier environments and fairer futures for all.

Research Question and Methodology

This chapter investigates how cross-border air pollution affects regional environmental health and what legal and institutional responses have emerged to address this challenge. It asks the central question: How can regional cooperation frameworks be strengthened to effectively manage transboundary air pollution and protect public health across borders?

To answer this, the study adopts a comparative legal and policy approach. It draws on regional case studies from Europe, Southeast Asia, North America, and East Asia to analyze both binding and non-binding mechanisms of cooperation. The focus is on how these regions have developed shared norms, implemented joint monitoring systems, and responded to health risks linked to air pollution that travels across national boundaries.

The methodology combines doctrinal legal research, policy analysis, and review of environmental health data from international and regional organizations. It includes qualitative evaluation of institutional practices, multilateral agreements, and coordination mechanisms among neighboring countries. Where relevant, the chapter incorporates examples of on-the-ground impacts drawn from peer-reviewed public health and environmental studies. This approach allows for a grounded understanding of how different regional frameworks function in practice, and where they succeed or fall short in responding to complex air quality challenges.

NORMATIVE AND LEGAL FOUNDATIONS OF TRANSBOUNDARY AIR GOVERNANCE

Air pollution does not respect national borders. What begins as industrial emission in one country often ends as respiratory illness in another. Over time, this reality has challenged traditional notions of environmental sovereignty. As cross-border pollution has grown in scope and severity, it has pushed legal systems to evolve beyond national frameworks. This chapter explores how legal norms have developed to govern shared atmospheric spaces, especially in regions where pollution is transboundary in character and consequence. It considers the shift from unilateral environmental regulation to cooperative legal regimes and examines both binding obligations and softer frameworks aimed at mitigating shared environmental harm. The discussion pays particular attention to how normative developments have been shaped by environmental justice concerns, unequal vulnerabilities, and calls for fairer, more inclusive regional cooperation.

Evolving Legal Norms for Shared Atmosphere

The development of legal norms around transboundary air pollution has followed a gradual yet significant evolution, shaped by scientific understanding, environmental crises, and political negotiations. The early foundations of this legal evolution are rooted in the principle that states have the sovereign right to exploit their own resources but also bear the responsibility not to cause environmental harm to other states. This principle, articulated in the 1972 Stockholm Declaration and reinforced in the 1992 Rio Declaration, laid the groundwork for legal thinking around cross-border environmental accountability.

The landmark 1941 Trail Smelter Arbitration between the United States and Canada is often cited as the first case to establish state responsibility for transboundary air pollution. Although focused on sulphur dioxide emissions from a single smelter, it introduced the core idea that a state can be held liable for environmental harm caused to another country. However, this ruling remained largely symbolic for decades, with few follow-up cases reinforcing its precedent.

By the late twentieth century, rising public awareness of acid rain in Europe and North America contributed to the emergence of more structured legal responses. Regional instruments such as the 1979 Convention on Long-range Transboundary Air Pollution (CLRTAP) reflected a new willingness among states to acknowledge shared responsibility for atmospheric degradation. CLRTAP introduced protocols targeting specific pollutants such as sulphur, nitrogen oxides, and persistent organic pollutants. Importantly, it also created a culture of regional dialogue and scientific cooperation that served as a legal model for other regions.

At the international level, there is no single treaty that governs transboundary air pollution globally. However, several multilateral environmental agreements touch upon the issue, including the 1992 United Nations Framework Convention on Climate Change (UNFCCC) and its related protocols. These agreements, although primarily climate-focused, implicitly recognize the interconnected nature of atmospheric harm and the need for collective responses.

Legal norms have also evolved through the work of international bodies and soft law instruments. For example, the United Nations Economic Commission for Europe has promoted environmental impact assessments that include cross-border pollution effects. Similarly, the ASEAN Agreement on Transboundary Haze Pollution, while limited in enforcement, represents a shift toward formalized regional obligations in Southeast Asia.

Recent years have witnessed increased calls for the legal recognition of the right to a clean, healthy, and sustainable environment (Spijkers, 2025). This right, now acknowledged in UN General Assembly resolutions and supported by human rights frameworks, places normative pressure on states to cooperate in preventing air pollution that affects populations across borders. It also aligns with broader commitments to environmental health, justice, and international solidarity.

While binding legal norms remain uneven and regionally concentrated, there is growing convergence on principles such as precaution, polluter pays, and public participation. These norms signal a move toward a shared legal understanding that the atmosphere is a collective resource requiring cooperative stewardship. The evolution of legal thinking in this domain continues to reflect both the urgency of the problem and the shifting ethical expectations of how states should behave toward one another in an increasingly interconnected world.

Binding and Non-Binding Legal Instruments

Efforts to govern transboundary air pollution have given rise to a range of legal instruments, which vary widely in their form, content, and enforcement potential. These instruments can be broadly categorized as either binding, those that create legal obligations enforceable under international or regional law, or non-binding, which rely on voluntary cooperation, political will, or soft power to achieve compliance. Together, they form a patchwork of governance mechanisms that address atmospheric challenges with differing degrees of authority and success.

Binding instruments

Binding legal instruments are formal agreements between states that typically include obligations, compliance procedures, and, in some cases, sanctions for non-performance. The most comprehensive and long-standing example in this area is the Convention on Long-range Transboundary Air Pollution (CLRTAP) adopted under the United Nations Economic Commission for Europe in 1979. CLRTAP now includes eight additional protocols that address specific pollutants and technical cooperation. These protocols have been instrumental in reducing sulphur dioxide and nitrogen oxide emissions across Europe and parts of North America. Importantly, they also include provisions on monitoring, scientific collaboration, and data-sharing, which help build trust and accountability among member states.

Another example is the ASEAN Agreement on Transboundary Haze Pollution (2002), a binding agreement among Southeast Asian nations designed to address the recurring problem of smoke haze caused by land and forest fires, particularly in Indonesia. While the agreement is legally binding, its enforcement mechanisms are limited, and it has faced criticism for the slow implementation of domestic regulations and monitoring practices. Nonetheless, it marks an important step in the evolution of regional legal cooperation on air pollution in the Global South.

In North America, the Canada–United States Air Quality Agreement (1991) stands out as a model of bilateral cooperation. The agreement focuses on acid rain and ground-level ozone, committing both countries to emission reductions, regulatory harmonization, and information exchange (Melious, 2014). It includes mechanisms for consultation, joint monitoring, and regular review, making it one of the more effective examples of a binding transboundary air pollution framework.

Non-binding instruments

Non-binding or soft law instruments also play a significant role, particularly where political sensitivities or institutional constraints limit the prospects for treaty-making.

These instruments include ministerial declarations, memoranda of understanding, codes of conduct, and guidelines issued by regional or international organizations. While they do not impose legal obligations, they can influence national policy, set standards, and prepare the ground for future binding commitments.

For instance, in East Asia, trilateral cooperation among China, Japan, and South Korea on air pollution has relied primarily on non-binding platforms such as the Tripartite Environment Ministers Meeting and associated joint research programs. These initiatives have helped foster mutual understanding, standardize monitoring practices, and create a basis for dialogue, even if no formal treaty has yet emerged.

Similarly, the United Nations Environment Programme's regional atmospheric programs have used soft law mechanisms to encourage collaboration in regions like Latin America and the Caribbean, where diverse legal and institutional capacities make binding agreements difficult. These programs often focus on capacity-building, best practices, and the development of regional action plans.

In some cases, non-binding instruments serve as precursors to binding law. They provide a testing ground for ideas, promote norm diffusion, and build the political momentum necessary for treaty formation. They are particularly valuable in areas of scientific uncertainty or where consensus is still emerging.

Interplay and Significance

The interplay between binding and non-binding instruments is a defining feature of contemporary transboundary air governance. Rather than viewing them as separate or competing, scholars increasingly recognize their complementary roles. Binding agreements offer legal certainty and enforcement, while soft law tools promote flexibility, inclusivity, and gradual norm internalization. Together, they reflect the layered and adaptive nature of environmental governance in a world where atmospheric interdependence is growing and where regional cooperation is essential to protect both public health and ecological stability.

This mixture of legal forms contributes to the broader global push for sustainable development and environmental justice. By enabling both immediate action and long-term legal evolution, these instruments play a key role in shaping fairer and more responsive regional approaches to shared environmental challenges.

Gaps in Coverage and Implementation

Despite the presence of both binding and non-binding instruments, significant gaps remain in the legal and institutional frameworks governing transboundary air pollution. These deficiencies are visible in multiple regions and manifest in several ways, including incomplete geographic coverage, weak enforcement, limited participation, and poor integration of scientific data into policy. Such gaps hinder the effective management of shared air resources and disproportionately affect vulnerable communities living near borders.

Uneven geographic coverage and participation

Many regions of the world, particularly in the Global South, lack robust transboundary air pollution agreements. While Europe and North America have established treaty-based frameworks, large parts of Africa, Latin America, Central Asia, and the Middle East do not have any comparable binding mechanisms. In regions where air pollution knows no political boundary, this absence of legal coverage results in unregulated emissions, fragmented responses, and limited regional solidarity. Even in regions with some cooperative instruments, such as Southeast Asia, participation may be incomplete. For example, the ASEAN Haze Agreement initially suffered from the non-ratification by key countries like Indonesia, despite being the major source of transboundary haze at the time of signing (Forsyth, 2014). Although this has since changed, early delays in ratification and implementation undermined regional efforts.

Weak enforcement and compliance mechanisms

A recurring challenge in both binding and non-binding instruments is the lack of effective enforcement. Many treaties lack clear sanctions or dispute resolution mechanisms. Instead, they rely heavily on reporting, monitoring, and peer pressure. While these mechanisms may work in contexts of high trust and strong institutional capacity, they are often inadequate in more fragile or politically sensitive regions. For instance, the enforcement provisions under the ASEAN Haze Agreement have been widely criticized as being too soft to compel national compliance. Even where national laws exist to implement treaty obligations, domestic enforcement agencies may lack the capacity or resources to monitor industrial sources, agricultural burning, or vehicular emissions that contribute to cross-border pollution.

Scientific uncertainty and data asymmetry

Another significant gap lies in the integration of scientific data and monitoring systems across regions. Transboundary air pollution is a highly technical issue that requires consistent data collection, atmospheric modelling, and pollutant tracking. In many regions, national governments maintain different standards for air quality measurement, emission inventories, and data sharing. This leads to disagreements about the source, extent, and impact of pollution episodes. In the East Asian context, for example, disagreements over data transparency between China, South Korea, and Japan have impeded progress toward a legally binding regional framework, despite growing public concern over PM2.5 levels.

Moreover, insufficient investment in transboundary monitoring infrastructure in lower-income countries creates data asymmetry that disadvantages weaker states in negotiations and dispute resolution. The absence of open and harmonized environmental data impedes mutual understanding, delays cooperative responses, and allows polluters to operate with limited scrutiny.

Lack of health-centered and equity-driven provisions

Existing legal frameworks often fail to fully integrate the public health dimensions of air pollution or address the environmental justice implications of unequal exposure. Legal texts may acknowledge the harmful effects of pollution on health, but enforcement and policy measures rarely prioritize affected communities or establish cross-border mechanisms to compensate or protect vulnerable populations (Rentschler & Leonova, 2023). Marginalized borderland communities, particularly those dependent on informal or agricultural economies, are frequently left without access to legal remedies or adequate public health protections.

Furthermore, the design of many regional instruments remains technocratic and state-centered, leaving little space for civil society, affected populations, or scientific communities to influence legal development and implementation. This exclusion contributes to a persistent gap between legal aspiration and on-the-ground outcomes.

Addressing these gaps is critical not only for environmental protection but also for advancing shared goals of regional stability, public health, and cooperative development. The failure to fill legal and institutional voids in transboundary air governance undermines broader aspirations for equity, sustainability, and inclusive growth. By identifying and analyzing these gaps, this study contributes to ongoing efforts to construct more responsive, participatory, and science-informed legal frameworks that support healthier and more resilient societies.

COMPARATIVE REGIONAL APPROACHES TO TRANSBOUNDARY AIR POLLUTION

Regional cooperation plays a crucial role in addressing the complex and borderless nature of air pollution. While the atmospheric movement of pollutants often defies national boundaries, legal and policy responses remain largely state-centric. Nevertheless, several regions have developed institutional frameworks and cooperative mechanisms to tackle transboundary air pollution through joint monitoring, standard-setting, and policy harmonization. These regional models vary widely in legal design, scope, and effectiveness. Some adopt structured compliance and enforceable legal obligations, while others rely on voluntary arrangements and diplomatic consensus. By examining how different regional frameworks function in practice, this chapter offers a comparative perspective on what works, what falls short, and what lessons can be drawn to inform future legal innovation. The case studies include Europe, Southeast Asia, North America, and East Asia, reflecting a diversity of political systems, legal traditions, and environmental challenges.

Europe: Legal Regionalism and Structured Compliance

Europe represents the most legally advanced and institutionally developed example of regional cooperation in transboundary air pollution control. The foundation of this framework lies in a dense web of binding legal instruments and regional organizations, including both the European Union (EU) and the United Nations Economic Commission for Europe (UNECE). These institutions have enabled the formulation, adoption, and enforcement of ambitious cross-border air quality standards grounded in scientific data and supported by compliance mechanisms.

The Convention on Long-Range Transboundary Air Pollution

One of the earliest and most influential instruments is the Convention on Long-Range Transboundary Air Pollution (CLRTAP), adopted in 1979 under the auspices of the UNECE. It created a regional legal regime for cooperation among European and some North American countries to address transboundary air pollution across large distances. Over time, the Convention has been supplemented by eight protocols targeting specific pollutants such as sulphur dioxide, nitrogen oxides, volatile organic compounds, and persistent organic pollutants. These protocols reflect a progressive tightening of emission ceilings, informed by evolving scientific understanding and technical capacity.

The CLRTAP framework is supported by monitoring programs such as the European Monitoring and Evaluation Programme (EMEP), which facilitates coordinated data collection, modelling, and information exchange. Parties are obligated to report emissions and implement national strategies aligned with the protocols. Importantly, the Convention has not remained static. Its Gothenburg Protocol, for instance, was amended in 2012 to incorporate health-related particulate matter and promote ecosystem protection, showing the regime's responsiveness to new environmental and public health challenges.

European Union Directives and Regional Implementation

In parallel with UNECE efforts, the European Union has developed its own legal regime that enforces air quality standards among member states. EU directives such as the Ambient Air Quality Directive (2008/50/EC) set legally binding limits on key pollutants, including fine particulate matter and nitrogen dioxide. Member states are required to develop air quality plans, monitor ambient conditions, and face potential infringement proceedings if they fail to meet the targets. The EU also links air pollution control to climate and energy policies, promoting integrated approaches to emissions reduction through its Green Deal and Fit for 55 packages.

One example of EU enforcement in action is the series of infringement cases brought by the European Commission against member states like Germany, France, and Italy for persistently exceeding nitrogen dioxide levels, largely due to diesel vehicle emissions. These legal actions underscore the seriousness with which the EU approaches compliance and the power of supranational institutions in enforcing environmental law.

Lessons from Europe's legal regionalism

Europe's experience highlights how legal regionalism, when backed by institutional capacity, public accountability, and political will, can yield measurable improvements in air quality. The combination of binding commitments, technical cooperation, public access to information, and judicial oversight creates a robust governance model. However, challenges remain. Differences in national implementation, resource constraints in some member states, and emerging concerns over ultrafine particles and cross-sectoral impacts require continuous adaptation. Nevertheless, the European model illustrates the potential of legally grounded regional cooperation to not only manage pollution but also safeguard public health and promote environmental justice.

Southeast Asia: ASEAN and the Haze Agreement

Transboundary air pollution in Southeast Asia, particularly the recurrent haze episodes caused by forest and peatland fires, has emerged as one of the region's most intractable environmental challenges. Originating primarily in Indonesia due to slash-and-burn practices used in clearing land for agriculture, these fires send dense clouds of smoke and harmful pollutants across borders, severely affecting the air quality in Malaysia, Singapore, Thailand, and beyond. The health, economic, and environmental consequences have been far-reaching. The 1997 haze crisis, for instance, led to thousands of premature deaths and over nine billion US dollars in estimated damages. The political fallout from such events catalysed regional efforts to formulate a collective legal and institutional response, eventually resulting in the 2002 ASEAN Agreement on Transboundary Haze Pollution, the first international treaty in the world to specifically address cross-border air pollution (Alvin, 2022).

The Agreement was envisioned as a framework for regional cooperation grounded in prevention, monitoring, and mitigation of haze-related disasters. It obliges signatory states to take active steps to reduce fire risk, enhance early warning and response systems, and share data and technical expertise. However, reflecting ASEAN's normative foundation rooted in non-interference and consensus, the Agreement does not include strong enforcement mechanisms. Instead, it relies on voluntary compliance, dialogue, and mutual support. This approach, often described as the "ASEAN Way," has allowed the Agreement to preserve diplomatic harmony, but has also limited its impact. Indonesia, the principal source country of haze pollution, did not ratify the treaty until 2014, a full twelve years after it was signed. During that time, multiple haze crises continued to affect the region, with limited accountability and inadequate response capacities on the ground.

Despite these limitations, the institutional architecture developed under the Agreement has played a pivotal role in laying the groundwork for coordinated action. The ASEAN Coordinating Centre for Transboundary Haze Pollution Control was established to serve as the hub for information exchange, capacity-building, and technical cooperation. The Sub-Regional Ministerial Steering Committee on Transboundary Haze Pollution and the ASEAN Ministerial Meeting on the Environment have provided high-level platforms for negotiation and review. At the operational level, ASEAN states have invested in satellite-based hotspot detection systems, air quality index harmonization, and joint research projects to understand the causes and impacts of haze. While these measures have strengthened the region's preparedness, they remain unevenly implemented across member states due to disparities in national capacities and governance frameworks.

The effectiveness of the ASEAN Haze Agreement is further constrained by deeply rooted political and economic factors. Many of the fires are set deliberately by commercial interests in palm oil and pulpwood plantations. These actors often operate with impunity in remote regions where regulatory oversight is weak and local enforcement is vulnerable to corruption or under-resourcing. The reluctance of states to confront powerful agribusiness lobbies or cede authority to a supranational enforcement mechanism perpetuates a gap between legal obligations and on-the-ground realities. Although public outrage in affected countries like Singapore has occasionally led to domestic legislative responses such as Singapore's Transboundary Haze Pollution Act of 2014 these national laws have limited reach beyond their borders and do not substitute for collective enforcement.

Nevertheless, there are incremental gains. Civil society participation and regional media attention have helped raise awareness and mobilize public support for haze mitigation. ASEAN's commitment to long-term haze-free targets, such as the Roadmap on ASEAN Cooperation towards Transboundary Haze Pollution Control with Means of Implementation (2016–2020), signals a growing willingness to move beyond aspirational rhetoric. Bilateral cooperation, such as Singapore's technical assistance to Indonesia and Malaysia's engagement with Indonesian authorities, has yielded practical benefits in fire management and community-level resilience. These efforts demonstrate that while formal legal enforcement remains weak, regional environmental diplomacy has fostered informal norms, political pressure, and cooperation pathways that can gradually shift state behavior.

The ASEAN model, in this context, reflects the potential of soft law in regions characterized by diverse political systems and sensitivities around sovereignty. Its focus on dialogue, trust-building, and incrementalism offers lessons for other regions grappling with transboundary environmental problems without robust legal integration. At the same time, it highlights the limits of cooperation that avoids legal accountability. As climate variability intensifies and land use pressures grow, the need for a stronger regional framework, one that balances sovereignty with responsibility, becomes increasingly urgent. The ASEAN haze regime, despite its challenges, continues to evolve and has contributed to regional capacity-building and environmental consciousness. It indirectly supports broader global goals linked to environmental health, sustainable ecosystems, and cross-border cooperation in the face of planetary challenges.

North America: U.S. - Canada Air Quality Agreement

The Air Quality Agreement signed between the United States and Canada in 1991 is widely regarded as a successful model of bilateral cooperation on transboundary air pollution. It was developed in response to growing concerns about acid rain caused by sulphur dioxide and nitrogen oxide emissions, which were affecting forests, lakes, and air quality in both countries. These pollutants, released from power plants and industrial facilities, were travelling across borders and creating shared environmental problems. The Agreement reflects a rare instance where two countries managed to create a binding legal framework based on scientific evidence, public pressure, and shared environmental goals. It requires both nations to reduce emissions of acid rain precursors and to report on progress regularly through joint reviews and public reports (McLean & Barton, 2008). A key feature of the Agreement is the establishment of the bilateral Air Quality Committee, which facilitates scientific cooperation, data sharing, and ongoing policy dialogue.

Over time, the Agreement expanded to include other pollutants, such as ground-level ozone, through additional annexes. It is notable that the Agreement has continued to function effectively despite political changes on both sides, in part because it is supported by strong domestic institutions like the U.S. Environmental Protection Agency and Environment and Climate Change Canada. These agencies have the technical capacity and legal authority to implement commitments and ensure compliance. The Agreement has led to significant reductions in cross-border pollution and helped restore ecosystems in affected regions. Its impact has also extended to public health, as cleaner air has meant fewer respiratory illnesses and related health costs for populations living near the border.

What sets this Agreement apart is the mutual trust and transparency it has fostered over decades. Both countries commit to comparable levels of ambition and regularly exchange data and updates. Unlike non-binding arrangements that depend on goodwill, the U.S.–Canada model is rooted in enforceable commitments and institutional follow-through. It has demonstrated that even without a supranational authority, neighboring countries can create and sustain effective environmental cooperation through law, science, and accountability. For regions struggling with weak legal frameworks or low trust, the U.S. - Canada experience offers valuable insights into how air quality governance can be made both practical and durable. In the broader context, it contributes to long-term goals of healthier environments, cooperative institutions, and regional stability.

East Asia: China–South Korea–Japan Trilateral Environmental Cooperation

East Asia has emerged as one of the most environmentally burdened regions in the world, with transboundary air pollution being a major source of political and public concern. The rapid industrialization of China, coupled with high energy demands and dense urban populations across China, South Korea, and Japan, has led to severe air quality degradation. Seasonal haze, fine particulate matter (PM2.5), and long-range transport of pollutants from mainland Asia to Korea and Japan have become recurrent issues. The public discourse in the region, particularly in South Korea and Japan, often links respiratory illnesses and reduced visibility to pollution originating from China. This has led to increasing pressure on national governments to pursue cooperative approaches that extend beyond domestic environmental policies.

The Trilateral Environmental Ministers Meeting (TEMM), established in 1999, has become the cornerstone of regional cooperation among China, South Korea, and Japan. Although not a binding legal agreement, TEMM serves as a high-level political forum where environmental ministers from the three countries meet annually to set shared priorities, develop joint projects, and promote information exchange. Through this mechanism, the countries have initiated various collaborative initiatives, such as the Acid Deposition Monitoring Network in East Asia and the Joint Research Project on Long-range Transboundary Air Pollutants (LTP). These programs aim to enhance scientific understanding of pollution flows, build regional data-sharing mechanisms, and promote dialogue on policy responses.

Despite the structured engagement, the cooperation has largely remained technical and diplomatic, with limited progress in developing enforceable legal instruments. Political sensitivities, historical tensions, and differing levels of economic development have made it difficult to align national priorities into a cohesive legal framework. Moreover, while China has made strides in domestic air pollution control, especially through its Air Pollution Prevention and Control Action Plan and recent efforts to reduce coal use, its commitment to regional frameworks often remains cautious. South Korea and Japan have relatively stronger domestic regulatory regimes and tend to push for greater transparency and mutual accountability, sometimes leading to friction in negotiations.

Nevertheless, the trilateral efforts have yielded some tangible benefits. Regular meetings have helped sustain environmental dialogue during times of geopolitical strain, and joint scientific monitoring has improved understanding of pollution patterns across borders. In recent years, the three countries have agreed to explore the possibility of a regional air quality forecasting system and to harmonize certain data standards. These steps indicate a gradual movement from soft cooperation toward more structured environmental governance, even if binding commitments remain elusive.

East Asia's experience illustrates the complexity of building cooperative frameworks in regions marked by historical mistrust and unequal capacities. Yet it also shows that persistent, science-based engagement can lay the foundation for deeper legal and policy coordination. By continuing to invest in institutional mechanisms, mutual data transparency, and incremental trust-building, the trilateral model could evolve into a more integrated regional approach to managing transboundary air pollution. The lessons from East Asia underscore the importance of regional solidarity, scientific diplomacy, and sustained multilateralism in tackling common environmental challenges.

HEALTH, JUSTICE AND INSTITUTIONAL DIMENSIONS OF CROSS-BORDER AIR POLLUTION

The consequences of cross-border air pollution are not merely environmental or technical; they are deeply human and political. Airborne pollutants do not stop at borders, but their impacts are felt most severely by populations with the fewest resources and the least voice in environmental governance. This chapter examines how transboundary air pollution creates unequal health outcomes, reinforces environmental injustice, and exposes institutional gaps in addressing cross-border harms. It highlights the need to center public health and equity within regional cooperation frameworks and calls for more accountable, inclusive, and transparent institutional responses.

Unequal Health Impacts and Borderless Burdens

Exposure to air pollution is a leading global environmental risk factor, but its health impacts are unevenly distributed. In many cases, the worst-affected populations live far from the original source of emissions. This disjuncture between polluters and sufferers creates a moral and legal dilemma, especially when pollutants cross borders and settle over communities that lack both the infrastructure to protect themselves and the political leverage to demand accountability.

In regions like South Asia, for instance, the seasonal movement of fine particulate matter and ozone precursors from one country to another results in health crises that are rarely addressed through joint public health planning. Northern India, with its heavy industrial base and biomass burning, contributes to smog episodes in neighboring Nepal and Bangladesh. However, the affected populations in these areas often lack access to air quality data, protective technologies, or even the basic right to information about the origin of the pollution that envelops their towns and cities. Rural children, elderly people, and those with preexisting health conditions are especially vulnerable, facing higher rates of asthma, bronchitis, and premature death.

Studies confirm that long-range transport of pollutants not only degrades air quality across regions but also disrupts public health systems in receiving countries (Liu et al., 2020). Transboundary exposure complicates disease attribution and medical treatment. Healthcare providers may misdiagnose pollution-related illnesses or fail to identify their source, and governments often hesitate to raise concerns with neighboring states for fear of political fallout. In such scenarios, the burden of proof and remediation falls on those least equipped to carry it.

These unequal health impacts reflect deeper structural inequalities embedded in global and regional governance systems. Communities situated near borders or in politically marginal zones are routinely excluded from decision-making processes on pollution management. Their needs rarely find space in regional air quality agreements, which tend to prioritize state sovereignty and industrial interests over local well-being. Even where health is mentioned, it is treated as a secondary concern, subordinated to economic calculations or diplomatic sensitivities.

Addressing these disparities requires more than improved emissions monitoring or transboundary dialogues. It calls for a rethinking of how public health is integrated into environmental policy and how regional cooperation frameworks can reflect the lived experiences of those who suffer most. Protecting the right to breathe clean air must be treated not as a secondary benefit but as a core objective of regional governance.

This approach aligns closely with international efforts to reduce health inequalities, promote access to essential services, and ensure inclusive decision-making. Recognising the differentiated health burdens of air pollution strengthens the case for shared responsibility and equitable policy interventions that protect the most vulnerable across borders.

Environmental Injustice and Structural Disparities

Environmental injustice in the context of cross-border air pollution refers to the disproportionate exposure of marginalized populations to environmental harms without corresponding influence over the policies that shape those risks. This asymmetry is neither accidental nor recent; it is the result of longstanding structural disparities rooted in economic inequality, political marginalization, and historical patterns of industrial development. While pollution itself may be airborne and borderless, its effects are sharply unequal. Poor and minority communities often situated near industrial zones, highways, or downstream of prevailing wind patterns tend to bear the greatest burden of toxic air. These communities rarely have access to adequate healthcare, lack political leverage, and are often excluded from regional decision-making bodies tasked with managing air quality.

In many parts of Southeast Asia, for instance, cross-border haze from large-scale agricultural burning in one country disproportionately affects indigenous communities and urban poor in neighboring nations. These populations are already grappling with limited access to clean water, reliable housing, or public health infrastructure, and the added burden of hazardous air quality leads to increased rates of asthma, heart disease, and premature death. Despite these realities, their voices are rarely included in regional negotiations or policy formulation. A similar pattern can be observed in North America, where indigenous communities near oil sands operations or downwind of coal-fired plants experience higher rates of respiratory and cardiovascular diseases, yet often struggle for recognition in cross-border environmental governance.

These disparities are further exacerbated by inadequate legal frameworks. Many regional agreements fail to incorporate procedural justice - meaning the right of affected individuals to participate meaningfully in environmental decision-making processes. Moreover, legal remedies for those harmed by transboundary pollution remain limited or inaccessible. Victims rarely have the standing to sue polluters located in another jurisdiction, and existing forums for cross-border environmental claims are often slow, fragmented, or unresponsive to the needs of vulnerable groups. This institutional silence perpetuates a cycle of exclusion where those who suffer most from pollution are least able to shape or challenge the systems that allow it.

Environmental injustice in transboundary contexts is not only about distributional inequalities but also about the lack of transparency and accountability within governance systems. In many cases, emission data are not disaggregated by locality or socioeconomic status, masking the reality of who is affected and to what extent. Regional bodies tasked with overseeing air quality may prioritize state sovereignty over local accountability, thereby sidelining communities with the greatest stake in cleaner air. Without mechanisms to address these structural disparities, cross-border air pollution will continue to entrench environmental inequalities.

Addressing these injustices requires more than technical solutions. It involves a reimagining of regional cooperation that puts equity at its core. This includes enabling meaningful public participation, ensuring fair access to information, and integrating human health concerns directly into the objectives of air pollution agreements. In doing so, regional frameworks can move closer to supporting inclusive institutions and healthier, more just societies.

Institutions, Transparency, and Regional Accountability

The effectiveness of cross-border air pollution governance hinges on the strength and integrity of regional institutions. These institutions act as the linchpin for monitoring, coordinating, and enforcing pollution control efforts that span national boundaries. Their credibility depends not only on legal mandates and scientific capacity but also on their ability to foster trust through transparency and responsiveness to the public. When institutions lack openness or suffer from weak mandates, regional cooperation often falters, and pollution control becomes fragmented or superficial.

One of the central challenges lies in the design of institutional frameworks that accommodate both state sovereignty and the need for collective action. In many regional arrangements, such as the ASEAN Agreement on Transboundary Haze Pollution, institutions operate through consensus-based models that prioritize diplomatic caution over firm commitments. While such models preserve regional harmony, they often result in vague obligations, limited enforcement mechanisms, and insufficient data-sharing practices. As a result, institutions tasked with addressing cross-border pollution are often unable to compel compliance or respond to acute public health risks with the necessary speed and coordination.

Transparency plays a crucial role in determining whether these institutions can function effectively. Public access to emissions data, air quality indicators, and mitigation reports empowers civil society to hold both states and industries accountable. In regions like Europe, the success of transboundary air quality regimes has been partly due to the integration of public reporting standards and independent scientific assessments into institutional practices. The European Environment Agency, for instance, routinely publishes air quality statistics accessible to the public, which strengthens confidence in governance processes and helps drive policy innovation (Laurent, 2022). In contrast, in other regions, air quality data may be withheld, selectively disclosed, or rendered inaccessible due to bureaucratic or political constraints. This opacity weakens institutional legitimacy and undermines opportunities for citizen engagement.

Accountability mechanisms are also uneven across regions. Some regional agreements include provisions for compliance review bodies or independent expert panels, while others rely solely on voluntary reporting. In many cases, institutions lack binding dispute resolution procedures or sanctions for non-compliance, limiting their capacity to address persistent violations or failures to act. Without consequences for inaction, cooperative frameworks risk becoming symbolic rather than substantive.

Importantly, regional institutions must be capable of adapting to evolving scientific evidence and shifting environmental conditions. Climate-induced changes in wind patterns, precipitation, and wildfire risks are altering traditional pollution flows and intensifying the urgency for effective institutional responses. This requires flexible legal mandates, consistent funding, and regular institutional assessments that can identify weaknesses and propose timely reforms. It also demands coordination not only among states but across sectors, including transportation, agriculture, health, and energy.

Building institutional resilience also involves ensuring that the needs and perspectives of affected communities are reflected in regional strategies. Mechanisms for public consultation, grievance redressal, and civil society participation should be embedded into institutional design. This participatory approach not only enhances legitimacy but also aligns regional governance with the broader objectives of justice and inclusion. Transparent institutions that are responsive to the needs of all stakeholders, especially those most affected, can better support long-term air quality improvements and foster a cooperative culture across borders.

By prioritizing transparency and accountability within regional institutions, countries can move closer to governance models that are equitable, responsive, and sustainable. This approach not only strengthens environmental outcomes but also supports more inclusive and effective decision-making structures, contributing meaningfully to broader goals of healthy ecosystems and strong institutions.

TOWARD RESILIENT AND COOPERATIVE REGIONAL GOVERNANCE

Transboundary air pollution is a shared problem that requires shared solutions. Countries often struggle to manage cross-border environmental harm because legal systems and institutions operate within national boundaries. This chapter focuses on how regional cooperation can be strengthened through innovative legal tools, policy frameworks, and institutional support. It explores ways to make regional governance more effective, fair, and adaptable in managing shared air resources. The goal is to move from fragmented actions to collaborative systems that protect public health and the environment.

Policy and Legal Innovations for Shared Airspaces

One of the key pathways toward resilient regional governance lies in the strategic use of legal and policy tools that promote cross-border coordination while respecting national sovereignty. Legal innovations include the incorporation of enforceable obligations, the recognition of extra-territorial environmental harm, and the development of procedural rights for affected communities. These tools help to move regional air governance from a primarily diplomatic framework to one that includes elements of accountability, rights, and duties.

For example, the European Union has pioneered binding regional standards on ambient air quality, along with mechanisms for monitoring and enforcement through supranational institutions such as the European Commission and the Court of Justice of the EU. These legal structures create a baseline of obligations that member states must comply with, while still allowing flexibility in domestic implementation. Crucially, the system enables legal action against non-compliant states and provides citizens with legal standing to challenge pollution-related breaches. This rights-based and rule-based structure has been instrumental in driving air quality improvements and fostering a culture of shared responsibility.

Outside of Europe, some regions are beginning to experiment with more robust policy instruments. In East Asia, for instance, the Trilateral Environmental Ministers Meeting between China, South Korea, and Japan has led to joint research initiatives, emission inventories, and policy dialogues that are gradually building a shared knowledge base (Ju, 2017). While these mechanisms are still largely non-binding, they reflect a growing recognition of the need for structured and cooperative approaches. More recent developments have included proposals for coordinated air quality targets, aligned reporting practices, and sector-specific action plans especially in transportation and energy.

At the national level, legal reforms can also contribute to regional solutions when countries adopt principles of environmental responsibility that extend beyond their borders. The inclusion of transboundary harm provisions in environmental laws or constitutional frameworks signals a willingness to be held accountable for cross-border impacts. This approach, while still rare, is gaining traction in environmental discourse and jurisprudence. Countries that embed such principles domestically can serve as leaders in regional efforts to strengthen collective commitments.

In parallel, policy innovations such as integrated air quality management (IAQM) are gaining relevance. IAQM frameworks encourage multi-sectoral planning that takes into account not only emissions but also meteorological data, land use, and public health indicators. These comprehensive strategies are better suited for regional application, especially in border regions where pollution sources and population centers are interlinked. Regional air sheds do not conform to political boundaries, and IAQM offers a model for managing them with precision and coordination.

Innovative governance also includes the adoption of environmental performance standards for industries operating in transboundary zones. These standards can be developed through regional consultation and harmonized to prevent regulatory evasion. Where countries fear competitive disadvantages, cooperation can take the form of technology-sharing agreements or joint regulatory pilots. Flexibility mechanisms such as cross-border emissions trading or joint investment in clean infrastructure can also make regional cooperation more attractive.

Ultimately, the move toward shared airspace governance requires a shift in perspective from managing pollution as a domestic issue to addressing it as a regional public good. Legal and policy frameworks must facilitate this shift by embedding principles of equity, scientific integrity, and intergovernmental trust. When done effectively, such innovations can lay the foundation for resilient institutions, empower affected populations, and reduce the unequal burdens of pollution exposure.

Technology, Financing, and Capacity Building

Tackling transboundary air pollution requires more than policy commitments and legal agreements. It calls for sustained investment in the tools, resources, and institutional capabilities that enable countries to monitor, understand, and respond to pollution across borders. Many regions facing acute air pollution challenges lack the technological infrastructure to accurately measure emissions, track pollutant movement, or predict health impacts. This technological gap often results in delayed responses, underreporting of pollution levels, and ineffective policy implementation.

In regions where air quality monitoring networks are sparse or uneven, data-sharing among neighboring countries becomes essential. Reliable data can form the basis of coordinated action, public accountability, and long-term trust between governments.

Technological solutions are evolving rapidly. Remote sensing, satellite-based monitoring, and low-cost air sensors can provide real-time data that helps in forecasting pollution episodes and identifying hotspots. These tools are particularly valuable for countries with limited environmental monitoring systems. Regional organizations and international partnerships can play a critical role in ensuring that such technologies are shared equitably and are tailored to local capacities. For example, the European Environment Agency has facilitated the use of integrated monitoring platforms across member states, supporting more consistent and accurate air quality assessments. Similar efforts are being explored in Southeast Asia and East Asia, although gaps remain in implementation and transparency.

Equally important is the question of financing. Addressing transboundary air pollution is a long-term commitment that demands sustained financial support. This includes funding for monitoring infrastructure, pollution control technologies, cross-border research initiatives, and public health interventions. Many low and middle income countries rely heavily on external funding for environmental programs. In such contexts, regional financing mechanisms or multilateral climate funds can help bridge resource gaps. The Green Climate Fund, while primarily focused on climate adaptation, offers useful lessons for designing cross-border financing instruments that can support cooperative air quality initiatives.

Capacity building is the third pillar of regional resilience. Legal frameworks and technology alone cannot deliver results without well-trained personnel, institutional memory, and public awareness. Many countries struggle with insufficient staffing in environmental agencies, lack of technical training, and poor coordination among different levels of government. Cross-border cooperation efforts must therefore prioritize institutional learning, knowledge exchange, and peer-to-peer training. Initiatives such as ASEAN's regional workshops on haze pollution, or the trilateral meetings among China, South Korea, and Japan, serve as platforms for capacity development and policy alignment (Charusombat, 2024). Strengthening the skills and knowledge of local and national actors ensures that environmental governance is not only reactive but anticipatory and adaptive.

Taken together, technology, financing, and capacity building form the backbone of effective and equitable regional air pollution governance. These components are not optional; they are foundational to achieving long-term improvements in air quality and health outcomes. Moreover, they reflect a deeper commitment to fairness, cooperation, and shared responsibility in addressing environmental challenges that do not respect borders. By investing in these pillars, regional alliances can move beyond reactive crisis management and build the infrastructure of resilience.

CONCLUSION AND WAY FORWARD

The crisis of transboundary air pollution reveals the deep interconnections between environmental integrity, public health, and regional cooperation. While air knows no borders, legal frameworks, political priorities, and institutional capacities often remain bounded by national jurisdictions. This misalignment between the scale of the problem and the scale of governance has contributed to regulatory fragmentation, reactive policymaking, and growing environmental inequities. The comparative review across regions demonstrates that there is no singular path to effective cross-border air quality governance. Rather, different models emerge from distinct institutional contexts, levels of economic integration, and political willingness to share sovereignty in the pursuit of common environmental goals.

Europe's legal regionalism, with its binding standards and supranational oversight, has allowed for meaningful environmental progress, but it remains heavily dependent on political cohesion and judicial enforceability. The ASEAN Haze Agreement presents a softer, consensus-driven approach rooted in regional diplomacy, but suffers from enforcement limitations and uneven national capacities. The U.S. - Canada model shows the potential of bilateral trust, technical collaboration, and institutional continuity, even without a supranational body. Meanwhile, East Asia's trilateral framework illustrates how environmental cooperation can gradually evolve even in the absence of strong legal commitments, though it also highlights the challenges of political rivalry and asymmetry. Together, these regional experiences underscore that effective cooperation is possible when built on trust, shared data, and mutual recognition of cross-border impacts.

Key lessons emerge from these diverse approaches. Legal instruments must be supported by credible implementation mechanisms. Monitoring technologies must be inclusive and interoperable across borders. Financing needs to be long-term and sensitive to disparities in national capabilities. Most importantly, institutional capacity and public engagement must be central to any regional environmental strategy. Without these foundations, legal texts risk becoming symbolic rather than transformative. The integration of public health concerns, transparency norms, and scientific cooperation can help anchor air governance within a broader regional public good framework. In doing so, it aligns efforts not only with legal obligations but also with a deeper sense of ecological interdependence and justice.

The way forward lies in strengthening regional resilience through layered cooperation. This includes promoting adaptive legal frameworks that respond to evolving scientific understanding and environmental threats. It involves investing in regional monitoring systems and early warning platforms that serve both governmental and community needs. It requires embedding equity and access at the heart of regional environmental governance, ensuring that vulnerable populations are not sidelined in decision-making or disproportionately burdened by pollution. Such an approach implicitly furthers the global pursuit of environmental justice and inclusive development, resonating with the broader international vision for sustainable and peaceful societies.

Cross-border air pollution control is not merely a technical or environmental issue. It is a matter of institutional design, legal imagination, and political ethics. Governing the air together demands more than cooperation—it demands solidarity. Through shared responsibilities and differentiated capabilities, regions can forge legal and institutional pathways that respect national sovereignty while upholding the right to a healthy environment for all. In doing so, they contribute meaningfully to a future where clean air is not a privilege constrained by geography, but a universal right anchored in law, cooperation, and collective stewardship.

REFERENCES

Alvin, S. D. (2022). Effectiveness of the ASEAN Agreement on Transboundary Haze Pollution. *Environmental Law Review*, *24*(4), 305–313. DOI: 10.1177/14614529221126335

Charusombat, P. (2024). Enhancing transboundary haze pollution cooperation in Southeast Asia (2003–2020): External actors and ASEAN supporting mechanisms. *Asia Social Issues*, *18*(1), e270363. DOI: 10.48048/asi.2025.270363

Forsyth, T. (2014). Public concerns about transboundary haze: A comparison of Indonesia, Singapore, and Malaysia. *Global Environmental Change*, *25*, 76–86. DOI: 10.1016/j.gloenvcha.2014.01.013

Grennfelt, P., Engleryd, A., Forsius, M., Hov, Ø., Rodhe, H., & Cowling, E. (2020). Acid rain and air pollution: 50 years of progress in environmental science and policy. *Ambio*, *49*(4), 849–864. DOI: 10.1007/s13280-019-01244-4 PMID: 31542884

Guttikunda, S., Lodoysamba, S., Bulgansaikhan, B., & Batdorj, D. (2013). Particulate pollution in Ulaanbaatar, Mongolia. *Air Quality, Atmosphere & Health*, *6*(3), 589–601. DOI: 10.1007/s11869-013-0198-7

Hadjiyianni, I. (2023). The extraterritorial reach of environmental law: Legitimacy concerns and the role of domestic courts in controlling transnational regulatory power. In Parrish, A., & Ryngaert, C. (Eds.), *Research handbook on extraterritoriality in international law* (pp. 372–388). Elgar Publishing., DOI: 10.4337/9781800885592.00031

Heilmann, D. (2015). After Indonesia's ratification: The ASEAN Agreement on Transboundary Haze Pollution and its effectiveness as a regional environmental governance tool. *Journal of Current Southeast Asian Affairs*, *34*(3), 95–121. DOI: 10.1177/186810341503400304

Jeffery, M. I. (2009). Book review: *International law and the environment*, by Patricia Birnie, Alan Boyle, and Catherine Redgwell. *Osgoode Hall Law Journal*, *47*(3), 595–601. DOI: 10.60082/2817-5069.1143

Ju, Y. (2017). Tracking the $PM_{2.5}$ inventories embodied in the trade among China, Japan and Korea. *Journal of Economic Structures*, *6*(1), 27. DOI: 10.1186/s40008-017-0089-4

Lan, R., Eastham, S. D., Liu, T., Saikawa, E., & Barrett, S. R. H. (2022). Air quality impacts of crop residue burning in India and mitigation alternatives. *Nature Communications*, *13*(1), 6537. DOI: 10.1038/s41467-022-34093-z PMID: 36376316

Laurent, É. (2022). Air (ine)quality in the European Union. *Current Environmental Health Reports*, *9*(2), 123–129. DOI: 10.1007/s40572-022-00348-6 PMID: 35347642

Liu, S., Xing, J., Wang, S., Ding, D., Chen, L., & Hao, J. (2020). Revealing the impacts of transboundary pollution on $PM_{2.5}$-related deaths in China. *Environment International*, *134*, 105323. DOI: 10.1016/j.envint.2019.105323 PMID: 31759275

McLean, B., & Barton, J. (2008). U.S.-Canada cooperation: The U.S.-Canada Air Quality Agreement. *Journal of Toxicology and Environmental Health. Part A.*, *71*(9–10), 564–569. DOI: 10.1080/15287390801997567 PMID: 18569627

Melious, J. (2014). The Canada-US Air Quality Agreement and its impact on air quality management in Canada. In Taylor, E., & McMillan, A. (Eds.), *Air quality management* (pp. 357–376). Springer., DOI: 10.1007/978-94-007-7557-2_16

Rentschler, J., & Leonova, N. (2023). Global air pollution exposure and poverty. *Nature Communications*, *14*(1), 4432. DOI: 10.1038/s41467-023-39797-4 PMID: 37481598

Rentschler, J., & Leonova, N. (2023). Global air pollution exposure and poverty. *Nature Communications*, *14*(1), 4432. DOI: 10.1038/s41467-023-39797-4 PMID: 37481598

Spijkers, O. (2025, April 4). The status of the right to a clean, healthy and sustainable environment under customary international law. *Verfassungsblog*. https://verfassungsblog.de/ip-hr2he-customary-international-law/ https://doi.org/10.59704/b4113d1f4013cea9

Spiteri, J., & von Brockdorff, P. (2021). Transboundary air pollution and respiratory disease mortality: Evidence from European countries. *Journal of Economic Studies (Glasgow, Scotland)*, *48*(7), 1371–1387. DOI: 10.1108/JES-05-2020-0210

Stevens, R., Poterlot, C., Trieu, N., Rodriguez, H. A., & Hayes, P. L. (2024). Transboundary transport of air pollution in eastern Canada. *Environmental Science. Advances*, *3*(3), 448–469. DOI: 10.1039/D3VA00307H

Underwood, A., Marcantonio, R., Wood, D., & Crippa, P. (2024). Interstate air pollution governance in the United States: Exploring Clean Air Act Section 126. *Environmental Management*, *74*(3), 401–413. DOI: 10.1007/s00267-024-02002-3 PMID: 38896241

Xia, L., Li, Y., Zhan, B., & Wu, C. (2025). Are regional environmental policies always effective? Evidence from China. *Environmental Research Letters*, *20*(7), 074067. Advance online publication. DOI: 10.1088/1748-9326/ade3f1

KEY TERMS AND DEFINITIONS

Atmospheric Governance: Legal and policy mechanisms aimed at regulating and managing the use of the atmosphere across jurisdictions.

Cross-Border Health Externalities: Public health impacts that occur in one country due to environmental actions or policies in another.

Environmental Health: The branch of public health concerned with how environmental factors impact human health and wellbeing.

Environmental Institutionalism: The role of formal organizations and frameworks in enforcing and shaping environmental governance.

Environmental Justice: The fair distribution of environmental benefits and burdens, ensuring no group bears disproportionate harm.

Regional Environmental Cooperation: Collaborative efforts among neighboring countries to address shared environmental challenges.

Shared Air Resources: Air as a common natural resource that transcends political boundaries and requires collective management.

Transboundary Air Pollution: The movement of air pollutants across national borders, affecting regions beyond their source.

Chapter 8
Data Sharing and Climate Monitoring Between Neighboring Territories

Divya Jyoti Thakur
https://orcid.org/0000-0002-0125-6118
University School of Business, Chandigarh University, Punjab

G. Chitra
Karpaga Vinayaga college of Engineering, India

Ravikumar Seenivasan
https://orcid.org/0000-0002-8235-8051
Woxsen University, India

Harita Ponnapalli
Woxsen University, India

Kasagani Durga
NBM Law College, India

P. Selvakumar
https://orcid.org/0000-0002-3650-4548
Department of Science and Humanities, Nehru Institute of Technology, Coimbatore, India

T. C. Manjunath
https://orcid.org/0000-0003-2545-9160
Rajarajeswari College of Engineering, India

ABSTRACT

In an era marked by escalating climate variability and increasingly frequent extreme weather events, the importance of robust and coordinated climate monitoring cannot be overstated. Climate change knows no borders; its effects transcend political, cultural, and geographical boundaries, impacting ecosystems, economies, and communities across nations. As such, the responsibility for observing, analyzing, and responding to climatic shifts must extend beyond the confines of individual states. Climate monitoring in neighboring territories plays a pivotal role in constructing a comprehensive understanding of regional and global climate systems. By pooling data and resources across borders, countries are better equipped to predict, adapt

DOI: 10.4018/979-8-3373-3972-6.ch008

to, and mitigate the multifaceted impacts of climate change. Neighboring territories often share interconnected ecosystems, river basins, coastlines, and atmospheric systems, making it crucial that climate data is collected and shared across these areas.

INTRODUCTION

In an era marked by escalating climate variability and increasingly frequent extreme weather events, the importance of robust and coordinated climate monitoring cannot be overstated. Climate change knows no borders; its effects transcend political, cultural, and geographical boundaries, impacting ecosystems, economies, and communities across nations. As such, the responsibility for observing, analyzing, and responding to climatic shifts must extend beyond the confines of individual states. Climate monitoring in neighboring territories plays a pivotal role in constructing a comprehensive understanding of regional and global climate systems. By pooling data and resources across borders, countries are better equipped to predict, adapt to, and mitigate the multifaceted impacts of climate change. Neighboring territories often share interconnected ecosystems, river basins, coastlines, and atmospheric systems, making it crucial that climate data is collected and shared across these areas. For instance, weather systems that develop in one country can affect its neighbors within hours or days, such as tropical storms, drought patterns, or sudden cold snaps. Likewise, gradual phenomena like desertification, sea-level rise, or shifts in biodiversity patterns often begin subtly in one region and extend to adjacent areas over time. Without collaborative monitoring, countries risk basing policy and emergency responses on incomplete or skewed data, which can lead to ineffective or even harmful decisions. Therefore, monitoring climate in adjacent regions is not merely beneficial; it is a strategic necessity for national and regional resilience.

Data sharing among neighboring territories significantly enhances the accuracy and timeliness of climate models. Regional climate models (RCMs), which are critical tools in predicting weather patterns, heatwaves, floods, and other climate-related risks, rely heavily on dense, high-resolution datasets. These datasets must encompass a broad geographical range to accurately simulate and project regional climate behavior. When neighboring countries openly share their meteorological, hydrological, and environmental data, they collectively improve the quality of these models. This collaboration allows for better forecasting of extreme weather events and more accurate long-term climate predictions, which are vital for infrastructure planning, agricultural strategy, water management, and public health preparedness. Moreover, collaborative monitoring and data sharing help harmonize early warning systems and disaster preparedness initiatives(Akiho & Raita, 2017). Climate-related disasters such as hurricanes, tsunamis, or wildfires often affect multiple countries simultaneously or sequentially. Shared data can provide early alerts to downstream or neighboring countries, allowing them to mobilize emergency resources more efficiently and potentially save lives and property. For example, if a country upstream detects unusual rainfall patterns that may lead to flooding, downstream countries can take preventive measures if they are promptly informed. This kind of regional cooperation enhances overall resilience and ensures that no country is left vulnerable due to a lack of information.

BACKGROUND OF DATA SHARING AND CLIMATE MONITORING

The benefits of climate monitoring in neighboring territories are not limited to risk reduction alone; they extend into economic, scientific, and diplomatic domains. Economically, shared climate data can optimize resource management across borders, such as joint water reservoirs, shared forests, and cross-border agricultural zones. It can reduce duplication of monitoring infrastructure and promote cost-sharing mechanisms, making climate surveillance more affordable for all parties involved. Scientifically, such collaboration fosters cross-border research initiatives, promotes the development of standardized monitoring techniques, and leads to more robust findings that are representative of broader climatic zones. Diplomatically, environmental cooperation often serves as a platform for broader peacebuilding and regional integration. Climate-related agreements, joint monitoring missions, and shared. In an era marked by escalating climate variability and increasingly frequent extreme weather events, the importance of robust and coordinated climate monitoring cannot be overstated. Climate change knows no borders; its effects transcend political, cultural, and geographical boundaries, impacting ecosystems, economies, and communities across nations. As such, the responsibility for observing, analyzing, and responding to climatic shifts must extend beyond the confines of individual states (Aragani V. M. et al. 2025). Climate monitoring in neighboring territories plays a pivotal role in constructing a comprehensive understanding of regional and global climate systems. By pooling data and resources across borders, countries are better equipped to predict, adapt to, and mitigate the multifaceted impacts of climate change. Neighboring territories often share interconnected ecosystems, river basins, coastlines, and atmospheric systems, making it crucial that climate data is collected and shared across these areas. For instance, weather systems that develop in one country can affect its neighbors within hours or days, such as tropical storms, drought patterns, or sudden cold snaps. Likewise, gradual phenomena like desertification, sea-level rise, or shifts in biodiversity patterns often begin subtly in one region and extend to adjacent areas over time. Without collaborative monitoring, countries risk basing policy and emergency responses on incomplete or skewed data, which can lead to ineffective or even harmful decisions. Therefore, monitoring climate in adjacent regions is not merely beneficial; it is a strategic necessity for national and regional resilience(Aronova E. 2017).

Moreover, collaborative monitoring and data sharing help harmonize early warning systems and disaster preparedness initiatives. Climate-related disasters such as hurricanes, tsunamis, or wildfires often affect multiple countries simultaneously or sequentially. Shared data can provide early alerts to downstream or neighboring countries, allowing them to mobilize emergency resources more efficiently and potentially save lives and property. For example, if a country upstream detects unusual rainfall patterns that may lead to flooding, downstream countries can take preventive measures if they are promptly informed. This kind of regional cooperation enhances overall resilience and ensures that no country is left vulnerable due to a lack of information(Biskaborn B. et al. 2019). The benefits of climate monitoring in neighboring territories are not limited to risk reduction alone; they extend into economic, scientific, and diplomatic domains. Economically, shared climate data can optimize resource management across borders, such as joint water reservoirs, shared forests, and cross-border agricultural zones. It can reduce duplication of monitoring infrastructure and promote cost-sharing mechanisms, making climate surveillance more affordable for all parties involved. Scientifically, such collaboration fosters cross-border research initiatives, promotes the development of standardized monitoring techniques, and leads to more robust findings that are representative of broader climatic zones(Boyd A. et al. 2019). Diplomatically, environmental cooperation often serves as a platform for broader peacebuilding and regional integration. Climate-related agreements, joint monitoring missions, and shared sustainability goals can open dialogue between neighboring nations, even those with otherwise strained relations. In conclusion, monitoring climate in neighboring territories is indispensable for understanding the complex and interconnected nature of our planet's climate system. It is a cornerstone of effective climate adaptation, disaster risk reduction, and sustainable development. The benefits of such efforts—ranging from improved forecasting and disaster resilience to scientific collaboration and diplomatic cooperation—far outweigh the challenges posed by cross-border coordination. As climate threats continue to intensify, regional partnerships for climate monitoring and data sharing are no longer optional but imperative(Buch E. 2019). They represent a proactive, informed, and united approach to safeguarding the well-being of current and future generations.

The Challenges of Climate Monitoring in Border Regions

While the importance of climate monitoring in border regions is increasingly acknowledged, the process is fraught with significant challenges that hinder its full potential. Border regions—those transitional spaces where two or more countries meet—are often marked by political sensitivities, jurisdictional complexities, and resource disparities, all of which complicate the implementation of effective and cooperative climate monitoring systems. The act of observing and interpreting climatic patterns in such areas is not merely a scientific endeavor; it is also a political, logistical, and diplomatic task. The success of climate monitoring in these zones relies heavily on data sharing, institutional coordination, trust-building, and technical interoperability—yet each of these elements faces distinct and persistent obstacles. One of the most prominent challenges in border-region climate monitoring is limited data sharing between countries. Despite the growing global emphasis on open data, many governments remain hesitant to share meteorological, hydrological, and environmental information with their neighbors. This reluctance may stem from concerns over national security, competitive advantage, or political mistrust. For instance, a country experiencing drought may be unwilling to disclose water scarcity data to a neighbor if it fears that this vulnerability could be exploited in water-sharing negotiations. Similarly, strategic infrastructure, such as dams or monitoring stations, may be considered sensitive and withheld from international collaboration. These barriers severely limit the creation of comprehensive datasets that are crucial for accurate climate modeling and forecasting across shared territories.

Compounding this issue is the lack of standardized data formats and methodologies. Even when data is shared, inconsistencies in how it is collected, processed, or reported can render it difficult to integrate or interpret. Different countries may use varied technologies, time intervals, or measurement units for collecting temperature, precipitation, or air quality data. In the absence of harmonized protocols, climate scientists and policymakers are often forced to navigate a fragmented and unreliable data landscape. This issue is particularly acute in regions with low institutional capacity, where outdated or poorly maintained monitoring equipment further reduces the reliability of climate observations. Coordination between national meteorological and environmental agencies also presents a significant challenge. Effective climate monitoring in border regions requires not only data sharing but also collaborative planning, regular communication, and shared decision-making processes. However, bureaucratic fragmentation, institutional silos, and varying national priorities can obstruct such coordination. Some countries may prioritize economic development over environmental monitoring, while others may lack the institutional mandates or technical expertise to engage in cross-border initiatives. Without mechanisms for synchronized observation and analysis, efforts to understand and respond to regional climate dynamics remain disjointed. Another layer of complexity arises from geopolitical tensions and historical conflicts between neighboring nations. In politically volatile border areas, cooperation on climate issues is often sidelined in favor of nationalistic or militarized agendas(Debnath P. et al. 2025). Distrust between governments can manifest as resistance to joint monitoring projects, refusals to engage in data exchange, or even sabotage of shared infrastructure. In some cases, the deployment of climate monitoring equipment near sensitive border areas may be interpreted as a security threat, leading to diplomatic standoffs or local opposition. These tensions not only stifle collaboration but also compromise the quality and coverage of climate data in strategically important regions.

Physical inaccessibility and environmental challenges can further complicate climate monitoring in border areas. Many border zones are located in remote or rugged terrains—such as mountain ranges, deserts, tropical forests, or coastal deltas—that are difficult to access and monitor effectively. Installing and maintaining equipment in such areas requires significant logistical planning and financial investment, often exceeding the capacities of local or national agencies. In some cases, monitoring stations in these regions are left unstaffed, neglected, or damaged due to extreme weather, vandalism, or conflict. These physical barriers create data blind spots that weaken the overall resolution and accuracy of climate assessments. Legal and regulatory inconsistencies between countries add another dimension of difficulty. Climate data collection and dissemination are often governed by national laws that vary widely in their openness, privacy considerations, and institutional mandates. While one country may support open-access environmental data policies, its neighbor may impose strict controls or bureaucratic hurdles that prevent timely information exchange(Hamblin J. 2011). In the absence of binding international frameworks or regional agreements, there is little recourse for enforcing data transparency or cooperation. Attempts to create shared legal platforms are often delayed by protracted negotiations or hampered by political instability.

Language barriers and cultural differences may also impede climate cooperation across borders. Miscommunications due to differing languages, scientific terminologies, or institutional cultures can delay or derail collaborative efforts. Moreover, communities living in border regions may be marginalized from national decision-making processes and lack access to climate information in a language or format they can understand. Bridging these cultural gaps requires not only translation but also community engagement strategies that are inclusive and context-sensitive. Despite these challenges, some positive examples of cross-border climate cooperation do exist. Initiatives such as the European Union's Copernicus program, the South Asian Climate Outlook Forum, and various UN-led efforts have shown that with sufficient political will, technical support, and institutional alignment, countries can overcome these obstacles. These models emphasize the need for regional governance structures, shared funding mechanisms, and joint capacity-building programs to enable effective border-region monitoring. However, scaling these successes globally requires a much stronger commitment from national governments, regional organizations, and the international community. In conclusion, the challenges of climate monitoring in border regions are multifaceted and deeply interwoven with political, institutional, and technical factors(Herzberg J. et al. 2018). The absence of reliable and integrated climate data in these critical zones undermines global efforts to address climate change, manage natural resources, and protect vulnerable communities. To move forward, there must be greater investment in cross-border infrastructure, harmonization of data standards, and diplomatic efforts to foster trust and transparency. Only through such collective action can we ensure that border regions are not left behind in the global pursuit of climate resilience and sustainable development.

The Role of Data Sharing in Climate Monitoring

In the fight against climate change and the effort to build climate resilience, data sharing has emerged as a cornerstone of effective climate monitoring. The global climate system is inherently interconnected—weather events, ocean currents, atmospheric pressures, and ecological shifts in one part of the world often influence conditions elsewhere. To monitor this complex and dynamic system, climate scientists and policymakers require access to vast quantities of diverse, high-resolution, and real-time data. This data must come from multiple sources, spanning geographical regions, climate zones, and timeframes. No single country, organization, or institution can generate all of this data alone. Therefore, data sharing across borders, institutions, and disciplines is not only beneficial but essential to the integrity and accuracy of climate monitoring efforts worldwide. The most immediate benefit of data sharing in climate monitoring is the improvement of climate models and forecasts(Jasny B.R. 2013). Predictive models—used for short-term weather forecasting and long-term climate projection—rely on comprehensive datasets that include atmospheric, oceanic, and terrestrial variables. These models become significantly more accurate and reliable when fed with data from a wide geographic area, including different altitudes, land covers, and seasonal cycles. Shared data ensures that models can account for transboundary phenomena such as El Niño, La Niña, atmospheric rivers, jet streams, and migratory patterns of flora and fauna. With improved models, scientists can better understand current climate trends and predict future scenarios with greater confidence, thereby informing policies on agriculture, urban planning, water management, public health, and disaster preparedness.

In addition to improving model accuracy, data sharing enhances early warning systems for extreme weather events such as hurricanes, floods, droughts, heatwaves, and wildfires. These hazards often affect multiple countries simultaneously or in sequence, especially in interconnected regions such as river basins, coastal zones, or mountain ranges. When countries and agencies share real-time data—such as satellite imagery, radar scans, and hydrological readings—they can detect emerging threats earlier and with more precision. This allows governments and communities to take preventive action, such as issuing evacuation orders, adjusting reservoir levels, or mobilizing emergency services. By acting on shared information, countries can save lives, reduce economic losses, and improve the coordination of humanitarian responses. Data sharing also facilitates international collaboration in scientific research, leading to stronger and more inclusive knowledge production. When climate data is accessible, it empowers researchers from different regions—including those from developing or under-resourced countries—to participate in global studies. This inclusivity leads to more diverse perspectives, localized insights, and region-specific solutions. Collaborative research projects based on shared data have produced important findings on global warming trends, biodiversity loss, carbon cycle dynamics, and climate feedback loops(Johri S. et al. 2025). They also foster trust among nations, enabling joint publications, shared funding opportunities, and the pooling of technical expertise. On a practical level, shared climate data supports decision-making at all levels of governance. Policymakers need reliable and timely climate information to draft and evaluate environmental regulations, plan infrastructure projects, manage energy grids, and allocate public resources. In regions where countries share ecosystems—such as the Amazon rainforest, the Mekong Delta, or the Sahel—joint data enables coordinated management of shared resources. For example, transboundary water management agreements often depend on shared hydrological data to determine fair water allocations and monitor compliance. Similarly, shared data on air pollution, deforestation, or soil degradation helps nations develop consistent environmental standards and align with international agreements such as the Paris Climate Accord or the Sendai Framework for Disaster Risk Reduction.

Moreover, data sharing contributes to transparency and accountability in climate action. As governments commit to reducing greenhouse gas emissions or enhancing adaptation capacities, shared data enables verification and peer review. Institutions such as the United Nations Framework Convention on Climate Change (UNFCCC) and the Intergovernmental Panel on Climate Change (IPCC) rely on openly shared data to track global progress toward climate goals. Transparent data also helps civil society, journalists, and the public hold governments and corporations accountable for their environmental impact and policy performance. In this sense, data sharing strengthens democratic oversight and builds public trust in climate institutions. Despite its many benefits, data sharing in climate monitoring is not without significant challenges. One major obstacle is the lack of standardization in data collection and formatting. Climate data is gathered by thousands of entities—

national meteorological agencies, research institutions, NGOs, private companies, and citizen scientists—using diverse technologies and methodologies. Differences in measurement units, data resolution, metadata descriptions, and file formats can make it difficult to aggregate or compare datasets. Inconsistent data practices may lead to duplication, gaps, or errors in analysis, undermining the value of shared information(Joseph M.P. 2002). Efforts to harmonize climate data standards are ongoing but require considerable coordination, training, and investment. Another challenge is data ownership and intellectual property concerns. Some institutions view their climate datasets as proprietary assets, especially when data collection requires substantial investment in infrastructure, expertise, or technology. There is often reluctance to share this data freely, particularly in the private sector or among research teams competing for recognition and funding. Governments may also withhold data due to national security concerns or geopolitical tensions, especially in strategic border areas. In some cases, sensitive information such as water table levels, land use changes, or carbon stock measurements is restricted from public access, despite its relevance to climate studies.

Funding and resource limitations further complicate the data-sharing landscape. Many developing countries lack the technical capacity to collect, store, process, and disseminate large volumes of climate data. Their national meteorological or environmental agencies may struggle with outdated equipment, insufficient staff, and unstable internet access. Even when willing to share data, these countries may be constrained by technical barriers or lack of support to maintain open-access platforms. Bridging this digital divide requires targeted international assistance, including investments in monitoring infrastructure, training programs, and cloud-based data services. Political and institutional fragmentation also poses significant barriers. Climate data governance often involves multiple ministries, departments, and agencies that operate in silos, with limited inter-agency communication or data exchange protocols. Even within a single country, data may be fragmented across departments responsible for weather, agriculture, forestry, energy, or health, each using different systems and standards. On the international level, the absence of binding agreements or enforceable standards limits the pressure to share data proactively(Kowarsch M. et al. 2016). Many regional climate data networks operate on a voluntary basis and suffer from inconsistent participation or funding shortfalls. Privacy and ethical considerations are emerging as new challenges in the era of big data and digital surveillance. Some climate-related data—such as satellite images, sensor readings, or geotagged field observations—can inadvertently reveal sensitive information about individuals, communities, or protected ecosystems. Indigenous groups, in particular, have raised concerns about data sovereignty and the need to respect traditional knowledge systems and land rights. Ethical data sharing requires protocols that ensure informed consent, benefit-sharing, and cultural sensitivity, especially in climate-vulnerable or marginalized regions. In conclusion, data sharing plays a foundational role in effective climate monitoring, enabling better forecasts, stronger research, improved governance, and more equitable international cooperation. Its benefits are far-reaching, touching nearly every aspect of climate adaptation, mitigation, and resilience-building(Mahadev Madgule N. et al. 2023). However, unlocking these benefits requires overcoming significant technical, political, institutional, and ethical challenges. To move forward, the global community must invest in standardization, capacity-building, trust-building, and legal frameworks that support open, secure, and inclusive data ecosystems. As the climate crisis deepens, the imperative for transparent, reliable, and widely accessible data grows stronger—data sharing is not a luxury but a lifeline for a sustainable future.

Data Sharing Mechanisms for Climate Monitoring

As climate change continues to intensify and manifest in increasingly unpredictable and destructive ways, the need for reliable, accessible, and timely climate data has become more urgent than ever. In this context, data sharing mechanisms have emerged as essential components of global and regional climate monitoring systems. These mechanisms, which include digital platforms, data portals, interoperability frameworks, and collaborative networks, facilitate the collection, dissemination, and use of climate-related data across diverse sectors and geopolitical boundaries(-Michener W.K. 2015). By enabling the exchange of meteorological, hydrological, oceanographic, and environmental information, these systems enhance the collective capacity to understand climate patterns, predict extreme events, assess environmental impacts, and implement effective mitigation and adaptation strategies. At the heart of most data sharing mechanisms for climate monitoring are climate data portals—web-based platforms designed to provide centralized access to diverse datasets from multiple sources. These portals often serve as user-friendly gateways where scientists, policymakers, planners, NGOs, educators, and the general public can retrieve climate information in various formats and resolutions. Examples of prominent global climate data portals include the World Meteorological Organization's (WMO) WMO Information System (WIS), the Global Climate Observing System (GCOS), and the NASA Earthdata portal, which aggregates satellite observations from missions like MODIS, Landsat, and Sentinel. These platforms allow users to visualize, download, and analyze data on temperature, precipitation, sea level, greenhouse gas concentrations, and land use changes over time.These platforms are often multilingual and designed with user-friendly interfaces that provide region-specific data products, climate projections, risk assessments, and decision-support tools. They help governments and stakeholders design targeted interventions, such as climate-resilient agriculture plans, coastal protection strategies, and water resource management frameworks(Paul C. et al. 2016). Beyond open-access portals, climate data is increasingly being exchanged through interoperable platforms and APIs (Application Programming Interfaces) that allow different databases and software systems to communicate seamlessly.

These platforms offer integrated computing resources, programming environments, and visualization tools, allowing users to perform complex analyses—such as land cover classification, anomaly detection, or trend analysis—at scale. By reducing technical barriers and infrastructure costs, cloud platforms democratize access to advanced climate data analytics, particularly for researchers and institutions in low- and middle-income countries(Porter J. 2010).Despite these advancements, the development and sustainability of data sharing mechanisms face several persistent challenges. One of the foremost issues is data fragmentation. Many climate data portals operate in isolation, with limited cross-platform integration or awareness(Ramakrishnan T. et al. 2022). Datasets are often siloed within institutional, sectoral, or national boundaries, making it difficult to assemble comprehensive views of regional or global climate systems. Efforts to address this require stronger coordination between institutions, harmonization of data standards, and incentives for interoperability. Another challenge is the digital divide that affects many developing countries and small island states. These regions often lack the technological infrastructure, stable internet connectivity, and trained personnel necessary to access and contribute to global data sharing mechanisms. This imbalance limits their participation in climate research and decision-making, and can exacerbate climate vulnerability due to information asymmetry. International support, including funding, capacity-building programs, and technology transfer, is essential to ensure equitable access to climate data resources.

Data quality and reliability remain critical concerns as well. While open-access platforms increase the volume of available data, they also risk disseminating low-quality, outdated, or unverified information. Users may struggle to assess the accuracy, resolution, or source credibility of datasets, particularly when metadata is incomplete or inconsistent. Establishing clear data validation protocols, peer-reviewed metadata standards, and user feedback mechanisms can help address these issues. Political and legal constraints can also impede the development of robust data sharing mechanisms. In some countries, climate data is considered a matter of national security or commercial confidentiality and is therefore restricted. Concerns over sovereignty, data misuse, or foreign surveillance may lead to resistance against open data policies. Overcoming these barriers requires building trust through bilateral agreements, regional data governance frameworks, and international conventions that respect national interests while promoting shared environmental stewardship. Finally, ethical and privacy considerations are becoming increasingly relevant as climate data intersects with social and personal information(Ran Y. et al. 2012). For example, data on climate impacts may inadvertently reveal the locations of vulnerable communities, endangered species, or culturally sensitive sites. Mechanisms for data sharing must include safeguards to protect privacy, respect indigenous knowledge, and ensure equitable benefit-sharing—especially in cases where local data informs global policy or commercial innovation.

Climate Monitoring Systems for Neighboring Territories.

Effective climate monitoring in neighboring territories is a critical component of regional and global climate resilience strategies. As climatic events frequently transcend political boundaries, the capacity to monitor, analyze, and respond to environmental changes must be rooted in systems that consider the interconnected nature of ecosystems, weather patterns, and human societies. Climate monitoring systems—comprising networks of ground-based weather stations, remote sensing technologies, data processing infrastructure, and collaborative frameworks—are essential for generating accurate, high-resolution, and timely climate data. When these systems are harmonized and coordinated across neighboring territories, they enable a shared understanding of regional climate dynamics, enhance early warning capabilities, and support the joint management of transboundary resources(Ran Y. et al. 2010). At the foundation of most climate monitoring systems are meteorological and hydrological observation stations that collect critical in-situ data on temperature, precipitation, humidity, wind speed and direction, solar radiation, soil moisture, river discharge, and other environmental parameters. These stations are often distributed across varying altitudes, latitudes, and land cover types to ensure comprehensive coverage of climatic zones. In neighboring territories, especially those that share physical geographies such as river basins, mountain ranges, or coastal regions, the distribution and density of weather stations play a key role in capturing localized climate behavior. For instance, countries sharing the Mekong River or the Alpine region rely on synchronized hydrometeorological networks to monitor seasonal changes, forecast floods, and plan water usage. When each nation within a shared ecological zone maintains a robust network of stations and shares their data in real time, the collective understanding of climate dynamics becomes far more precise and actionable.

Complementing ground-based systems are a wide range of remote sensing technologies—primarily satellite-based sensors—that provide continuous, large-scale coverage of the Earth's surface and atmosphere. Remote sensing platforms operated by organizations such as NASA, ESA (European Space Agency), ISRO (Indian Space Research Organisation), and JAXA (Japan Aerospace Exploration Agency) offer high-frequency, spatially detailed observations that are particularly valuable for monitoring transboundary phenomena. These include land surface temperatures, sea surface heights, vegetation indices, snow cover, glacier movement, cloud formation, and aerosol concentrations. Unlike ground stations, which are often affected by terrain limitations or maintenance issues, satellites can observe inaccessible or politically sensitive border areas without physical intrusion. This makes them indispensable for climate monitoring in neighboring territories where direct access might be constrained by geography or international relations(Rasmussen L.H. et al. 2018). Remote sensing also supports change detection and trend analysis over time, allowing scientists and decision-makers to assess long-term shifts in regional climates. Similarly, ASEAN countries cooperate on climate monitoring through the ASEAN Specialized Meteorological Centre (ASMC), which tracks transboundary haze and other regional climate issues. These frameworks enhance interoperability between national systems and foster mutual trust, joint training, and investment in shared infrastructure. Automatic Weather Stations (AWS) and smart sensor networks are increasingly deployed in border regions to support real-time climate data collection. These stations, often solar-powered and internet-connected, can automatically transmit observations to centralized databases without requiring constant human oversight. They are especially useful in remote or conflict-prone border zones where manual data collection may be risky or logistically challenging (Shende A. D. et al. 2025).

Figure 1. Measuring and Monitoring Urban Impacts on Climate Change

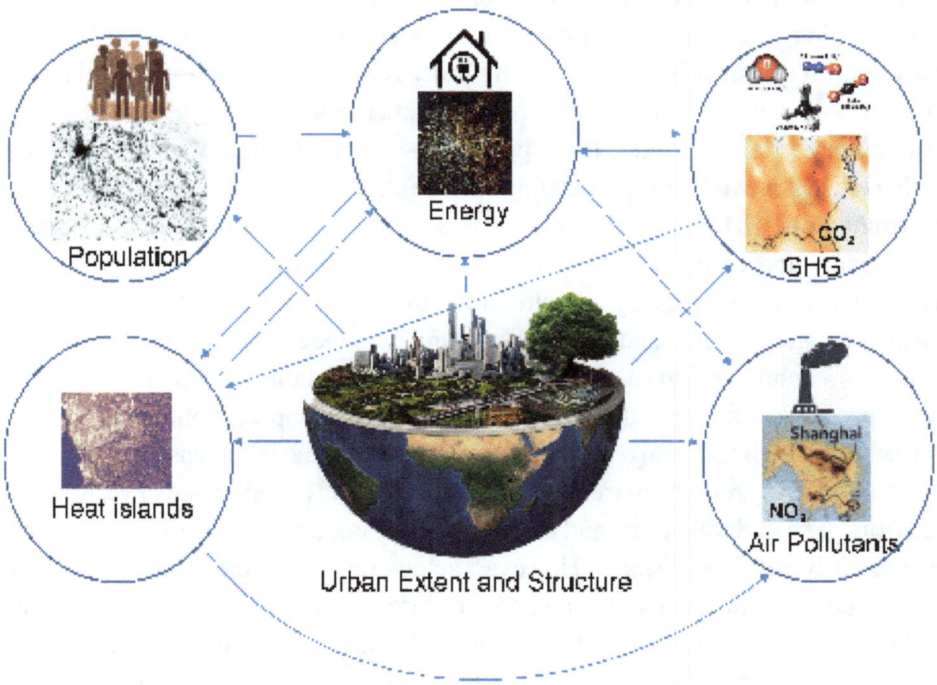

These systems not only save lives and reduce economic losses but also promote regional solidarity in the face of shared climate threats. However, despite the increasing sophistication of climate monitoring systems, there are several technical and institutional challenges to their effective deployment across neighboring territories. First, there is often a disparity in the capacity of countries to invest in and maintain monitoring infrastructure(Sreenivasulu A. et al. 2022). While some nations operate advanced satellite ground stations and dense weather observation networks, others may lack the resources to install or sustain even basic meteorological equipment. This imbalance can result in uneven data coverage and limit the reliability of regional climate assessments. Addressing this requires international support, funding mechanisms, and regional cooperation to build capacity and close data gaps. Overcoming these challenges requires diplomacy, transparency, and confidence-building measures. Multilateral climate agreements and science diplomacy initiatives can help establish neutral platforms where countries can collaborate on monitoring without compromising their sovereignty or security.By combining ground-based weather stations, advanced remote sensing technologies, automated sensors, and regional cooperation frameworks, these systems provide the foundation for timely, accurate, and actionable climate information(Stanford Encyclopedia of Philosophy 2020).

Data Quality and Validation for Climate Monitoring

In the realm of climate monitoring, data quality and validation represent foundational pillars that determine the reliability, usefulness, and credibility of climate analyses, forecasts, and decision-making processes. Climate data—derived from a wide array of sources such as ground-based weather stations, satellites, ocean buoys, remote sensors, and citizen science platforms—serves as the raw material for detecting environmental trends, modeling climate change, assessing vulnerability, and informing policy. However, the value of this data is only as strong as its accuracy, consistency, and completeness. Without rigorous quality control and validation procedures, climate data may contain errors, biases, or anomalies that can mislead scientists, policymakers, and the public, ultimately undermining efforts to address climate risks and develop sustainable solutions. Data quality in climate monitoring encompasses several key attributes, including accuracy, precision, completeness, consistency, timeliness, and representativeness. These dimensions ensure that climate observations truly reflect environmental conditions, can be replicated or verified, are free from gaps or duplication, and are delivered promptly for real-time applications such as early warning systems. However, achieving and maintaining high-quality data is a complex task due to the diversity of instruments, collection environments, methodologies, and human factors involved in the climate observation process. Weather stations may encounter equipment malfunctions, sensor drift, calibration issues, or human recording errors. Satellite observations, while extensive in coverage, may suffer from limitations in spatial resolution, cloud interference, or algorithmic uncertainties. Ocean-based sensors must contend with harsh conditions, biofouling, and maintenance challenges. These realities necessitate robust frameworks for data quality assurance and validation at every stage of the data lifecycle.Common data cleaning techniques include outlier detection, missing value imputation, duplicate removal, and smoothing of time-series data. For example, missing temperature records might be estimated using statistical interpolation based on surrounding time points or adjacent stations. Advanced methods, such as machine learning-based imputation, are also being explored to enhance data recovery from sparse or noisy datasets(Sugumaran K. M. et al. 2024). Data cleaning is particularly important in large, long-term climate datasets where even small inconsistencies can accumulate and bias trend analyses over decades.

Quality control (QC) systems in climate monitoring are typically structured in multi-tiered levels. The first level, known as automated quality checks, uses algorithms to detect gross errors, such as values outside the physical range (e.g., negative precipitation, excessively high wind speeds), instrument failure signals, or duplicated timestamps. These initial checks are fast and scalable, making them suitable for real-time or near-real-time data streams. The second level, or statistical quality control, applies more nuanced techniques such as moving averages, regression analysis, and climatological range checks to detect subtler anomalies. The third level often involves manual review and expert validation, especially for datasets used in high-impact applications such as national climate assessments or international climate reports. Human analysts play a crucial role in interpreting complex anomalies that might result from rare weather events, sensor upgrades, or changes in station location. Maintaining data quality also depends on metadata documentation, which includes detailed information about the data source, measurement instruments, calibration procedures, sampling frequency, and any modifications made during data processing. Proper metadata ensures that datasets can be interpreted correctly and reused by other researchers or institutions without ambiguity. It also helps track the provenance of data, making it easier to audit and reproduce analyses. For instance, if a sudden shift in recorded temperatures corresponds with a station relocation, metadata can explain the discontinuity and guide data homogenization efforts.

Homogenization is a specialized process within climate data quality assurance aimed at removing non-climatic biases introduced by changes in observation practices. Over time, climate records may be affected by station relocations, instrument replacements, urban development (e.g., the urban heat island effect), or changes in observation schedules. These factors can introduce artificial trends or jumps in the data that do not reflect actual climate variability. Homogenization techniques use statistical models to detect and adjust for such discontinuities, ensuring that long-term climate records remain consistent and suitable for trend analysis. This is particularly important in climate change studies, where the detection of subtle signals over time requires high-precision data. While climate monitoring organizations invest heavily in quality control infrastructure, several challenges persist. First, data quality varies widely between regions, particularly between developed and developing countries. Some national meteorological services lack the resources, training, or equipment needed to implement rigorous quality assurance procedures. Their datasets may suffer from frequent gaps, inconsistencies, or limited temporal coverage. International cooperation, technical assistance, and funding support are essential to enhance global data equity and ensure that climate monitoring systems worldwide adhere to minimum quality standards (Tanopir C. et al. 2011). Creating incentives for data producers to adhere to these standards—such as funding conditions, data repository rankings, or recognition for data stewardship—can enhance compliance and accountability. Finally, emerging data sources such as citizen science, mobile sensors, and Internet of Things (IoT) devices introduce new complexities to data quality management. While these sources can fill spatial or temporal gaps in traditional monitoring networks, their data may be

Remote Sensing Technologies for Climate Monitoring

Remote sensing technologies, which rely on satellite-based observations, airborne sensors, and unmanned aerial vehicles (UAVs), have become indispensable tools for climate monitoring in the Himalayan region. These technologies offer several advantages over traditional ground-based monitoring systems, especially given the challenging topography of the Himalayas, where many areas are difficult to access due to high altitudes, dense forests, and harsh weather conditions. Remote sensing allows for continuous, large-scale observation of climate variables such as temperature, precipitation, snow cover, glacial dynamics, and land use change, providing crucial data that would otherwise be difficult or impossible to collect in real time. One of the most significant remote sensing applications in the Himalayas is glacier monitoring. Glaciers in the region, which serve as essential freshwater reservoirs, are highly sensitive to temperature fluctuations and changes in precipitation patterns. Satellites equipped with synthetic aperture radar (SAR), such as Sentinel-1 from the European Space Agency (ESA), allow for the monitoring of glacier movements, ice mass loss, and changes in surface characteristics. These sensors can penetrate cloud cover, which is a common challenge in mountainous regions, enabling year-round monitoring even during the monsoon season. Additionally, optical and infrared sensors, such as those on the Landsat satellites, provide high-resolution imagery that can be used to track snow cover extent, seasonal variations, and the rapid retreat of glaciers, which has been a key indicator of climate change in the region.

Remote sensing technologies also play a vital role in monitoring vegetation, deforestation, and land use change, which are increasingly influenced by both natural and anthropogenic climate drivers. Satellite images from the Moderate Resolution Imaging Spectroradiometer (MODIS) and Suomi NPP (National Polar-Orbiting Partnership) can be used to track changes in vegetation health, forest cover, and land degradation (The Intergovernmental Panel on Climate Change 2019). Given the transboundary nature of the Himalayan ecosystem, data sharing is essential for effective climate monitoring and response. The Himalayan region spans several countries, each with its own climate monitoring capabilities, policies, and governance structures. A fragmented approach to climate data collection, analysis, and distribution risks duplicating efforts, wasting resources, and creating data silos that limit the region's ability to address shared climate challenges. Collaborative data sharing mechanisms are therefore crucial to pooling resources, harmonizing data collection methods, and enabling informed decision-making at the regional level. One of the key platforms for regional cooperation in climate monitoring is the Hindu Kush Himalaya Monitoring and Assessment Program (HIMAP), a project led by the International Centre for Integrated Mountain Development (ICIMOD), a regional intergovernmental organization based in Nepal. HIMAP facilitates the sharing of data on climate trends, glacier dynamics, water resources, and biodiversity, providing a common platform for regional governments, research institutions, and non-governmental organizations (NGOs) to access and contribute to the data pool. This platform promotes collaboration across national boundaries and ensures that all stakeholders have access to up-to-date, high-quality data for regional climate adaptation and disaster risk management.

In addition to HIMAP, several other regional initiatives are focused on enhancing climate data sharing. The South Asian Association for Regional Cooperation (SAARC) and the Asian Development Bank (ADB) have worked to improve regional climate monitoring infrastructure and foster data-sharing agreements between countries in the Himalayan region. Programs like SERVIR-Himalaya, a collaboration between NASA and ICIMOD, support the use of satellite data and geographic information systems (GIS) to enhance decision-making in climate adaptation, disaster management, and resource allocation. SERVIR-Himalaya facilitates data sharing and capacity building in countries such as Nepal, Bhutan, India, and Bangladesh, helping to build regional resilience to climate change by providing open access to critical climate and environmental datasets(Turchetti & Roberts, 2014). Despite the progress made in regional collaboration, several challenges remain when it comes to data sharing in the Himalayan region. Political barriers often complicate cross-border data sharing, particularly in areas where national interests, territorial disputes, or concerns over sovereignty are involved. For example, data on water resources, such as glacial meltwater or river flow data, is highly sensitive in regions with shared river basins, and countries may be reluctant to share such data due to fears about resource control or potential mismanagement.

Technical challenges also present significant obstacles to data sharing. The lack of standardized data formats, metadata protocols, and interoperable systems across the countries in the region makes it difficult to combine and compare data from different sources. Many national climate monitoring systems operate in isolation, using proprietary technologies and formats that are not easily adaptable to regional or international standards. In some areas, limited internet connectivity or lack of data infrastructure in remote regions hampers the ability to store, share, and analyze data in real time. Additionally, financial and institutional constraints often limit the capacity of countries in the Himalayan region to invest in advanced climate monitoring technologies, maintain consistent data collection networks, or participate in large-scale regional data-sharing initiatives. In many instances, national and local governments face competing priorities, and climate monitoring may not receive adequate funding or attention. The reliance on external funding for regional initiatives, while essential, can sometimes lead to issues of sustainability and continuity. To overcome these challenges, a series of strategic approaches can help enhance climate monitoring and data sharing in the Himalayan region. Standardization of data formats and the adoption of common protocols for data sharing will help ensure that climate data from different countries and organizations is compatible and can be easily integrated into shared regional platforms. International organizations like the World Meteorological Organization (WMO) and the Group on Earth Observations (GEO) can provide leadership in developing these common standards and encouraging their adoption by national meteorological and hydrological agencies. Additionally, building technical capacity in the region is essential to ensuring that local and national institutions can not only collect high-quality data but also analyze and share it effectively..

Blockchain for Transparent and Secure Data Sharing

Blockchain technology, which underpins cryptocurrencies like Bitcoin, is gaining attention for its potential applications in climate data sharing. Blockchain's core feature—its ability to create an immutable, transparent, and decentralized ledger—offers solutions to some of the most pressing challenges in climate data sharing. These include data security, accountability, and trust among stakeholders. In the context of climate monitoring, blockchain could be used to create a transparent and secure system for sharing data between countries, organizations, and institutions. Another emerging trend is the deployment of Internet of Things (IoT) devices for real-time climate monitoring. IoT technology enables the installation of sensors and smart devices that collect real-time data on a range of environmental factors, including temperature, humidity, soil moisture, air quality, and water levels. These devices, often deployed in remote and rural areas, can provide localized, granular climate data that may not be captured by satellites or traditional monitoring stations. For instance, IoT-enabled weather stations and river monitoring sensors can provide real-time insights into extreme weather events, such as floods, heatwaves, and droughts, enabling early warning systems that are critical for disaster preparedness. The data collected by these IoT devices can be automatically transmitted to centralized platforms via cloud computing, where it can be shared among government agencies, research institutions, and local communities(Vijayakumar G. et al. 2024). The key benefit of IoT devices is their ability to provide continuous, high-frequency data streams, which can be integrated with other climate monitoring datasets to improve forecasting accuracy and risk assessment. These devices, often part of larger smart city or smart agriculture initiatives, also offer the potential for bottom-up data sharing, empowering local communities to participate in climate monitoring and adaptation efforts. One of the most promising developments in climate data sharing is the continued push for open data initiatives and the creation of global data platforms. Open data platforms allow free and open access to climate data for governments, researchers, and communities. These platforms serve as repositories where datasets from satellites, weather stations, IoT sensors, and climate models can be accessed by anyone with the proper tools and permissions.

Several initiatives are already paving the way for open climate data sharing. The Global Climate Observing System (GCOS), a collaborative initiative by major international organizations, aims to ensure the availability of high-quality climate data. Similarly, the World Climate Research Programme (WCRP) and the United Nations Framework Convention on Climate Change (UNFCCC) have worked to establish data repositories and collaborative platforms that promote the sharing of climate data among scientists, governments, and other stakeholders. As open data policies continue to evolve, data interoperability and standardization will be key factors in ensuring that climate data from different sources can be easily integrated and used for climate modeling, analysis, and decision-making. Moreover, the role of cloud computing will be crucial in managing the vast amounts of data generated by these initiatives, enabling efficient storage, processing, and real-time dissemination. The future of climate data sharing will also see increased involvement of citizen science and crowd-sourced data. As more people become aware of the impacts of climate change, individuals and communities are increasingly contributing to climate monitoring efforts. Through smartphone apps, sensor networks, and online platforms, citizens are helping collect data on local weather conditions, air quality, water levels, and biodiversity.

Future Research Direction of Emerging Technologies in Climate Monitoring.

Climate change is one of the most pressing challenges facing humanity today. Its impacts, ranging from rising temperatures and shifting weather patterns to extreme weather events such as floods, droughts, and storms, have profound implications for ecosystems, human health, food security, and economic stability. As the global community strives to mitigate and adapt to these impacts, accurate and timely climate data becomes indispensable for informed decision-making. Monitoring climate variables—such as temperature, precipitation, sea levels, and atmospheric composition—requires advanced, scalable, and cost-effective technologies capable of processing vast amounts of data. In this context, emerging technologies, including artificial intelligence (AI), blockchain, and Internet of Things (IoT), are revolutionizing the landscape of climate monitoring by offering new ways to collect, analyze, share, and validate data. These technologies are enabling more precise, real-time insights into climate trends, improving early warning systems, and enhancing the effectiveness of climate adaptation strategies. This section delves into the role of these emerging technologies in advancing climate monitoring, exploring their applications, benefits, and potential for transforming the way we track and respond to climate change.Artificial intelligence (AI) is quickly becoming a cornerstone of climate monitoring, enabling the analysis of large and complex datasets that are essential for understanding climate dynamics. AI can process and interpret vast amounts of climate-related data far more efficiently than traditional methods, providing deeper insights into the causes and effects of climate change. The role of AI in climate monitoring can be grouped into several key areas: Climate data, generated through various sources such as satellite imagery, weather stations, and sensor networks, can be vast and unstructured. AI, particularly machine learning (ML) and deep learning algorithms, can help automate the processing of this data, transforming it into actionable insights. For example, AI can analyze satellite images to identify patterns of deforestation, land use change, or glacial retreat, all of which have significant implications for climate change. Deep learning models are particularly effective in detecting subtle trends or anomalies that may not be immediately apparent to human analysts, enabling quicker identification of climate changes.

AI-enhanced climate models also have the potential to simulate the impacts of various climate mitigation strategies, such as carbon capture and storage or renewable energy deployment, helping policymakers assess the potential effectiveness of different approaches in reducing greenhouse gas emissions. Another area where AI is making a significant impact is in the automation of data collection and data integration. In the past, climate data collection has been labor-intensive and time-consuming, often requiring manual input and data reconciliation from disparate sources. AI-powered systems, particularly those combined with Internet of Things (IoT) sensors and satellite-based monitoring, can automatically collect real-time data from multiple sources and integrate it into centralized databases. This automation helps streamline the process, reducing the risk of human error and ensuring the timely availability of accurate climate data. AI is also becoming indispensable in climate risk assessment, a critical component of climate adaptation strategies. By analyzing historical data and real-time inputs, AI can help predict the likelihood of extreme events like floods, droughts, wildfires, and heatwaves. These insights allow governments, businesses, and local communities to implement targeted mitigation measures and preparedness strategies. For example, AI-powered platforms can monitor changing weather conditions, provide early warning systems for natural disasters, and help with disaster relief coordination.

The Importance of International Cooperation in Climate Monitoring

Climate change is a global phenomenon that transcends borders, with its impacts affecting every corner of the Earth, regardless of nationality, economic status, or geographic location. From rising sea levels and altered weather patterns to the increased frequency of extreme events such as floods, droughts, and wildfires, no country is immune to the effects of climate change. As the world continues to grapple with the escalating challenges of a warming planet, climate monitoring has become an essential tool for understanding the scale and scope of these changes and for developing effective mitigation and adaptation strategies. However, climate monitoring is an inherently complex and global endeavor that requires the coordinated efforts of multiple stakeholders, including national governments, international organizations, research institutions, and the private sector. International cooperation in climate monitoring is therefore vital for ensuring that data is shared, analyzed, and utilized in ways that can address the pressing challenges of climate change (Wang Z. et al. 2018). The role of data sharing and coordinated action is central to advancing climate monitoring on a global scale. In this discussion, we explore the importance of international cooperation in climate monitoring, its benefits, and the challenges that come with it, as well as strategies for fostering collaboration to address the urgency of the climate crisis. Climate change is a collective problem that no single country or region can solve alone. Greenhouse gases (GHGs) emitted in one part of the world can affect ecosystems and communities in distant areas.

Accurate and timely climate data is the cornerstone of effective climate monitoring, decision-making, and policymaking. However, data collection and analysis for climate monitoring are expensive, resource-intensive, and time-consuming. Developing nations, in particular, often lack the technical capacity, financial resources, or infrastructure to conduct comprehensive climate monitoring and data collection on their own. In this context, data sharing between countries and institutions is crucial to ensuring that comprehensive, high-quality climate data is available for all nations, regardless of their wealth or development status. Data sharing allows countries to access critical climate information that can inform national climate policies, disaster risk reduction strategies, and adaptation plans. For instance, data on precipitation patterns, sea-level rise, and temperature trends is essential for formulating climate adaptation strategies in regions vulnerable to droughts, floods, or coastal erosion. The ability to share data across borders ensures that countries have access to the most up-to-date and accurate information about climate change impacts and can develop more effective, evidence-based responses.Moreover, climate change is not just an environmental issue; it is also an economic and social challenge. Sectors such as agriculture, fisheries, energy, water resources, and transportation are all vulnerable to the effects of climate change. Access to climate data enables policymakers to assess the risks and vulnerabilities of these sectors and make more informed decisions about how to build resilience and reduce the economic costs of climate change. For example, timely data on drought conditions can inform irrigation practices in agricultural areas, while data on sea-level rise can guide the development of coastal infrastructure. Furthermore, global climate models—which are used to project future climate conditions—rely on a vast array of data from different regions.

CONCLUSION

One of the most powerful tools in climate monitoring is the ability to develop global climate models that simulate the potential impacts of climate change on a global scale. These models rely on vast amounts of data from a variety of sources, including weather stations, satellites, ocean buoys, and ground-based sensors. International cooperation in sharing these data sources is critical for improving the accuracy of climate models and ensuring that they are able to simulate the potential impacts of climate change on different regions, ecosystems, and sectors. For example, seasonal forecasting of climate conditions, such as El Niño and La Niña events, requires data from various countries and regions, as these phenomena have far-reaching effects on weather patterns worldwide. By sharing data on ocean temperatures, atmospheric pressure, and wind patterns, countries can develop better predictive models that allow them to prepare for extreme weather events, such as droughts, floods, and hurricanes, and take appropriate action. In addition, early warning systems for natural disasters—such as cyclones, floods, and wildfires—rely on accurate, real-time climate data. Satellite imagery and IoT-based sensors are particularly important in providing real-time monitoring of climate hazards. Through international cooperation, data from these systems can be shared across borders to provide alerts in advance of natural disasters, allowing affected regions to take preventive measures and minimize damage. For example, early warnings of an approaching cyclone or flood can allow governments and communities to evacuate at-risk populations, secure infrastructure, and take other proactive steps. The Sendai Framework for Disaster Risk Reduction (2015-2030), an international agreement adopted by the United Nations, emphasizes the need for data sharing and international cooperation in disaster risk reduction, including climate-related disasters. Climate monitoring is not solely about collecting and sharing data; it also involves the ability to analyze and interpret this data in meaningful ways. Effective climate monitoring requires a robust scientific and technical foundation, which often necessitates collaborative research and capacity building across borders. Research institutions, universities, and non-governmental organizations (NGOs) from around the world collaborate on climate studies, sharing knowledge, expertise, and resources to better understand the impacts of climate change. For example, the Intergovernmental Panel on Climate Change (IPCC), an international body that assesses the science related to climate change, brings together climate scientists from all over the world to produce comprehensive assessment reports that inform international climate negotiations and policymaking. These reports rely on data from diverse regions, and the collective expertise of scientists from different countries is crucial in interpreting this data and making sense of complex climate systems. In addition to research collaborations, capacity-building programs are essential for helping developing countries enhance their climate monitoring capabilities. Many low- and middle-income nations lack the technical expertise, infrastructure, and funding needed to implement robust climate monitoring systems. By fostering international partnerships and providing training and resources, developed countries and international organizations can help build the capacity of these nations to collect and analyze climate data. Initiatives such as the Green Climate Fund (GCF), which supports climate adaptation and mitigation projects in developing countries, are an example of how international cooperation can help strengthen the global response to climate change.

REFERENCES

Akiho, S., & Raita, M. (2017). An Agreement on Enhancing International Arctic Scientific Cooperation: Only for the Eight Arctic States and Their Scientists? In *The Yearbook of Polar Law Online* (8th ed., pp. 129–162). Brill Publishers.

Aragani, V. M., Anumolu, V. R., & Selvakumar, P. (2025). Democratization in the Age of Algorithms: Navigating Opportunities and Challenges. In Wongmahesak, K., Wekke, I., Seftyono, C., & Nurdin, N. (Eds.), *Democracy and Democratization in the Age of AI* (pp. 39–56). IGI Global Scientific Publishing., DOI: 10.4018/979-8-3693-8749-8.ch003

Aronova, E. (2017). Geophysical Datascapes of the Cold War: Politics and Practices of the World Data Centers in the 1950s and 1960s. *Osiris, 32*(1), 307–327. DOI: 10.1086/694094

Biskaborn, B., Smith, S., Noetzli, J., Matthes, H., Vieira, G., Streletskiy, D. A., Schoeneich, P., Romanovsky, V. E., Lewkowicz, A., Abramov, A., Allard, M., Boike, J., Cable, W. L., Christiansen, H. H., Delaloye, R., Diekmann, B., Drozdov, D., Etzelmüller, B., Grosse, G., & Lantuit, H. (2019). Permafrost is warming at a global scale. *Nature Communications, 10*(1), 264. DOI: 10.1038/s41467-018-08240-4 PMID: 30651568

Boyd, A., Gatewood, J., Thorson, S., & Dye, T. (2019). Data Diplomacy. *Science & Diplomacy, 8*, 1. Retrieved June 2, 2021, from http://sciencediplomacy.org/article/2019/data-diplomacy PMID: 31598426

Buch, E. (2019). Arctic In Situ Data Availability. In Copernicus Programme Services; European Environment Agency: Kobenhavn, Denmark; p. 86.

Debnath, P., Deshmukh, S. M., Laroia, M., Selvakumar, P., Manjunath, T. C., & Bhattacharya, S. (2025). Challenges Faced by Countries in Renewable Energy Adoption. In Özsungur, F., Chaychi Semsari, M., & Küçük Bayraktar, H. (Eds.), *Geopolitical Landscapes of Renewable Energy and Urban Growth* (pp. 295–320). IGI Global Scientific Publishing., DOI: 10.4018/979-8-3693-8814-3.ch010

Hamblin, J. (2011). *Oceanographers and the Cold War*. University of Washington Press. DOI: 10.1515/9780295801858

Herzberg, J., Kerht, C., & Torma, F. (2018). *Ice and Snow in the Cold War: Histories of Extreme Climatic Environments*. Berghahn. DOI: 10.3167/9781785339868

Jasny, B. R. (2013). Realities of data sharing using the genome wars as case study—An historical perspective and commentary. *EPJ Data Science*, *2*(1), 1–15. DOI: 10.1140/epjds13

Johri, S., Selvakumar, P., Satyanarayana, P., Das, A., Gupta, S., & Bhattacharya, S. (2025). Urbanization and Energy: The Nexus of Global Challenges. In Özsungur, F., Chaychi Semsari, M., & Küçük Bayraktar, H. (Eds.), *Geopolitical Landscapes of Renewable Energy and Urban Growth* (pp. 453–478). IGI Global Scientific Publishing., DOI: 10.4018/979-8-3693-8814-3.ch016

Joseph, M. P. (2002). The Canadian Cryospheric Information Network: Facilitating Access to Sea Ice, Lake Ice, Snow Cover, Permafrost, and Glacier Data. In *Proceedings of the IEEE International Geoscience and Remote Sensing Symposium*, Toronto, ON, Canada, 24–28 June.

Kowarsch, M., Garard, J., Riousset, P., Lenzi, D., Dorsch, M. J., Knopf, B., Harrs, J.-A., & Edenhofer, O. (2016). Scientific Assessments to Facilitate Deliberative Policy Learning. *Palgrave Communications*, *2*(1), 16092. DOI: 10.1057/palcomms.2016.92

Madgule, M., Vinayaka, N., Sonkhaskar, Y. M., Patel, D. R., Karthikeyan, R., & Selvakumar, P. (2023). Mechanical properties and microstructure of activated TIG welded similar joints of Inconel alloys by desirability approaches. *Materials Today: Proceedings*, *77*, 528–533.

Michener, W. K. (2015). Ecological Data Sharing. *Ecological Informatics*, *29*, 33–44. DOI: 10.1016/j.ecoinf.2015.06.010

Paul, C., Oliver, K., & Wellstead, A. (2016). To Bridge the Divide between Evidence and Policy: Reduce Ambiguity as Much as Uncertainty. *Public Administration Review*, *76*(3), 399–402. DOI: 10.1111/puar.12555

Porter, J. (2010). A Brief History of Data Sharing in the U.S. Long Term Ecological Research Network. *Bulletin of the Ecological Society of America*, *91*(1), 14–20. DOI: 10.1890/0012-9623-91.1.14

Ramakrishnan, T., Mohan Gift, M. D., Chitradevi, S., Jegan, R., & Subha Hency Jose, P. Nagaraja, Rajneesh Sharma H.N., Selvakumar P., Sintayehu Mekuria Hailegiorgis, (2022). "Study of Numerous Resins Used in Polymer Matrix Composite Materials", Advances in Materials Science and Engineering, vol. 2022, Article ID 1088926, 8 pages, 2022. https://doi.org/DOI: 10.1155/2022/1088926

Ran, Y., Li, X., Cheng, G., Zhang, T., Wu, Q., Jin, H., & Jin, R. (2012). Distribution of permafrost in China: An overview of existing permafrost maps. *Permafrost and Periglacial Processes*, *23*(4), 322–333. DOI: 10.1002/ppp.1756

Ran, Y., Xin, L., Guodong, C., Tingjun, Z., Qingbai, W., Huijun, J., Rui, J., Romanovsky, V. E., Smith, S. L., & Christiansen, H. H. (2010). Permafrost thermal state in the Polar Northern Hemisphere during the international polar year 2007–2009: A synthesis. *Permafrost and Periglacial Processes*, *21*(2), 106–116. DOI: 10.1002/ppp.689

Rasmussen, L. H., Zhang, W., Hollesen, J., Cable, S., Christiansen, H. H., Jansson, P.-E., & Bo, E. (2018). Modelling present and future permafrost thermal regimes in Northeast Greenland. *Cold Regions Science and Technology*, *146*, 199–213. DOI: 10.1016/j.coldregions.2017.10.011

Shende, A. D., Tekade, S. A., Deshmukh, A. A., Tembhurkar, S. P., & Selvakumar, P. (2025). Water Quality Monitoring and Control With AI. In Ruiz-Vanoye, J., & Díaz-Parra, O. (Eds.), *Smart Water Technology for Sustainable Management in Modern Cities* (pp. 51–80). IGI Global Scientific Publishing., DOI: 10.4018/979-8-3693-8074-1.ch003

Sreenivasulu, A., & Manimozhi, T. Satheesh kumar P., Selvakumar P. and Singh P.,(2022). "The Model of Smart Sensing Device For Sensitive Nanoclusters Modification in Sensing Properties," 2022 2nd International Conference on Advance Computing and Innovative Technologies in Engineering (ICACITE), Greater Noida, India, 2022, pp. 1043-1047, DOI: 10.1109/ICACITE53722.2022.9823846

Stanford Encyclopedia of Philosophy Stanford University. (2020). Models in Science. Available online: https://seop.illc.uva.nl/entries/models-science/

Sugumaran, K. M. D., Pandiyarajan V., Y. B S, Selvakumar P. and M. S, (2024). "Analysis of Electromyography Signals for Control of Mechanical Prosthesis using Machine Learning Techniques," 2024 2nd International Conference on Intelligent Data Communication Technologies and Internet of Things (IDCIoT), Bengaluru, India, 2024, pp. 1059-1067, DOI: 10.1109/IDCIoT59759.2024.10467707

Tanopir, C., Allard, S., Gouglass, K., Aydinoglu, A. U., Wu, L., Read, E., Manoff, M., & Frame, M. (2011). Data Sharing by Scientists: Practices and Perceptions. *PLoS One*, *6*(6), 6. DOI: 10.1371/journal.pone.0021101 PMID: 21738610

The Intergovernmental Panel on Climate Change (IPCC). (2019). United Nations Intergovernmental Panel on Climate Change's Special Report on the Ocean and Cryosphere. Available online: https://www.ipcc.ch/srocc/

Turchetti, S., & Roberts, P. (2014). *The Surveillance Imperative: Geosciences during the Cold War and Beyond*. Palgrave. DOI: 10.1057/9781137438744

Vijayakumar, G., Rajkumar, M., Rajiv Chandar, N., Selvakumar, P., & Duraisamy, R. (2024). Muniyandi Rajkumar, Rajiv Chandar N, Selvakumar P and Ramesh Duraisamy, (2024). Environmentally friendly TDS removal from waste water by electrochemical ion exchange batch-type recirculation (EIR) technique. *Environmental Science. Water Research & Technology, 10*(4), 826–835. DOI: 10.1039/D3EW00793F

Wang, Z., Kim, Y., Seo, H., Um, M.-J., & Mao, J. (2018). Permafrost response to vegetation greenness variation in the Arctic tundra through positive feedback in surface air temperature and snow cover. *Environmental Research Letters*, **•••**, 14.

KEY TERMS AND DEFINITIONS

Climate Monitoring: The systematic collection, analysis, and interpretation of climate data (e.g., temperature, rainfall, humidity) over time to understand trends and inform policy decisions.

Data Sharing;: The practice of making data accessible between different organizations, regions, or countries to enable collaboration and more effective decision-making.

Interoperability: The ability of different data systems and software to communicate, exchange, and use information effectively across organizations or countries.

Meteorological Data: Data collected on atmospheric conditions including temperature, precipitation, wind speed, and pressure, typically used in weather forecasting and climate analysis.

Cross-Border Collaboration: Cooperative efforts between neighboring countries or regions to address shared challenges such as climate change and environmental monitoring.

Early Warning Systems (EWS): Technologies and communication networks used to detect and provide timely information on climate-related hazards like floods, droughts, or storms.

Climate Resilience:The capacity of communities: ecosystems, or systems to anticipate, prepare for, and respond to hazardous climate events and long-term changes.

Open Data: Data that is freely available for anyone to access, use, and share without restrictions, often promoted to enhance transparency and collaboration.

Standardized Data Protocols: Agreed-upon rules and formats for collecting, storing, and exchanging data to ensure consistency and usability across borders.

Geospatial Information Systems (GIS): Technology that captures, stores, and analyzes spatial or geographic data, crucial for visualizing climate impacts and coordinating responses.

Chapter 9 Environmental Diplomacy and Cross-Border Negotiations

Ramesh Safare
https://orcid.org/0000-0001-5045-6338
Marwadi University, Rajkot, India

S. Guru Basava Aradhya
https://orcid.org/0000-0002-3335-2681
Chandigarh Group of Colleges, Jhanjeri, India

E. Nirupama
https://orcid.org/0009-0003-2915-0678
Saveetha School of Management, Saveetha Institute of Medical and Technical Sciences, India

P. Selvakumar
https://orcid.org/0000-0002-3650-4548
Department of Science and Humanities, Nehru Institute of Technology, Coimbatore, India

Kannan Kunnathully
https://orcid.org/0000-0001-6467-2146
Vinayaka Mission's Law School, Vinayaka Mission's Research Foundation, India

T. C. Manjunath
https://orcid.org/0000-0003-2545-9160
Rajarajeswari College of Engineering, India

A. Ranjithkumar
https://orcid.org/0000-0002-6744-7603
Vel Tech Rangarajan Dr. Sagunthala R&D Institute of Science and Technology, India

ABSTRACT

Environmental diplomacy signifies the collective effort of nations to tackle pressing environmental challenges that transcend individual borders. As the global community confronts increasingly complex and interconnected issues like climate change, pollution, and deforestation, the importance of fostering international cooperation

DOI: 10.4018/979-8-3373-3972-6.ch009

has grown exponentially. Unlike traditional diplomacy, which primarily centers on matters of politics, economics, and security, environmental diplomacy strives to encourage collaboration to safeguard and sustain the natural world. The health of the environment directly impacts global stability and security, making it imperative for nations to engage in multilateral approaches to address these concerns. Through agreements, knowledge-sharing initiatives, and resource allocation, environmental diplomacy aims to develop sustainable solutions that benefit humanity as a whole. The interconnected nature of environmental challenges demands a collaborative approach that takes into account diverse cultures, perspectives, and interests.

INTRODUCTION TO THE IMPORTANCE OF ENVIRONMENTAL

Environmental diplomacy signifies the collective effort of nations to tackle pressing environmental challenges that transcend individual borders. As the global community confronts increasingly complex and interconnected issues like climate change, pollution, and deforestation, the importance of fostering international co-operation has grown exponentially. Unlike traditional diplomacy, which primarily centers on matters of politics, economics, and security, environmental diplomacy strives to encourage collaboration to safeguard and sustain the natural world. The health of the environment directly impacts global stability and security, making it imperative for nations to engage in multilateral approaches to address these concerns. Through agreements, knowledge-sharing initiatives, and resource allocation, environmental diplomacy aims to develop sustainable solutions that benefit humanity as a whole. The interconnected nature of environmental challenges demands a collaborative approach that takes into account diverse cultures, perspectives, and interests. This type of diplomacy is not merely about preventing environmental degradation but is also about building stronger international bonds. It emphasizes the idea that preserving the Earth is not just a responsibility of individual countries but a shared duty that holds immense value for collective progress.For instance, greenhouse gas emissions originating from one country contribute to climate-related consequences in regions halfway across the world. This underscores the critical need for global agreements, such as the Paris Agreement, which provides a framework for nations to unite in mitigating climate change. Similarly, the contamination of shared water bodies like rivers and seas highlights the importance of coordinated action between neighboring countries(Agrawala, S. and Andresen, S. 1999). Environmental diplomacy reinforces the realization that interdependence is a defining characteristic of these global challenges. No country is immune to the effects of environmental degradation; therefore, working together to address these concerns is not only desirable but essential. The repercussions of unsustainable practices often extend beyond geographical boundaries, affecting biodiversity, ecosystem health, and societal stability. Environmental diplomacy, thus, serves as a bridge, connecting nations in pursuit of shared goals and sustainable futures.

The interconnectedness of the modern global economy further amplifies the significance of environmental diplomacy. When environmental degradation occurs in one region, the ripple effects can impact trade, migration patterns, and social stability on a global scale. For example, deforestation in one part of the world can lead to disruptions in weather patterns, negatively affecting agriculture in distant regions. Similarly, pollution from industries can contaminate air and water supplies that are crucial for neighboring countries' populations. The interlinked nature of economies means that environmental issues can have cascading effects, influencing economic growth and human well-being worldwide. Environmental diplomacy acknowledges this interconnectedness and strives to address potential risks comprehensively. Through innovative collaborations, nations can not only mitigate damage but also develop adaptive strategies that enhance resilience. This approach allows countries to balance economic development with environmental sustainability, ensuring that progress does not come at the cost of ecological health(Aragani, V. M., et al. 2025). Additionally, by fostering international partnerships, environmental diplomacy paves the way for shared technological advancements and effective policy-making that promotes long-term solutions.

The urgency of taking action through environmental diplomacy is underscored by the accelerating pace of environmental degradation. Scientific studies and reports continue to signal alarming trends in climate change, habitat destruction, and the loss of biodiversity. From rising global temperatures to frequent natural disasters, the consequences of unchecked human activities are becoming increasingly evident. Environmental diplomacy emerges as a vital mechanism for addressing these challenges before they escalate further. By bringing together policymakers, scientists, and activists, it facilitates discussions that lead to actionable outcomes(Atkinson, R. 1995). This collaborative approach enables nations to share knowledge, resources, and strategies that effectively combat environmental issues. However, the role of environmental diplomacy is not limited to crisis management. It also includes preventive measures that reduce the likelihood of future problems. Promoting clean energy alternatives, encouraging sustainable agricultural practices, and conserving natural resources are some of the ways through which environmental diplomacy supports global efforts toward sustainability. The urgency of such measures is reflected in the growing awareness and commitment to addressing environmental issues at both national and international levels.In a world marked by diverse cultures, values, and interests, environmental diplomacy serves as a unifying force that transcends individual differences. It encourages nations to recognize the interconnectedness of their actions and embrace cooperative solutions. By doing so, environmental diplomacy contributes to building trust and strengthening relationships among countries. Moreover, it highlights the intrinsic value of environmental preservation as a foundation for societal well-being. As global citizens, humanity must prioritize the health of the Earth as a fundamental aspect of its existence. Environmental diplomacy not only offers practical pathways for achieving this but also inspires collective action that benefits future generations. It embodies the belief that cooperation and collaboration can overcome even the most daunting challenges(Barnett, M. and Sikkink, K. 2008). As environmental crises continue to evolve, the commitment to engaging in diplomacy remains essential. Ultimately, the success of environmental diplomacy depends on the willingness of nations to work together, guided by the shared vision of a sustainable and harmonious world.

THE ROLE OF CROSS-BORDER NEGOTIATIONS IN ENVIRONMENTAL DIPLOMACY

Environmental diplomacy embodies international cooperation to tackle complex environmental challenges that transcend borders. As the world faces mounting crises, including climate change, pollution, and deforestation, the significance of collaborative global efforts has never been more evident. Unlike traditional diplomacy, which often prioritizes issues of politics, economics, and national security, environmental diplomacy is dedicated to creating solutions that protect and sustain the planet. These efforts involve fostering partnerships, sharing knowledge, and allocating resources to achieve outcomes that benefit humanity as a whole. The health of the environment is intricately linked to global stability and security, demanding multilateral approaches where nations come together for collective action. The interconnected nature of environmental concerns necessitates an understanding that actions in one country invariably affect others. Environmental diplomacy thus operates on the principle of shared responsibility, emphasizing that sustainable development is a universal goal. It serves not only to address ecological crises but also to strengthen international ties and promote mutual respect(Berry, M. 2002). Through unified efforts, countries can establish frameworks that balance national interests with global priorities, ensuring the well-being of both current and future generations.

The transboundary nature of environmental issues highlights the necessity of cooperation between nations. Greenhouse gas emissions in one region inevitably contribute to climate changes affecting distant territories, making global agreements like the Paris Agreement crucial in combating these challenges. Similarly, the contamination of shared water bodies calls for coordinated action among neighboring countries to safeguard vital ecosystems. Environmental diplomacy plays a pivotal role in recognizing interdependence and addressing these transnational concerns collaboratively. Its effectiveness lies in the ability to unite diverse nations under shared objectives, fostering partnerships that transcend differences(Betsill, M. 2008). This collaborative approach ensures that resources are allocated equitably while innovative solutions are prioritized. By addressing environmental issues collectively, diplomacy prevents isolated efforts and promotes comprehensive strategies that reflect the interconnectedness of global challenges. Recognizing that no nation is immune to the ripple effects of environmental degradation reinforces the need for unity in the pursuit of sustainability. With the proper frameworks, countries can transition toward practices that are both ecologically sound and economically viable.

The global economy's interconnectedness further underscores the importance of environmental diplomacy. Environmental degradation in one region can have cascading effects on trade, migration patterns, and social stability worldwide. For instance, deforestation not only disrupts local ecosystems but can also lead to broader consequences, such as altered weather patterns that negatively impact agricultural productivity across continents. Pollution generated by industrial activities may contaminate air and water sources, affecting populations far beyond national borders. Environmental diplomacy serves as a bridge to mitigate these risks and develop adaptive strategies that enhance resilience. Through international collaborations, nations can devise innovative approaches that balance economic development with ecological conservation. Diplomatic efforts allow for the exchange of technologies, expertise, and resources, fostering progress that aligns with sustainability goals. By integrating environmental considerations into economic policies, countries can create pathways that support both growth and stewardship(Betsill, M. and Corell, E. 2008). Environmental diplomacy, therefore, becomes an essential tool for addressing issues that are deeply intertwined with human development. As nations collaborate, they forge a unified vision that not only addresses current challenges but also anticipates future needs, ensuring long-term stability and prosperity.

The urgent need for environmental diplomacy is driven by the accelerating pace of degradation and the growing scientific evidence of its consequences. Rising temperatures, habitat destruction, and biodiversity loss are alarming trends that demand immediate attention. Through diplomatic channels, countries can come together to implement solutions that address both immediate crises and long-term challenges. Effective environmental diplomacy facilitates the development of frameworks that prioritize sustainable practices, enabling nations to adopt cleaner energy sources, promote conservation, and enhance resource efficiency. Preventive measures, such as reforestation efforts and investments in renewable technologies, are integral to these initiatives. By fostering dialogue among policymakers, scientists, and stakeholders, environmental diplomacy accelerates the transition toward resilience and sustainability. It also acknowledges the inequities inherent in environmental challenges, advocating for inclusive approaches that elevate vulnerable communities. Nations with limited capacities benefit from shared expertise and funding mechanisms, reducing disparities and ensuring progress is equitable(Bexell, M., et al. 2010).

Figure 1. The role of environmental diplomacy and economic factors on environmental degradation

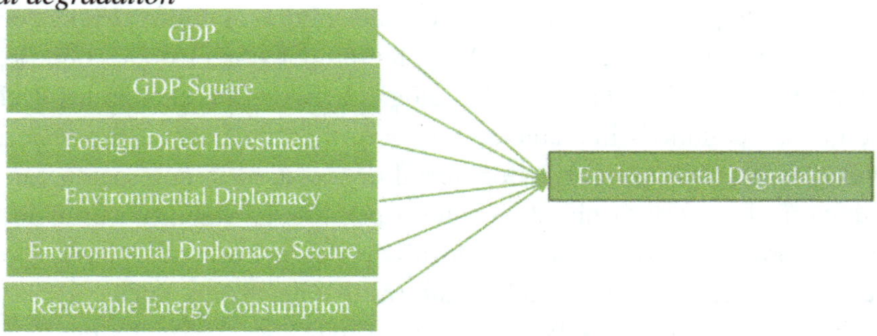

Environmental diplomacy extends beyond addressing crises to envisioning solutions that inspire collective progress. Its transformative potential lies in fostering a sense of shared responsibility among nations, bridging cultural differences to unite under common objectives. Diplomacy strengthens trust, paving the way for agreements that safeguard essential resources like freshwater, forests, and fisheries while mitigating potential conflicts. Additionally, innovative approaches like green finance align economic incentives with environmental priorities, encouraging sustainable practices across sectors. Diplomatic platforms facilitate the implementation of projects that promote renewable energy and sustainable agriculture, enhancing the capacity of nations to adapt to environmental challenges. In the face of rising natural disasters and extreme weather events, diplomacy supports resilience-building efforts, including early warning systems and coordinated disaster response strategies. By addressing root causes and promoting sustainable management, environmental diplomacy contributes to peacebuilding in regions affected by resource scarcity(Braithwaite, J. and Drahos, P. 2000). Its success, however, hinges on the political will and collective action of governments, corporations, and communities. As the world navigates environmental crises, the commitment to diplomacy remains a beacon of hope, guiding humanity toward a harmonious and sustainable future.

THE EVOLUTION OF ENVIRGOVERNMENTS, INTERNATIONAL ORGANIZATIONS, NONGOVERNMENTAL ORGANIZATIONS (NGOS), AND THE PRIVATE SECTOR

Cross-border negotiations play a pivotal role in environmental diplomacy, providing a platform for nations and regions to collaboratively address shared environmental concerns that extend beyond national boundaries. Complex issues like climate change, water resource management, deforestation, pollution, and the conservation of transboundary ecosystems demand cooperative efforts across borders due to their global or regional impact. These negotiations are critical as they bring countries together to exchange knowledge, negotiate terms, and establish binding agreements aimed at resolving common environmental challenges. Their significance lies not only in fostering cooperation but also in creating frameworks that enable nations to achieve mutual goals while addressing pressing ecological crises. However, the process is not without its complexities. Cross-border negotiations often encounter challenges, such as differing national priorities, economic disparities, and geopolitical tensions, which can hinder progress. Despite these obstacles, such negotiations remain indispensable for securing long-term, sustainable environmental outcomes. By bridging divides and encouraging shared responsibility, they pave the way for innovative and inclusive solutions that benefit both humanity and the planet as a whole. Cross-border negotiations are a cornerstone of international environmental agreements, providing an organized framework where nations collaborate to address shared environmental risks. These negotiations play a crucial role in establishing cooperative strategies aimed at safeguarding ecosystems, mitigating climate change, and preventing environmental degradation that could have far-reaching impacts on neighboring states(Carpenter, C. 2001). In addition, cross-border negotiations are instrumental in the management of shared resources like rivers, lakes, and oceans(Debnath, P., et al. 2025). Such negotiations are essential in preventing conflicts over resource access, ensuring the fair distribution of natural assets, and fostering joint efforts to protect ecosystems that are critical to human livelihoods, biodiversity, and economic prosperity. Moreover, cross-border negotiations enhance multilateral environmental cooperation in addressing global challenges. The Montreal Protocol, which brought nations together to phase out ozone-depleting substances, stands as a testament to the power of diplomacy in achieving tangible environmental results. These negotiations often involve diverse stakeholders, including governments, scientific experts, technical advisors, and civil society organizations, all of whom contribute invaluable knowledge and insights.

Cross-border negotiations are vital for fostering collective action among nations to tackle environmental challenges that no single country can resolve on its own. Environmental problems, particularly those related to climate change and pollution, are inherently global. For example, greenhouse gas emissions originating in one country contribute to climate impacts that affect the entire planet, while biodiversity loss in a specific region can trigger a chain reaction of consequences that cross national boundaries. Through cross-border negotiations, countries can work together to set shared goals and establish actionable plans to address these pressing concerns(Downie, C. 2012). These collaborative discussions enable nations to agree on strategies for reducing emissions, conserving ecosystems, and controlling pollution. Such agreements culminate in collective commitments that promote environmental sustainability and ultimately benefit the global community as a whole One of the key advantages of cross-border environmental negotiations is the opportunity they provide for countries to share resources and expertise. Nations facing similar environmental challenges can work together to exchange technical knowledge, financial support, and best practices. For instance, developing countries often require assistance to transition to renewable energy systems or implement measures to adapt to climate change. Developed nations, with their advanced technological capabilities and greater financial resources, can offer support through initiatives such as climate finance, technology transfer, and capacity-building programs. This collaborative model ensures that less-resourced countries are not left behind in global environmental efforts. By fostering equitable partnerships, cross-border negotiations promote progress that benefits all nations, creating a unified approach toward tackling environmental challenges on a fair and inclusive basis.

Environmental issues, particularly those involving shared natural resources, often become sources of tension or even outright conflict between neighboring countries. Cross-border negotiations offer an essential mechanism for preventing or resolving disputes over resources such as rivers, forests, and fisheries. Despite the historical tensions between the two nations, this treaty has successfully facilitated cooperative water management. Such agreements highlight how cross-border negotiations can shift potential conflict into opportunities for collaboration and peacebuilding, demonstrating the transformative power of diplomacy in fostering stability and environmental sustainability. Cross-border environmental negotiations play a crucial role in strengthening international cooperation and promoting multilateralism. Since environmental challenges often demand coordinated action on a global scale, these negotiations pave the way for establishing international frameworks such as conventions, treaties, and protocols to address these pressing issues. Such frameworks not only provide mechanisms for tracking progress and enforcing compliance but also create opportunities for the exchange of information and resources among nations(Downie, C. 2013). Moreover, the inclusion of multiple stakeholders—governments, international organizations, non-governmental organizations (NGOs), and the private sector—ensures that a diverse array of perspectives is considered, resulting in solutions that are both comprehensive and inclusive.

This collaborative approach builds trust and fosters solidarity among countries, ultimately contributing to greater geopolitical stability and a united global effort toward environmental sustainability.

While cross-border negotiations offer numerous advantages, they are not without significant challenges. The complexities inherent in environmental diplomacy, combined with the diverse interests of participating nations, often make these negotiations both arduous and time-consuming. Countries may have conflicting priorities driven by their distinct political, economic, and social circumstances, which can make it difficult to reach consensus. Furthermore, the dynamics of international relations—such as disparities in power and economic capabilities—add another layer of difficulty to the process. These elements underscore the intricate nature of achieving agreements that not only address pressing environmental concerns but also meet the varying needs of all parties involved. Below are some of the primary obstacles that countries face during cross-border negotiations in the context of environmental diplomacy(Downie, C. 2014). One of the most significant challenges in cross-border negotiations is the divergence of national interests among participating countries. Nations differ in terms of their economic development, industrial capacity, and environmental priorities, which often leads to conflicting stances during negotiations. For instance, developed nations are typically more inclined to adopt ambitious measures to combat climate change, as they possess the financial resources and advanced technologies needed to implement such actions. In contrast, developing countries may prioritize poverty alleviation and economic growth, perceiving environmental agreements as potential constraints on their ability to industrialize and expand. These differing priorities can result in stalemates or compromise agreements that fall short of addressing the full scope of the issues at hand.

Principles of Environmental Diplomacy

The evolution of environmental diplomacy represents a dynamic and progressive shift in international relations, highlighting the growing acknowledgment that environmental challenges transcend national borders and demand coordinated, multilateral solutions. Over the past century, environmental diplomacy has transitioned from being a peripheral issue to becoming a central focus of global political and economic agendas. This transformation has been driven by an enhanced understanding of environmental challenges, the establishment of international legal frameworks, the active participation of non-state actors, and the significant role of scientific research in shaping policy decisions(Gough, C. and Shackley, S. 2001). By examining key milestones and developments, this analysis underscores how environmental diplomacy has evolved into an essential mechanism for tackling some of the world's most urgent environmental challenges. The origins of environmental diplomacy can be traced back to the late 19th and early 20th centuries, when environmental concerns started gaining international attention, albeit in a limited and often localized context. Early efforts in this field were primarily focused on the management and preservation of shared natural resources, such as rivers, lakes, and forests, along with protecting endangered species. During its initial stages, environmental diplomacy was largely characterized by bilateral agreements between neighboring countries, addressing specific regional concerns.

One of the earliest examples of formal environmental diplomacy was the International Sanitary Conferences in the mid-19th century, which aimed to curb the spread of infectious diseases through international cooperation. Although these conferences were not directly focused on environmental conservation, they laid the foundation for future collaboration on cross-border issues. However, these early initiatives were often narrow in scope and did not fully encompass the broader environmental challenges that would later emerge on the world stage. The post-World War II era witnessed a surge in global environmental movements, driven by increasing industrialization, rapid urbanization, and growing awareness of environmental degradation. By the 1960s and 1970s, the negative impacts of unchecked industrial activities—such as pollution, habitat destruction, and resource depletion—became more apparent, spurring demands for international cooperation(Hilzenrath, D. 1993). This period marked a significant transition from a narrow focus on managing resources to a broader understanding of environmental sustainability, laying the groundwork for more comprehensive and coordinated approaches to environmental diplomacy. The 1970s marked a transformative decade in the evolution of environmental diplomacy, with a series of groundbreaking events and developments that laid the foundation for its institutionalization on a global scale. Bringing together delegates from 113 countries, the conference heralded a new era where environmental concerns were officially recognized as central to international relations and sustainable development. The Stockholm Conference also produced the Stockholm Declaration, a landmark document that outlined key principles for environmental protection and sustainability. UNEP's creation underscored the growing recognition that environmental challenges demanded focused attention and resources from the international community(Joachim, J. 2003). During the same decade, significant strides were made in developing international legal frameworks for environmental conservation.

This agreement tackled rising concerns over ocean pollution and emphasized the importance of safeguarding marine ecosystems. These early treaties and agreements set crucial precedents for future international environmental initiatives. They demonstrated a growing willingness among nations to engage in cooperative efforts to address pressing ecological challenges, ultimately strengthening the foundation of environmental diplomacy for decades to come. The 1980s marked a pivotal period in the global recognition and engagement with environmental issues, heralding deeper awareness and action(Johri, S., et al. 2025). The decade also witnessed a significant intensification of international initiatives aimed at addressing emerging environmental threats. This protocol stands as the first universally ratified environmental agreement and successfully addressed the global issue of ozone depletion by implementing measures to phase out ozone-depleting chemicals. The success of the Montreal Protocol highlighted the power of international cooperation in tackling complex environmental problems and set an enduring precedent for future agreements on issues such as climate change and biodiversity conservation(Keck, M. and Sikkink, K. 1998). Public awareness of environmental concerns surged during the 1980s, fueled by the efforts of environmental NGOs, scientists, and media outlets.

International Environmental Agreements and Frameworks

Environmental diplomacy is a complex and multifaceted discipline that lies at the intersection of international relations, environmental science, and sustainable development. Its primary aim is to promote international cooperation and address pressing global environmental issues that transcend national boundaries. These guiding values ensure that environmental diplomacy remains effective, equitable(Khagram, S. 2006). A core principle of environmental diplomacy is sovereignty, which refers to a nation's right to govern itself, manage its territory, and make decisions free from external interference. This principle is central to international relations, including environmental diplomacy, as it ensures that each country has the autonomy to determine its environmental policies and priorities. However, sovereignty also brings with it a responsibility to manage natural resources and the environment in ways that do not harm other countries or the shared global commons. Environmental diplomacy emphasizes this responsibility, highlighting the need for nations to strike a balance between exercising sovereignty and addressing the global impacts of their actions.

This balance is particularly important because many environmental challenges are inherently transboundary in nature. The actions of one country—such as emitting greenhouse gases or polluting a river—can have significant consequences for neighboring nations and the planet as a whole. In this context, sovereignty must coexist with collective responsibility. While countries maintain the freedom to set their environmental policies, they also have an obligation to consider the broader international implications of their decisions(Lake, D. 2008). For instance, greenhouse gas emissions in one nation contribute to global climate change, while pollution from a river in an upstream country can affect water quality downstream. Recognizing this interconnectedness is essential for effective environmental diplomacy. Such agreements exemplify how environmental diplomacy can respect national sovereignty while fostering collaboration to address urgent global challenges. Another key principle of environmental diplomacy is cooperation, which is essential given the complexity and scale of global environmental challenges. Issues such as climate change, deforestation, ocean pollution, and freshwater resource depletion are inherently transboundary problems that cannot be effectively addressed by any single country or entity. The principle of cooperation underscores the need for nations to work together, share resources, exchange knowledge, and build partnerships to create solutions that are not only effective but also equitable and sustainable(Lake, D. 2010). Environmental diplomacy encourages countries to embrace multilateralism, wherein collaboration occurs through international frameworks, treaties, and organizations. Similarly, the Convention on Biological Diversity (CBD) unites countries in efforts to protect ecosystems and biodiversity. These agreements exemplify how cooperation enables nations to address complex, global challenges by pooling their resources and expertise, facilitating knowledge-sharing, and aligning their policies for greater impact.

Businesses can contribute to sustainability by adopting green technologies and environmentally friendly practices, while NGOs play a critical role in advocating for stronger environmental protection measures, raising awareness, and holding governments accountable for their commitments. This multi-stakeholder approach reinforces the growing recognition that environmental diplomacy requires collaborative efforts across various sectors to achieve shared environmental goals. The principle of mutual benefit is another cornerstone of environmental diplomacy. It emphasizes that all parties involved in negotiations or agreements should derive advantages from the outcome. This principle challenges the notion of environmental diplomacy as a zero-sum game, where one nation's gain comes at the expense of another's loss. Instead, it seeks to identify solutions that create win-win scenarios, benefiting all stakeholders—whether developed or developing nations—through cooperation(Leggett, J. 1999). For example, in the context of climate change, both mitigation and adaptation measures often yield mutual benefits. In return, developing countries gain access to vital resources and technologies that enhance their resilience, while contributing to the global effort to curb emissions. Furthermore, environmental diplomacy can foster opportunities for green economies, where investments in renewable energy, sustainable agriculture, and clean technologies generate economic benefits for all participating nations. The concept of mutual benefit also manifests in the management of shared natural resources. For example, transboundary water treaties like the Indus Water Treaty between India and Pakistan aim to ensure equitable access to water resources while promoting cooperation for the sustainable management of the river system. Similarly, the Convention on the Law of the Sea (UNCLOS) seeks to balance coastal states' rights to use marine resources with the need for global collaboration in protecting oceans and their ecosystems.

The Role of International Organizations in Environmental Diplomacy

The adoption and implementation of such frameworks often involve intricate negotiations that balance national interests with global priorities(Levy, D. 2005). This analysis examines the principles, objectives, structure, and impact of these agreements while highlighting how they have evolved to address emerging environmental challenges. The Paris Agreement, adopted in December 2015 under the UNFCCC, stands out as one of the most ambitious and widely supported international environmental agreements to date. Its central aim is to limit the global temperature rise to well below 2°C above pre-industrial levels, with a further goal of capping the increase at 1.5°C(Lund, E. 2013). This flexible framework acknowledges the need for every country, irrespective of its economic development level, to actively participate in addressing climate change.

This flexibility has broadened participation, ensuring the inclusion of key emitters such as China, India, and the United States, which were not fully committed under the Kyoto Protocol. This global inclusiveness is essential for tackling climate change, a challenge that demands collective action from all nations. The Paris Agreement also incorporates a robust transparency mechanism to track progress(Mahadev Madgule, et al. 2023). Countries are required to submit regular reports on their emissions and their efforts to meet their NDCs, which are then reviewed internationally. This system enhances accountability and motivates nations to raise their ambitions over time. Moreover, the agreement emphasizes climate finance, with developed countries pledging to provide $100 billion annually to support the mitigation and adaptation efforts of developing nations—a critical element for ensuring equitable progress. Despite its successes, the Paris Agreement faces significant challenges. One major obstacle is the gap between current emission reduction commitments and the level of action needed to meet the 1.5°C target. If fully implemented, existing pledges are still insufficient to avert the worst effects of climate change. Consequently, much of the agreement's focus is now on urging countries to strengthen their NDCs and adopt more aggressive policies to cut emissions. Another pressing issue is climate finance. While the $100 billion annual goal is a significant milestone, many developing nations argue that this amount falls short of what is necessary to meet their adaptation and resilience needs. Despite these challenges, the Paris Agreement remains a landmark in international climate governance. Its adaptability, inclusivity, and emphasis on financial support provide valuable insights for future environmental agreements(Newell, P. 2000). Recognizing that biodiversity is a shared concern for humanity and essential for the planet's well-being, the CBD operates on the principle of sustainable development, integrating conservation with economic and social progress. Its broad scope encompasses terrestrial, marine, and aquatic ecosystems while addressing threats such as habitat loss, invasive species, pollution, and climate change(Odell, J. and Sell, S. 2006). The flexibility of the CBD allows member nations to tailor their actions to suit their unique circumstances and priorities.

A notable milestone under the CBD is the adoption of the Aichi Biodiversity Targets in 2010 at the Nagoya Conference in Japan. The Nagoya Protocol, adopted in 2010 as part of the CBD, established a legally binding framework for ensuring that the benefits of utilizing genetic resources and traditional knowledge are fairly shared, particularly benefiting indigenous communities whose contributions have often been exploited without compensation. However, implementing the CBD's goals remains challenging at the national level(Pitzl, J. and Stewart, E. 2000). Competing priorities, such as economic development and resource extraction, can undermine conservation efforts. The global biodiversity crisis, exacerbated by accelerating habitat loss, invasive species, and climate change, necessitates coordinated and urgent action. Recent discussions under the post-2020 Global Biodiversity Framework have emphasized the role of nature-based solutions in combating climate change, such as ecosystem restoration to absorb carbon dioxide. This integrated approach underscores the interconnectedness of biodiversity and climate change, advocating for a holistic model of environmental governance.

Beyond the Paris Agreement and the CBD, numerous other international agreements play crucial roles in protecting the global environment. The Montreal Protocol is often celebrated as one of the most successful environmental treaties, having nearly eliminated ozone-depleting substances and contributing significantly to the recovery of the ozone layer(Porter, G., et al. 2000). Similarly, CITES has been instrumental in regulating the international trade of endangered species to prevent their extinction. While the Kyoto Protocol faced challenges in compliance, it laid the groundwork for the more inclusive Paris Agreement by introducing concepts like carbon markets and emissions trading. The UNCCD, addressing land degradation and desertification, focuses on sustainable land management, rehabilitating degraded areas, and fostering global cooperation to combat desertification. These agreements collectively demonstrate the international community's commitment to addressing environmental challenges through cooperation, innovation, and shared responsibility.

Cross-Border Negotiations

Among the most prominent international organizations involved in environmental diplomacy are the United Nations Environment Programme (UNEP), the European Union (EU), and various specialized agencies within the UN system. Each of these entities plays a unique and critical role in shaping global environmental governance(Putnam, R. 1988). This discussion explores the contributions of these organizations to environmental diplomacy, examining their structures, mandates, achievements, challenges, and evolving influence in global environmental governance. The European Union (EU) is another significant actor in international environmental diplomacy. Although it is a regional organization, the EU has consistently taken a leadership role in shaping global environmental policy, both within its member states and on the international stage. The EU's commitment to environmental sustainability is deeply embedded in its legal and policy frameworks, making it one of the most influential entities in global environmental diplomacy. Its role is multifaceted, encompassing internal policy development and external diplomatic efforts aimed at addressing global environmental issues. One of the EU's most notable contributions to international environmental diplomacy is its leadership on climate change(Ramakrishnan T. et al. 2022). The EU's success in implementing these policies has enhanced its leverage in international negotiations and solidified its position as a global leader in the fight against climate change. For instance, the EU has committed to protecting 30% of its land and seas by 2030, a target that is consistent with the global biodiversity framework under the CBD. Additionally, the EU promotes sustainable agricultural and forestry practices both within its borders and through international cooperation. Its efforts in securing the Paris Agreement illustrate its ability to bring together diverse nations and stakeholders to reach consensus on critical environmental actions(Risse-Kappen, T. 1995). The EU's environmental diplomacy extends to its trade policies as well. Through its trade agreements, the EU integrates environmental sustainability principles and encourages the adoption of high environmental standards among its trade partners. The EU also employs trade sanctions to enforce compliance with international environmental norms. For example, the European Union Timber Regulation (EUTR) aims to combat illegal logging by regulating timber imports into the EU, ensuring that environmental and ethical standards are upheld. In addition to UNEP and the EU, numerous other international organizations play pivotal roles in advancing environmental diplomacy on a global scale (Schattschneider, E.E. 1960). Each of these bodies brings its expertise and coordination capabilities to address specific environmental issues, reinforcing the collective effort to tackle global environmental challenges.

The World Health Organization (WHO) plays a vital role in environmental diplomacy by addressing issues related to environmental health and focusing on the impacts of pollution, climate change, and ecosystem degradation on human health. WHO collaborates with countries to develop and implement policies that mitigate health risks stemming from environmental factors, such as air pollution, unsafe water, and chemical exposure. Its efforts are crucial in ensuring that environmental challenges are addressed in ways that protect human well-being on a global scale. The Food and Agriculture Organization (FAO) is another key player, operating at the intersection of environmental diplomacy and food security. FAO's initiatives emphasize the importance of sustainable agricultural practices, land-use management, and ecosystem protection, all of which are essential for global food production. The organization actively supports the Global Soil Partnership and advocates for sustainable fisheries and forest management practices, highlighting its commitment to promoting environmental sustainability alongside food security. The World Meteorological Organization (WMO) contributes significantly to environmental diplomacy by providing critical scientific data and assessments related to climate change, weather patterns, and natural disasters. The WMO facilitates international cooperation through early warning systems for extreme weather events, enhances understanding of climate change impacts, and equips countries with the tools and knowledge needed to adapt to these changes. Its work is indispensable in fostering resilience and preparedness against environmental threats(Sell, S. 2003). These NGOs provide technical expertise, advocate for more robust environmental policies, mobilize public awareness and support, and hold governments accountable for their environmental commitments. Their efforts complement those of international organizations, creating a comprehensive and collaborative approach to addressing global environmental challenges.Although international organizations play a vital role in environmental diplomacy, they face a number of significant challenges. One of the most pressing issues is ensuring the effective implementation of environmental agreements. While these organizations are adept at providing frameworks, offering guidance, and extending financial support, the actual enforcement of environmental policies and commitments typically falls under the jurisdiction of national governments(Sewell, G. 1996). This decentralized approach often leads to inconsistencies, as some nations struggle to fulfil their obligations or lack the resources and capacity to take meaningful action. These challenges highlight the need for stronger mechanisms to bridge the gap between international guidance and national-level execution.

Negotiation Strategies and Tactics

Cross-border negotiations in environmental diplomacy are inherently complex, presenting both significant challenges and remarkable opportunities. Environmental issues such as climate change, pollution, and resource depletion do not adhere to national boundaries, making effective governance dependent on collaboration between nations that often have differing political, economic, and social priorities. Negotiating solutions to these global challenges—whether through treaties, multilateral agreements, or regional frameworks—requires careful balancing of competing interests, navigating geopolitical tensions, addressing economic disparities, and ensuring that all stakeholders have a voice in the process. Although such negotiations can be slow and difficult, they also create opportunities for transformative cooperation and collective action(Shende, A. D., et al. 2025). This discussion explores the key challenges and opportunities within cross-border environmental negotiations, emphasizing the critical roles played by diplomacy, international institutions, and national interests in shaping effective outcomes. Furthermore, the effects of these issues often extend far beyond the borders of individual nations, influencing neighboring countries and even global systems. Addressing these interconnected challenges requires robust diplomatic frameworks that foster cooperation while navigating the complexities of diverse interests and perspectives.

Climate change serves as a clear example of a global environmental issue that necessitates cross-border collaboration. Greenhouse gas emissions produced in one country can impact the climate of another, while the broader effects of climate change—such as rising sea levels, extreme weather events, and shifts in agricultural patterns—are not limited by national borders. Similarly, pollution, whether in the form of air, water, or plastic waste, often crosses borders and creates regional or even international environmental crises that demand coordinated responses. For instance, the transboundary nature of air pollution calls for cooperation among countries sharing air and water bodies, as sources of pollution frequently originate from multiple locations. Shared water resources, such as river basins, lakes, or aquifers, also present unique challenges for cross-border negotiations. These resources often traverse multiple national boundaries, making their management contentious. Nations may have differing priorities when it comes to utilizing water for agriculture, energy production, or industrial purposes. A prominent example is the Nile River basin, where Egypt, Sudan, and Ethiopia have conflicting interests concerning the construction of the Grand Ethiopian Renaissance Dam (GERD). This situation underscores the complexities involved in negotiating the use of shared water resources effectively.

Biodiversity conservation introduces further cross-border challenges, as ecosystems and wildlife do not adhere to political boundaries. Migratory species, such as birds, whales, and fish, rely on international cooperation for their protection. Moreover, biodiversity loss is often driven by activities that span multiple countries, including deforestation, overfishing, and illegal wildlife trade. To safeguard biodiversity, nations must reach agreements on policies that protect habitats, designate protected areas, and prevent unsustainable practices that threaten shared ecosystems. These challenges highlight the critical need for cross-border negotiations that bring together a wide range of stakeholders with diverse interests to forge mutually beneficial solutions. While the shared nature of these problems underscores the necessity of cooperation, the process of achieving common ground is fraught with difficulties that can impede effective action(Sreenivasulu A., et al. 2022). Diplomacy and collective commitment remain vital in overcoming these obstacles and advancing global environmental governance. One of the most significant challenges in cross-border environmental negotiations is the divergence of national interests.

In climate change negotiations, for example, industrialized nations like the United States, China, and European Union member states often face pressure to take more aggressive actions to reduce their emissions. Meanwhile, developing countries such as India and Brazil argue that their relatively low historical contributions to global emissions mean they should not bear the same level of responsibility. This disparity creates tension during multilateral climate negotiations, making it difficult to achieve binding global agreements, as evidenced during the Kyoto Protocol and Paris Agreement discussions. Conflicts over transboundary water resources provide another example of divergent national interests in environmental diplomacy. Shared water resources—such as river basins, lakes, or aquifers—frequently become points of contention, as neighbouring countries may have competing priorities regarding how water is allocated and utilized. A prominent example is the ongoing dispute among Turkey, Syria, and Iraq over the allocation of water from the Tigris and Euphrates rivers. Each nation seeks to secure its interests in using the water for agriculture, industry, and energy, highlighting the complexities of managing shared resources. Cross-border environmental negotiations are further complicated by significant power imbalances between countries(Sugumaran K. M, D., et al. 2024). Wealthier, developed nations typically have greater financial resources, advanced technology, and technical expertise, which provides them with a considerable advantage during negotiations. In contrast, developing countries often lack the financial and technical capacity to effectively address environmental challenges, leaving them at a disadvantage. This imbalance can lead to inequitable agreements, where wealthier nations push for policies that align with their interests, while less-developed countries may be pressured into compromises that fail to adequately meet their needs. For instance, the Paris Agreement on climate change required all countries to commit to Nationally Determined Contributions (NDCs). However, it also recognized the need for developed nations to provide financial and technological support to help developing countries meet their targets. Despite this acknowledgment, the commitment by developed nations to deliver $100 billion annually in climate finance remains unmet, leaving many vulnerable nations without the resources needed to implement climate adaptation and mitigation measures effectively. In biodiversity conservation negotiations, similar imbalances arise. Wealthier nations often advocate for ambitious global conservation goals, but developing countries argue that they should be compensated for preserving ecosystems that benefit the global community(Tallberg, J. and Jönsson, C. 2010). However, implementing this protocol has proven challenging, as many countries struggle to enforce benefit-sharing agreements effectively. These examples underscore the complexities and inequities that characterize cross-border environmental negotiations.

Role of Science Technology in Cross-Border Negotiations

Negotiation strategies and tactics in cross-border environmental diplomacy are crucial for achieving sustainable and effective solutions to global environmental challenges. These strategies are influenced by a variety of factors, including the diversity of stakeholders, the complexity of the issues at hand, the asymmetry of power and resources between negotiating parties, and the need to strike a balance between national interests and global priorities. Environmental negotiations often involve not only governments but also international organizations, non-governmental organizations (NGOs), the private sector, and local communities, making the process highly dynamic and multifaceted. Understanding the strategies and tactics employed in these negotiations is essential for advancing environmental goals, building consensus, and ensuring that agreements are fair and enforceable. This discussion delves into the key negotiation strategies and tactics commonly used in cross-border environmental diplomacy, offering insights into how negotiators navigate the intricate balance of interests, power dynamics, and geopolitical considerations. Cross-border environmental negotiations inherently involve diverse stakeholders with varying priorities, political systems, economic conditions, and levels of environmental vulnerability.

Whether addressing climate change, transboundary pollution, biodiversity conservation, or the management of shared resources, negotiators must craft agreements that are not only effective in resolving specific issues but also acceptable to all involved parties. The ultimate aim is to reach agreements that foster international cooperation, enhance environmental protection, and promote sustainable development. The negotiation process requires balancing competing interests, managing conflicting values, and navigating complex legal and institutional frameworks. Strategies and tactics are employed to achieve these objectives while ensuring that outcomes are equitable, practical, and long-lasting. These strategies are dynamic and evolve over time in response to changing circumstances, priorities, and power dynamics. Effective negotiators must, therefore, be skilled in assessing situations, understanding the positions of all parties, and using the most appropriate tactics to secure favorable results. One of the most widely used strategies in cross-border environmental negotiations is multilateralism, which involves the participation of multiple countries or parties in a single negotiation process. Multilateral approaches are particularly effective for addressing global environmental challenges, such as climate change or ocean pollution, that transcend national borders.

Negotiations often occur within the framework of international institutions like the United Nations Framework Convention on Climate Change (UNFCCC), the Convention on Biological Diversity (CBD), or the Global Environment Facility (GEF), which provide platforms for multilateral discussions and agreement-making.

Coalition building is a critical tactic within multilateral negotiations. Countries with shared interests or concerns often form alliances to strengthen their negotiating positions and advocate for common objectives(United States Senate 1997). Similarly, developing nations with shared concerns about the economic and social impacts of environmental policies may form coalitions to advocate for financial assistance and technological transfers from developed countries. By building coalitions, negotiators can leverage collective strength to influence negotiation outcomes, enabling smaller or less powerful countries to pool resources and amplify their voices. Issue linkage is another strategic approach often used in cross-border environmental negotiations. This tactic involves connecting unrelated or loosely related issues to create a package deal that offers mutual benefits. By linking multiple issues, negotiators can incentivize parties to reach agreements on one issue by offering concessions on another. This approach is particularly effective in scenarios with conflicting interests, as it provides negotiators with opportunities to trade concessions in ways that benefit all involved parties.

In environmental diplomacy, issue linkage often combines environmental goals with economic or development-related objectives. For example, a country might commit to stricter emissions reduction targets in exchange for financial aid, technological transfers, or capacity-building support from wealthier nations. Similarly, biodiversity conservation negotiations might be tied to trade agreements, where conservation measures are incorporated into broader economic frameworks to encourage countries to fulfill their environmental commitments. By expanding the range of potential solutions, issue linkage allows negotiators to find compromises that satisfy all stakeholders. However, the challenge lies in identifying appropriate issues to link and ensuring that the resulting trade-offs are perceived as fair and equitable by all parties.This approach proves particularly valuable in cross-border environmental negotiations, where parties often have differing perspectives but share common goals, such as addressing climate change or conserving biodiversity. The primary aim of principled negotiation is to create win-win solutions that address the fundamental interests of all stakeholders, rather than resulting in a win-lose outcome. In the context of environmental diplomacy, this often involves crafting solutions that balance environmental protection with economic growth, social equity, and political feasibility. For instance, in climate change negotiations, the objective might be to establish ambitious yet achievable emission reduction targets for both developed and developing nations. This would include providing financial assistance and technological transfers to help developing countries meet their targets while advancing global climate goals. Principled negotiation also stresses the importance of joint problem-solving and collaboration over confrontation or adversarial tactics. By focusing on shared interests and exploring innovative solutions, negotiators can uncover pathways to agreement that might not have been evident at the outset. This collaborative, interest-based approach fosters trust and improves the likelihood of achieving sustainable and equitable outcomes in environmental diplomacy(Vijayakumar G, et al. 2024).

Recommendations for Enhancing Environmental Diplomacy

The future of environmental diplomacy stands at a crucial juncture, where the urgency of global environmental crises converges with evolving geopolitical dynamics, technological advancements, and shifting social expectations. As the world confronts escalating challenges such as climate change, resource depletion, and pollution, environmental diplomacy will increasingly be called upon to create innovative, multi-faceted solutions that balance national interests with the global good. Moving forward, the trajectory of environmental diplomacy will be shaped by several interconnected trends, challenges, and opportunities, emphasizing the need for collaborative, inclusive, and forward-thinking approaches to secure the planet's future. Multilateral cooperation will remain a cornerstone of environmental diplomacy. Indigenous knowledge and local perspectives will remain critical in shaping sustainable and culturally appropriate solutions to environmental challenges(Yin, R. 2009). Global challenges like climate change and biodiversity loss transcend national borders, yet international efforts to address these crises have often been hindered by conflicting national interests and power imbalances. As the planet's environmental systems grow more interconnected and fragile, the need for collective action will intensify.

The future of environmental diplomacy will demand a nuanced understanding of equity, justice, and fairness, with special attention to vulnerable communities—often located in the Global South—whose contributions to environmental crises are minimal but who face the greatest impacts. Regional cooperation and international alliances will also play an essential role in addressing region-specific environmental issues(Wapner, P. 1995). Regional environmental diplomacy is expected to grow, with nations and sub-regions working together to develop tailored solutions for shared challenges, such as transboundary pollution and natural resource management. The integration of emerging technologies will be a defining feature of future environmental diplomacy. Innovations in areas such as artificial intelligence (AI), blockchain, remote sensing, and bioengineering have the potential to revolutionize how environmental challenges are monitored, managed, and mitigated. These technologies will enhance data collection and predictive modeling while enabling more transparent, accountable, and efficient governance frameworks for addressing global environmental issues. AI, for instance, will improve climate modeling accuracy, facilitate real-time biodiversity monitoring, and strengthen natural disaster forecasting. The future of environmental diplomacy will increasingly rely on open, data-driven approaches that engage a wide range of stakeholders in real-time discussions, ensuring decisions are informed by the best available evidence and that affected communities are empowered to hold decision-makers accountable(White, R. 1998).

CONCLUSION

Blockchain technology, with its secure and decentralized nature, will provide the foundation for more efficient carbon markets, sustainable supply chains, and reliable tracking of environmental impacts. Meanwhile, emerging geoengineering technologies, such as solar radiation management (SRM) and carbon capture and storage (CCS), may become central to climate diplomacy as potential tools for managing global warming. However, these technologies pose significant ethical and environmental risks, meaning their future deployment will require rigorous scientific evaluation, international regulation, and broad public consent to avoid exacerbating existing inequalities or creating new challenges. Technological advancements will also democratize data access, enabling local communities and civil society organizations to participate more meaningfully in environmental negotiations. As environmental crises grow more severe, the future of environmental diplomacy will be increasingly characterized by inclusive, participatory governance processes. Traditional diplomatic frameworks, often dominated by state actors, will no longer suffice to address the complexity of global environmental issues. Civil society, indigenous groups, local communities, and businesses must play integral roles in shaping environmental policies at both national and international levels, ensuring solutions are equitable, effective, and sustainable. Indigenous peoples, who often find themselves on the frontlines of environmental degradation, possess invaluable knowledge of ecosystem management, biodiversity conservation, and sustainable resource use. Recognizing and respecting indigenous rights will be a cornerstone of future environmental diplomacy, ensuring that their voices are heard in international negotiations and that their land stewardship practices are incorporated into global environmental governance. Furthermore, the engagement of young people and youth movements, such as Fridays for Future, is expected to grow even more influential in the coming years. The future of environmental diplomacy must prioritize intergenerational equity by including young people—those who will face the long-term impacts of today's environmental decisions—in key decision-making processes. The rising climate justice movement will play an integral role in advocating for the rights of marginalized communities and future generations, driving transformative actions that address both social inequality and environmental sustainability.

REFERENCES

Agrawala, S., & Andresen, S. (1999). Indispensability and indefensibility? The United States in the climate treaty negotiations. *Global Governance*, *5*(4), 457–482. DOI: 10.1163/19426720-00504004

Aragani, V. M., Anumolu, V. R., & Selvakumar, P. (2025). Democratization in the Age of Algorithms: Navigating Opportunities and Challenges. In Wongmahesak, K., Wekke, I., Seftyono, C., & Nurdin, N. (Eds.), *Democracy and Democratization in the Age of AI* (pp. 39–56). IGI Global Scientific Publishing., DOI: 10.4018/979-8-3693-8749-8.ch003

Atkinson, R. (1995). Climate debate ends on cool note greenhouse gas cutbacks stalled 2 years. The Washington Post, 8 April, p. 15.

Barnett, M., & Sikkink, K. (2008). From international relations to global society. In Reus-Smit, C., & Snidal, D. (Eds.), *The Oxford handbook of international relations* (pp. 62–83). Oxford University Press.

Berry, M. (2002). Validity and reliability issues in elite interviewing. *PS, Political Science & Politics*, *35*(4), 679–682. DOI: 10.1017/S1049096502001166

Betsill, M. (2008). Environmental NGOs and the Kyoto Protocol negotiations: 1995 to 1997. In Betsill, M., & Corell, E. (Eds.), *NGO diplomacy: The influence of nongovernmental organizations in international environmental negotiations* (pp. 43–66). The MIT Press.

Betsill, M., & Corell, E. (2008). Introduction to NGO diplomacy. In Betsill, M., & Corell, E. (Eds.), *NGO diplomacy: The influence of nongovernmental organizations in international environmental negotiations* (pp. 1–18). MIT Press.

Bexell, M., Tallberg, J., & Uhlin, A. (2010). Democracy in global governance: The promises and pitfalls of transnational actors. *Global Governance*, *16*(1), 81–101. DOI: 10.1163/19426720-01601006

Braithwaite, J., & Drahos, P. (2000). *Global business regulation*. Cambridge University Press. DOI: 10.1017/9780521780339

Carpenter, C. (2001). Businesses, green groups and the media: The role of non-governmental organisations in the climate change debate. *International Affairs*, *77*(2), 313–328. DOI: 10.1111/1468-2346.00194

Debnath, P., Deshmukh, S. M., Laroia, M., Selvakumar, P., Manjunath, T. C., & Bhattacharya, S. (2025). Challenges Faced by Countries in Renewable Energy Adoption. In Özsungur, F., Chaychi Semsari, M., & Küçük Bayraktar, H. (Eds.), *Geopolitical Landscapes of Renewable Energy and Urban Growth* (pp. 295–320). IGI Global Scientific Publishing., DOI: 10.4018/979-8-3693-8814-3.ch010

Downie, C. (2012). Toward an understanding of state behavior in prolonged international negotiations. *International Negotiation*, *17*(2), 295–320. DOI: 10.1163/157180612X651458

Downie, C. (2013). Transnational actors: Nongovernmental organizations, civil society and individuals. In Harris, P. (Ed.), *Routledge handbook of global environmental politics* (pp. 176–186). Routledge. DOI: 10.4324/9780203799055.ch14

Downie, C. (2014). *The politics of climate change negotiations: strategies and variables in prolonged international negotiations*. Edward Elgar. DOI: 10.4337/9781783472116

Gough, C., & Shackley, S. (2001). The respectable politics of climate change: The epistemic communities and NGOs. *International Affairs*, *77*(2), 329–345. DOI: 10.1111/1468-2346.00195

Hilzenrath, D. (1993). Miscalculations, lobby effort doomed BTU tax plan. The Washington Post, 11 June, p. 1.

Joachim, J. (2003). Framing issues and seizing opportunities: The UN, NGOs, and women's rights. *International Studies Quarterly*, *47*(2), 247–274. DOI: 10.1111/1468-2478.4702005

Johri, S., Selvakumar, P., Satyanarayana, P., Das, A., Gupta, S., & Bhattacharya, S. (2025). Urbanization and Energy: The Nexus of Global Challenges. In Özsungur, F., Chaychi Semsari, M., & Küçük Bayraktar, H. (Eds.), *Geopolitical Landscapes of Renewable Energy and Urban Growth* (pp. 453–478). IGI Global Scientific Publishing., DOI: 10.4018/979-8-3693-8814-3.ch016

Keck, M., & Sikkink, K. (1998). *Activists beyond borders: advocacy networks in international politics*. Cornell University Press.

Khagram, S. (2006). Possible future architectures of global governance: A transnational perspective/prospective. *Global Governance*, *12*(1), 97–117. DOI: 10.1163/19426720-01201008

Lake, D. (2008). The state and international relations. In Reus-Smit, C., & Snidal, D. (Eds.), *Oxford handbook of international relations* (pp. 41–61). Oxford University Press.

Lake, D. (2010). Rightful rules: Authority, order, and the foundations of global governance. *Studia Quaternaria, 54*(3), 587–613.

Leggett, J. (1999). *The carbon war: dispatches from the end of the oil century.* Allen Lane.

Levy, D. (2005). Business and the evolution of the climate regime. In Levy, D., & Newell, P. (Eds.), *The business of global environmental governance* (pp. 73–104). The MIT Press.

Lund, E. (2013). Environmental diplomacy: Comparing the influence of business and environmental NGOs in negotiations on reform of the clean development mechanism. *Environmental Politics, 22*(5), 739–759. DOI: 10.1080/09644016.2012.737253

Madgule, M., Vinayaka, N., Sonkhaskar, Y. M., Patel, D. R., Karthikeyan, R., & Selvakumar, P. (2023). Mechanical properties and microstructure of activated TIG welded similar joints of Inconel alloys by desirability approaches. *Materials Today: Proceedings, 77*, 528–533.

Newell, P. (2000). *Climate for change: non-state actors and the global politics of the greenhouse.* Cambridge University Press. DOI: 10.1017/CBO9780511529436

Odell, J., & Sell, S. (2006). Reframing the issue: The WTO coalition on intellectual property and public health, 2001. In Odell, J. (Ed.), *Negotiating trade: developing countries in the WTO and NAFTA* (pp. 85–114). Cambridge University Press. DOI: 10.1017/CBO9780511491610.003

Pitzl, J., & Stewart, E. (2000). *The US position on the Kyoto Protocol: Senate ratification or not?* Georgetown University., Available from http://faculty.guisd.org/product.cfm?ProductID=193&MenuCategory=2

Porter, G.. (2000). *Global environmental politics.* Westview Press.

Putnam, R. (1988). Diplomacy and domestic politics: The logic of two-level games. *International Organization, 42*(3), 427–460. DOI: 10.1017/S0020818300027697

Ramakrishnan, T., Mohan Gift, M. D., Chitradevi, S., Jegan, R., & Subha Hency Jose, P. Nagaraja, Rajneesh Sharma H.N., Selvakumar P., Sintayehu Mekuria Hailegiorgis, (2022). "Study of Numerous Resins Used in Polymer Matrix Composite Materials", Advances in Materials Science and Engineering, vol. 2022, Article ID 1088926, 8 pages, 2022. https://doi.org/DOI: 10.1155/2022/1088926

Risse-Kappen, T. (1995). Bringing transnational relations back in: Introduction. In Risse-Kappen, T. (Ed.), *Bringing transnational relations back in: non-state actors, domestic structures and international institutions* (pp. 3–34). Cambridge University Press. DOI: 10.1017/CBO9780511598760.002

Schattschneider, E. E. (1960). *The semisovereign people: a realist's view of democracy in America*. Holt, Rinehart and Winston.

Sell, S. (2003). *Private power, public law: the globalization of intellectual property rights*. Cambridge University Press. DOI: 10.1017/CBO9780511491665

Sewell, G. (1996). Conflicting beliefs: National implementation of the United Nations framework convention on climate change. *Environmental Impact Assessment Review, 16*(3), 137–150. DOI: 10.1016/0195-9255(96)00031-5

Shende, A. D., Tekade, S. A., Deshmukh, A. A., Tembhurkar, S. P., & Selvakumar, P. (2025). Water Quality Monitoring and Control With AI. In Ruiz-Vanoye, J., & Díaz-Parra, O. (Eds.), *Smart Water Technology for Sustainable Management in Modern Cities* (pp. 51–80). IGI Global Scientific Publishing., DOI: 10.4018/979-8-3693-8074-1.ch003

Sreenivasulu, A., & Manimozhi, T. Satheesh kumar P., Selvakumar P. and Singh P.,(2022). "The Model of Smart Sensing Device For Sensitive Nanoclusters Modification in Sensing Properties," 2022 2nd International Conference on Advance Computing and Innovative Technologies in Engineering (ICACITE), Greater Noida, India, 2022, pp. 1043-1047, DOI: 10.1109/ICACITE53722.2022.9823846

Sugumaran, K. M. D., Pandiyarajan V., Y. B S, Selvakumar P. and M. S, (2024). "Analysis of Electromyography Signals for Control of Mechanical Prosthesis using Machine Learning Techniques," 2024 2nd International Conference on Intelligent Data Communication Technologies and Internet of Things (IDCIoT), Bengaluru, India, 2024, pp. 1059-1067, DOI: 10.1109/IDCIoT59759.2024.10467707

Tallberg, J., & Jönsson, C. (2010). Transnational actor participation in international institutions: where, why, and with what consequences? In Jönsson, C., & Tallberg, J. (Eds.), *Transnational actors in global governance: patterns, explanations, and implications* (pp. 1–21). Palgrave Macmillan. DOI: 10.1057/9780230283220_1

United States Senate. (1997). Senate resolution 98. 105th Congress. 1st Session. Available from: http://thomas.loc.gov/cgi-bin/bdquery/z?d105:S.RES.98: [Accessed 5 December 2013].

Vijayakumar, G., Rajkumar, M., Rajiv Chandar, N., Selvakumar, P., & Duraisamy, R. (2024). Muniyandi Rajkumar, Rajiv Chandar N, Selvakumar P and Ramesh Duraisamy, (2024). Environmentally friendly TDS removal from waste water by electrochemical ion exchange batch-type recirculation (EIR) technique. *Environmental Science. Water Research & Technology, 10*(4), 826–835. DOI: 10.1039/D3EW00793F

Wapner, P. (1995). Politics beyond the state: Environmental activism and world civic politics. *World Politics, 47*(3), 311–340. DOI: 10.1017/S0043887100016415

White, R. (1998). Kyoto and beyond. *Issues in Science and Technology, 14*(3), 59–65.

Yin, R. (2009). *Case study research design and methods*. Sage Publications.

Chapter 10
India–UAE Bilateral Cooperation to Combat Climate Change:
Strategic Synergies and Future Pathways

Tushar Bhatia
https://orcid.org/0009-0000-7532-0488
Department of Humanities and Social Sciences, Graphic Era University, Dehradun, India

S. Srinivasan
https://orcid.org/0009-0002-0179-9849
Department of Humanities and Social Sciences, Graphic Era University, Dehradun, India

Gaurav Dimri
Department of Humanities and Social Sciences, Graphic Era University, Dehradun, India

ABSTRACT

India and the United Arab Emirates (UAE) have evolved their long-standing bilateral relationship into a strategic climate partnership. Both countries are committed to combating climate change through cooperation in renewable energy, green hydrogen, and climate finance, aligned with their nationally determined contributions under the Paris Agreement. India's Green Hydrogen Mission and the UAE's Net Zero 2050 strategy form the foundation for collaborative energy transition goals. Joint initiatives, MOUs, and investment agreements have expanded clean energy

DOI: 10.4018/979-8-3373-3972-6.ch010

development, grid connectivity, and support for developing nations. This study assesses the effectiveness of these bilateral efforts, their influence on global climate negotiations, and the shared vision for a sustainable, low-carbon future.

INTRODUCTION

Over the last two decades, India-UAE bilateral relations have expanded in various dimensions, such as trade, energy, oil and gas, defense, agriculture, culture, and space, etc. The bilateral trade and investments reached $85 billion in 2023, compared to $7 billion in 2003 (Quamar et al., 2023). Both countries have a Comprehensive Economic Partnership Agreement to reach US$100 billion in non-oil trade sectors in the next five years. Moreover, the third-largest trading partner of India is the UAE, while the second-largest trading partner of the UAE is India. India exports to the UAE over US$35.62 billion in FY 2023–2024, but the UAE exports over US$48 billion in FY 2023–2024 (Saber, 2024). India and the UAE strengthened their relationship through a Comprehensive Strategic Partnership in 2017. In recent years, many different kinds of business opportunities have emerged that have increased trade relations between them, such as energy, food security, defense, climate change, maritime security, AI, and infrastructure development (Voyer et al., 2018).

The bilateral trade is projected to reach US$250 billion by 2030. Additionally, the UAE is financially supporting India's renewable energy program, which aims to achieve 500 gigawatts of renewable energy capacity by 2030. Apart from this, both countries are working together on a low-carbon future (Dutta, 2021). Global warming due to climate change is driven by human activities such as the burning of fossil fuels, deforestation, industrial processes, and the use of non-renewable energy over many decades, which release CO_2 in large amounts and produce more heat in the Earth's atmosphere (Swagatika Priyadarshini et al., 2025). Apart from this, both countries are working together on a low-carbon future. Global warming, due to climate change, is driven by human activities such as the burning of fossil fuels, deforestation, industrial processes, and the use of non-renewable energy over many decades, which release CO_2 in large amounts and produce more heat in the Earth's atmosphere. To reduce global warming, various multilateral platforms and global frameworks have focused on limiting the Earth's temperature rise, such as the UN Framework Convention on Climate Change (1992), the UN-SDGs (2015), and the Paris Agreement (2015) (Amin et al., 2022; Leung & You, 2023).

Global countries came together to conclude that broad categories of action are to limit global temperature to no more than 1.5 degrees Celsius, achieve net-zero emissions by 2050, ensure commitments by many countries to use renewable energy like solar and wind to reduce CO_2 emissions, and recognize that financing is required for adjustments within the countries (Shahzad et al., 2024). To achieve these aims and targets at the global level, India has set its own goals, objectives, and missions at the national level and through bilateral partnerships with countries (Anika et al., 2022). At the national level, the Green Hydrogen Mission was launched in 2023 to establish India as a green hydrogen market hub that produces, uses, and exports low-carbon hydrogen derived from renewable energy sources. Therefore, New Delhi has set a target to produce 5 million metric tonnes of green hydrogen per annum by 2030 (Verma et al., 2024). Moreover, the country targets achieving net-zero emissions by 2070, as announced at COP 26 in Glasgow, and aims to generate 40% of its electricity from non-fossil sources by 2030; however, this target was achieved earlier in 2021 (nine years ahead of schedule). However, India's climate leadership was observed at the G20 Summit and COP 28, which emphasized three main agendas: advancing climate action domestically, focusing on finance and climate equity, and requiring global support for climate action (Henderson, 2021).

In light of the significance of global support in addressing climate change, India-UAE bilateral relations are driven by mutual interests in energy security, technological partnership, and sustainable development (Shahnawaz & Mir, 2025). The reason for both countries coming together is the huge potential for market access, expanded trade relations in non-oil trade sectors, and the emerging decarbonization market of green hydrogen, green ammonia, and renewable energy at the regional and global levels (Sim, 2024). Over the past few years, India and GCC member states have strengthened relations in the renewable energy sector in a general manner, and the India-UAE commitment in particular (Ashwarya, 2023). As a result, the UAE allocated US$75 billion in funds to India for promoting clean energy. Even the UAE envisages becoming the global hub of green hydrogen and achieving net-zero emissions by 2050. Indeed, India and the UAE signed an MOU in January 2023 for the development of green hydrogen and the installation of undersea cable connectivity under the initiative of "One Sun, One World, and One Grid," announced by PM Modi.

Apart from this, India and the UAE face several challenges while developing renewable energy, green hydrogen, and grid connectivity. Such challenges include the high production cost of green hydrogen, high infrastructure costs for storage, transportation, and distribution of hydrogen, and a lack of technological advancement for improving the efficiency and affordability of green hydrogen. Moreover, both countries are still lacking in policy reforms, innovations, investments, supply chains, and investment capital (Khan & AI-Ghamdi, 2023).

India and the UAE have had strong bilateral trade and diplomatic relations since the 1970s, during the Gulf oil boom. The relationship gained momentum when both countries began cooperating on climate change through bilateral and multilateral platforms (Baomar, 2024). The countries have national goals and targets in the renewable energy sector and aim to be green hydrogen producers at the global level.

But their common vision is to combat climate change through bilateral partnership while respecting the founding principles of the UN Framework Convention on Climate Change, nationally determined contributions under the Paris Agreement, and the UN Sustainable Development Goals (Sattar, 2022). India and the UAE are also leading members of the multilateral platform, like the International Solar Alliance (2016), having a commitment to advance solar energy deployment globally in a general manner but with bilateral commitment in a particular manner. The study examines the evolving bilateral cooperation between India and the UAE in addressing climate change (Pradhan, 2010; Jha, 2021; Khan et al., 2021) . It explores joint initiatives and collaborations in renewable energy, green hydrogen, and climate finance and investment. However, while addressing climate change, India and the UAE also jointly support discussions in favor of Global South countries' demand for adequate climate finance and technology transfer from developed countries on world forums. By analyzing joint statements and comments, official documents of both countries, and signed agreements between them, the study aims to assess the effectiveness of this bilateral cooperation in achieving climate goals (Alkhaldi et al., 2023; Tibni, 2024).

Significance of the study

The significance of this study lies in the evolving bilateral dynamics of relations between India and the UAE regarding climate change cooperation beyond borders. The study provides a comprehensive analysis of bilateral relations while addressing environmental challenges. Their bilateral collaborations and implications support policy reforms, technical feasibility, enhanced infrastructure, and investment flows in the fields of renewable energy and green hydrogen while combating climate change. In addition to this, the relations also expand South-South cooperation not only to reduce global emissions but also to build regional stability and achieve sustainable development goals, namely SDG 7, SDG 13, and SDG 17.

Research Objectives

1. To analyse the scope and depth of India–UAE bilateral cooperation in combating climate change.
2. To evaluate the effectiveness of joint initiatives in renewable energy, green hydrogen, and climate finance.
3. To assess the role of India–UAE collaboration in influencing global climate policies and negotiations.
4. To identify challenges and opportunities in enhancing bilateral climate cooperation.

Research Questions

What are the key areas of collaboration between India and the UAE in addressing climate change?
How effective have joint initiatives been in achieving their intended climate objectives?
In what ways does the India–UAE partnership influence global climate policy frameworks?
What are the potential avenues for strengthening bilateral cooperation in the future?

Background of the study

Climate change is a critical global issue requiring joint efforts across nations. India and the United Arab Emirates (UAE), as emerging economies, have recognized the need for sustainable development and clean energy transitions. Their bilateral relationship has expanded beyond trade to include strategic cooperation on climate action. Driven by shared commitments under the Paris Agreement and platforms like COP and the International Solar Alliance (ISA), both countries are investing in renewable energy, green hydrogen, and carbon reduction initiatives. India's "Panchamrit" goals and the UAE's Net Zero by 2050 strategy reflect their strong climate agendas. The India-UAE High-Level Joint Task Force and recent events such as COP28 in the UAE have further strengthened this cooperation. As both nations face climate-related risks such as heatwaves, water stress, and rising sea levels, this study explores how their partnership can be enhanced for greater climate resilience and sustainable growth.

LITERATURE REVIEW

The strategic partnership between India and the UAE has significantly evolved over the years. Both countries are leading members of the International Solar Alliance (2016), committed to upholding the Paris Agreement. India supported the outcomes of COP 28, which was held in Dubai in 2022. Similarly, the UAE also supported India's hosted G20 conference, which addressed global climate issues with a focus on limiting global temperature rises (Mallah et al., 2024). The cornerstone of such cooperation is the MOUs and agreements on climate change that began in 2014. Recently, an MOU on investment and cooperation in renewable energy was signed in 2024. UAE investment in India's green energy sector is made through ventures like Masdar and the Abu Dhabi Investment Authority (Diaz et al., 2023). Moreover, both share a common vision for low-carbon energy solutions. Additionally, PM Modi's announcement of the "One Sun, One World, and One Grid" initiative highlights a mutual commitment to solar energy proliferation. Both countries have their own nationally determined contributions under the Paris Agreement of 2016. India's net-zero emission target by 2070, hydrogen production capacity of 10 million metric tonnes per annum by 2030, manufacturing of water-based hydrogen, and the global hydrogen market target to reach over 100 MMT by 2030 (Prasad et al., 2024). Whereas, the UAE's net-zero emissions will be achieved by 2050; it aims to be a low-emission hydrogen producer and supplier by 2031, producing 1.4 million tons per annum by 2031 and 15 million tonnes per annum by 2050. Their national policies and targets will be achievable through the bilateral partnership in long-term goals such as the global hydrogen market in Asia and Europe (Shibata et al., 2024).

India-UAE Hydrogen Task Force

Both countries recognize the urgent need for climate action; therefore, political leadership has pledged to intensify cooperation to accelerate the energy transition under the MoU on green hydrogen development and investment. In this regard, they have agreed to establish a joint Hydrogen Task Force with a focus on scaling up clean technologies in the production of green hydrogen. Both agreed on UAE investments in India's clean technology programs, which are underpinned through business-to-business and public-private partnerships. With the launch of the task force in January 2023, India's ambitious goal to become a global hub for green hydrogen by 2030 will be realized also fulfilling the sustainable climate development goals (Akomea-Frimpong, 2022).

Last year, on the occasion of Indian Independence Day, the Indian Ambassador to the UAE highlighted that 'India-UAE relations have reached new heights'. Moreover, Ambassador Sunjay Sudhir emphasised that the bilateral relationship has undergone a significant shift in recent years because of a shared national vision to mitigate climate change on both sides; as a result, 'India and the UAE are closer than ever before' DD News. It has been argued that India's Green Hydrogen Mission and the UAE National Hydrogen Strategy both have complementary long-term goals and mutual benefits as nations to become global green hydrogen producers (Almzeedi, 2024). They have common global goals but different national strategies to achieve them. For instance, the UAE's low-carbon hydrogen production target is 1.4 million metric tonnes per annum, which encompasses both blue hydrogen (from natural gas with carbon capture and utilisation technologies) and green hydrogen from renewable sources (Olabi & Jouhara, 2024).

At present, the UAE's cost competitiveness of hydrogen is \$2.7/kg, which is projected to fall to \$1.7/kg by 2030. Whereas India's global green hydrogen landscape involves large-scale production, domestic usage, and global exportation (from renewable sources such as wind and solar energy), with cost-competitive production (Stern & Lankes, 2022). Moreover, India's green hydrogen production capacity target is 5 million metric tonnes per annum by 2030 (produced from the water electrolysis process). Both collaborative stakeholders have become reliable and interdependent partners through technological advancements and innovations aimed at enhancing efficiency, reducing costs, improving storage capacities, and investing in research and development (Abdin, 2024).

India-UAE transnational Green Credit Initiative

In the 28th session of the UN Climate Change Conferences (COP 28) held in the UAE, PM Modi of India and President Sheikh Mohamed Al Nahyan hosted the Green Card Initiative. The aim of this initiative is to promote environmental conservation and preservation by encouraging individuals, groups, and private companies to proactively participate in greening the planet through activities like tree plantation, water conservation, sustainable agriculture, and waste management beyond the legal requirements (Odeyemi et al., 2024). Indeed, this initiative is a part of the Green Credit Programme, which is related to the mission of "LIFE – Lifestyle for the Environment," introduced under the UNFCCC (COP 26 in 2021) (Selvakumar et al., 2025).

However, while addressing the COP 28 summit, PM Shri Narendra Modi mentioned that "India has set a perfect example of balancing both the economy and ecology to the world." Nevertheless, India has 17% of the global population but has emitted less than 4% of carbon emissions. Moreover, the Indian government also identified Green Ammonia projects for sustainable climate-economic cooperation in the future between India and the UAE while addressing the COP 28 World Summit (Joyce & Stanford, 2023). For the significance of this Green Credit Initiative, PM Modi also emphasised, "As a human being gives importance to his/her healthy life, in a similar manner, global member countries must start thinking about environmental health. Also, we will have to look at what is to be included as positive for the Earth's health. Additionally, UAE President Sheikh Mohamed Al Nahyan expressed his appreciation and welcomed the step of the "Green Credit Initiative" as a web portal for global climate cooperation against climate change (Chang, 2025).

India-UAE as joint stakeholders in Renewable Energy sector

Renewable energy forms the cornerstone of the India–UAE climate partnership to address climate change. In October 2021, the UAE pledged US$75 billion to support India's clean energy transition. Building on this, both nations signed a landmark agreement in January 2023 focused on green hydrogen investment, potentially adding 60 GW to India's renewable energy capacity. Beyond clean energy, cross-border electricity grid connectivity is crucial for energy sustainability. These initiatives stem from the Comprehensive Strategic Partnership established in January 2017. Bilateral cooperation, which expanded beyond hydrocarbons since the first MoU in 2014, is supported by robust policy frameworks emphasizing energy storage technologies and innovative undergrid connectivity.

At COP28, hosted by the UAE, India reaffirmed its stance on "common but differentiated responsibilities," with support from the UAE. India has consistently urged developed nations to honor climate finance commitments, emphasizing the disparity between high-emission nations and those with limited resources. Prime Minister Modi reiterated that wealthy nations must expand their net-zero goals and commit to a new global climate target (Leal-Arcas, 2024). India's pledge to reach net-zero by 2070 was announced at COP26 in 2021. The UAE has actively invested in India's green sector through entities like the Abu Dhabi Investment Authority and Masdar, which have formed joint ventures with Indian companies to build green infrastructure. The UAE remains India's largest investor (Matz, 2018), with over 70% of GCC foreign investments originating from the UAE. The signing of the India–UAE Bilateral Investment Treaty on August 31, 2024, further enhances investment.

Recent initiatives also promote hybrid renewable systems. Ren Serv Global Pvt. Ltd., a ReNew Energy subsidiary, now operates in both countries to optimize wind and solar project efficiency (Hartmann et al., 2021). ReNew is expanding its footprint in the UAE, while Indian firm Avaada Group, backed by Brookfield, is in talks to establish a green hydrogen and ammonia plant with a 1 million-tonne capacity. Discussions with the Abu Dhabi government have focused on enabling energy banking for surplus solar and wind power. Such facilities could enhance long-term off-take agreements, ensuring revenue stability and closer cooperation in energy security strategies.

These long-term agreements also open doors to tax incentives, green loans, and financial transparency in cross-border clean energy ventures. In response to evolving global energy challenges, the UAE revised its Energy Strategy 2050 under President Sheikh Mohamed bin Zayed Al Nahyan, declaring 2023 the "Year of Sustainability". This aligns with India's investment target of AED 150–200 billion by 2030 under the Paris Agreement's Nationally Determined Contributions.

The strategic partnership also extends to innovation, such as the collaboration between Indian startup LOHUM Clean Tech and UAE authorities to establish the UAE's first EV battery recycling facility—showcasing effective cross-border cooperation in addressing resource scarcity. However, challenges persist, including bureaucratic delays, supply chain inefficiencies, carbon pricing issues, infrastructure gaps, and lack of innovation in materials and logistics (Oshilalu, 2024). Both nations continue to support policy reforms at global climate forums. India's Environment Minister, Mr. Bhupendra Yadav, emphasized that global climate action must be grounded in equity and climate justice, calling on developed countries to lead with ambitious commitments (Thompson, 2025). In this context, the following section explores gaps in global climate policy and outlines actionable recommendations for reform, jointly proposed by India and the UAE (Al-Sarihi & Mason, 2020).

India-UAE joint supportive voice: On changes and transformations in global climate policies execution

Over the past 30 years, more than 3,000 climate change policies have been developed globally. Under the Paris Agreement, countries—regardless of income, population, or carbon footprint—have committed to climate action (Makarov et al., 2020). Surprisingly, many low-emission developing countries have taken significant steps, while wealthy nations with high emissions have been criticized for inadequate financial support, lack of technology transfer, and weak contributions to climate funds. Despite their numerous climate declarations, developed countries have often failed to demonstrate tangible progress. During 2010–2020, several emerging economies, including India, Colombia, Mexico, Peru, and South Africa, successfully implemented impactful policies. For example, Peru promoted bicycle transport, South Africa's Bus Rapid Transit eased traffic and emissions, and Colombia introduced a Green Building Code. India rapidly expanded solar power initiatives, with strong support from its UAE partnership in capital investment and innovation. These examples underscore the need for globally coordinated, equitable policies benefiting all nations.

However, global consensus on a common climate policy framework remains elusive. While emerging economies have led valuable reforms, broader implementation requires politically supported, well-designed policies (Ibrahim, 2024). As World Bank Vice President Juergen Voegele noted, "with the right mix of political support and policy design, we can decarbonize development". PM Modi echoed similar sentiments at COP26 in Glasgow: "Let's move together, let's interact together, and everyone's mind should be one." India has become a proactive global climate actor, aligning domestic initiatives—such as the National Action Plan on Climate Change (2008)—with international collaboration. Similarly, the UAE's National Climate Action Plan (2017–2050) emphasizes emissions reduction and resilience through bilateral partnerships. Both nations go beyond traditional multilateral platforms like the UN-COP model, collaborating to strengthen global climate governance (Sebbagh et al., 2024).

India and the UAE have criticized the lack of consensus on carbon pricing, weak enforcement, and vague emission targets under current global frameworks. Together, they promote universal climate principles beyond national targets and advocate for inclusive climate action (Shah et al., 2025). Frustration with the limitations of the Paris Agreement has led many Global South nations to pursue mini-lateral, bilateral, and triangular forums. India and the UAE are active in climate alliances such as the International Solar Alliance (ISA), I2U2 (India, Israel, UAE, USA), and the UAE-France-India trilateral (UFI). Their joint leadership emphasizes a just energy transition, ensuring universal access to affordable and sustainable energy. Moving forward, India and the UAE advocate for substantial reforms in climate policy under the UN-COP system, which the next section elaborates.

Rich nations imposed carbon tariff upon the developing nations

In 2023, the European Union announced its Carbon Border Adjustment Mechanism (CBAM), set to impose carbon taxes from January 1, 2026, on carbon-intensive imports such as steel, cement, and fertilizers (Jakhar, 2024). This decision has raised concerns among major exporters like India, China, and the USA. India's Finance Minister Nirmala Sitharaman criticized the move as "unilateral and arbitrary," warning of negative impacts on Indian exports. Commerce Minister Piyush Goyal also stated that India would consider retaliatory tariffs. Many Global South countries oppose CBAM, arguing it violates the "Common but Differentiated Responsibilities" principle under the UNFCCC. Rich countries have largely failed in providing adequate climate finance and technology transfer. While renewable energy costs have dropped, emissions continue to rise. The IPCC (2021) warned that temperatures could exceed 2°C by the end of the century.

According to environmental economists, developing countries need $6 trillion by 2030 to meet their climate goals. However, instead of funding, developed nations are imposing carbon tariffs. In contrast, carbon trading offers a flexible, market-driven approach (Yang, 2025). One carbon credit equals one tonne of CO_2 reduced. Markets are either compliance-based (regulated) or voluntary (private sector-led). Many Global South nations, including India and the UAE, are now collaborating on carbon credit markets and clean energy transitions (Dagher et al., 2020).

Under Article 6 of the Paris Agreement, countries can trade carbon credits to meet their NDCs. For instance, if one country exceeds its emissions reduction target, it can sell the surplus to another (Article 6.2). India and the UAE are actively participating in this mechanism. India is building a national carbon market framework that includes both compliance and voluntary systems. Since 2015, Indian industries have saved 106 million tonnes of CO_2. India also aims to cut 1 billion tonnes of CO_2 by 2030, in line with PM Modi's "Panchamrit" targets announced at COP26.

The Energy Conservation (Amendment) Act, 2022, enables issuance of carbon credit certificates, while India's Bureau of Energy Efficiency enforces emissions targets for high-carbon sectors like steel, cement, and petrochemicals (Shee Weng, 2025). A robust Measurement, Reporting, and Verification (MRV) system ensures transparency and prevents double counting. India is also working with the UAE to build a reliable carbon trading ecosystem, supporting mutual goals for climate resilience and energy transition (Sreekanth, 2024). The UAE updated its NDC in 2024 and launched a National Register for Carbon Credits (NRCC), applicable to entities emitting over 0.5 million tonnes of CO_2 annually. It aims to cut emissions by 47% by 2035, from 196.3 to 103.5 million tonnes of CO_2e. The NRCC enables both domestic and international carbon credit exchange, aligned with Article 6(4) of the Paris Agreement.

While the UAE's approach differs slightly from India's, both countries are committed to measurable and verifiable emission reductions. To address sectors with high emissions, India should also invest in carbon capture and storage (CCS) and collaborate with the UAE on green industrial technologies. This would help India protect its exports from EU carbon tariffs while enhancing its global climate leadership. Ultimately, the India–UAE carbon credit partnership supports mutual net-zero ambitions and provides a model for Global South countries seeking climate finance, technology transfer, and access to global carbon markets (Dsilva et al., 2024).

India-UAE approach to Developed countries: Remove barriers to enable climate action

Developing countries continue to face major barriers in climate sustainability efforts due to limited access to advanced technology, financial resources, and R&D infrastructure. These challenges hinder their progress toward achieving Nationally Determined Contributions (NDCs) by 2030. External support through international cooperation remains critical, as countries vary in vulnerability to climate change impacts and tailor their own mitigation, adaptation, and resilience strategies under the Paris Agreement's 1.5–2°C goal. Many developing nations urgently require technology to address climate-related losses and damages. Although technology development and transfer are enshrined in Article 4(7) of the UNFCCC and Articles 10(5) and 10(6) of the Paris Agreement, Global South and small island states have repeatedly criticized developed countries for failing to uphold their commitments. This inaction is seen as a major obstacle to achieving NDCs.

Emerging economies like India, Brazil, South Africa, Indonesia, Turkey, and Mexico are building bilateral and multilateral partnerships with both developing and developed nations to meet their climate goals more effectively (Suri et al., 2024). Countries like India and the UAE uphold the principle of "Equity and Common but Differentiated Responsibilities," emphasizing that wealthier nations must offer financial and technological support. At COP28 (UAE, 2023), countries agreed on the first Global Stocktake (GST) to assess collective progress toward the 1.5°C target MEA, India. The GST emphasized key actions: scaling up renewable energy, reducing emissions of methane and other harmful gases, phasing down fossil fuels, submitting updated NDCs for the 2025–2035 period, and enhancing loss and damage funding for vulnerable nations.

India PM Narendra Modi and UAE President Mohamed Al Nayan highlighted the GST's balanced approach and urged developed countries to ensure fair climate finance distribution. Both leaders stressed the need to support developing nations—especially for long-term plans like solar energy corridors—under the principle of equity and justice. Additionally, India and the UAE called for doubling adaptation finance for developing countries by 2025 compared to 2019 levels MEA, India, 2023. The following table presents year-wise climate finance contributions from 2018 to 2022 (Manohar, 2024).

Table 1. Finance Contributions

Financial Year	Climate finance contributions
2018	USD 79.9 Billions
2019	USD 80.4 Billions
2020	USD 83.3 Billions
2021	USD 89.6 Billions
2022	USD 115.9 Billions

Source: OECD, www.oecd.org

As per the World Resources Institute report, in 2023, developing countries received a record USD 125 billion in financial support. The climate finance budget has increased, and Multilateral Development Banks' (MDBs) contributions to low- and middle-income countries have also increased. Although India and the UAE appreciated the financial incremental contribution, at the same time they were criticized for the improper mobilization of climate funds. Both reliable climate partners emphasized the need to reform climate funding through the importance of highly concessional finance and non-debt instruments to support developing countries. Similarly, at COP 29 (held in Baku, Azerbaijan, 2024), India and the UAE both strongly opposed the climate finance deal of USD 300 billion annually by 2035. This insufficient amount could not meet the updated NDC targets of developing states. Furthermore, developing nations are not financially capable of building green technology or even able to buy carbon credits from developed nations. Overall, both India and the UAE reiterated the USD 1.3 trillion annual climate deal discussed last year in 2023, which is seen as more realistic to deal with climate action. Indian Chief Economic Advisor V. Anantha Nageswaran has strongly condemned rich nations for their ‘double standards’ on climate action (The Hindu, 2024).

Indeed, this insufficient climate budget with a lack of clarity affects the India-UAE climate cooperation projects due to the lack of financial resources. Moreover, India-UAE energy transition projects have attracted other climate-struggling developing countries desiring to become a part of this cooperation because it provides assistance for ambitious NDC targets. Along with this, both countries exploring energy transition projects have brought a paradigm shift to the global connectivity corridor. As both countries are members of the International Solar Alliance, they therefore have more responsibilities to support the climate goals of Global South countries. Further, both are emerging renewable markets, so there are more opportunities for foreign investments in both ways, according to their requirements. Currently, India’s annual finance flow for green energy is only one-fourth of the amount required to achieve its nationally determined contributions as per the Paris Agreement.

India-UAE strategic climate action: Opportunities, Challenges and Solutions

The bilateral relationship between India and the UAE has been evolving in diversified directions. As mentioned earlier, both are strategic partners, which rejuvenates the relationship to be more strengthened, reliable, and remarkable. By virtue of comprehensive friendship, both nations have pledged to combat climate change through energy transition. In order to become a global energy market, India and the UAE signed a Comprehensive Economic Partnership Agreement. It provides great opportunities for exploring their national climate goals as a paradigm shift toward global sustainability goals. For this purpose, renewable energy is the most crucial weapon to reduce carbon emissions through the collaboration of India and the UAE in different fields.

Indeed, through climate collaboration, many possibilities arise for opening new foreign trades, ensuring energy security, and reducing dependence on fossil fuels. However, beyond the bilateral engagements, India-UAE is represented as a lead member in multi-aligned groups like I2U2 (India, Israel, UAE, and USA), the International Solar Alliance, the trilateral collaboration of India, UAE, and France on clean energy projects, and the Coalition for Disaster Resilient Infrastructure. As the India-UAE climate partnership focuses on the perception of net-zero emissions, sustainable development, and positioning as a future global energy trade exporter, it also addresses domestic economic priorities.

Further, their collaboration spans many agreements and MOUs, joint ventures, and long-term strategic initiatives. Recently, the Crown Prince of Dubai, Sheikh

Hamdan bin Mohammed Al Maktoum, emphasized "the family relationship between the two nations". This vision promotes multi-cooperation, which strengthens its climate partnership in the following pathways –

India-UAE energy deal: Scopes and Advantages

India and the UAE aim to boost non-oil bilateral trade to USD 100 billion by 2030, with energy cooperation as a core driver of economic and climate goals. UAE investments and technological collaboration are helping India emerge as a global hub for solar photovoltaic (PV) manufacturing. Leveraging public-private partnership models, UAE capital is supporting solar and wind projects in India by ensuring risk-sharing, structured project implementation, and private sector participation. During Crown Prince Khaled bin Mohammed bin Zayed Al Nahyan's visit to India in September 2024, multiple agreements were signed across clean energy, nuclear energy, agriculture, and food security. A key agreement involved Abu Dhabi National Oil Company supplying over 1 million metric tons of LNG annually to India for 15 years—highlighting their ongoing energy collaboration.

The UAE is India's seventh-largest investor, with USD 18 billion invested as of 2023. Given India's net-zero ambition by 2070, requiring USD 10 trillion in investments, entities like Abu Dhabi Investment Authority have backed leading Indian renewable firms such as Greeko and ReNew. UAE companies—ADNOC, Masdar, and Taqa—along with India's Avaada Group, are investing in renewable projects, including large-scale solar and wind energy storage. Avaada's projects manage storage capacities from 50,000 to over a million units, adjusted to energy demand. However, the intermittency of renewables and limited lending by commercial banks hinder growth due to the absence of a robust energy banking mechanism. To address this, both nations plan to collaborate on developing energy technologies, small grids, IoT applications, and R&D in energy efficiency (Biyik et al., 2021).

Beyond energy, the two countries are expanding into sustainable industrial development. A 2023 MoU between UAE's Minister of Industry and India's Commerce Minister promotes industrial decarbonization. A green hydrogen strategy is also underway. Both countries are leveraging water electrolysis-based green hydrogen to become global producers (Horri & Ozcan, 2024). Their initiatives align with growing opportunities in the low-carbon hydrogen market. This collaboration is strategic and integrated—clean hydrogen is generated using renewable sources and powered by grid connectivity, as seen in ADNOC's green hydrogen production. Although their national targets differ, India and the UAE share a unified approach to reducing GHG emissions via reliable and visionary energy transition strategies (Zaiter, 2024). Together, they are exploring the use of green hydrogen in key sectors like steel, transport, industry, and shipping. This is crucial for India, particularly in light of the potential EU carbon tax. Reducing green hydrogen costs remains a joint priority (Sambasivam & Sarma, 2024).

India-UAE strategic technology development in Renewable energy

India and the UAE share a strong alignment toward a successful and action-driven COP28 climate summit. Both Prime Minister Modi and President Sheikh Mohamed Al Nahyan recognize COP28 as a pivotal platform for inclusive climate action. They emphasize the need for renewable energy solutions that include storage technologies, low-carbon innovations, and energy efficiency to support global sustainability.

At COP26, PM Modi introduced the "One Sun, One World, One Grid" initiative, advocating for a global solar grid. This initiative began during the International Solar Alliance event in 2018. Echoing this vision, President Al Nahyan praised India's G20 leadership and reiterated a joint commitment to enhancing climate collaboration through strategic technology sharing and innovation.

Their collaboration extends to renewable energy technology and the logistics sector, focusing on smart routing, advanced data exchange, and predictive forecasting. However, several challenges persist:

- Transporting large components like wind turbine blades or sensitive solar parts over difficult terrain increases logistics costs.
- Solar power intermittency necessitates robust energy storage.
- Undersea renewable energy connectivity requires advanced infrastructure and storage systems.
- Administrative delays and bureaucratic inefficiencies slow project implementation.

Addressing these challenges requires a multi-stakeholder approach involving governments, industry leaders, technocrats, logistics experts, and administrative bodies. Joint India-UAE investments in infrastructure upgrades, green innovation, and strategic technological cooperation are key. Innovative supply chain solutions and integrated logistics systems are essential to streamline clean energy development and accelerate the transition to a green future.

RESEARCH METHODOLOGY

The research study adopts a qualitative approach to understand the bilateral climate cooperation between India and the UAE. In this qualitative study, secondary data includes bilateral agreements and MOUs, climate action reports from both countries, and policy documents, which will provide a comprehensive understanding of the existing framework. The data analysis will be employed to identify shared interests, technological partnerships, and policy alignment. The study will assess strategic relations, policy frameworks, and the outcomes of climate cooperation over the period. This research is descriptive, exploratory, and analytical. It endeavours to describe the existing framework, analyse the bilateral cooperation, and explore the potential future of climate cooperation between the two countries.

RESULT

India-UAE Green Hydrogen Development

India and the UAE have been strengthening their energy partnership in trade, production, supply, and storage capacity of green hydrogen in the recent year, 2023. It is important to note here that the demand for clean energy is expected to increase by nearly 35 percent by 2035 in South Asian countries, according to the International Energy Agency (Abdallah et al., 2024). As mentioned earlier, India-UAE bilateral strategic cooperation in green hydrogen production was a cornerstone to foster economic growth in low-carbon global trade, giving rise to job creation and sustainable energy solutions against climate change. In a nutshell, the India-UAE formidable and lucrative partnership, recognized by PM Modi, congratulated the UAE as the host country of the Conference of All Parties (COP) in 2023. In a reciprocal manner, His Highness President Sheikh Mohamed bin Zayed Al Nahyan congratulated India for its leadership role in the G20 presidency India-UAE: Joint Statement on Climate Change. In this context, at COP28, the UAE launched 6.6 GW of clean energy capacity in India, including the production of 10 MW of wind and solar energy. These projects are expected to be financed by the UAE's $30 billion climate fund, Alterra. The finance stakes contributed by both countries under the specific projects related to green hydrogen (Tyson & Weiss, 2025).

Disappointment of COP 29 Climate Finance Deal

India and other Global South nations have consistently criticized the Paris Agreement for lacking adequate financial and technological support from developed Global North countries. At the UN-COP29 summit in Baku New Delhi strongly opposed a climate finance deal that was adopted by consensus, despite disappointment among Global South countries, including India and the UAE. These nations had been demanding $1.3 trillion in climate support over the past three years, but the agreed amount was only $300 billion—far below expectations. Indian delegate Chandni Raina described the agreement as "nothing more than an optical illusion (Bhat, 2025).

Despite contributing far less to global emissions than countries like the EU, USA, and China, Global South countries remain committed to climate action. At COP28 in the UAE, PM Modi emphasized this, stating that developing nations "have not contributed to greenhouse gas emissions but are willing to find global solutions". India has repeatedly stressed that without climate finance and technology transfer from wealthy nations, Global South countries cannot meet their climate goals. As a result, global climate change demands a unified response. In this context, the India-UAE climate partnership plays a critical role in advancing sustainable development, driving green technology, supporting renewable energy, and promoting green credit initiatives—especially for the Global South.

Table 2. Source: Ministry of Environment, Forest and Climate Change of India

NDC targets and its update	India
In the year 2015 NDC target	Reduced emissions of GDP by 33-35% by 2030 from 2005 level.
2022 NDC update	40% electricity from non-fossil fuel by 2030

https://www.pib.gov.in

However, indeed, many Global South nations have committed themselves to reducing greenhouse gas emissions, and these countries are using bilateral and multilateral collaborations to achieve their national climate targets. Like this, India and the UAE have updated their NDC targets. For India, the updated NDC in August 2022 set a target to reduce carbon emissions intensity of its GDP by 45% by 2030 from 2005 levels and to achieve 50% of cumulative electric power installed capacity from non-fossil-based energy resources by 2030. Previously, in the 2015 NDC, the target was to reduce carbon emissions intensity by 33-35% by 2030 from 2005 levels and achieve 40% of cumulative electricity from non-fossil fuels by 2030. The 2015 targets were achieved well before the deadline.

On the other hand, the UAE also updated its target to reduce its greenhouse gas (GHG) emissions by 47% by 2035. In 2023, the target was to reduce GHG emissions by 40% by 2030 (as mentioned in Table 3).

Table 3. Source: Ministry of Climate change and Environment of UAE

NDC targets and its update	UAE
Present target of NDC	Reduce GHG emissions by 47% by 2035.
Former target in July, 2023	To reduce its GHG emissions to 40% by 2030.

https://unfccc.int.

India and other Global South nations have shown strong commitment to reducing greenhouse gas (GHG) emissions, using bilateral and multilateral cooperation to meet climate goals. Under the Paris Agreement, India's 2015 NDC targeted a 33–35% reduction in emissions intensity of GDP by 2030 (from 2005 levels) and 40% electricity from non-fossil sources—both achieved ahead of time. In 2022, India updated its targets to a 45% reduction in emissions intensity and 50% installed power capacity from non-fossil fuels by 2030. Similarly, the UAE raised its emissions reduction target from 40% by 2030 (set in 2023) to 47% by 2035. These updates highlight both countries' growing leadership in global climate action and commitment to sustainable development.

India-UAE joint initiatives of Renewable energy, Green hydrogen and Climate finance: To combat climate change

Human activities, particularly the usage of fossil fuels such as oil, gas, and petrol, have released a huge amount of greenhouse gases since the mid-20th century, leading to global warming on the Earth. The global countries are affected by climate change at the national, regional, and global levels (Jogdand, 2020). Due to climate change, Earth's temperature rises, sea levels rise, and glaciers warm and melt. The global countries came together to conclude the reduction of carbon dioxide, the primary gas driving global warming on the Earth. At the world level, the prominent climate change agreements are the UN Framework Convention on Climate Change, the Kyoto Protocol, and the Paris Agreement (a legally binding international treaty) (Gao et al., 2017). But on the bilateral level, India and the UAE jointly committed to mitigate climate change. This has happened due to their strong and strategic relationship in various sectors over a period of time (Krzymowski, 2020).

As a result of their historic bilateral foreign relations, both countries have enjoyed a deep-rooted partnership and enhanced cooperation. As a result, they share common interests in many areas like trade and investment, energy, defence, and technology. Even both countries elevated their relations to reach a "Comprehensive Strategic and Economic Partnership" in 2017 and 2022, respectively. Similarly, in the arena of climate change, there have been many bilateral MOUs and agreements signed over the last ten years. Specifically, over the last three years, energy security has been the mainstay of India-UAE ties (Siyech, 2020). Apart from this, Indian PM

Narendra Modi mentioned that "the two countries have vigorous cooperation on multiple pillars" during an interview with the Arabic newspaper *Al-Ittihad* on 30 November 2023 (NRI Focus, 2023). As per the concerned government records, PM Modi visited the UAE six times from 2014 to 2024. Moreover, PM Modi emphasized that "India and the UAE joined to enhance each other's strength and provide support to the International Solar Alliance for a global solar facility".

India's National goal and mission to mitigate Climate change

India is continuously working to reduce its dependence on fossil fuels for generating electricity. The transformation is shifting to renewable energy sources like solar, wind, and hydropower, as well as supporting technologies such as green hydrogen and biogas (Ramesh, 2025). According to the Central Electricity Authority, India's remarkable achievement in renewable energy capacity reached 203.18 gigawatts. Over the past decade, India's renewable energy installed capacity increased from 76.37 GW in 2014 to 209.44 GW in 2024. This successful achievement was historic and remarkable not only because of mitigating climate change but also because it provided approximately 1.02 million jobs and employment opportunities. At present, India is in 4th position globally in the renewable energy sector, and holds the 4th and 5th positions in wind and solar power capacity, respectively (International Renewable Energy Agency, 2024) (Erdoğdu et al., 2025).

Moreover, at the national level, many programs and schemes are running to help attain national targets, foster sustainable energy for the future, and secure energy security. The main programs are as follows: National Green Hydrogen Mission, PM-KUSUM, Green Energy Corridor, and PLI schemes (Dubey et al., 2023). As an emerging global renewable power, India's electricity generation capacity has touched 452.69 GW as per government records of October 2024. Furthermore, renewable energy capacity contributed 46.3% of the total energy capacity, with solar power providing 90.76 GW, wind power close to 47.36 GW, hydropower contributing 46.92 GW (large) and 5.07 GW (small), and bio-power providing 11.32 GW.

Figure 1. Installed Renewable Energy Capacity (in MW) In India

On the other hand, green hydrogen is significant in the fight against climate change, particularly in heavy industries like steel, iron, cement, and chemicals, where large amounts of carbon emissions have resulted in global warming. Replacing fossil fuels with green hydrogen has decreased CO_2 emissions. At the global level, hydrogen-based energy is rapidly evolving. However, most countries used grey hydrogen, which had an impact on the global climate. After that, many countries adopted the blue economy and then shifted to green hydrogen. According to the International Renewable Energy Agency, approximately 25% of green hydrogen production will be traded at the world level by 2050 (Elegbeleye et al., 2025). The bar chart shows a steady rise from 7,548.11 in 2020–2021 to 28,723.65 in 2024–2025, reflecting nearly a fourfold increase over five years. Moderate annual growth was observed until 2023–2024, followed by a sharp jump of over 10,000 units in 2024–2025, indicating a significant surge in investment, production, or output.

Figure 2. Sector-wise contribution in India in GW

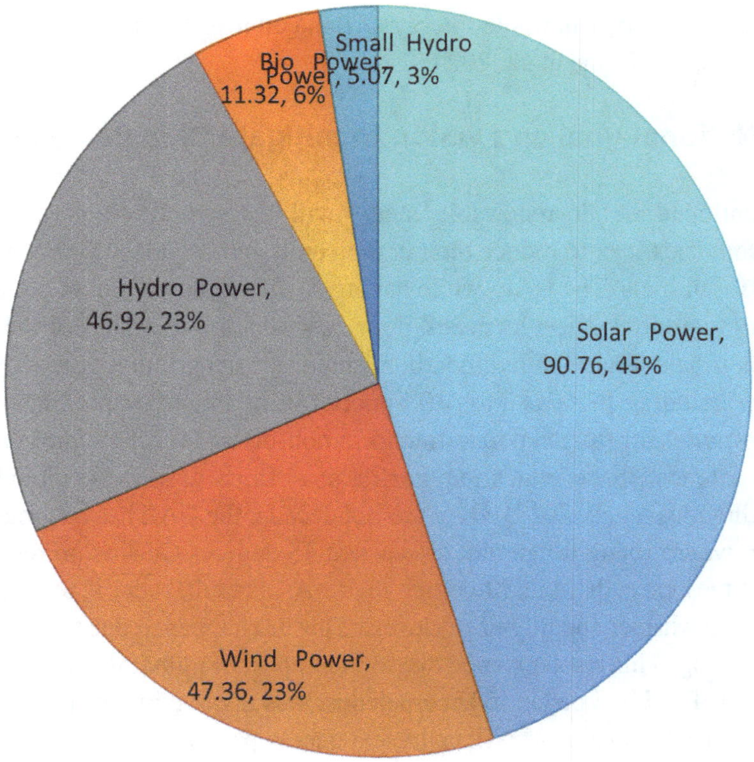

The pie chart displays the composition of renewable energy sources, with Solar Power accounting for the largest share at 45% (90.76 units). Wind Power and Hydro Power contribute equally at 23% each, with values of 47.36 and 46.92 units respectively. Bio Power makes up 6% (11.32 units), while Small Hydro Power contributes the least at 3% (5.07 units). This distribution indicates a strong reliance on solar energy within the renewable mix, followed by significant contributions from wind and hydro sources. The relatively smaller shares of bio and small hydro power suggest areas with potential for future expansion in the renewable energy sector. The Indian government initiated the National Green Hydrogen Mission on 4 January 2023, with an initial commercial investment of Rs 19,744 crore (Ministry of New and Renewable Energy, 2024). The expected outcome of NGHM is to produce a minimum of 5 million metric tonnes of green hydrogen per annum by 2030. To attain national targets and to become a global market leader in the renewable energy sector and green hydrogen, the Indian government has started certain initiatives and engagements for the promotion of clean energy with both developed and developing countries at bilateral and multilateral platforms. For instance, a Memorandum of Understanding was signed between the Ministry of Investment of the UAE and the Ministry of New and Renewable Energy of India on investment and cooperation in renewable energy (Raina et al., 2024).

UAE's National goal and vision to mitigate Climate change

As mentioned earlier, renewable energy will become the global trade market because countries seek to reduce their carbon emissions and lessen dependence on imported fossil fuels. The UAE, while having an abundant amount of oil, coal, and natural gas to not only meet its needs, demands, and developments of the country, also enables the spillover of hydrocarbons into trade exportation to more than 150 countries (including India as a top 10 importer of hydrocarbon products), providing opportunities for financial investments in non-oil trade sectors like agriculture, maritime defence, space, education, health, and efforts to mitigate global climate change (Ghoddusi et al., 2022). Over the last decade, the UAE has become the hottest place, where rising temperatures reached 45 degrees Celsius on regular days and sometimes exceeded the 50-degree mark. As a result, there is less rain, more droughts, and warmer and higher sea levels. This has happened due to the impact of global warming. This has happened due to the impact of global warming (Subraelu et al., 2022). The UAE began to become conscious about climate change in 2007, when it decided to develop a civilian nuclear energy program to generate electricity by reducing the carbon footprint.

Figure 3. Installed Renewable Energy Capacity (in MW) in UAE

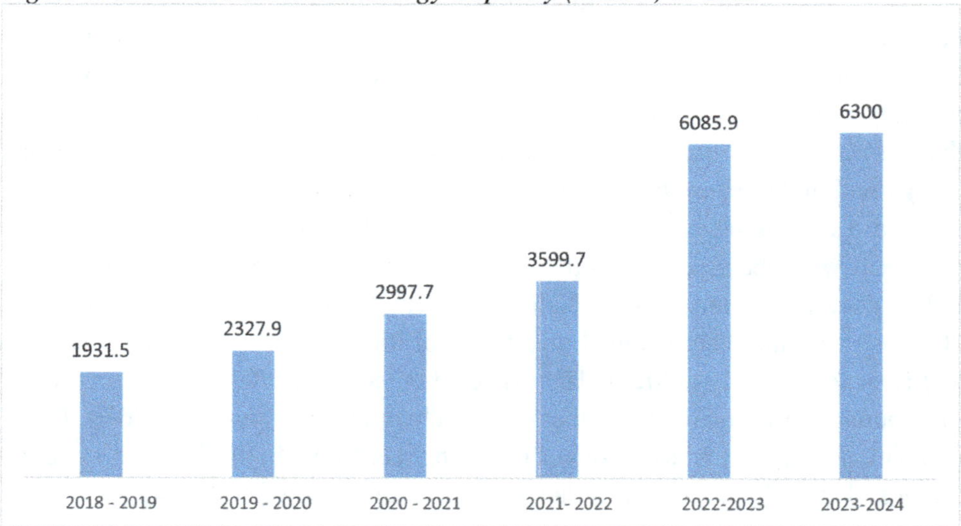

The bar chart illustrates a steady and significant rise in values from 2018–2019 to 2023–2024. Starting at 1,931.5 in 2018–2019, the value increases incrementally each year, reaching 2,327.9 in 2019–2020 and 2,997.7 in 2020–2021. A moderate rise continues in 2021–2022 to 3,599.7, followed by a sharp surge to 6,085.9 in 2022–2023. The upward trend peaks at 6,300 in 2023–2024. The data highlights a more than threefold growth over the six-year period, with the most substantial jump occurring between 2021–2022 and 2022–2023—an increase of over 2,400 units, likely signaling a major boost in investment, capacity, or output in that timeframe. Later on, the country gradually shifted towards clean energy and green hydrogen for achieving the targets of UN Sustainable Goals, and being a leading member of the International Solar Alliance (2016) ensures that the country is responsible and resilient against global climate change (Olabi et al., 2023). In addition to this, the UAE is interested in diversifying its energy sector at both regional and world levels. Therefore, the UAE is actively investing in clean energy to spread out its energy mix and reduce reliance on fossil fuels. According to the International Energy Agency, natural gas is the primary source of electricity generation in the UAE under the National Energy Strategy (Eveloy & Gebreegziabher, 2019).

Figure 4. Electricity generation contribution in UAE (2022)

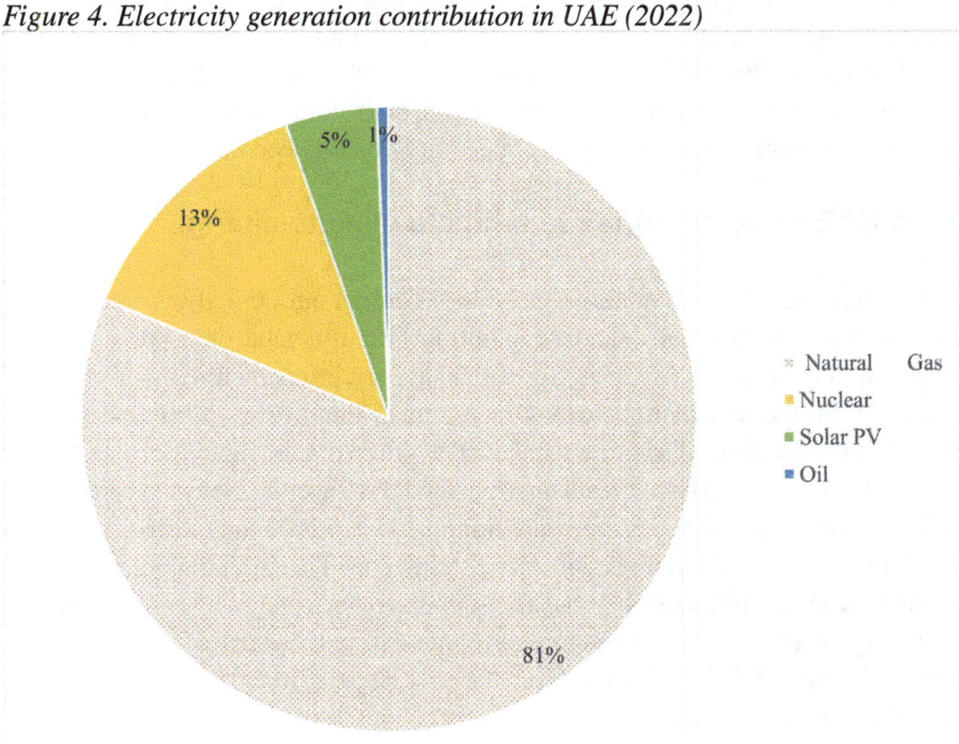

By 2050, the country set its target for alternative energy sources to meet economic requirements and environmental goals as follows – 44 percent alternative energy (like solar and wind), 38 percent natural gas, 12 percent clean coal, and 6 percent nuclear UAE Embassy in Washington. Furthermore, when the UAE hosted COP 28 in 2023, it announced an updated strategy which includes – job creation likely to be 50,000 by 2030, becoming zero-carbon emission by 2050, tripling renewable energy capacity to 14 GW by 2030, and raising the percentage of alternative energy in the total energy mix to 30 percent by 2031. In parallel to this, the UAE also envisions green hydrogen production under the National Hydrogen Strategy 2050, which strives to enhance its role as a low-emission hydrogen producer and supplier by Green Hydrogen Organisation. While the global hydrogen economy is concerned, the strategic planning is in a phased manner — firstly to reach 1.4 million tons per annum production by 2031, in the second phase it will be 7.5 mtpa by 2040, and the third phase target will be 15 mtpa by Green Hydrogen Organisation. Actually, the UAE's alternative energy production is helpful at the domestic and regional level by enabling decarbonisation across sectors. But speaking almost prophetically, these long-term national targets create internal employment and will make the country a global trade hub in the energy and hydrogen economy. For the lynchpin of the global trade, the UAE is also expanding collaborations with other countries on energy and green hydrogen projects such as Germany, the Netherlands, Norway, Denmark, and with India also. The UAE and India have signed landmark agreements aimed to be the global leaders in green hydrogen production; both aim to boost renewable energy capacity, moreover, undersea cable connectivity ties between them.

India-UAE joint synergies to mitigate climate change

India and the UAE's strategic partnership has grown substantially, encompassing defense, economic ties, and a shared commitment to sustainable development. Their cooperation spans renewable energy, green technology, infrastructure, environmental conservation, and contributions to the Green Climate Fund. Prime Minister Narendra Modi has visited the UAE six times since 2015—marking the first visit by an Indian PM in 34 years—while UAE President Mohamed Bin Zayed Al Nahyan visited India in 2017. These high-level engagements culminated in a 2022 joint vision statement on advancing green technology and energy transition. Their collaboration not only aligns with national climate goals but also supports other developing nations through knowledge sharing and infrastructure development. India has also deepened clean energy ties with GCC countries. Over the past decade, India and the UAE have signed several key agreements, including:

MoU on Renewable Energy Cooperation (2014)

General Framework Agreement on Renewable Energy Cooperation (2016)
Comprehensive Strategic Partnership (2017)
International Solar Alliance Framework Agreement (2017)
Comprehensive Economic Partnership Agreement (2022)
MoU on Climate Action (2022)
MoU on Green Hydrogen Development and Investment (2023)
MoU on Investment Cooperation in Renewable Energy (2024)
MoU on Electricity Interconnection and Trade (2024)

These agreements reflect strong potential for joint action on clean energy and climate change through a multi-faceted strategy. Both nations adhere to the UNFCCC and Paris Agreement (2015), aiming to meet their nationally determined contributions (NDCs). In line with global goals to limit temperature rise to well below 2°C, preferably 1.5°C, India and the UAE are proactively strengthening their alliance under the banner of 'technology for green development' to drive carbon emission reductions.

DISCUSSION

India and the UAE have significantly strengthened their partnership in green hydrogen development, especially in 2023, through cooperation in trade, production, storage, and clean energy investments. With South Asia's clean energy demand projected to rise by 35% by 2035 (IEA), this bilateral partnership is pivotal for low-carbon economic growth and climate resilience (Matar et al., 2023). At COP28, the UAE committed $30 billion through its Alterra fund, launching 6.6 GW of clean energy projects in India, including wind and solar power. Despite this progress, disappointment loomed at COP29 when developing nations, including India and the UAE, criticized the climate finance deal as inadequate—$300 billion versus the $1.3 trillion demanded—highlighting persistent inequities between Global North and South (Rao & Saad, 2025). India's NDC update targets a 45% emissions intensity reduction and 50% electricity from non-fossil fuels by 2030, while the UAE raised its GHG reduction goal to 47% by 2035. Both nations lead initiatives in renewable energy and green hydrogen, with India launching the National Green Hydrogen Mission (aiming for 5 MMT annual production by 2030) and scaling renewables to 209.44 GW by 2024, creating over a million jobs. Similarly, the UAE's Net-Zero Strategy aims for 14 GW renewable capacity and 15 MMT green hydrogen by 2050, while diversifying energy sources to reduce its carbon footprint (Hilali, 2025). Over a decade of bilateral agreements—including MoUs on renewable energy (2014–2024), CEPA (2022), and interconnection projects—have deepened cooperation, aligned with UNFCCC and Paris Agreement goals (Hurlbert & Akpan, 2023). The India-UAE green alliance is a model of Global South leadership, advancing clean energy, sustainable development, and global decarbonization through strategic, high-level collaboration (Suri et al., 2024).

CONCLUSION

India-UAE bilateral cooperation on climate change exemplifies a strategic partnership through agreements on renewable energy, green hydrogen, grid connectivity, climate action, trade, and investments by public-private players in the solar and wind sectors. Both countries are working on the reduction of greenhouse gases that have an impact at the global level through their own national goals and targets. To achieve its targets, India-UAE bilateral collaborative relations on renewable energy projects are crucial; therefore, both have a strategic partnership in many sectors like economy, defence, food, and climate change mitigation. Hence, India and the UAE have good compatibility to work together on policy reforms, incentivise investments, and enhance infrastructure to become global traders in green hydrogen and renewable energy. There are several challenges, such as the essential role of policy reforms and frameworks on carbon trade, technological hurdles, and the need for more finance and investment in transportation networks, pipelines, ports, and storage facilities for green hydrogen. All these obstacles will eventually be reduced through environmental diplomacy, technology transfer, incentivised investments, and dialogue processes to increase trade in renewable energy and green hydrogen on a large scale at the global level for the future.

ABBREVIATIONS

AED - Automated External Defibrillator
AI - Artificial Intelligence
CBAM - Carbon Border Adjustment Mechanism
CEPA - Comprehensive Economic Partnership Agreement
CO2 - Carbon-dioxide
COP - Conference of the Parties
CCS - Carbon capture and storage
FY - Financial Year
GHG - Greenhouse gas
GCC - Gulf Cooperation Council
GST - Global Stocktake

GW - Giga Watt
G20 - Group of Twenty
IEA - International Energy Agency
IPCC - Intergovernmental Panel on Climate Change
ISA - International Solar Alliance
MEA - Ministry of External Affairs
MoU - Memorandum of Understanding
MMT - Million metric ton
MW - Mega Watt
NDC - Nationally Determined Contributions
NGHM - National Green Hydrogen Mission
PLI - Production Linked Incentives
PM - Prime Minister
PM-KUSUM - Pradhan Mantri Kisan Urja Suraksha Evam Utthan Mahabhiyan
SDG - Sustainable Development Goals
UAE - United Arab Emirates
UFI - UAE-France-India trilateral
UNFCCC - United Nations Franework Convention on Climate Change
UN-SDG - United Nation- Sustainable Development Goals
US - United Nations

REFERENCES

Abdallah, I., Alhosin, H., Belarabi, M., Chaouki, S., Mahmoud, N., & Tayah, J. (2024). A Pan-Asian energy transition? The new rationale for decarbonization policies in the world's largest energy exporting countries: A case study of Qatar and other GCC countries. *Energies*, *17*(15), 3776. DOI: 10.3390/en17153776

Abdin, Z. (2024). Empowering the hydrogen economy: The transformative potential of blockchain technology. *Renewable & Sustainable Energy Reviews*, *200*, 114572. DOI: 10.1016/j.rser.2024.114572

Akomea-Frimpong, I., Jin, X., & Osei-Kyei, R. (2022). Managing financial risks to improve financial success of public—private partnership projects: A theoretical framework. *Journal of Facilities Management*, *20*(5), 629–651. DOI: 10.1108/JFM-03-2021-0036

Al-Sarihi, A., & Mason, M. (2020). Challenges and opportunities for climate policy integration in oil-producing countries: The case of the UAE and Oman. *Climate Policy*, *20*(10), 1226–1241. DOI: 10.1080/14693062.2020.1781036

Alkhaldi, M., Moonesar, I. A., Issa, S. T., Ghach, W., Okasha, A., Albada, M., Chelli, S., & Takshe, A. A. (2023). Analysis of the United Arab Emirates' contribution to the sustainable development goals with a focus on global health and climate change. *International Journal of Health Governance*, *28*(4), 357–367. 357-367. DOI: 10.1108/IJHG-04-2023-0040

Amin, M., Shah, H. H., Fareed, A. G., Khan, W. U., Chung, E., Zia, A., Rahman Farooqi, Z. U., & Lee, C. (2022). Hydrogen production through renewable and non-renewable energy processes and their impact on climate change. *International Journal of Hydrogen Energy*, *47*(77), 33112–33134. DOI: 10.1016/j.ijhydene.2022.07.172

Anika, O. C., Nnabuife, S. G., Bello, A., Okoroafor, E. R., Kuang, B., & Villa, R. (2022). Prospects of low and zero-carbon renewable fuels in 1.5-degree net zero emission actualisation by 2050: A critical review. *Carbon Capture Science & Technology*, *5*, 100072. DOI: 10.1016/j.ccst.2022.100072

Ashwarya, S. (2023). India's national role conception and relations with GCC countries under Modi: A focus on Saudi Arabia. *Journal of Asian and African Studies*, *58*(4), 535–555. DOI: 10.1177/00219096231162104

Baomar, A. A. H. A. (2024). *The GCC'S Complex Interdependence with China and its Implications on Western Alliances*. Sultan Qaboos University.

Bhat, G. M. (2025). "Drill, Baby Drill" for Oil and Gas: COP29 New Climate Finance. *Journal of the Geological Society of India*, *101*(3), 287–290. DOI: 10.17491/jgsi/2025/174094

Biyik, C., Allam, Z., Pieri, G., Moroni, D., O'fraifer, M., O'connell, E., Olariu, S., & Khalid, M. (2021). Smart parking systems: Reviewing the literature, architecture and ways forward. *Smart Cities*, *4*(2), 623–642. DOI: 10.3390/smartcities4020032

Chang, W. C. (2025). A grand plan for health equity. *International Journal for Equity in Health*, *24*(1), 1–15. DOI: 10.1186/s12939-025-02536-w PMID: 40542351

Dagher, L., Heeter, J., & Mansour, M. H. (2020). What can GCC countries learn from well-established green power markets in other countries? In *Gulf Conference on Sustainable Built Environment* (pp. 309-327). Cham: Springer International Publishing. https://doi.org/DOI: 10.1007/978-3-030-39734-0_19

Diaz, S., Al Hammadi, N., El Nasr, A. S., Villasuso, F., Prakash, S., Baobaid, O., Gracias, D., & Mills, R. (2023, October). Green Corridor: A Feasible Option for the UAE Decarbonization Pathway, Opportunities & Challenges. In *Abu Dhabi International Petroleum Exhibition and Conference* (p. D011S017R003). SPE. DOI: 10.2118/216033-MS

Dsilva, J., Shaikh, M. I., & Usmani, F. Z. A. (2024). Net Zero Policies and Climate Resilience: A Focus on Emerging Technologies and Implementation Strategies in the Middle East. *Transition Towards a Sustainable Future: Net Zero Policies and Environmental Sustainability*, 119-138. DOI: 10.1007/978-981-97-5756-5_6

Dubey, B., Agrawal, S., & Sharma, A. K. (2023). India's renewable energy portfolio: An investigation of the untapped potential of RE, policies, and incentives favoring energy security in the country. *Energies*, *16*(14), 5491. DOI: 10.3390/en16145491

Dutta, A. (2021). Unpacking the India-EU economic relationship. *A 2030 Vision for India's Economic Diplomacy*, 28. https://www.orfonline.org/public/uploads/posts/pdf/20230424134334.pdf#page=28

Elegbeleye, I., Oguntona, O., & Elegbeleye, F. (2025). Green Hydrogen: Pathway to Net Zero Green House Gas Emission and Global Climate Change Mitigation. *Hydrogen*, *6*(2), 29. DOI: 10.3390/hydrogen6020029

Erdoğdu, A., Dayi, F., Yanik, A., Yildiz, F., & Ganji, F. (2025). Innovative solutions for combating climate change: Advancing sustainable energy and consumption practices for a greener future. *Sustainability (Basel)*, *17*(6), 2697. DOI: 10.3390/su17062697

Eveloy, V., & Gebreegziabher, T. (2019). Excess electricity and power-to-gas storage potential in the future renewable-based power generation sector in the United Arab Emirates. *Energy, 166*, 426–450. DOI: 10.1016/j.energy.2018.10.088

Gao, Y., Gao, X., & Zhang, X. (2017). The 2 C global temperature target and the evolution of the long-term goal of addressing climate change—From the United Nations framework convention on climate change to the Paris agreement. *Engineering (Beijing), 3*(2), 272–278. DOI: 10.1016/J.ENG.2017.01.022

Ghoddusi, H., Moghaddam, H., & Wirl, F. (2022). Going downstream–an economical option for oil and gas exporting countries? *Energy Policy*, *161*, 112487. DOI: 10.1016/j.enpol.2021.112487

Hartmann, J., Inkpen, A. C., & Ramaswamy, K. (2021). Different shades of green: Global oil and gas companies and renewable energy. *Journal of International Business Studies*, *52*(5), 879–903. DOI: 10.1057/s41267-020-00326-w

Henderson, J. (2021). COP 26—Examining the balance between ambitious pledges and realistic expectations. In *Oxford Energy Forum: A Quarterly Journal for Debating Energy Issues and Policies* (No. 129). Oxford: The Oxford Institute for Energy Studies. https://www.econstor.eu/handle/10419/306991

Hilali, I. (2025). Accelerating the transition to a hydrogen economy: Achieving carbon neutrality. In *Accelerating the Transition to a Hydrogen Economy* (pp. 105–127). Elsevier., DOI: 10.1016/B978-0-443-24002-7.00009-4

Horri, B. A., & Ozcan, H. (2024). Green hydrogen production by water electrolysis: Current status and challenges. *Current Opinion in Green and Sustainable Chemistry, 47*, 100932. DOI: 10.1016/j.cogsc.2024.100932

Hurlbert, M. A., & Akpan, J. (2023). Dialectic narratives, hostile actors, and Earth's resources in Saskatchewan, Canada. *Sustainability Science*, *18*(1), 285–301. DOI: 10.1007/s11625-022-01214-y PMID: 36157091

Ibrahim, A. (2024). Transitioning Emerging Economies Toward Sustainability by Implementing Targeted Policy Interventions. *ESTIDAMAA, 2024*, 53-59. https://peninsula-press.ae/Journals/index.php/ESTIDAMAA/article/view/130

Jakhar, T. C. A. T. R. (2024). Beyond Borders: CBAM's Revolutionary Potential and Challenges in Achieving Carbon Neutrality. *policy, 3*, 4.

Jha, V. (2021). 'Soft Law in a Hard Shell': India, International Rulemaking and the International Solar Alliance. *Transnational Environmental Law*, *10*(3), 517–541. DOI: 10.1017/S2047102520000400

Jogdand, O. K. (2020). Study on the effect of global warming and greenhouse gases on environmental system. In *Green chemistry and sustainable technology* (pp. 275–306). Apple Academic Press. DOI: 10.1201/9780367808310-12

Joyce, C., & Stanford, J. (2023). Manufacturing the Energy Revolution. *Centre for Future Work, August,* https://futurework. org. *au/report/manufacturing-the-energy-revolution.*

Khan, M., Mfitumukiza, D., & Huq, S. (2021). Capacity building for implementation of nationally determined contributions under the Paris Agreement. In *Making climate action more effective* (pp. 95–106). Routledge., https://www.taylorfrancis.com/chapters/edit/10.4324/9781003162407-8/capacity-building-implementation-nationally-determined-contributions-paris-agreement-mizan-khan-david-mfitumukiza-saleemul-huq DOI: 10.4324/9781003162407-8

Khan, M. I., & Al-Ghamdi, S. G. (2023). Hydrogen economy for sustainable development in GCC countries: A SWOT analysis considering current situation, challenges, and prospects. *International Journal of Hydrogen Energy, 48*(28), 10315–10344. DOI: 10.1016/j.ijhydene.2022.12.033

Krzymowski, A. (2020). The European Union and the United Arab Emirates as civilian and soft powers engaged in Sustainable Development Goals. *Journal of International Students, 13*(3), 41–58. https://www.ceeol.com/search/article-detail?id=979401. DOI: 10.14254/2071-8330.2020/13-3/3

Leal-Arcas, R. (2024). Global Energy Transition: Reflecting on COP 28 and Case Studies of Qatar and Morocco. *Forthcoming in Florida Journal of International Law, 35*(1).

Leung, T. C. H., & You, C. S. X. (2023). Climate change in Hong Kong with the integration of UN SDG 13 climate action. In *SDGs in the Asia and Pacific Region* (pp. 1–26). Springer International Publishing., DOI: 10.1007/978-3-030-91262-8_18-1

Makarov, I., Chen, H., & Paltsev, S. (2020). Impacts of climate change policies worldwide on the Russian economy. *Climate Policy, 20*(10), 1242-1256. ttps://DOI: 10.1080/14693062.2020.1781047

Mallah, A. Q., Gayen, M. H., & Basu, P. P. (2024). India's Environmental Diplomacy: In Pursuit of a Safer Planet. In *75 Years of India's Foreign Policy: Bilateral, Conventional and Emerging Trends* (pp. 463-483). Singapore: Springer Nature Singapore. DOI: 10.1007/978-981-97-6054-1_20

Manohar, H. (2024). Exploring the nexus of renewable energy and international investment law in India: Opportunities and issues. *Jindal Global Law Review, 15*(2), 635–670. DOI: 10.1007/s41020-024-00239-x

Matar, A., Fareed, Z., Magazzino, C., Al-Rdaydeh, M., & Schneider, N. (2023). Assessing the co-movements between electricity use and carbon emissions in the GCC area: Evidence from a wavelet coherence method. *Environmental Modeling and Assessment, 28*(3), 407–428. DOI: 10.1007/s10666-022-09871-0

Matz, C. A. (2018). State-sponsored sustainability within the Emirate of Abu Dhabi. *The Palgrave Handbook of Sustainability: Case Studies and Practical Solutions*, 803-833. DOI: 10.1007/978-3-319-71389-2_44

Odeyemi, O., Usman, F. O., Mhlongo, N. Z., Elufioye, O. A., & Ike, C. U. (2024). Sustainable entrepreneurship: A review of green business practices and environmental impact. *World Journal of Advanced Research and Reviews, 21*(2), 346–358. DOI: 10.30574/wjarr.2024.21.2.0461

Olabi, A. G., Abdelkareem, M. A., Mahmoud, M. S., Elsaid, K., Obaideen, K., Rezk, H., Wilberforce, T., Eisa, T., Chae, K., & Sayed, E. T. (2023). Green hydrogen: Pathways, roadmap, and role in achieving sustainable development goals. *Process Safety and Environmental Protection, 177*, 664–687. DOI: 10.1016/j.psep.2023.06.069

Olabi, V., & Jouhara, H. (2024). An assessment of current hydrogen supply chains in the Gulf Cooperation Council (GCC). *Energy, 131576*, 131576. Advance online publication. DOI: 10.1016/j.energy.2024.131576

Oshilalu, A. Z. (2024). Sustainability meets scalability: Transforming energy infrastructure projects into economic catalysts through supply chain innovation. *International Journal of Research Publication and Reviews, 5*(12), 762–779. DOI: 10.55248/gengpi.5.1224.3417

Pradhan, S. (2010). India and the Gulf Cooperation Council (GCC): An economic and political perspective. *Strategic Analysis, 34*(1), 93–103. DOI: 10.1080/09700160903351340

Prasad, S., Yadav, K. K., Kumar, S., Pandita, P., Bhutto, J. K., Alreshidi, M. A., Ravindran, B., Yaseen, Z. M., Osman, S. M., & Cabral-Pinto, M. M. (2024). Review on biofuel production: Sustainable development scenario, environment, and climate change perspectives– A sustainable approach. *Journal of Environmental Chemical Engineering, 12*(2), 111996. DOI: 10.1016/j.jece.2024.111996

Quamar, M. M., Hameed, S., & Kumaraswamy, P. R. (2023). United Arab Emirates. In *Persian Gulf 2023: India's Relations with the Region* (pp. 335-382). Singapore: Springer Nature Singapore. https://doi.org/DOI: 10.1007/978-981-99-6380-5_9

Raina, S., Shandilya, S., Kothari, R., & Pathania, D. (2024). *Green Hydrogen: International Collaborators, Agreements, Opportunities, and Challenges, and Their Impact on India's Research and Development Sector. Green Hydrogen Economy for Environmental Sustainability* (Vol. 2). Applications, Challenges, and Policies., DOI: 10.1021/bk-2024-1474.ch014

Ramesh, S. (2025). Renewable Energy Technologies. In *The Political Economy of Contemporary Human Civilisation, Volume II: From Quantum Computing and Nuclear Fusion to War and Conflict* (pp. 183-218). Cham: Springer Nature Switzerland. DOI: 10.1007/978-3-031-84185-9_6

Rao, U., & Saad, M. (2025). Economic and financial aspects of renewable energy environment. In *Renewable Energy Projects and Investments* (pp. 243–258). Elsevier., DOI: 10.1016/B978-0-443-29869-1.00013-1

Saber, A. (2024). *Sovereign wealth funds in a post-pandemic landscape: a uk perspective* (Doctoral dissertation, University of Sussex). https://www.researchgate.net/profile/Ali-Saber-17/publication/384635695_Sovereign_wealth_funds_in_a_post-pandemic_landscape_a_UK_perspective/links/67006342869f1104c6c65638/Sovereign-wealth-funds-in-a-post-pandemic-landscape-a-UK-perspective.pdf

Sambasivam, B., & Sarma, R. N. (2024). India's National Green Hydrogen Mission: An analysis of the strategies, policies for net-zero emissions and sustainability. *Environmental Research: Energy*, *1*(4), 045015. DOI: 10.1088/2753-3751/ad8fb2

Sattar, U. (2022). A conceptual framework of climate action needs of the least developed party countries of the Paris agreement. *International Journal of Environmental Research and Public Health*, *19*(16), 9941. DOI: 10.3390/ijerph19169941 PMID: 36011575

Sebbagh, T., Şahin, M. E., & Beldjaatit, C. (2024). Green hydrogen revolution for a sustainable energy future. *Clean Technologies and Environmental Policy*, *26*(12), 4017–4040. DOI: 10.1007/s10098-024-02995-9

Selvakumar, P., Seenivasan, S., Patel, S., Das, A., & Ravikumar, S. (2025). Environmental Sustainability. In *Renewable Energy and the Economic Welfare of Society* (pp. 103–132). IGI Global Scientific Publishing., DOI: 10.4018/979-8-3693-7580-8.ch005

Shah, G. M., Sapkota, K., & Spakota, M. (2025). *A Comprehensive Review of Theories, Models and Approaches Governing Climate Change Discourse and Global Climate Finance*. Models and Approaches Governing Climate Change Discourse and Global Climate Finance. DOI: 10.2139/ssrn.5287468

Shahnawaz, M., & Mir, R. Y. (2025). Strategic Convergence in the Arabian Gulf: Evaluating India-UAE Partnership Post 2014. [IJSSL]. *Indian Journal of Social Science and Literature*, *4*(4), 19–24. DOI: 10.54105/ijssl.C1166.04040625

Shahzad, S., Faheem, M., Muqeet, H. A., & Waseem, M. (2024). Charting the UK's path to net zero emissions by 2050: Challenges, strategies, and future directions. *IET Smart Grid*, *7*(6), 716–736. DOI: 10.1049/stg2.12185

Shee Weng, L. (2025). *The EU's Carbon Border Adjustment Mechanism (CBAM): Balancing Climate Action and Global Trade. The EU's Carbon Border Adjustment Mechanism (CBAM)*. Balancing Climate Action and Global Trade.

Shibata, Y., Nian, V., Bhandari, A., & Roychoudhury, J. (2024). Using hydrogen for decarbonization, industrial development, and energy security: Shared ambitions for Japan, ASEAN Member States, and India. In *The Clean Hydrogen Economy and Saudi Arabia* (pp. 329–373). Routledge., https://www.taylorfrancis.com/chapters/oa-edit/10.4324/9781003294290-14/using-hydrogen-decarbonization-industrial-development-energy-security-yoshiaki-shibata-victor-nian-amit-bhandari-jitendra-roychoudhury DOI: 10.4324/9781003294290-14

Sim, L. C. (2024). The Arab Gulf states in the Asian energy market: Is the Russia-Ukraine war a game changer? *Policy Studies*, *45*(3-4), 633–652. DOI: 10.1080/01442872.2023.2279986

Siyech, M. S. (2020). Understanding India's increased counter terrorism relations with Saudi Arabia and the UAE. *India Review*, *19*(4), 351–375. DOI: 10.1080/14736489.2020.1797316

Sreekanth, K. J. (2024). Energy Policy Analyses, Energy Transition and Sustainability. In *Energy Efficiency Improvements with Emission Abatement for Energy Sustainability: Energy Statistics* (pp. 135-167). Singapore: Springer Nature Singapore. DOI: 10.1007/978-981-97-9480-5_8

Stern, N., & Lankes, H. P. (2022). *Collaborating and delivering on climate action through a climate club*. Grantham Research Institute on Climate Change and the Environment, London School of Economics & Political Science.

Subraelu, P., Sefelnasr, A., Yagoub, M. M., Sherif, M., Ebraheem, A. A., Raj Sekhar, A., & Nageswara Rao, K. (2022). Global warming climate change and sea level rise: Impact on land use land cover features along uae coast through remote sensing and gis. *Journal of Ecosystem & Ecography*, *12*, 1–8. DOI: 10.4172/2157-7625.1000329

Suri, N., Ghosh, N., Taneja, K., Patil, S., & Mookherjee, P. (2024). India-Middle East-Europe Economic Corridor: Towards a New Discourse in Global Connectivity. *Observer Research Foundation, April, 9*. https://www.orfonline.org/research/india-middle-east-europe-economic-corridor-towards-a-new-discourse-in-global-connectivity?trk=public_post_comment-text

Swagatika Priyadarshini, L. A., Kaushik, V., & Kataria, R. (2025). The Role of Anthropogenic Activities, Energy Demand, and Global Climate Change. In *The Intersection of Global Energy Politics and Climate Change: A Comprehensive Analysis of Energy Markets and Economics* (pp. 35–61). Springer Nature Singapore., DOI: 10.1007/978-981-96-0535-4_2

Thompson, A. (2025). From rivals to partners: The cooptation of emerging powers into the climate regime. *Contemporary Security Policy*, *46*(1), 126–149. DOI: 10.1080/13523260.2024.2420477

Tibni, H. (2024). Assessing the Effectiveness of the 10 United Nations Global and Multilateral Climate Treaties. *European Journal of Development Studies*, *4*(4), 41–55. DOI: 10.24018/ejdevelop.2024.4.4.366

Tyson, L. D., & Weiss, D. (2025). Climate Finance: Progress, Challenges, and Opportunities on the Path to a Sustainable Planet. *California Management Review*, *67*(2), 5–39. DOI: 10.1177/00081256241296902

Verma, P., Sharma, B., & Mehta, P. (2024). Green Hydrogen Production: Impacts on Environmental Sustainability Scale. In *Green Hydrogen Economy for Environmental Sustainability. Volume 1: Fundamentals and Feedstocks* (pp. 29-48). American Chemical Society. DOI: DOI: 10.1021/bk-2024-1473.ch002

Voyer, M., Schofield, C., Azmi, K., Warner, R., McIlgorm, A., & Quirk, G. (2018). Maritime security and the Blue Economy: Intersections and interdependencies in the Indian Ocean. *Journal of the Indian Ocean Region*, *14*(1), 28–48. DOI: 10.1080/19480881.2018.1418155

Yang, B. (2025). Current Development and Practice: The Economics of Climate Change. In *The Carbon Emission Liability Mechanism: A New Solution for Global Carbon Neutrality* (pp. 45-114). Singapore: Springer Nature Singapore. https://doi.org/DOI: 10.1007/978-981-96-3612-9_3

Zaiter, I. (2024). *Optimizing the UAE's Energy Sector through Hydrogen Integration for Cost-Effectiveness and Net-Zero Carbon Emissions* (Doctoral dissertation, Khalifa University of Science).

Chapter 11
Integrated Cross-Border Renewable Energy Infrastructure for a Green Transition

R. N. Ravikumar
https://orcid.org/0009-0009-3705-1681
Marwadi University, Rajkot, India

S. Aarthi
https://orcid.org/0009-0006-9064-2091
Marwadi University, Rajkot, India

ABSTRACT

This research explores the critical role of cross-border renewable energy integration in facilitating a sustainable and just green transition. It highlights how regional collaboration across legal, technical, and policy domains enhances energy security, grid stability, and climate resilience. The manuscript evaluates successful models like the North Sea Wind Power Hub and India-Bhutan Hydropower Cooperation, illustrating the benefits of shared infrastructure and joint governance. Emphasis is placed on harmonized regulations, stakeholder inclusion, smart grid technologies, and robust environmental safeguards. The work concludes that cross-border systems are not only technologically feasible but geopolitically essential for achieving net-zero goals, economic integration, and equitable energy access.

DOI: 10.4018/979-8-3373-3972-6.ch011

1. INTRODUCTION TO CROSS-BORDER ENERGY COLLABORATION

As the world shifts the gears in combating the altering climate, the transition to renewable energy has become an immediate agendum. However, since the effects of climate have trans-boundaries, the solutions must also be extra-national. The article dwells on the problem of cross-border renewable energy infrastructure, which is one of the strategic solutions to green transition. Cross border energy collaboration involves co-operation, arrangement of power systems among surface neighbours, thus approving of the sharing of resources, increased stability of energy resources, and de-carbonization of schemes within the area. The presence of high imbalances in the supply of renewable energy resources such as the potential of the offshore wind in the North Sea or in the potential of hydro in the Himalayas renders cross border integration a requirement in the delivery of a fair and effective access to energy (Balsalobre-Lorente et al., 2018). Such a stakeholder partnership also helps minimise overlaps, minimise spending, and promotes energy diplomacy that cuts the geopolitical level and economic integration perpendicularly and horizontally. However, technology is not the only condition of such transition but also coordinated laws, sources of funding, representative governments, and institutional support. In this Research work, the author discussed policy, technical, and social components of transnational renewable energy infrastructure with reference to the successful examples of the same cases worldwide. It observes that not only is it possible, but also necessary to create a resilient, low-carbon future, and a sustainable and equitable energy transition to everyone, through cross-border systems.

1.1 Background Study

International integration of renewable energy sources has been pegged to the growing awareness that global warming is a global problem and must be dealt with congruently on an international scale. Previously energy systems could be formed on the borders of the country and they have relied considerably on the fossil structure and infrastructures is centralized (Nag, 2021). The disadvantages of such solution, though, in the aspect of resource incompetence, grid instability, and geopolitical risks, have brought the requirement of more specific, interconnected energy system. Multiple international accords, including the Paris Agreement and Sustainable Development Goals (SDGs), have indicated the importance of renewable energy in relation to the process of introducing net-zero goals and climate resilience. Meanwhile, the regional blocs such as the European Union (EU), SAARC, and the ASEAN have similarly implemented their own structures and technical agreements in order to allow trading of energy across borders and construction of infrastructural facilities.

The recent trends in technology sector such as the high-capacity transmission lines, smart grids and energy storage system have enabled the cross-border integration to become even more achievable. Moreover, it has been further enhanced by economic interdependence and reliance and joint feebleness of the environmentally to encourage the convergence of resources and alignment of strategies in planning energy by nations (N. Naval & J.M Yusta, 2024). Against this, there are still immense obstacles with regard to harmonization of regulation, finance, involvement of stakeholders and protection of the environment. The Research work uses the above context to discuss policies, technologies, institutional mechanisms, and real models describing the increasing perceptual and practical opportunities surrounding the hybridization of cross-border renewable energy cooperation.

1.2. Literature Review

Transboundary renewable energy integration has become one of the fundamental elements of transforming the world into sustainable energy systems. The literature and policy makers concur that interconnections across countries can enhance the utilization of renewable sources through transfer of power across nations who have surplus and those who have deficit hence minimising the cost of the entire system as well as the levels of emissions. Initial researches focused on economic and technical advantages of regional connectivity (Axon et al., 2017). TEN-E (Trans-European Networks for Energy) produced by the European Union is an example, which demonstrated the benefits of joint infrastructure planning as well as integration of markets, to make electricity flow, and stabilise cross-border grids. Equally, a study conducted on this by Zetland on Nordic electricity market showed that the synergy effect of collaborative planning and mutual infrastructure minimized costs of operation on such projects and made optimum use of hydropower and wind power generation (Poque González et al., 2023). The latest literature has further accommodated institutional, regulatory and geopolitical factors.

The harmonization of grid codes, equal treatment of transmission pricing as well as issues related to ownership of sovereignty have been singled out to be vital to success (Wu et al., 2015). The ASEAN Power Grid and the West African Power Pool are examples of projects that give guidance regarding the possibilities of energy cooperation across countries, along with many of the issues that must be considered such as a sense of trust, harmonisation of investment processes, and a more inclusive approach to stakeholders. Technologically there is an increase in the development of smart grids, an HVDC (High Voltage Direct Current) transmission line, and regional SCADA (Supervisory Control and Data Acquisition) systems which have enhanced the possibility of cross-border integration (Nag, 2021). Research has depicted that the use of digital mediums and real-time surveillance is a very important way of enhancing fault capture, power dispatch, and flexible grid. In addition, scholars have started discussing the social and environmental aspects of such initiatives. The impact analysis is usually done to show that a major infrastructure project can invoke land-use conflicts, indigenous people, or ecosystem (Nasr et al., 2020). The literature puts greater emphasis on the subject of sustainability and community-based planning in cross border renewable energy projects. Mechanisms such as climate policies and governance structures, including the multilateral treaties, green bonds, as well as carbon markets are considered to be crucial in mitigating risks and coordinating businesses with the national net-zero efforts. Generally, effective cross-border integration has been noted on the aspect of multi-dimensionality in terms of technical, economic, political, and environmental coordination (Ekechukwu & Simpa, 2024). The upcoming studies will put more emphasis on evaluating resilience, adaptive governance, and optimising systems through AI and digital twins.

1.3. The Urgency of Green Transition in a Climate-Constrained World

With the increasing urgency of climate change, driven by the rising global temperatures, occurrence of extreme weather phenomena, and other forms of destruction of ecosystems, a green transition is urgently needed. Countries all over the world are under pressure to decarbonize their economy and honour their promises under international obligations such as the Paris Accord. The green transition that is based on the adoption of renewable energy, energy efficiency, as well as sustainable development, is now not a choice but rather a requirement. The energy systems based on fossil fuels are in a way not sustainable, harmful to the environment, and to say the least, they are even geopolitically dangerous. The most rapid transition to the use of renewable energy sources, i.e., solar, wind, hydro and geothermal energy, is substantial to reduce the risks of climate change, conserve the biodiversity, and give people sustainable living conditions (Ogino et al., 2019). This transformation should though be comprehensive, powerful and consistent in the world. It needs more than a technological change, it needs a systemic change it will entail policy change, infrastructure change, finance change, and behavior change. Any latent reaction can put the sustainability of the environment and social-economic future in jeopardy.

1.4. The Role of Cross-Border Energy Links in Renewable Energy Expansion

The main aspect of incorporating renewable energy systems into scale with efficiency and equity involves cross border integration. The renewable energy sources are not evenly distributed in the countries and hence some sources of wind energy (the North Sea), solar energy (North Africa) or even hydro energy (the Himalayas). Without the coordination between countries, not only can some countries not enjoy the entire benefits of the energy potential that nature has to offer, but it is also not able to ensure a stable supply of the energy (Pu et al., 2021a). The interconnections and harmonizations between energy grids can facilitate the sharing of the surplus energy generation, minimize discrepancies in renewable energy generation and to stabilize the grid. The cross-border projects reduce the redundancies and the cost of investments is reduced as a result of the economy of scale and shared infrastructure. Integration fosters energy diplomacy, which encourages political cooperation, regional and shared economic growth. Moreover, ambitious decarbonization strategies are already being sought by the nations; this situation is an opportunity to seek a win-win situation by using transnational structures to organize climate work (Utelbayev et al., 2024). Therefore, cross-border integration is not only a technical project but also a strategic and a geopolitical project to have a just, inclusive, and fast energy transition. The figure 1 underscores the major advantages of the Cross-Border Integration in renewable energy. It increases effectiveness with the help of Energy Efficiency, Grid Stability, Cost Reduction, Political Collaboration, and Climate Coordination.

Figure 1. Key Benefits of Cross-Border Renewable Energy Integration

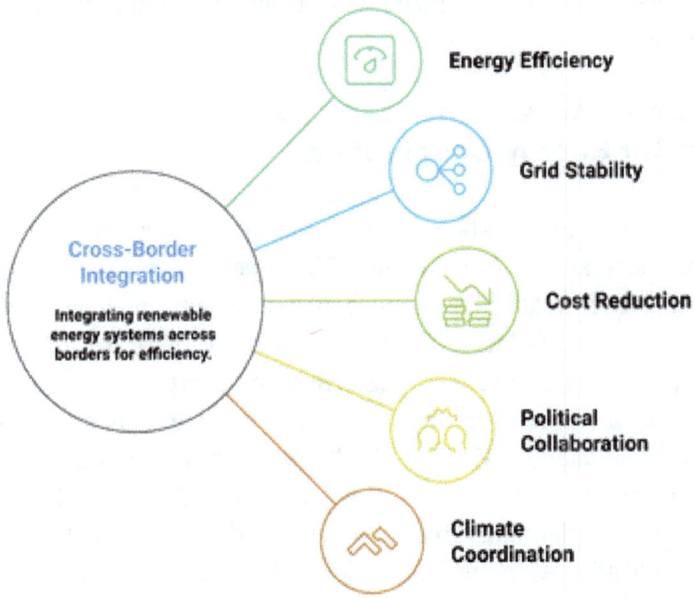

2. POLICY AND LEGAL FRAMEWORKS FOR CROSS-BORDER RENEWABLE PROJECTS

The establishment of a commendable policy and law support structure is an ingredient of the success of cross border renewable-energy projects. These frameworks provide guidelines, commitments and incentives of the interested participating nations and stakeholders who are guaranteed of cooperation, equity and sustainability. The harmonization of the regulatory standards introduced by the grid codes, tariff structure and environmental standards will be needed at the national level in order that the exchange of energy and implementation of project can take place with similarity. The case of regional measure, e.g. the creation of a regional policy, such as the example of European Union or SAARC Energy Charter, also encompasses the case of coordinated planning, permitting and compliance process (Ballo et al., 2022). The legal framework also assumes a crucial role in establishment of forms of ownership, Power Purchase Agreements (PPAs), protection of investments and their dispute resolution models. Joint ventures, funding modes and technology sharing are usually established on the basis of treaties and bilateral or multilateral agreements. Notably, policies should embrace the principle of climate justice to ensure that the benefits of energy can be distributed equally to all people and people should be meaningfully engaged in this matter. To conclude, the cross-border renewable cooperation can be optimized with strong legal and policy coordination as its requirement.

2.1. International Agreements and Legal Instruments for Energy Cooperation

The basis of governing cross-border renewable energy cooperation lies in the international agreements and legal instruments. These mechanisms make commitments between the countries formalised, provide regulatory clarity and legal predictability to the investors and stakeholders. The most prominent ones are Energy Charter Treaty (ECT), providing a multilateral agreement on energy trade and investment and Paris Agreement where everyone is invited to take up collective actions and pledges in renewable energy (Meus et al., 2019). Technical coordination, market integration and infrastructure planning Cooperation through technical coordination, market integration and planning through bilateral and regional treaties such as the Nordic Electricity Market Agreement or the South Asia Regional Initiative on Energy Integration (SARI/EI) can be very useful as well. Such contracts usually contain the clauses on the common possession of the projects and the unification of tariffs, synchronization of the grids, as well as the resolution of the disputes. They also encourage exchange of technological and ecological standards. Legal tools should also be dynamic, to accommodate changing energy demands, and consistent, to avoid the occurrence of conflicts and to guarantee compliance. They play a vital part in creating trust, openness, and continuity within long-term inter-border energy projects.

2.2. Financing Mechanisms and Regulatory Harmonization

Harmonization financing and regulation is an important facilitator of cross border renewable energy infrastructure. Multi-source financing usually involves the combination of money provided by governments, the private sector and development banks, and climate finance provided by such tools as the Green Climate Fund. They have to de-risk by government, by having clear policies, sovereign guarantees and incentives such as feed-in tariffs and carbon credits or carbon tax exemptions and once this is done governments will have an attractive return on their sustainable investment (Carlos Muñoz & Oliva, 2024). Equally important is the harmonization of the regulatory environment of the involved countries. The various policies on the tariffs, energy prices and grid codes and environment compliance might hinder the project process and the investors might be scared out of the project. Well synchronized structures make the process of permitting simplified, provide interoperability in energy systems, and also the coordination of grid functions. Institutions like African Union or the ASEAN are a way of achieving regional power pools and policies to enable such ambitions. The complex of efficient financial models and harmonization of the regulations is enough to decrease the prices and reduce delays but also create the confidence of the investors and social accountability in such a manner that cross-border energy projects are not only economically viable but also environmentally safe.

3. INFRASTRUCTURE AND TECHNOLOGY FOR GRID SYNCHRONIZATION

The basis of the cross border renewable energy cooperation is the concept of Grid synchronization that enables it to share energy on any case needed in a stable and steady fashion across the borders. It involves aligning of the frequency, voltage and the operating standard of the corresponding power grids in a way that it flows effectively and dependably. It involves much investment in modern infrastructure and modern technologies to be synchronized. One of them is High Voltage Direct Current (HVDC) line, the most appropriate in the context of long-distance energy transportation with losses reduced to a minimum and Flexible AC Transmission Systems (FACTS) that enhance grid stability and control (Do et al., 2023). Smart grids like intelligent standards like digital tracking, load balancing, and automated fault monitoring systems (a.k.a. smart grid technologies) are considered to be necessary due to the unpredictable nature of renewable renewables like wind and solar.

Other steps to stabilize the grid include energy storage e.g. battery banks, pumped hydro and thermal storage, which stores excess energy that is released when peak demand is experienced. The Supervisory Control and Data Acquisitions (SCADA) systems also provide the operators to monitor and control the activities of grid across boundaries. The country-to-country technical standards, robust cybersecurity and business operations should be aligned to enable successful grid synchronization also. The potential of transnational renewable energy systems can be achieved with grid synchronization potentially, by synergizing infrastructure development with technological innovation to achieve the potential and encourage a resilient, green and connected energy future (Pu et al., 2021b). Figure 2 represents eight key technical elements, starting with HVDC lines and SCADA systems and then moving on to cybersecurity and smart grid systems that enable the creation of safe, stable, and efficient integration of renewable energy across borders.

Figure 2. Key Technical Enablers for Secure Cross-Border Renewable Energy Integration

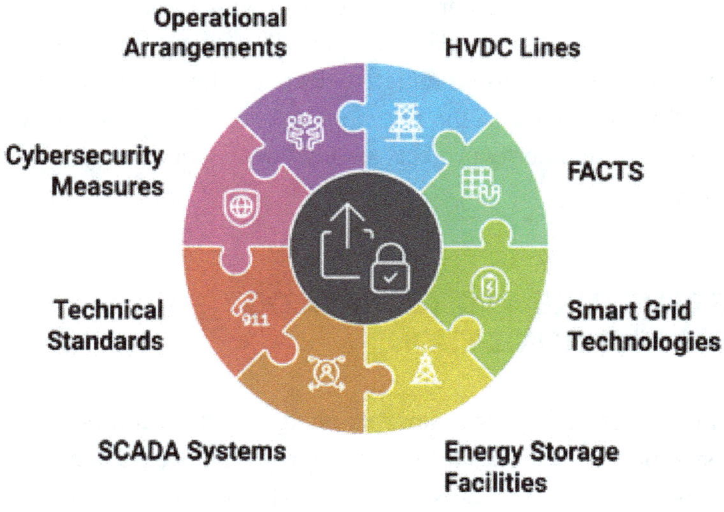

3.1. Technical Requirements for Cross-Border Energy Integration

Energy integration across borders requires the use of a vast amount of technical coordination whereby the various national grids exercise the capacity to work together in an efficient, secure, and reliable manner. The most important thing is the frequency synchronisation, i.e. all grids connected with each other should always have the same frequency of operation (e.g. 50 or 60 Hz). There is also the need of voltage compatibility and reactive power management to prevent the transmission losses and to prove system stability (Roldan-Fernandez et al., 2018). Such interconnection infrastructure as High Voltage Direct Current (HVDC) and Flexible Alternating Current Transmission Systems (FACTS) are essential to energy transfer over long distances through minimal losses. There is also a need to have a language translation grid codes, interoperable communications protocols and on-line tools to monitor the system to perform the task of managing the load dispatch, demand and supply and contingencies. Nations also need to invest on joint protection systems, load flow analysis equipment, or emergency response to be resilient with regard to faults or outages (Singh et al., 2018). These technical requirements are the basis of a safe, efficient and a combined cross-border renewable energy system.

3.2. Role of Smart Grids, Storage, and Digital Platforms

The smart grids, energy storage, and digital platforms are the technological components that constitute the modern cross-border renewable energy. Smart grids use sensors, AI-directed automation and analytics to monitor energy flow to and staying within the power grid in real-time to ensure that dynamic load balancing and fault isolation can be achieved (Zidan et al., 2025). This is particularly essential to the organization of the moving and de-centralization of the renewable energy sources in inter-relational areas. Energy storage Systems Lithium-ion batteries Flow batteries Pumped hydro Energy storage systems may be significant since they may receive additional energy produced at their peak of generation and discharge it to the generation and demand of their systems. This has the effect of stabilizing grid frequency and ensures that power exchange between countries remains constant (Axon et al., 2017). Digital platforms and Internet of Things enhancements that are featured in digital devices and platforms offer observability and efficiency of the grid and contribute to the realization of predictive maintenance, demand forecasting, and remote control of assets. The blockchain and cloud-based systems would also assist in transacting trading in energies and their bills transparently across the borders between the partners. Together, such technologies realize adaptive energy resilient and efficient energy integration, therefore, a fully integrated green energy future. The digital workflow of efficient energy management includes four steps shown in the figure 3. The first one is the Smart Grid Monitoring to track energy in real-time and the second is Energy Storage to stabilize demand and supply. Then there are Digital Platforms that strengthen the control of the grid, which is followed by Transparent Trading that allows energy to be billed and exchanged. These aspects combined make it possible to have an intelligent, sensitive, and transparent energy ecosystem.

Figure 3. Role of Smart Energy Management

4. ROLE OF REGIONAL INSTITUTIONS AND INTERGOVERNMENTAL COLLABORATION

The success of cross-border renewable energy programs is promoted by regional institutions as well as intergovernmental cooperation. The bodies serve as forums of communication, organization and resolution of conflicting interests by countries that share common energy objectives and geographic positions. they do much more than simply provide administration; they provide integration in the alignment of policies, coordination of shared infrastructure development and sponsorship of transnational energy schemes. Regional energy integration through power pools, renewable energy corridors and uniform regulatory boards has been promoted by the

European Union (EU), African Union (AU), South Asian Association of Regional Cooperation (SAARC) and Association of Southeast Asian Nations (ASEAN) organizations. Such organizations assist in eliminating political, legal, and technical obstacles that impede multinational energy collaboration most of the time (Chileshe, 2023). In addition, long-term trust and accountability can be developed with intergovernmental relationships and alliances by build-up of legally contracting and binding intergovernmental agreements, common governing policies and laws and fair benefit sharing process. They also create a scheme of environmental impact control, community involvement and disputing of cross border differences. Overall, regional organizations and multi-level cooperation play a crucial role in developing a favourable environment in which cross-border renewable energy systems will operate, thus enhancing sustainable development, decarbonization in the region, and geopolitical stability.

4.1. Contributions of Regional Blocs (e.g., EU, SAARC, ASEAN)

Regional blocs which have played significant roles in the cross-border cooperations in the field of renewable energy include the European Union (EU), South Asian Association of Regional Cooperation (SAARC) and Association of Southeast Asian Nations (ASEAN). The said bodies ensure convergence of national policies, construction of shared infrastructure and harmonization of laws and regulations in the member states. The Framework like Trans-European Energy Networks (TEN-E), which secures energy security, grid interoperability, and decarbonization in Europe have been an ambitious global leader and project like the North Sea Wind Power Hub by the EU (Ishola et al., 2024). The SAARC has promoted the SAARC Energy Ring and SAARC Framework Agreement on Energy Cooperation that has offered the platform on which electricity trade in South Asia can be encompassed. The Power grid feasibility study that ASEAN has put forward is an effort to set up multilateral electric power trading with an aim of utilizing the available potentials of widely distributed renewable energy capacity within the Asian countries. The blocs also provide technical support, financial backing and blueprints of policy that give rise to sustainable, cross-border energy cooperation as they facilitate peace and economic integration.

4.2. Local Government and Stakeholder Involvement in Joint Energy Planning

Although the international and regional institutions offer macro-level structures, the final success and sustainability of the cross-border renewable energy projects rely on the local governments and stakeholders being involved. Local government is important in issuing of permits, acquisition of land, environmental compliance, and community mobilization. Their involvement will make both the national and regional energy plans consistent with local realities and development's needs (Menon, 2024). Furthermore, inclusion of the stakeholders comprising of the local community, civil society organization, indigenous communities, and the private sector players in the process of planning ensures transparency, equity and acceptance in the society. Some strategies to allow the community and participatory decision-making with benefit-sharing models can reduce opposition, land-use conflicts, and environmental justice. There is also need to have capacity-building initiatives and local knowledge-sharing forums to ensure that stakeholders have the capacity to be part of effective cross-border energy governance. Said in other words, inclusive, bottom-up method helps to make sure that the renewable energy projects are not only technologically feasible but also socially viable and politically promoted.

5. ENVIRONMENTAL AND SOCIAL CONSIDERATIONS

Even though cross border renewable energy projects are essential in decarbonization, they should be planned to include solid environmental and social protective measures to provide long-term sustainability and acceptance amidst the communities. Such projects usually traverse ecologically sensitive areas and use people of different backgrounds; hence, there is need to do an environmental and social impact assessment (ESIA) in the planning and implementation phases. On an environmental level, the huge infrastructure such as transmission lines and hydroelectric dams may disrupt the ecosystem, wildlife routes and waterways (Nasr et al., 2020). The ecological impact minimization is highly urgent and that is why it is necessary to introduce the mitigation procedures, e.g. biodiversity offsets, eco-sensitive routing, international environmental standards. In terms of social, such projects can affect the question of livelihoods, land rights, and cultural heritage; especially, in the rural and indigenous communities. Social injustices and oppositions should be avoided using clear communication, remunerations, and involvement of all stake holders. Gender responsive planning and consideration of local governance systems also increases legitimacy (Omenge et al., 2019). In conclusion, the joint effort to merge environmental ethics and social justice into cross-border renewable activities guarantees sustainable cross-border projects besides being compliant with the regulations.

5.1. Environmental Assessments and Ecological Safeguards

Ecological safeguards and Environmental Assessments (EAs) are the essential sections of accountable trans-boundary renewable energy development. Such tests indicate the possible environment effects of facilities like wind farms, sun fields, transmission lines, or water power plants before launching the implementation of the project. Environmental Impact Assessment (EIA), usually referred to as Strategic Environmental Assessment (SEA), can aid in the identification, forecast and alleviation of any negative impact on biodiversity, water systems, forests and migrating routes. Since cross-border projects cut across several ecological zones, management of transboundary environmental governance becomes crucial (Rountree & Baldwin, 2018). The countries must come together to study together to be in a position to know the cumulative impacts on the environment and checking on the international conventions on cumulative impacts on the environment, like the Espoo Convention. Some ecological protection mechanisms are possible to be wildlife-friendly design of infrastructure, green corridors, reforestation and restoring the habitats projects. The measures would not only reduce the environmental degradation rates, but also enhance the legitimacy of the projects by making them compatible with the sustainability agenda. This will keep energy infrastructure healthy in terms of resilience, but it will not adversely affect the ecological integrity of common natural systems.

5.2. Inclusive Development: Community Benefits and Social Acceptance

The fourth critical success factor is social inclusion and community engagement-in the long-term cross-border renewable energy projects depend on the ability of the project developer and the partners to secure sufficient social inclusion and community engagement. Other projects that fail to consider the local needs or lock out the people in whose name projects are initiated are likely to be delayed, meet resistance, or fail altogether (Wu et al., 2015). Inclusive development requires accommodation to the social aspects of energy infrastructure, including land rights, displacement, cultural heritage and entitlements to benefits. Goodwill and local ownership can be supported using community benefits like local employment, revenue-sharing, corruption-free infrastructure and access to reasonable energy. The vulnerable groups, such as the indigenous population and women, should receive special attention; since underrepresented in planning processes, they usually represent a disproportionate disadvantage. Participatory planning frameworks, Free, Prior, and Informed Consent (FPIC) and consultation of the stakeholders early on are imperative to trust and transparency. Social acceptance grows when the communities feel involved in the development process and actually gain benefits out of energy projects implementation, since this approach changes the opposition to the collaboration and strengthens the sustainability of the cooperation in regions.

6. CASE STUDIES: SUCCESSFUL MODELS OF CROSS-BORDER RENEWABLE PROJECTS

Case studies of practical cross-border cooperation in terms of renewable energy present a strong inspiration about the practicality of cross-border collaboration, its economic rewards, and its strategic advantages. Such initiatives as the North Sea Wind Power Hub and the India-Bhutan Hydropower Cooperation are examples of innovative decision-making that prove that shared resources, unanimous policies, and technological collaboration can result in a win-win situation (Agostini et al., 2017). North Sea Wind Power Hub can be seen as a demonstration of a bold strategy of offshore wind-generation and grid-connection not only between a couple of countries in the EU but also decarbonization and energy landscape security on a large scale. At that, there is the example of the India-Bhutan cooperation which exemplifies the effectiveness of bilateral relations on the hydroelectric energy infrastructure development that has aided the growth of the economy and sustainable energy resource base of Bhutan and India. The two situations highlight the need to have a political will, coordination, and inclusive governance in carrying out transnational projects (Poque González et al., 2023). They will provide future cases of renewable projects that can be replicated in efforts of going beyond borders in a bid to provide sustainability, economic integration, as well as resilience to the changing climate.

6.1. North Sea Wind Power Hub (Denmark, Germany, Netherlands, Belgium)

The North Sea Wind Power Hub (NSWPH) is a revolutionary form of renewable energy collaboration, which is to complete the exploitation of the huge potential of offshore wind in the North Sea, coupled with interconnection of the electricity grid of various European nations. Over the direction of Denmark, Germany, the Netherlands, and Belgium, the vision project plans on constructing one or more artificial energy islands to become the collection, converted and transmitted hub of wind-generated electricity across borders. It is estimated that by 2050 the NSWPH will generate 70 to 150 GW of offshore wind power that will provide electricity to more than 150 million households in Europe (Lehmann et al., 2022). By 2024, the 4 participating countries already possess over 60% of the current offshore wind capacity found in Europe and Denmark and Germany are the pioneers in offshore deployment. The hub structure provides the possibility of sharing and balancing loads between interconnected countries enhancing the energy security and lowering the usage of fossil fuels.

It is estimated that the infrastructure costs will be lowered by up to 30-40% as compared to constructing individual country infrastructure, as facilities like the converters, platforms and transmission systems will be shared. The utilisation of the High Voltage Direct Current (HVDC) transmission lines implies that there is no loss of efficiency up to long transmission distances. The central part of the project is environmental and economic sustainability. The NSWPH is anticipated to play a huge role in meeting the EU objective of cutting down the emissions of greenhouse gases by 55% by 2030. Besides, the project would produce more than 100,000 indirect and direct jobs in the offshore construction and operations as well as maintenance in countries taking part in the project (Lim et al., 2024). In addition to technical innovation, the NSWPH establishes a previous in regional rule and conformity of guidelines. Shared policy frameworks, collaborative funding mechanism, and aligned environmental assessments are in the process of being developed in order to harmonise their implementation and level distribution of benefits. Essentially, the North Sea Wind Power Hub is not only an example of an affordable, sustainable, and transnational wind energy establishment, but it is also a scalable model that will define how international collaboration will accelerate the European green transition, establishing a standard globally concerning the integration of offshore energy sources.

The table provides a concise overview of the North Sea Wind Power Hub (NSWPH) a pioneering multilateral initiative involving Denmark, Germany, the Netherlands, and Belgium. The project exemplifies large-scale cross-border cooperation in offshore renewable energy, targeting up to 150 GW capacity by 2050. It highlights key elements such as energy capacity, technology use, economic and environmental benefits, and governance structure (Magnani & Cittati, 2022). The table 1serves as a reference for understanding how integrated infrastructure, shared investment, and policy alignment can accelerate regional green transitions and set global benchmarks for transnational energy collaboration.

Table 1. Summary of North Sea Wind Power Hub – A Cross-Border Renewable Energy Model

Category	Details
Participating Countries	Denmark, Germany, Netherlands, Belgium
Project Type	Offshore wind energy with cross-border transmission
Target Capacity	70–150 GW by 2050
Household Coverage	Power supply for over 150 million European households
Current Offshore Share	>60% of Europe's offshore wind capacity (as of 2024)

continued on following page

Table 1. Continued

Category	Details
Infrastructure Benefit	30–40% cost reduction due to shared platforms, converters, and HVDC systems
Technology Used	Artificial energy islands, HVDC transmission, offshore wind turbines
GHG Reduction Goal	Significant contribution toward EU's 55% emissions reduction target by 2030
Job Creation	Estimated 100,000+ direct and indirect jobs
Energy Benefits	Grid balancing, increased energy security, reduced fossil fuel dependency
Governance Model	Joint policy frameworks, coordinated funding, shared regulatory structures
Significance	A scalable, cost-effective, and environmentally sustainable model for regional energy cooperation and integration

The Figure 4 illustrates the estimated current offshore wind capacity share (in %) and the projected contribution (in GW) by 2050 from each of the four participating countries Denmark, Germany, the Netherlands, and Belgium in the North Sea Wind Power Hub. The chart highlights the collective leadership of these countries, especially Denmark and Germany, in current offshore capacity, and their expected role in scaling up to a combined 150 GW by mid-century through a shared, cross-border energy island model.

Figure 4. North Sea Wind Power Hub: Offshore Wind Capacity Contributions

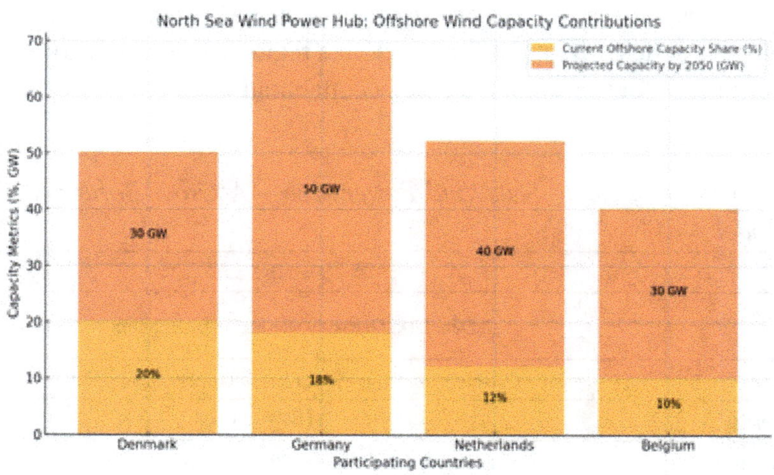

6.2. India-Bhutan Hydropower Cooperation

In our opinion, the India-Bhutan Hydropower Cooperation acts as one of the oldest and successful examples of bilateral cooperation in cross-border renewable energy project. Based on trust, common geography, and common economic good, this nexus has made excellent relations to Bhutan in terms of developing as an economy and to India in terms of clean energy conversion. Around 95% of Bhutanese energy production is hydropower; it is the basis of the renewable energy policy of the country and a big export product. Beneath government-to-government financing and technical support, India has helped in building up some large hydropower projects in Bhutan. Good examples of the projects are Chukha (336 MW), Tala (1,020 MW) and Mangdechhu (720 MW) that export a significant share of generated power to India. Bhutan sells more than 75% of its hydro-generation capacity to India and with this revenue, it generates almost a quarter of Bhutan GDP and above 40% of the total government revenue as of 2024.

In the case of India, the collaboration is important to achieve its low-emission energy targets in the Paris Agreement. The imported clean energy in the form of a hydropower plant in Bhutan is lessening the carbon payload, reinforcing neighboring grids, and growing India in its intention to accomplish half of cumulative electric power capacity of non-fossil fuel sources by 2030. The import of hydro power is especially useful when electricity demand is high during summer, and supplements the Indian solar dominated energy mix. Institutional coordination is also witnessed in the partnership (Nakahara et al., 2019). The Hydropower Committee and bilateral tariff agreement on setting, transmission and regulation of project run operations with no exchange of conflicts and unsmooth business transactions. Moreover, the two countries carry out common environmental impact studies, engage the local communities in the planning process and make sure that environmental protection is extended. Some new projects such as the 1,200 MW Punatsangchhu-II, as well as the prospects of joint ventures in power trade and grid network development are the most recent developments that are causing an increase in this relationship. To sum up, the example of the India-Bhutan hydropower model shows how long-term cross-border renewable energy cooperation, strategic in nature, can provide both mutual economic benefits, projections on sustainable development and support the low-carbon agenda in the framework of regional stability and energy security. The table 2 outlines the key features of the India-Bhutan Hydropower Cooperation, a bilateral renewable energy partnership focused on hydropower development and energy trade.

Table 2. Summary of India-Bhutan Hydropower Cooperation – A Bilateral Renewable Energy Model

Category	Details
Participating Countries	India and Bhutan
Project Type	Bilateral hydropower development and electricity trade
Hydropower Share in Bhutan	~95% of total electricity generation
Key Projects Supported by India	Chukha (336 MW), Tala (1,020 MW), Mangdechhu (720 MW), Punatsangchhu-II (1,200 MW – upcoming)
Electricity Export to India	>75% of Bhutan's hydropower output (as of 2024)
Contribution to Bhutan's Economy	~25% of GDP and >40% of total government revenue
India's Energy Goals	Supports India's target of 50% electric power from non-fossil sources by 2030
Environmental Measures	Joint Environmental Impact Assessments (EIAs), community engagement, and ecological safeguards
Institutional Mechanism	Hydropower Committee, bilateral agreements on tariffs, governance, and transmission
Strategic Benefits	Energy security for India, economic development for Bhutan, regional stability, and shared low-carbon development

The Figure 5 showcases the installed and planned generation capacities of key hydropower projects developed through the India-Bhutan energy partnership. Projects like Tala (1,020 MW) and Mangdechhu (720 MW) have played a crucial role in Bhutan's energy export and India's clean energy import strategy. The upcoming Punatsangchhu-II project, with a planned capacity of 1,200 MW, signifies the continued strengthening of this bilateral cooperation, reinforcing both nations' commitment to sustainable and mutually beneficial renewable energy development.

Figure 5. Major Hydropower Projects under India-Bhutan Cooperation

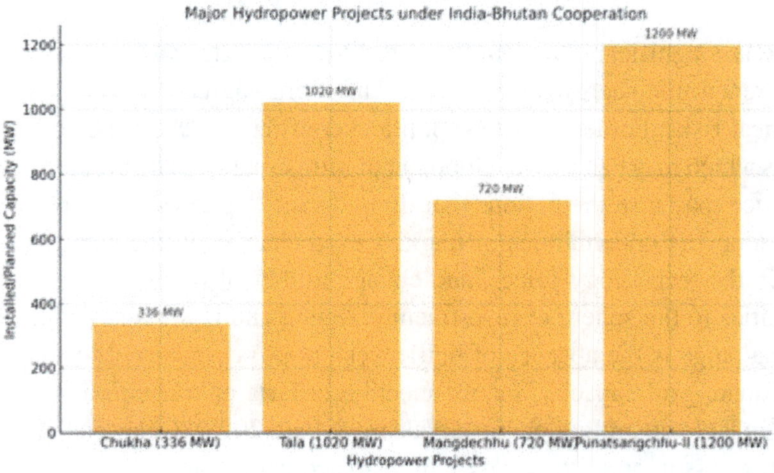

7. CHALLENGES AND FUTURE DIRECTIONS

The potential of cross-border renewable energy projects is tremendous but has been challenged by various issues such as lack of regulatory alignment, complicated permitting process, and lack of consistent energy policy nationally. The risk of dependency and the geopolitical tensions as well as the technical barriers including the grid incompatibility, the absence of standardization, and cybersecurity risks are additional barriers to improvement. The implementation is also postponed by social resistance, environmental issues, and the poor stakeholder involvement (Reynolds, 2024). The next generation relationships should focus on the legal framework, cooperation, and regional institutions. As smoother transfers can be made possible with the help of digital technologies, AI-based load balancing, and advanced energy storage. The key to success lies in the universal governance, equitable distribution of benefits, infrastructure that will resist climate change, strengthened multilateral treaties, and even cooperation at the community level, which will guarantee the decarbonization process and the attainment of global energy equity (Narain et al., 2021).

7.1. Geopolitical Risks, Market Uncertainty, and Infrastructure Bottlenecks

There are significant challenges related to geopolitics and markets that international renewable energy projects face. The energy agreements are vulnerable to geopolitical risks, border conflicts, political conflicts, or change of bilateral relationships which may halt the development of infrastructure. The problems of energy dependence may turn the nations that do not want to participate in the long-term cooperation. The uncertainties of markets even changes in the energy prices and change in the technology prices, and change in the subsidies are also adding the complexities to the sphere of investment planning and risk assessment. The other major challenge is the absence of bottlenecks in infrastructure. Most of the areas lack adequate grid capacity, interconnection criteria or transmission facilities to transfer, in large quantities, energy (Wakrim & Fait, 2023). Delay can be expensive due to slowness in permitting, delays in land acquisition and poor regulatory environments. Besides, insufficient in-region technical resources and insufficient local process can impose additional pressure on the project dependence on the external means. To overcome these challenges, it is required to have risk-sharing regimes, a well-developed system of energy trading relations, and cooperation between the government and the business community (Waris et al., 2019). Regional integration should be aided by infrastructural investments, diplomacy of trust and institutional cooperation within the region.

7.2. Strategic Roadmap: Scaling Cross-Border Green Energy Networks

Transnational green energy networks require a longitudinal scheme of growth, many-sided. Firstly, nations ought to harmonize the energy policies, grid codes and green standards to harmonize the development of transnational projects. Governance The regional power pools and the multi-lateral arrangements such as those arranged by the EU, ASEAN or the SAARC may facilitate governance as well as collaborative planning, conflict settlement, and share of benefits. Second, the level of investments in the infrastructure of the next generation is high (Atalay, 2018). These include intelligent grids, super capacity transmission lines (including HVDC) as well as solutions that combine energy storage to address the variable element of renewables. The main agents of the smooth operations will be digital technologies such as forecasting based on AI, trading in energy based on blockchain, and real-time monitoring. Third, it should develop a sound financial framework to the extent of creating a pool of finance, green bonds and climate funds that will facilitate the movement of finance between countries. The policies should also be designed in a way that they will draw in an inclusive growth to empower the locals and bring a social-environmental sustainability (Adegboyega et al., 2024). The last but not the least is strategic imperative of cross-border scalability of renewable networks capacity not only to future-proof energy supply, on the one hand, but also to future-proof climate-resilience in the region and decarbonization in the long term. There are four pillars Policy Harmonization, Financial Architecture, Infrastructure Investment and Strategic Imperatives that can support the development of cross-border green energy systems as indicated in the figure 6.

Figure 6. Pillars of Cross-Border Green Energy Networks

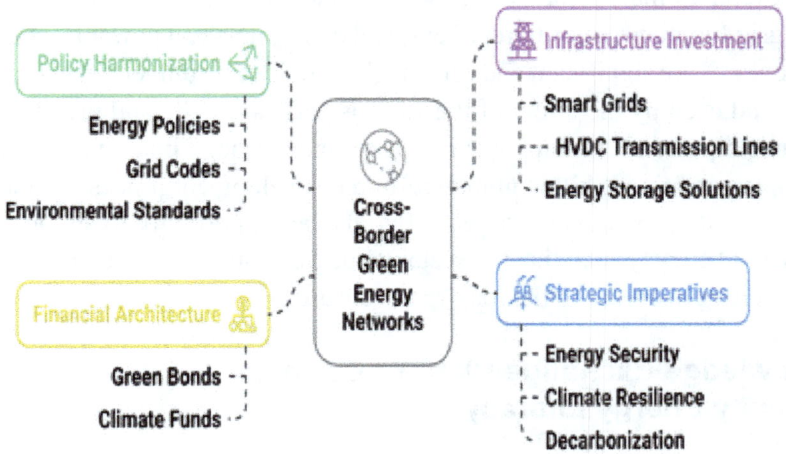

8. STAKEHOLDER ENGAGEMENT AND CAPACITY BUILDING

The participation as well as the capacity building of the stakeholders is also important in sustainable cross border renewable energy projects. Besides infrastructure and policies, the preparedness of local actors is another ability to achieve success. Capacity building should also target grid operators, engineers, policy makers and the community leaders and equip them with a technical knowledge base as well as contextual background. The shortage in skills can be addressed by training in smart grid management, cross-border regulation and clean energy technologies. Through knowledge exchange platforms, i.e. work-shops, digital and taskforces, innovation and shared learning are discovered. Energy literacy helps individuals to empower and exploit projects. Public-private partnerships can be used to fund the resultant socially encompassing, resilient and institutionally sustainable initiatives through education, research funding and curriculum alignment with industry needs, and the result (Dewi, 2024).

8.1 Training and Technical Skill Development for Local Actors

The success of transnational renewable energy projects entails both the expertise and the desire of the local stakeholders; the grid managers, engineers, policy makers and the community leaders. These actors are expected to be very well trained to operate and maintain complex cross-boundary power systems. The capacity-building exercises must prioritize the sector-related abilities that are applied such as smart grid functions, inter-border laws, and emerging clean power technologies. These programs are significant and can only be provided by universities, vocational institutes and even by the union of the various parties (public and privately). The local ownership and operational viability is attained through linking training to the actual implementation. Skilled human resource development process is not solely a means of dealing with technical gaps but also an opportunity to strengthen the engagement of society such that the sustainability and success of incorporating the renewable energy sources in the long run is ensured.

8.2 Knowledge Exchange Platforms and Community Energy Literacy

It is believed that platforms of knowledge exchange play a pivotal role in increasing the effectiveness in cross border renewable energy project collaboration, innovation, and shared learning among the countries that deal with the projects. The stakeholders in such platforms, such as those of the regional workshops, intergovernmental task forces, and the online forums, enable the stakeholders to share the best practices and lessons learned, and technological actions ahead. At the same time, the community energy literacy needs to be also increased as a method of gaining local support and involvement. People can be informed about the benefits and duties of the renewable energy systems by conducting the educational program of publicity, by school courses and community contacts. The communities are likely to involve themselves in a positive and sustainable way with a clear understanding of how the energy systems work and with the impact of these energy systems to their livelihoods. Overall, such efforts provoke inclusivity, transparency and social acceptance of cross border green energy initiatives.

9. MONITORING, EVALUATION, AND IMPACT ASSESSMENT

It could only be able to add long-term value in as far as there is a system of organized Monitoring, Evaluation, and Impact Assessment (MEIA) framework to such cross-border renewable energy projects. These structures see initiatives fulfil their goals whether it is reducing emissions, enhancing energy security, or acceptable improvement (Ekechukwu & Simpa, 2024). The performance indicators are to be presented on the table. These may be Greenhouse Gas (GHG) emission targets (e.g. 40-55% by 2030), improving grid reliability (up to 30% less blackouts) or reducing electricity importation and exportation expenses (1525% with infrastructure sharing). In the case of real-time energy flow, fault monitoring and performance benchmarking of national grids, technical monitoring systems such as Supervisory Control and Data Acquisition (SCADA) system and regional dashboards may be used to monitor. Social and environmental impact assessment on the post-implementation is also important and will determine their impact on the livelihood of the locals, the ecologies and accessibility of the resources. Such tests can be used to identify such problems as unfair benefit sharing or unexpected eco-stress (Kularathna et al., 2019). Mechanisms of feedback, i.e., multi-stakeholder review boards, consultations to the population, and annual progress reports should be institutionalized to maintain improvement. Such mechanisms provide the guarantee that previous projects use knowledge in another phase to develop the next one, which results in the enhanced responsiveness of policies and transparency in operations. Essentially, strong MEIA processes increase accountability, build trusts and make effective, adaptive and sustainable cross-border energy projects in a country. The figure 7 shows key performance indicators for cross-border renewable energy projects. It highlights improvements in GHG emission reduction (55%), grid reliability (30%), and cost savings (25%). These metrics guide impact assessment and policy adjustments.

Figure 7. Performance Metrics for Cross-Border Renewable Energy Projects

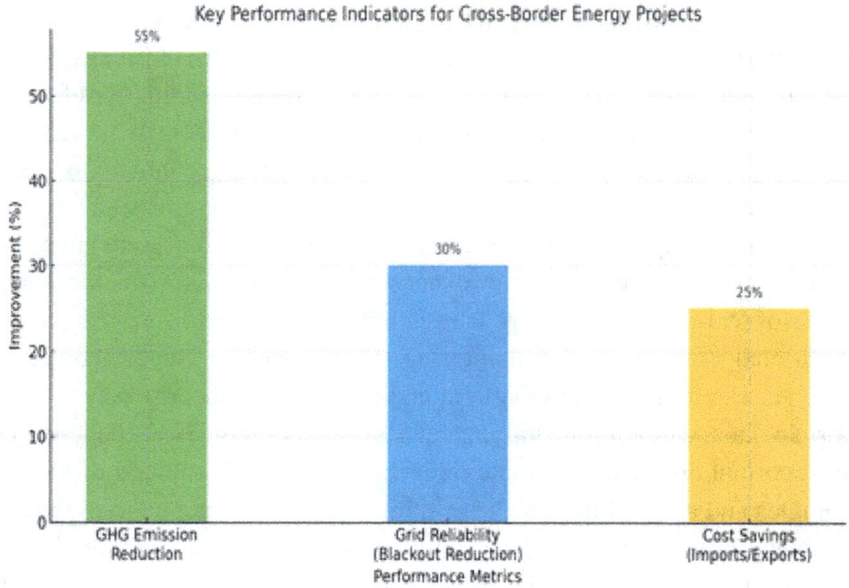

9.1 Performance Indicators and Real-Time Monitoring Tools

It is imperative to set performance indicators to assess how successful cross-border renewable energy projects can be. Measures of success, like reductions in GHG emissions (e.g., 40-55% by 2030), grid stability (e.g. 30% reduction in blackouts), and cost savings (e.g., 15-25% of shared infrastructure) are concrete measures. Real-time monitoring instruments, such as Supervisory Control and Data Acquisition (SCADA) systems and local dashboards, can keep track of flows of energy, animate faults, and view performance in connected grids. These are the tools that facilitate evidence-based decision making, improvement of transparency, and achieve environmental and economic objectives, thus, enhance accountability and long-term operational performance.

9.2 Social and Environmental Impact Evaluation and Feedback Mechanisms

In order to make sure that cross-border renewable energy projects are fair and sustainable, social and environmental impact assessment is essential. These assessments look at the impact which projects have on the local livelihood and biodiversity, and ecosystems and the availability of resources. Important instruments are post-implementation surveys, ecological monitoring and impact assessment of stakeholders (Adegboyega et al., 2024). In order to achieve the transparency and a flexible improvement, systematic feedback systems including community consultations, multi-stakeholder review boards and a regular progress report should be institutionalized. The mechanisms can detect such problems as inequitable sharing of benefits or unintentional ecological pressure and correct them. The integration of feedback loops leads to an improvement in policy responsiveness, increases the level of trust within the communities and facilitates the involvement of society in energy transformations, inclusively and in ways that are socially responsible.

10. POLICY RECOMMENDATIONS AND WAY FORWARD

In order to turn vision of cross-border renewable energy cooperation into reality, there should be specific and implementable policy recommendations to be processed as a short-term, medium-term, and short-term attempts to turn this vision into reality. These suggestions give decision-makers the way forward to help in breaking down practical impediments, lowering interests, and promoting accommodative and sustainable development (Maqbool et al., 2020). The short-term goals that the countries should work on are compatibility in the energy tariffs, the recognition of each other's technical standards and establishing pilot bilateral agreements in power trading. These initial steps can create institutional confidence, minimize regulatory friction as well as prove proof-of-concept to additional collaboration. They should also invest in their joint feasibility studies, capacity-building programs and fast-track permitting cross-border infrastructure.

Moving to the medium term, to support a more significant integration, regional transmission corridors should be created with the help of co-financed projects in the sphere of infrastructure and energy storage systems shared. There can be regional regulatory read out bodies or coordination bodies that regulate operation, resolve disputes, guarantee stability of the grids. Convergence in policy in environmental compliance as well as planning cycle synchronization will also play an important role at this level (School of Media et al., 2021). The activities should progress ultimately to the signing of formal treaty commitments, the establishment of common climate and green infrastructure funds and the establishment of transnational carbon markets. By integrating the energy cooperation into the regional climate governance, it is possible to ensure continuity and resilience. Finally, through a staged and policy-based solution, the cross-border renewable energy pipelines end up being transformative and sustainable.

10.1 Short and Medium-Term Strategic Interventions

An efficient implementation of integration of renewable energy cross-border starts with specific short- and medium-term measures. The short-term step that these nations should put on their agenda is the harmonisation of energy tariffs, harmonisation of technical standards, and initiating pilot bilateral power trading agreements. Such preliminary work imports an element of trust amongst institutions, eases regulatory burdens, and serves as a proof-of-concept to future partnership. The development of the regional transmission corridors, co-financing of the shared infrastructure, and the establishment of the regulatory coordination bodies should become the focus in the medium term (Atalay, 2018). Through these interventions, energy circulation is smoother, and the interoperability of the grid increases, and a coordinated planning is encouraged. In combination, these gradualist strategies can establish the basis of extensive, sustainable, and politics-supported cross-border energy collaboration.

10.2 Long-Term Vision: Treaty-Level Commitments and Climate Governance

In order to achieve long term multinational cooperation in renewable energy, it is necessary to have the long-term vision anchored on formal treaty-based undertaking. These agreements are supposed to form long-time frontiers of shared infrastructures, planned policies, and shared responsibility. Establishing the regional climate and green infrastructure funds, and transnational carbon markets shall help to achieve large-scale financing and fair distribution of benefits. An embedded energy cooperation within the wider climate governance systems guarantees continuity, resilience, and stability of the law (Ekechukwu & Simpa, 2024). This strategy enhances the geopolitical consensus but also serves as confidence to the international trends on decarbonization. The long-term governance should rest on the principles of inclusivity, transparency and environmental sustainability to give fair and a future-proof energy transition.

11. CONCLUSION

The conversion to sustainable energy systems cannot be affected through individual national effort. The relevance of regional partnership as discussed in relation to the renewable energy partnership between countries also points out the consistency of institutions and technology fit in an attempt to solve the global climatic crises in a decent manner. Climate change is transnational and its resolutions ought to be extraterritorial especially in the energy sector. Not only that cross-border energy infrastructure is able to contribute to higher affordability and energy security but it is also a trigger of the main motivation behind decarbonization, regional economies and climatic adaption integration. Policy and legal alignment is one such success that the initiatives might need. Regulations, cross border fragmented regulations and climate justice principles should be harmonized to instill transparency, fairness and sustainability. The climate in which investment and innovations are to be done is conducive to green bonds and multilateral climate funds financing that have harmonised rules.

The critical factor of technology is also highlighted in the chapter. Such infrastructure as High Voltage Direct Current (HVDC) direct current transmission lines, smart grids, and developed energy storage facilities is needed to develop the renewable sources beyond the borders of nations. By these tools, it will be possible to balance in real time, reduce the losses in transmission and make interconnected grids resilient. This will lead to the increase of the level of transparency, predictive maintenance and safe energy trading made possible by the appearance of a new generation of digital environments and AI and blockchain technologies, which will render cross-border energy transactions technologically and economically feasible. Their ability to make policies to fit together, plan and conflict resolution are the key to regional organizations like the EU, ASEAN, and SAARC. The renewable energy joint ventures can be re-established through the assistance of common vision, trust, and regulation which proved effective in the such projects like the North Sea Wind Power Hub or the India-Bhutan Hydropower Cooperation currently they still have serious problems. It has geopolitical tensions, fragmentation in regulations, bad infrastructure and uncertainties in the market that cannot be focused on rapid development. Inclusive planning, environmental protection and fair distribution of benefits ought to be made by other factors like social and environmental problems like displacement, environmental devastation, and inclusion of stakeholders. The coordination of policy changes, the robustness of infrastructure investments and empowerment of the local communities through the participatory governance will be required in future to develop the cross-border green energy networks. The institutional ability-building and technological innovation will facilitate the multilateral cooperation. Last but not least, the politically integrated cross-border renewable energy systems are not only technically possible but also are geopolitically necessary. They are a roadmap of a zero-carbon future, energy justice, and a more collaborative and greener global future.

REFERENCES

Adegboyega, A. W., Ani, E. C., Oladunni, S. O., & Pramanik, A. S. (2024). Project management tools in renewable energy integration: A review of U.S. perspectives. Eng. Sci. Technol. j., 5(7), 2364–2378.

Agostini, C., Silva, C., & Nasirov, S. (2017). Failure of energy mega-projects in Chile: A critical review from sustainability perspectives. *Sustainability (Basel)*, *9*(6), 1073. DOI: 10.3390/su9061073

Atalay, Y. (2018). Understanding input and output legitimacy of environmental policymaking in the gulf cooperation council states. *Environmental Policy and Governance*, *28*(1), 39–50. DOI: 10.1002/eet.1794

Axon, S., Chapman, A., & Light, D. (2017). Stakeholder engagement in coastal sustainability transitions: An emerging research agenda. *Reg. Mag.*, *308*(4), 20–22. DOI: 10.1080/13673882.2017.11958673

Ballo, A., Valentin, K. K., Korgo, B., Ogunjobi, K. O., Agbo, S. N., Kone, D., & Savadogo, M. (2022). Law and policy review on green hydrogen potential in ECOWAS countries. *Energies*, *15*(7), 2304. DOI: 10.3390/en15072304

Balsalobre-Lorente, D., Shahbaz, M., Roubaud, D., & Farhani, S. (2018). How economic growth, renewable electricity and natural resources contribute to CO2 emissions? *Energy Policy*, *113*, 356–367. DOI: 10.1016/j.enpol.2017.10.050

Carlos Muñoz, J., & Oliva, H. (2024). Effect of combining carbon policies and price controls in cross-border trade of energy on renewable generation investments. *The Energy Journal (Cambridge, Mass.)*, *45*(1), 81–103. DOI: 10.5547/01956574.45.1.jmun

Chileshe, M. (2023). International renewable energy finance mechanisms and the role of local stakeholders in project design and implementation: Perceptions from Zambia. *Open J. Bus. Manag.*, *11*(06), 3066–3080. DOI: 10.4236/ojbm.2023.116168

Dewi, L. (2024). Exploring stakeholder perspectives on supply chain sustainability in renewable energy industries. In *Research Square*. DOI: 10.21203/rs.3.rs-4301842/v1

Do, T. N., Burke, P. J., & Lu, B. (2023). Harnessing solar and wind for sustainable cross-border electricity trade in the Greater Mekong Subregion. *Frontiers in Environmental Science*, *11*, 11. DOI: 10.3389/fenvs.2023.1188335

Ekechukwu, D. E., & Simpa, P. (2024). A comprehensive review of renewable energy integration for climate resilience. Eng. Sci. Technol. j., 5(6), 1884–1908.

Ishola, A. O., Odunaiya, O. G., & Soyombo, O. T. (2024). Stakeholder communication framework for successful implementation of community-based renewable energy projects. *Int. J. Front. Eng. Technol. Res.*, *7*(2), 25–43. DOI: 10.53294/ijfstr.2024.7.2.0047

Kularathna, A. H. T. S., Suda, S., Takagi, K., & Tabeta, S. (2019). Evaluation of co-existence options of marine renewable energy projects in Japan. *Sustainability (Basel)*, *11*(10), 2840. DOI: 10.3390/su11102840

Lehmann, J., Bouillass, G., Fofack-Garcia, R., & Pérez-López, P. (2022). Towards social Life Cycle Assessment of Energy Systems: a case study on offshore wind farms from companies' perspective. E3S Web Conf., 349, 12002.

Lim, A. B., Poelzer, G., & Noble, B. (2024). Social value of renewable energy in remote northern Indigenous communities. *Journal of Aboriginal Economic Development*, *14*(1), 74–91. DOI: 10.29173/jaed8

Magnani, N., & Cittati, V.-M. (2022). Combining the multilevel perspective and Socio-technical imaginaries in the study of community energy. *Energies*, *15*(5), 1624. DOI: 10.3390/en15051624

Maqbool, R., Deng, X., & Rashid, Y. (2020). Stakeholders' satisfaction as a key determinant of critical success factors in renewable energy projects. *Energy, Sustainability and Society*, *10*(1), 28. DOI: 10.1186/s13705-020-00259-0

Menon, S. (2024). Fuzzy cooperative game model for Renewable Energy advancement: A case study of offshore wind farms. In *Research Square*. DOI: 10.21203/rs.3.rs-5720102/v1

Meus, J., den Bergh, K., Delarue, E., & Proost, S. (2019). On international renewable cooperation mechanisms: The impact of national RES-E support schemes. *Energy Economics*, *81*, 859–873. DOI: 10.1016/j.eneco.2019.05.016

Nag, T. (2021). Barriers to cross-border energy cooperation and implications on energy security: An Indian perspective with reference to energy trade in south Asia. *Global Business Review*, *22*(6), 1530–1552. DOI: 10.1177/0972150919826380

Nakahara, Y., Tabata, T., Ohno, T., Furukawa, F., Inokuchi, K., Katagiri, K., & Hirayama, Y. (2019). Discussion on regional revitalization using woody biomass resources as renewable energy. *International Journal of Energy and Environmental Engineering*, *10*(2), 243–256. DOI: 10.1007/s40095-019-0300-5

Narain, D., Teo, H., Lechner, A., Watson, J., & Maron, M. (2021). Biodiversity risks and safeguards of China's hydropower investments in Belt and Road Initiative (BRI) Countries. In *Research Square*. DOI: 10.21203/rs.3.rs-778318/v1

Nasr, A. K., Kashan, M. K., Maleki, A., Jafari, N., & Hashemi, H. (2020). Assessment of barriers to renewable energy development using stakeholders approach. *J. Entrep. Sustain. Issu.*, *7*(3), 2526–2541. DOI: 10.9770/jesi.2020.7.3(71)

Naval, N., & Yusta, J.M. (2024). Technical versus socio-economic and environmental criteria in power transmission projects. Renew. Energy Power Qual. J., 21(2).

Ogino, K., Nakayama, M., & Sasaki, D. (2019). Domestic socioeconomic barriers to hydropower trading: Evidence from Bhutan and Nepal. *Sustainability (Basel)*, *11*(7), 2062. DOI: 10.3390/su11072062

Omenge, P. M., Eshiamwata, G. W., Makindi, S. M., & Obwoyere, G. O. (2019, September). Public participation in environmental impact assessment and its substantive contribution to environmental risk management: Insights from eia practitioners and other stakeholders in Kenya's renewable energy sub-sector. Energy and Sustainability VIII.

Poque González, A. B., Masip Macia, Y., Ferreira, L. D. C., & Valdes, J. (2023). Socio-ecological controversies from Chilean and Brazilian sustainable energy transitions. *Sustainability*, *15*(3), 1861.

Pu, Y., Li, Y., & Wang, Y. (2021a). Structure characteristics and influencing factors of cross-border electricity trade: A complex network perspective. *Sustainability (Basel)*, *13*(11), 5797. DOI: 10.3390/su13115797

Pu, Y., Li, Y., & Wang, Y. (2021b). Structure characteristics and influencing factors of cross-border electricity trade: A complex network perspective. *Sustainability (Basel)*, *13*(11), 5797. DOI: 10.3390/su13115797

Reynolds, S. (2024). Stakeholder engagement and its impact on supply chain sustainability in the context of renewable energy. In *Preprints*.

Roldan-Fernandez, J. M., Gómez-Quiles, C., Merre, A., Burgos-Payán, M., & Riquelme-Santos, J. M. (2018). Cross-border energy exchange and renewable premiums: The case of the Iberian system. *Energies*, *11*(12), 3277. DOI: 10.3390/en11123277

Rountree, V., & Baldwin, E. (2018). State-level renewable energy policy implementation: How and why do stakeholders participate? *Frontiers in Communication*, *3*, 3. DOI: 10.3389/fcomm.2018.00006

School of Media, Communication Universities of Douala, C., Leicester, U. K., & Nyoh, I. (2021). Communicating climate change and energy in rural Africa: A case analysis to explain how participatory communication can support transition to renewables and adoption of solar technologies in rural Africa. Ciências e Pol\'\ iticas Públicas / Public Sciences &, 183–200.

Singh, A., Jamasb, T., Nepal, R., & Toman, M. (2018). Electricity cooperation in South Asia: Barriers to cross-border trade. *Energy Policy*, *120*, 741–748. DOI: 10.1016/j.enpol.2017.12.048

Utelbayev, B. T., Azhigali, A. S., & Amangeldin, A. G. (2024). Environmental sustainability in cross-border energy trade: The case of the Kazakhstan – China pipeline. *Jour*, *21*(1), 140–148. DOI: 10.55452/1998-6688-2024-21-1-140-148

Wakrim, A., & Fait, S. (2023). The future of hydropower: Civil engineering and renewable energy integration. *Int. J. Hydropower Civ. Eng.*, *4*(2), 12–15. DOI: 10.22271/27078302.2023.v4.i2a.30

Waris, M., Sufia Azlan, Z., Fadzline Muhamad Tamyez, P., Ullah, M., & Khan, A. (2019). *Analyzing the constructs of stakeholder engagement towards renewable energy projects success in Malaysia: A PLS approach.* KnE Soc. Sci. DOI: 10.18502/kss.v3i22.5090

Wu, G. C., Deshmukh, R., Ndhlukula, K., Radojicic, T., & Reilly, J. (2015). *Renewable energy zones for the Africa clean energy corridor. Office of Scientific and Technical Information.* OSTI. DOI: 10.2172/1328753

Zidan, H., Malik, M. T., Rafique, U., Azeem, F., Manzoor, T., & Manzoor, H. U. (2025). Exploring the potential of cross-border energy trade in SAARC countries for achieving Sustainable Development Goals (SDGs). *Energy Science & Engineering*, *13*(4), 2063–2081. DOI: 10.1002/ese3.70032

KEY TERMS AND DEFINITIONS

Benefit-Sharing Mechanism: Policy tools or agreements ensuring equitable distribution of economic and social benefits from energy projects to affected communities.

Climate Resilience: The ability of energy systems and communities to adapt to and recover from climate-related disruptions.

Cross-Border Energy Cooperation: Collaborative development and operation of energy infrastructure between countries to enhance energy security, efficiency, and sustainability.

Digital Platforms: Software systems (e.g., IoT, AI, blockchain) used for real-time monitoring, forecasting, billing, and energy trading across borders.

Energy Charter Treaty (ECT): An international agreement that provides a legal framework for energy cooperation, investment protection, and dispute resolution.

Energy Storage Systems: Technologies like batteries or pumped hydro used to store excess energy and release it during high demand, ensuring grid stability.

Environmental Impact Assessment (EIA): A process to evaluate the ecological consequences of energy projects before implementation, ensuring mitigation of adverse effects.

Free, Prior, and Informed Consent (FPIC): A principle ensuring communities, especially indigenous ones, have the right to approve or reject projects affecting their land or livelihoods.

Grid Synchronization: Alignment of technical parameters like frequency and voltage across different countries' power grids to allow seamless and stable energy transfer.

High Voltage Direct Current (HVDC): A long-distance, low-loss electric transmission system essential for transporting electricity across national boundaries.

Inclusive Energy Planning: A participatory approach involving local governments, communities, and stakeholders in the planning and execution of energy projects.

Power Purchase Agreement (PPA): A contract between an energy producer and buyer outlining the sale, pricing, and delivery of electricity, often used in cross-border deals.

Regional Power Pool: A shared electricity network allowing countries to trade and balance power for regional grid stability and economic benefits.

Renewable Energy Integration: Incorporation of renewable energy sources such as wind, solar, and hydro into power systems, especially across interconnected national grids.

Smart Grid: An advanced energy system using digital communication, automation, and AI to manage electricity flow dynamically and efficiently.

Chapter 12
The Cross-Border Influence of Climate Change on the Current Reality of Environmental Risks in Spain

Ignacio Sotelo Pérez
University Institute of Environmental Sciences. Complutense University of Madrid, Spain

Maria Sotelo Pérez
https://orcid.org/0000-0002-5541-7941
King Juan Carlos University, Spain

José Antonio Sotelo Navalpotro
University Institute of Environmental Sciences. Complutense University of Madrid, Spain

ABSTRACT

This research highlights the impact and repercussions of so-called "environmental risks" on the future of territories marked by the cross-border impacts generated by the evolutionary processes of the renowned climate change. In this way, a concise study of the most northwestern region of Spain is carried out, based on the analysis of extreme meteorological phenomena -such as extratropical cyclonic low pressure systems-, on the physical, human and economic environment in the coastal areas of the Province of Pontevedra, focus attention on how these atmospheric processes constitute a key element in the configuration of the climate regime, where their

DOI: 10.4018/979-8-3373-3972-6.ch012

frequency and intensity generate episodes of high disturbance, with significant implications for territorial dynamics and the safety and activity of its inhabitants. Likewise, it emphasizes the need for climate risk planning tools, based on action and performance measures by local governments that promote cooperation to mitigate or eradicate cross-border risks and impacts, thus strengthening resilience in border regions.

1. INTRODUCTION[1]

In the processes of analysis and assessment of cross-border strategies for adaptation and mitigation of climate change, with the aim of specifying and establishing Territorial Planning and Development Models, through the design and application of Environmental Policies that regulate the complex environmental management of the territory -especially when addressing inter- and intra-border realities-, it is necessary to start from the maxim that climate change is a reality as long as it is conceived as a process of climate transformation that has become more pronounced in recent decades, consolidating the idea of a crisis. Thus, natural changes in carbon dioxide concentrations and Earth's orbital parameters have interacted with cyclical freezing and thawing processes in the Arctic, as well as with ocean current systems, intensifying their effects on the global climate. All of this has had and continues to have an impact on the environmental reality of the planet in general, and of Spain in particular, since it strengthens and intensifies the countless natural risks, especially those of a climatic nature, which have increased since the middle of the last century -in many cases linked to anthropogenic processes such as the increase in urban areas-, and which define Spain as one of the areas with the highest levels of natural risk on the European continent. Therefore, throughout this research, the incidence and impact of extreme meteorological phenomena on the physical, human and economic environment in the study area of the territorial reality of the coastal areas of the Province of Pontevedra are highlighted, highlighting how these atmospheric processes constitute a key element in shaping the climate regime of the Galician coast, where their frequency and intensity generate episodes of high disturbance, with significant implications for territorial dynamics and the safety of its inhabitants; which is why climate risk planning measures are needed, based on action and performance measures by local governments that promote cooperation to mitigate or eradicate cross-border risks and impacts (Sotelo Pérez, M. & Sotelo Pérez, I., 2017), as well as to strengthen resilience in border regions.

So, we start from the epistemological assumption that the contemporary understanding of the so-called "risk society" requires an interdisciplinary approach based on contributions from multiple fields of knowledge. In this sense, it is essential to pay attention to the studies developed by specialists who, with solid training in disciplines such as Geography, Economics, and Law, contribute to unravelling the conceptual dimensions that shape this paradigm. These works also provide a body of empirical data and analytical frameworks that allow other researchers to address, from a critical and rigorous perspective, a complex reality characterized by its dynamism and the challenges posed by social action. In this context, it is imperative to develop theoretical and practical responses that address emerging problems linked to socio-spatial transformation processes and risk management in its many forms.

In the analysis of environmental risks linked to climate change, especially in its cross-border dimension within the Spanish context, it is important to consider the contributions of specialists who warn of the need to guarantee minimum conditions of material well-being for the population that makes up the social fabric of the State. This requirement, formulated in terms of equity and sustainability, translates into the identification of economic indicators that allow us to differentiate between two interrelated, but conceptually distinct, processes: economic expansion and structural progress. Economic expansion refers to the quantitative increase in productive activity, accompanied by an intensification of investments that favor the consolidation of strategic infrastructure in key sectors such as agriculture and industry. In contrast, structural progress implies a qualitative transformation of the socioeconomic fabric, assessed through macroeconomic parameters such as employment rates, resilience to financial shocks, and the system's capacity to absorb and overcome crises. This analytical framework allows us to address environmental challenges not only as physical phenomena, but also as expressions of territorial and social imbalances that require integrated responses based on planning, governance, and environmental justice.

It is essential to recognize that the progress of a community is not limited to the accumulation of resources or economic growth, but is manifested in its capacity to intervene innovatively and efficiently in the configuration of its physical, symbolic, and social environment. This transformative power also extends to the way in which this community establishes links and dynamics with other territorial and political entities, in an increasingly complex network of interdependencies. In this context, it is crucial to rethink the guidelines that guide progress policies (both planning and final, as well as specific (instrumental and sectoral)), both nationally and internationally, promoting approaches that prioritize equity, sustainability, and multi-scale cooperation as fundamental pillars of a contemporary development strategy.

From this perspective, it is legitimate to argue that the aforementioned economic development constitutes, in essence, a direct derivation of economic growth-among other variables, such as social progress and environmental quality-. Such a statement inevitably leads to an analysis of the various risks, or more precisely, the various impacts resulting from the increase in infrastructure and industrial production.

These processes contribute to the realization of one of the fundamental objectives of contemporary society, politically structured, settled in a specific territory and spatially delimited. This phenomenon gives rise to a series of circumstances that generate significant repercussions on what, in the collective imagination, has been conceptualized under the theoretical model of the subjective natural heritage environment. The case demonstrates how events that unfold within the borders of the same state and in the same historical period can have profoundly different origins and characteristics. In certain contexts, these origins lie in the social fabric, in the dynamics of what might be called social forces. In other cases, however, these phenomena emerge as a result of minority, voluntary, and structured actions, guided by explicit and deliberate interests, which are identified as processes of manipulation.

In the field of risk epistemology, it is essential to adopt an integrative approach that transcends the traditional dichotomy between environmental crisis and social crisis, recognizing the existence of a single crisis of a structural and multifactorial nature: *the socio-environmental crisis*. This configuration demands a critical reinterpretation of current development models, in which poverty eradication and the preservation of ecological systems are not conceived as separate objectives, but rather as interdependent components of a single sustainability strategy (Sotelo Navalpotro, J.A., 2000).

This research focuses on the study of risk vectors affecting the northwesternmost territory of Spain, whose territorial and socioeconomic configuration presents a high degree of heterogeneity and a complex historical structure. In this context, the precautionary principle takes on a central regulatory role, establishing that, in the face of potential threats of serious and irreversible damage, the absence of conclusive scientific evidence should not constitute an obstacle to the adoption of preventive measures. This logic is justified by the growing incidence of negative externalities arising from the dominant economic system, characterized by a neoliberal market capitalism on a global scale, in which market self-regulation does not in itself guarantee distributive equity or territorial cohesion, and where certain economic conglomerates exert a capacity for political pressure that sometimes exceeds the sovereignty of States. Although the global scale constitutes the essential analytical horizon for a comprehensive understanding of the phenomenon, transformative praxis must be deployed from the local level. Consequently, it is pertinent to investigate the existence of *risk regions* within the national territory, using a multi-scale approach that allows for the identification of territorial patterns of vulnerability and the development of mitigation strategies tailored to the socio-environmental specificities of each context.

2. ENVIRONMENTAL RISKS: CROSSING BOUNDARIES

The investigation into the risks inherent in development processes, in the territorial sphere of borders, requires an epistemological reconfiguration that incorporates new ecological frameworks as essential interpretive vectors for understanding the evolution of the Social Sciences. In particular, disciplines such as Geography and Economics are challenged by the need to integrate systemic approaches that transcend traditional dichotomies between nature and society and that allow us to address the complexity of territorial and economic phenomena from a relational and multi-scalar perspective. Economists such as John Kenneth Galbraith and Amartya Sen have called for the economy to be restored to its original function, based on the rational articulation of policies aimed at ensuring future well-being. This requirement takes on particular relevance in a context characterized by the proliferation of theoretical models and regulatory proposals that, although innovative, remain fragmented and lack a consolidated paradigmatic coherence. In this sense, contemporary economics is immersed in a process of paradigmatic transition, in which attempts are emerging to redefine its ontological and methodological foundations. In light of Thomas Kuhn's "Theory of Scientific Revolutions," it can be said that we are in a phase of structural dissonance, in which the dominant paradigm shows signs of exhaustion, while the emerging one has not yet reached sufficient maturity to constitute itself as a hegemonic framework. This situation invites critical reflection on inherited conceptual limitations and the possibilities of opening up to new ways of thinking about the economy, more sensitive to the ecological, social, and territorial interdependencies that shape the contemporary world.

So, it becomes crucial to examine the interrelationship between the conceptual architectures of development—understood as an integral process beyond economic growth—and the spatial configurations that accompany it. Although the notion of local development has recently been consolidated as an analytical category, its roots can be traced back to historical forms of social organization, revealing a persistent intellectual concern that has spanned eras and disciplines. In this sense, the transition to modernity, especially between the late 19th and early 20th centuries, gave rise to a profound theoretical confrontation. Two paradigms clashed: one oriented toward institutional centralization and state cohesion, and another that championed community structures and ties of proximity. This dichotomy was reflected in the famous distinction between *Gesellschaft* and *Gemeinschaft* proposed by the sociologist Ferdinand Tönnies in German thought, as well as in the divergence between the structuralist sociology of the sociologist Émile Durkheim and the humanist geography of Vidal de la Blache in France, each defending profoundly different models of sociability and territoriality.

In contrast, to Durkheim's sociological vision, which focused on social structures as determinants of collective behaviour, Paul Vidal de la Blache offers a counter from the perspective of geography, stating that this discipline does not deal with people in the abstract, but with concrete places: the *science des lieux*. In his conception, "place" is not merely a physical location, but a singular territorial entity, where natural and human elements intertwine through everyday practices and ways of life that he calls genres de vie. This perspective inaugurates a relational reading of space, in which territory is configured as a product of the interaction between environment and society. In parallel, Ebenezer Howard introduces a visionary proposal in the face of the ravages of industrialization and uncontrolled urban expansion: the *Garden Cities*. These garden-cities, conceived as self-sufficient centers on the outskirts of metropolises, sought to reconcile the benefits of the rural environment with urban opportunities. Their design aspired to create integrated communities where work, housing, and social life coexisted harmoniously within a humanized territorial framework. The logic underlying this model responds to a dual need: on the one hand, to mitigate the adverse effects of a natural environment altered by industrial dynamics (so-called natural risks), and on the other, to prevent the generation of new threats derived from human action (anthropogenic and technological risks). In this sense, territorial planning becomes a strategic tool for rebalancing the relationship between society and the environment, promoting habitable spaces that not only protect the population but also foster a conscious and sustainable transformation of the environment.

Throughout this study, a methodological strategy based on the deliberate selection and rigorous analysis of specific cases has been adopted, based on the premise that case studies constitute a research tool with a high epistemological density in the field of human and social sciences. This approach allows for a systematic and in-depth exploration of unique social entities, facilitating the understanding of complex phenomena in their territorial, historical, and structural context. The research is articulated through a methodological dialectic that combines the hypothetico-deductive method with the inductive method, favoring a heuristic approach that transitions between the formulation of theoretical propositions and empirical inference. Within this framework, the concept of risk is defined as a potential contingency, whether natural, anthropogenic, or hybrid, capable of generating significant disruptions to the physical, patrimonial, or ecological integrity of socio-environmental systems. In accordance with the previously established methodological guidelines, a tripartite classification of risks is proposed, based on their etiology and the dynamics that shape them. Natural risks originate in physical, geodynamic, or bioclimatic processes inherent to the natural environment. Their occurrence is modulated by the geographic, geomorphological, and ecological conditions of each region, determining their spatial distribution, intensity, and recurrence. Anthropogenic risks, on the other hand, emerge from human practices accumulated over time, linked to patterns of territorial occupation, production, consumption, and social behaviour. They reflect an unbalanced interaction between society and the environment that often leads to structural damage. Finally, technological risks arise from the use, implementation, or failure of human-developed technical systems. These risks range from industrial accidents to malfunctions in critical infrastructure, posing complex challenges in terms of risk governance, territorial resilience, and the ethics of innovation.

This analytical framework allows risks to be addressed not as exogenous or isolated threats, but as manifestations of the socio-environmental relationships that shape the territory. Consequently, the possibility arises of designing mitigation, adaptation and transformation strategies that respond to local specificities and are guided by principles of sustainability, equity and spatial justice, which allow territorial cooperation and development to be addressed from a cross-border perspective that promotes optimization, by creating economically viable, socially acceptable, and environmentally responsible situations, minimizing -if not eradicating-, the risks, impacts, or damage to the environment.

3. THE CROSS-BORDER REALITY OF CLIMATE CHANGE

Climate change is defined as a sustained modification in the state of the Earth's climate system, composed of interrelated subsystems: the atmosphere, hydrosphere, cryosphere, lithosphere, and biosphere. This variation, which extends over time scales long enough to induce a new state of dynamic equilibrium, requires a comprehensive methodological approach that considers both the characterization of the phenomenon and the conceptual delimitation of the term "climate" and its constituent elements. From a geophysical perspective, climate can be understood as the temporal succession of spatial configurations of fundamental meteorological variables -primarily air temperature and precipitation-, distributed over the Earth's surface. This definition allows us to analyse climate as a complex system, governed by multiscale interactions between physical, chemical, and biological processes, whose variability responds to both internal and external forcings. Human evolution took place during a relatively recent geological period, corresponding to the Late Pleistocene and Holocene, during which climatic conditions remained relatively stable, as Waters et al., 2016 explain. However, in the contemporary context, a significant acceleration in climate variability patterns is observed, largely attributable to the intensification of human activities. Greenhouse gas emissions, land-use changes, massive deforestation, and the expansion of energy-intensive technological systems have acted as anthropogenic drivers of disturbance, generating imbalances in the energy and biogeochemical flows of the climate system. This phenomenon poses substantial challenges for predictive modelling, climate risk management, and the formulation of public policies aimed at mitigation and adaptation. Consequently, climate change must be addressed as a systemic problem that requires interdisciplinary responses, integrating knowledge from climatology, ecology, environmental economics, and territorial planning, in order to promote resilient strategies in the face of scenarios of increasing climate uncertainty (Sotelo Pérez, M. & Sotelo Pérez, I., 2017).

This way, the Earth's climate system is composed of several interrelated subsystems (atmosphere, oceans, cryosphere, and biosphere), in which the energy that drives the physical and biological processes within these components comes primarily from radiation emitted by the Sun, an active star that generates energy through nuclear fusion reactions. Although nuclear fission energy has limitations for its practical application in the terrestrial context, solar energy constitutes the dominant source that supports phenomena such as atmospheric circulation, ocean currents, plant development, and the metabolic functions of living organisms. In addition, a small proportion of thermal energy originates in the interior of the planet, a product of nuclear reactions of radioactive decay, which contribute to geological dynamics, including tectonic activity.

The global mean temperature is defined as the annual average of air temperatures recorded on the Earth's surface at different geographic locations. Throughout the planet's evolution, this parameter has undergone significant variations. According to the principles of physics, everybody with a temperature emits electromagnetic radiation. In the case of Earth, this emission is primarily in the infrared spectrum, whose wavelength is determined by surface temperature. Some of this radiation is absorbed by certain gases present in the atmosphere at very low concentrations, known as trace gases, which have infrared absorption properties and play a key role in the greenhouse effect.

The presence of low-concentration gaseous compounds in the atmosphere, such as carbon dioxide, methane, and water vapor, plays a fundamental role in modulating the energy balance of the climate system. These gases, known as trace greenhouse gases, have the ability to absorb and re-emit infrared radiation emitted by the Earth's surface, resulting in a partial retention of energy in the troposphere. This phenomenon contributes to the increase in air temperature in the lower layers of the atmosphere, through a mechanism analogous to the thermal insulation provided by porous materials, which limit heat loss from a body to the environment. Air temperature, even under similar global average temperature conditions across geological eras, is determined by multiple structural and dynamic factors. These include the spatial configuration of the continents and oceans, the dynamics of ocean currents, and the concentration of radiatively active gases in the lower atmosphere. The spatiotemporal distribution of atmospheric temperature during the last ice age was influenced by the existence of a polar ocean basin without island interruptions, as well as by the arrangement of three major ocean systems: the Atlantic, the Indo-Pacific, and the Southern Ocean. In this context, the Arctic Ocean, surrounded by continental landmasses, can be conceptualized as a physical system with nonlinear behavior, susceptible to metastable states. This dynamic can be represented by mathematical bifurcation models, which describe transitions between distinct climate regimes (see Figure 1).

Figure 1. Phase diagram of water as a log-lin chart and spatio-temporal distribution of atmospheric temperature.

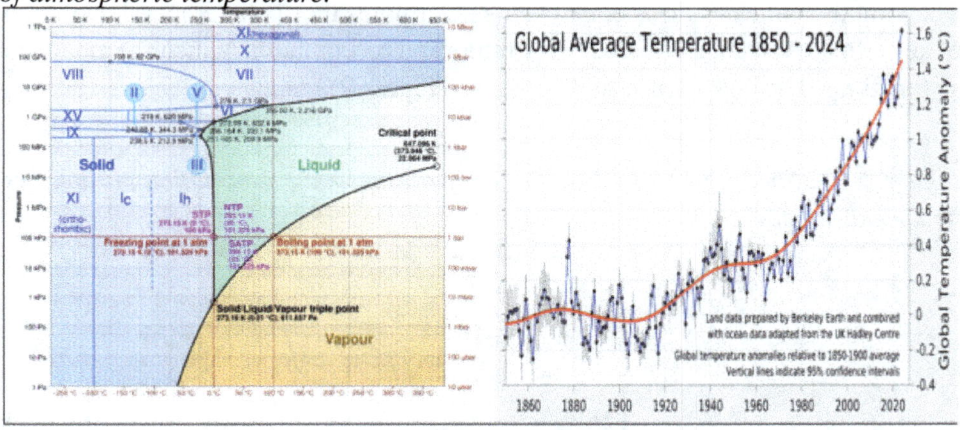

Source: U.S. Geological Survey (USGS) and Berkeley Earth System (2025).

Variations in atmospheric carbon dioxide concentrations, along with oscillations in Earth's orbital parameters, are key factors modulating the climate system. These processes have been amplified by mechanisms internal to the Earth system, such as ocean thermohaline circulation and the expansion and contraction cycles of ice masses at polar latitudes, particularly in the Arctic. Human sociocultural evolution has been closely linked to the climatic dynamics of the Quaternary. Various paleo-environmental and archaeological records suggest that transitions between glacial and interglacial states have influenced settlement patterns, social organization, and technological development. Some mythological narratives that refer to catastrophic events could represent cultural interpretations of extreme climate phenomena occurring during these transitions. At the same time, significant transformations in land use and agricultural productivity associated with regional climate variations have been documented.

The concept of "climate change" requires contextualization within the intrinsically dynamic nature of the climate system. While variations have been a constant throughout geological history, the trends observed today show an acceleration and direction that do not correspond to the patterns of natural variability recorded in previous periods. This uniqueness suggests the intervention of additional forcings, the magnitude and persistence of which are altering the balance of the global climate system. Thus, the inherent complexity of the climate system generates multiple levels of uncertainty, the exploration of which requires advanced analytical tools. An effective approach consists of the application of dynamic models based on non-linear differential equations, capable of representing the interaction between five interdependent subsystems: the atmosphere (analysed both from its thermodynamic and mechanical behaviour and its gaseous composition and relationship to electro-magnetic radiation), water bodies, ice masses, the biosphere, and the Earth's surface. The joint dynamics of these components determines the spatiotemporal evolution of key climate variables, such as temperature and precipitation, which are modelled as functions that vary in time and space (see Figure 2).

Figure 2. Spatio-temporal modeling of temperature and precipitation parameters in a scenario marked by climate change.

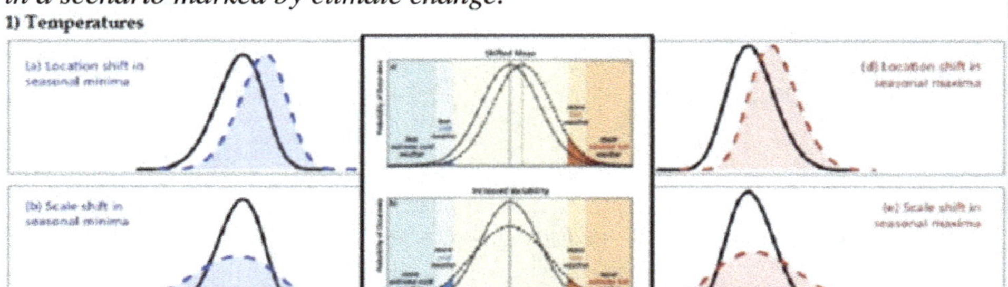

Source: Kodra, E. & Ganguly, A.R. (2014) & Hoylman, ZH, Bocinsky, RK y Jencso, KG. (2022).

As seen in the temperature graph in the outer panel (a), the impact of changes in location parameters on the distribution of extreme values is illustrated. Similarly, panels (b) and (c) show how adjustments to scale and shape parameters affect the distribution. Changes in the location parameter influence the average or typical magnitude of extreme events, while changes in the scale parameter affect the dispersion or width of the distribution. Regarding the shape parameter, variations in this parameter have a direct effect on the behaviour of the upper tail, that is, on the frequency and severity of the most intense extreme events. The reference distributions, which correspond to the family of Generalized Extreme Value distributions, are represented in black. The shifted distributions are shown in blue and red, corresponding to the simulated statistics of the seasonal minimum and maximum values, respectively. For a more detailed description of the procedure used to generate the six side plots, please refer to the Supplementary Information (SI) section, which explains the process performed with randomly simulated data under generalized extreme value models. For its part, in the precipitation graph, a theoretical data set of accumulated precipitation is presented on the horizontal axis, while the associated probability density function is shown on the vertical axis. On the left, a conceptual model is presented illustrating how probability distributions can change over time when conditions shift from a wetter, less variable state to a drier, more variable state. *On the right, a demonstration of how this change can generate a dry bias during dry seasons and a wet bias during periods of higher precipitation is shown, applied to a theoretical distribution* (Hoylman, ZH, Bocinsky, RK & Jencso, KG., 2022).

Thus, the sustained increase in average global temperature, linked to the rising concentration of greenhouse gases in the atmosphere, represents a significant transformation in the Earth's climate system. Although life on the planet will continue in various forms, the focus is on the conditions that will allow the persistence and adaptation of the human species and its social structures. Throughout history, humanity has inhabited extremely diverse climatic environments, from tropical zones to hot deserts and polar regions. This adaptive capacity suggests that the species can survive substantial changes in the global climate. However, new environmental conditions imply a profound reorganization of social, economic, and territorial systems. Human migrations, historically associated with variations in resource availability, could intensify in response to the redistribution of freshwater in different regions of the planet.

The global average temperature is an aggregate measure that reflects temperature variations across different geographic areas. In the coming decades, an increase in the amount of solar energy absorbed by the Earth's surface is expected. In tropical regions, where ocean cover predominates, this energy increase translates primarily into increased evaporation, without necessarily implying a significant increase in temperature. In contrast, the polar regions, covered by ice, accumulate thermal energy without changing their physical state, which generates more pronounced warming at high latitudes. A direct consequence of this energy redistribution is the poleward shift of the precipitation systems currently affecting temperate regions. In the case of the Iberian Peninsula, this phenomenon is particularly significant. Its geographic location determines a rainfall pattern characterized by seasonal rains and periods of drought. The shift of the polar jet stream toward more northern latitudes would intensify episodes of drought, interrupted by sporadic torrential rains, significantly altering the region's water balance and ecological stability.

And it is that, Climate change is generating profound transformations in the planet's natural systems. One of the most evident manifestations is the accelerated disappearance of glaciers, a phenomenon that has intensified in recent decades. In the Pyrenees, the Monte Perdido glacier has practically disappeared in the last hundred years. This trend is replicated around the world: Nepal has lost almost a third of its glacial volume in thirty years, and European glaciers could shrink by half in the next hundred years; or, on remote Heard Island in the Southern Ocean, the loss of approximately 25% of its ice surface has been documented since the mid-1950s. This territory, considered one of the last pristine ecosystems on the planet, has suffered significant glacial retreat, despite its isolation and limited direct human intervention.

The decline in winter snowpack compromises freshwater reserves available in summer, especially in regions that depend on melting snow to supply reservoirs and aquifers. Furthermore, rising average temperatures, even by as little as two degrees, increase evapotranspiration from soils and vegetation, worsening periods of drought and reducing water availability for agriculture and human consumption. In tropical and Caribbean areas, global warming has additional implications. Rising ocean temperatures intensify relative air humidity, which enhances the formation and development of hurricanes. These phenomena are fueled by water vapor in rising air columns, and a warmer atmosphere provides greater energy for their intensification. Recent research confirms that hurricanes are strengthening more rapidly and leaving more intense rainfall, with an 8% increase for every additional degree in sea temperature. Furthermore, the weakening of the polar jet stream in certain regions favors atmospheric conditions more conducive to the formation of persistent tropical cyclones.

A paradigmatic case of the impact of global warming was recorded in Greenland in 2024, when a subglacial flood was documented that released approximately 100 billion litters of water. This phenomenon was caused by the fracturing of the ice from below, challenging traditional models of ice sheet behaviour. The massive release of subglacial water not only contributes to sea level rise but also alters the internal dynamics of the ice sheet, reducing its structural stability and accelerating its displacement into the ocean. Likewise, during the 2023–2024 annual cycle, Greenland has again recorded a net loss of ice mass, showing a glacial retreat that has continued over the past thirty years. Despite abundant winter snowfall that delayed the onset of the melt season, intense melting episodes during the summer resulted in a total loss estimated at around 80 billion tons of ice. This phenomenon has been intensified by the *calving* of large blocks at the edges of the ice shelf, a process known as calving, which generates icebergs and contributes significantly to sea level rise. Furthermore, it has been observed that much of the mass loss comes from areas below sea level, especially in deep fjords, further complicating the stability dynamics of the ice sheet.

Figure 3. Changes in the Greenland ice sheet in 2019 and 2022.

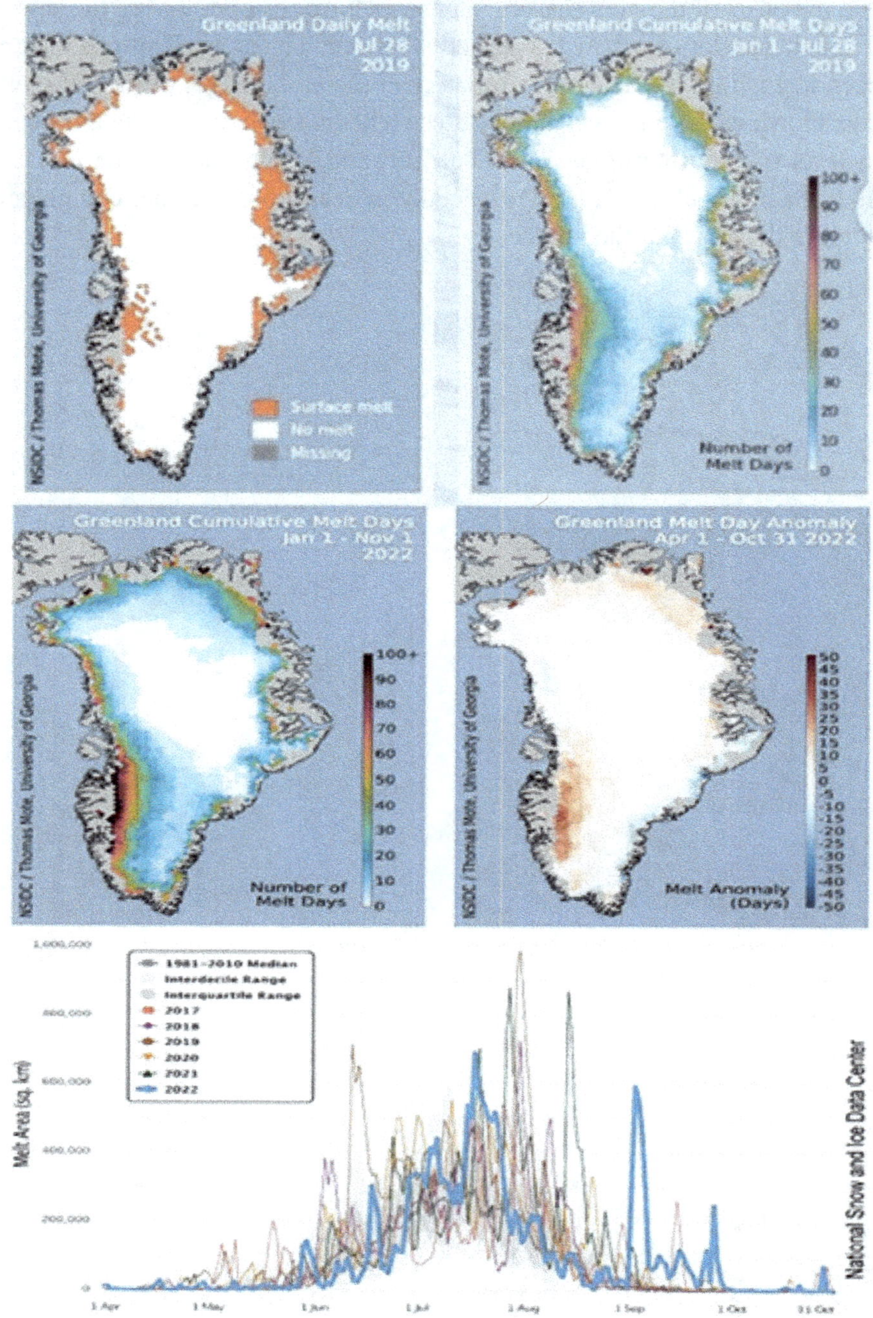

Source: NSIDC (2019 y 2022).

A recent study published in Nature has revealed that Greenland's glaciers have lost 20% more mass than previously estimated. This revision, based on satellite imagery and big data analysis, has corrected previous measurements that did not account for glacial retreat at the island's edges. In total, the ice sheet is estimated to have lost more than one billion metric tons. Although this loss has not had an immediate direct impact on global sea level, it does pose a threat to ocean circulation and global heat distribution. Ocean warming, which reached new records in 2024, continues to be the main driver of accelerated melting, affecting both the surface and base of the ice sheet. This is certainly not an isolated incident, in 2012, almost the entire surface of Greenland melted in a short period of time (more than half of the ice sheet was reduced in just four days). This happened again in August 2019, a year that left an indelible mark on Greenland's climatic history (see Figure 3). On a day that defied expectations, the island lost nearly two billion tons of ice. Temperatures soared, reaching levels more than twenty-three degrees above normal. What would normally be a transitional period toward cooling unexpectedly turned into a day of massive melting. From space, satellites captured a transformed landscape: pools of liquid water spread across the white surface, like open wounds on a glacier. On land, the village of Qaanaaq experienced an unprecedented day, with the thermometer reading temperatures above seventeen degrees Celsius, an unthinkable figure in such a northern region.

Such events, once considered exceptional, are no longer rare. Researchers from the Geological Survey of Denmark and Greenland warn that global warming is creating the perfect conditions for ice to yield more frequently. Even the sea ice surrounding the island showed signs of weakening, affected by record-breaking ocean temperatures. The consequences of this phenomenon go beyond the landscape. Melting ice in Greenland not only threatens to raise sea levels by several meters, but could also alter global weather patterns. Furthermore, the water that accumulates on the ice reduces its ability to reflect sunlight, further accelerating the melting process. What happened that day was not just a meteorological anomaly. It was a silent warning, a sign that the climate balance is changing faster than we imagine, as similar unusual situations were recorded again in 2021 and 2022, maintaining a continuous temporal dynamic that, as we can see, is intensifying over increasingly shorter periods of time. Available data show that Greenland is on a critical trajectory with respect to climate change. If current trends continue, they could lead to a tipping point, where glacial mass loss becomes irreversible, triggering a significant rise in global sea level. This phenomenon is not a hypothetical scenario, but a tangible consequence of ongoing climate change. Increasingly frequent and intense extreme weather events underscore that climate change is already a tangible reality, not a future threat (in fact, ice cores indicate that such melting events typically occur approximately every 150 years; in fact, prior to 2012, a similar event was recorded at the end of the 19th century). However, if similar phenomena continue to occur in the coming years, it could signal a worrying shift in climate conditions, and it is in this context that urgent and coordinated global responses are required to mitigate their effects and prevent irreversible damage to our ecosystems and societies.

One of the most notable phenomena in the melting process is the alternation between freezing and thawing. In the higher elevations, most of the meltwater refreezes quickly and almost immediately, while in coastal regions, a large amount is lost directly to the ocean. This process contributes to an additional phenomenon: the calving of large blocks of ice, which intensifies after these events, as is the case with the "Spalte" ice mass, a section of the 79N glacier. This phenomenon is the second of its kind and is characterized by the collapse of enormous masses of floating water, the magnitude of which is extraordinary. Although, at first, researchers did not consider these collapses to be alarming in themselves, the combination of these events with accelerated melting raises questions about possible changes in the climate system. In this context, satellite observations have become crucial, as they allow us to study how these events might be interrelated and how they might reflect a broader alteration in global climate patterns (see Figure 4).

Figure 4. Ice mass calving "Spalte" section of glacier 79N in Greenland.

Source: ESA, Copernicus (2019).

During these melting processes, which have occurred annually since July 2012, Greenland's surface ice cover has melted over a larger area than at any time in more than 30 years of satellite observations. Nearly all of Greenland's ice sheet, from the thinnest coastal areas to 2 kilometers deep inland, has experienced some degree of surface melting annually by the middle of this month due to warmer temperatures, according to measurements from three independent satellites analysed by scientists from the National Aeronautics and Space Administration and universities.

On average, during the summer, about half of the Greenland ice sheet's surface melts naturally. At high altitudes, most of this meltwater quickly refreezes in place. Near the coast, some of the meltwater is retained by the ice sheet, and the rest is lost to the ocean. But this year after year, the degree of ice melting at the surface has increased significantly. Researchers have not yet determined whether this widespread melting event will affect the overall volume of ice loss in the summer, keeping in mind that such pronounced melting on the ice sheet in this area has not occurred since 1889, according to the ice cores analysed.

In this way, climate change, understood within the framework of natural, anthropogenic, and technological risks, represents a complex challenge that must be analysed from diverse perspectives. Despite the increasing uncontrolled emission of greenhouse gases, scientific theory suggests that global warming will reach a natural limit at which the global average temperature will stabilize, regardless of further increases in the concentrations of these gases. This phenomenon, derived from the absorption by the atmosphere of thermal radiation emitted by the Earth's surface, will cease when the atmosphere has reached its maximum absorption capacity. Currently, almost 95 percent of Earth's radiation is already absorbed, with the remainder escaping directly into outer space through the so-called "atmospheric window." Therefore, although climate change predictions are varied and, in many cases, point to catastrophic scenarios, we must recognize that a moderate increase in global temperature, although significant, is not sufficient in principle to drastically transform the Earth's climate. Indeed, it is estimated that the upper limit of warming will not exceed two and a half degrees, which may ultimately offer a more nuanced and less alarmist view of future climate projections.

However, it is no less true that, in recent decades, there has been a notable increase in the vulnerability and exposure of human and natural systems to threats of climatic and geophysical origin. This phenomenon manifests itself in the intensification of risks associated with extreme events, such as intense rainfall resulting from rapidly evolving atmospheric processes, including so-called explosive cyclogenesis. The climate regime exerts a differentiated influence on various types of territories - including natural, rural, urban, and coastal areas-, conditioning their stability and resilience. This dynamic is particularly observable in regions such as Galicia, and more pronounced in its coastal strip, which is the focus of this research.

Spain is highly exposed to multiple natural risks, especially those of climate origin. This situation is exacerbated by the accelerated and, in many cases, disorderly expansion of urban centers, as well as by land occupation processes that have lacked adequate planning in various regions of the country. As a result, Spain is among the most vulnerable in Europe in terms of natural risk, with significant implications at the territorial, economic, and social levels. So-called natural risks -such as floods, droughts, heat waves, and extreme weather events-, directly affect the resident population, generating impacts that extend beyond the physical realm, affecting public health, infrastructure, and social cohesion. This situation is intensified by the complexity of Spain's physical environment, characterized by a strategic geographic location, diverse topography, and marked climate variability. Furthermore, the demographic dynamics -with a heterogeneous, constantly growing and mobile population-, together with the economic structure of many regions, contribute to consolidating Spain as a risk territory, exposed not only to natural threats, but also to anthropogenic and technological factors that amplify its systemic vulnerability.

So, the assessment of natural risks, when considered within the context of territorial dynamics, is embedded in the processes that shape geographical realities. These processes, in turn, materialize in the repercussions derived from the inherent complexity of spatial differentiation caused by human activities, specifically those related to the use of land for recreational purposes. Interaction with the physical environment, particularly with meteorological phenomena, entails an inherent exposure to risks that intensify when the areas involved are classified as "vulnerable territories". While it is true that the study of risks and their corresponding impacts, both natural and anthropogenic, has gained relevance in recent decades, it is necessary to observe how, in the Spanish context, and more specifically in Galicia, significant transformations have occurred. These changes are largely attributable to the modernization and expansion of transportation infrastructure, as well as the reconfiguration of the population's lifestyle. This phenomenon can be related to the so-called "rural exodus," the consequences of which have been described with nostalgia by some authors when recalling the characteristics of rural landscapes prior to their abandonment. The evolution of lifestyles and the territorial transformations associated with these processes have been the subject of reflection in various studies, which underscore the socioeconomic and environmental significance of these changes. However, what has the greatest impact is the current state of urban areas, large cities crowded with people who live, and, more especially, in recent decades, specifically since the late fifties of the last century, in which the progressive disappearance of villages and towns, or at least of their cultures and essences, has been observed, marking the maximum stipulated repeatedly in recent times in which it is projected that, in a time horizon of approximately twenty years, ninety percent of the world's population will reside in urban areas. This fact has significant implications from a geographical perspective, especially considering the recent focus of studies on "natural risks", which underscores the urgency of adopting a comprehensive geographic approach. This approach must encompass both the physical and human dimensions of the territory, with the aim of understanding the complexity of interactions between humans and their environment, while recognizing that these processes are dynamic and constantly evolving.

Throughout this analysis, we examine the territorial situation of the province of Pontevedra, whose climate is profoundly influenced by its proximity to the sea. The region is under the influence of the Gulf Stream, giving it a temperate and humid climate, particularly noticeable in coastal areas, where winters are milder compared to the interior. Wind patterns vary considerably throughout the year, and notable differences are observed between different areas. In Vigo, for example, the prevailing winds come from the west, heading toward the estuaries, and from the south, while in Pontevedra the most common winds are from the north and southwest. This climate, characterized by its mildness and variability, represents a crucial factor in territorial dynamics, influencing the characteristics of the environment and human activities in the region.

Nevertheless, it is important to note that, despite the relatively benign conditions in the area, extreme weather events, such as extratropical cyclonic low-pressure systems, also occur, and are considered natural risks. These meteorological events, by altering the usual pattern of atmospheric conditions, can have significant repercussions on both climate stability and territorial management and regional development strategies. The occurrence of these phenomena highlights the need to understand the risks inherent to the local climate system in order to anticipate and mitigate their potential impacts on the region.

Within the framework of territorial management dynamics, it is evident that the diversification of human activities beyond traditional periods of peak demand has become one of the key strategies for ensuring the sustainability and stability of regional development. This phenomenon takes on particular relevance in coastal areas that, due to their geographic location and environmental characteristics, are subject to a series of climatic conditions that directly influence patterns of land use and occupation. In this sense, integrated territorial planning, which addresses both the physical and social aspects of the environment, is presented as an essential tool for balancing human demands with the capabilities of the natural environment (Sotelo Pérez, I. & Sotelo Pérez, M., 2023).

The climate of the region in question, defined by its proximity to the ocean and the associated atmospheric influences, is a determining factor in the organization of the space. Meteorological variations, especially those related to atmospheric pressure fluctuations, tend to be more pronounced during the winter months, although temperature fluctuations remain within a moderate range, with no significant thermal extremes recorded. The low altitude of the terrain limits the presence of meteorological phenomena associated with low temperatures, such as snow, which creates a series of limitations and possibilities for the occupation of the territory.

Despite the generally mild climate, the area's meteorological characteristics pose challenges to the stability of human activities throughout the year. The seasonality of climate phenomena generates periods in which atmospheric conditions can become unstable, increasing the likelihood of extreme weather events that disrupt territorial equilibrium. Among these phenomena, extratropical cyclonic low-pressure systems stand out as a relevant example of meteorological events that can disrupt normal weather conditions and, consequently, affect the sustainability of infrastructure and systems that depend on a stable climate.

A thorough understanding of these risks, along with the identification of the risks inherent to the prevailing climate, is a fundamental element for proper territorial management. Understanding how climate change impacts development and spatial occupancy is key to developing mitigation strategies that reduce the vulnerability of areas affected by natural phenomena. In this context, the study of interactions between natural elements and human activities must be approached comprehensively, in order to anticipate and manage the impacts of climate variations on the territory.

4. TERRITORIES VULNERABLE TO THE CROSS-BOUNDARY RISKS OF CLIMATE CHANGE

In the contemporary context, the study of societies characterized by their vulnerability to various risks (natural, anthropogenic, and/or technological) requires a holistic approach, supported by the convergence of multiple scientific disciplines. Research conducted by specialists in fields such as Geography, among others, provides an essential conceptual framework for elucidating the nature of this phenomenon, which often manifests itself as a highly complex and abstract theoretical category. The interdisciplinary perspective underlying such studies enables a more rigorous interpretation of the risks emerging from current social dynamics, while providing crucial analytical tools for a deeper understanding of the socio-territorial reality in which we are immersed. These studies not only offer a conceptual understanding of the risks inherent in contemporary society, but also provide data and methodologies that allow other researchers to approach a complex reality characterized by difficulties inherent in social, economic, and technological processes, as well as their impacts on cross-border situations. Given this reality, the need to find answers to problems arising from human activities becomes a priority in the study of cross-border risks. When analysing the aforementioned "cross-border risks", many of these experts emphasize the importance of ensuring that the people who make up the social structure can achieve a minimum level of material development, regardless of border boundaries. This objective is realized through a series of factors, among which economic aspects stand out. Regional economic growth, increased per capita income, and the availability of advanced technologies are fundamental elements for improving competitiveness and productivity in interregional industrial systems. These factors allow us to distinguish between economic growth, understood as a quantitative increase in production, and territorial development, which advocates an equitable distribution of resources and a qualitative improvement in living conditions. All of this is part of a macroeconomic analysis that seeks to understand the dynamics of progress without losing sight of the risks it entails, both locally and globally.

In this context, cross-border risks take on particular relevance, as they involve phenomena whose nature, impact, and management transcend national borders, simultaneously affecting multiple territories and populations. These risks can arise from shared natural disasters (such as forest fires, floods, or droughts), negative externalities of industrial processes (such as air or water pollution), or systemic crises (such as pandemics, cyberattacks, or failures of critical infrastructure). Addressing them requires not only technical and political cooperation between states, but also a reconfiguration of regulatory and environmental governance frameworks at the regional and international levels.

Within this analytical framework, economic growth can be conceived as an empirical manifestation of sustained growth in production and the consolidation of infrastructure. This phenomenon not only drives territorial expansion but also significantly impacts the political and social makeup of the communities that inhabit these spaces. It is therefore imperative to examine the impacts and risks associated with the intensification of infrastructure and the growth of industrial activity, as these processes generate structural transformations in territories and influence collective perception and environmental management. Likewise, contemporary social systems tend to be oriented toward the conservation of the spaces they occupy, although this objective is hampered by the limited availability of viable alternatives that enable the implementation of effective territorial strategies, especially when addressing the most complex border realities. This inadequacy gives rise to a series of dynamics that directly impact social consciousness, affecting what has been conceptualized as the "subjective natural heritage environment", a notion that refers to the way in which human practices impact the territory, understood as the setting where both the benefits of economic development and the risks inherent in this process materialize. Economic activities, in general terms, are organized around the principles of progress and modernization, with the aim of satisfying social demands through production and consumption. In a development-oriented society, it is essential to have an economic system capable not only of ensuring the coverage of basic needs but also of providing the necessary resources to confront and mitigate the effects of natural risks, whose occurrence is characterized by their unpredictability. In this sense, development must be accompanied by forward-looking mechanisms that allow for anticipating these risks and minimizing their impacts in the medium and long term.

In the contemporary context, a territory is considered to have reached an advanced stage of development when its population not only enjoys decent and sustainable living conditions, but also benefits from a consolidated, efficient, and technologically sophisticated productive structure. This level of development is reflected in the availability of capital goods that facilitate strategic investment, contributing to the reduction or elimination of socioeconomic inequalities between different population groups and between different regions (regardless of their borders). Likewise, the institutional capacity of the territory to design and implement monitoring, early warning, and preventive management systems for natural risks is recognized, with the aim of mitigating their adverse effects on society and regional economic activities. In this sense, a territory can be categorized as developed when these conditions are met within its geographic boundaries, giving rise to a capitalist economic system characterized by its structural maturity and the incorporation of environmental protection as a cross-cutting axis of its public policies. This paradigm has evolved in recent decades into the concept of sustainable development, understood as a model that balances the economic, social, and ecological dimensions of progress. Within this framework, territories that have managed to harmoniously integrate economic growth with social equity and environmental responsibility are considered developed, especially in contrast to those regions that still struggle to achieve this balance. This development process entails the assumption of differentiated responsibilities both at the legal level (through the State's capacity to guarantee the effective recognition of rights and obligations), and at the level of civil society, whose active participation is essential to ensure that development translates into tangible and equitable benefits for all citizens.

It is essential to consider the complex, yet invaluable, environmental reality that underlies all human activity in the region. In the current context, this environment faces a situation that can be described as one of legal and environmental risk, marked by industrial and productive infrastructures fully regulated by rigorously structured legal systems and regulations, the implementation of which falls to technically competent jurisdictional bodies. However, this idealized vision is often contrasted with practice, in which these systems sometimes appear to disregard the effective fulfillment of fundamental rights, as well as the protection of public health and collective well-being, despite the fact that these guarantees are formally recognized in legal codes. This apparent dysfunctionality generates the perception that the institutions responsible for ensuring regulatory compliance do not fully perform their functions or that they are sometimes subordinated to political interests or particular economic models (Sotelo Pérez, I. & Sotelo Pérez, M., 2023). Despite the growing regulatory landscape and the proliferation of environmental regulations, the natural environment continues to show signs of deterioration, pollution, and increasing vulnerability, reflecting a critical situation of structural risk. Furthermore, the development of new territorial areas is often determined by the confluence of social, political, and economic interests, which frequently minimize or even ignore the risks inherent in these spaces. This dynamic reinforces the need to adopt a comprehensive approach to risk analysis, which would allow for the design of territorial planning strategies oriented toward sustainability and environmental resilience. In this context, and taking as reference previous studies on the so-called "regions at risk" -such as the Mediterranean, Cantabrian and Canary Islands coasts-, it is pertinent to extend the analysis to other vulnerable areas, such as the Galician Atlantic coast, and, in particular, when addressing the study of the coastal areas of the Province of Pontevedra, which requires specialized attention within the framework of environmental and legal management, given its growing vulnerability, marked by the risks that hit said coast, from a cross-border perspective.

5. THE REALITY OF CROSS-BOUNDARY ENVIRONMENTAL RISKS IN THE COASTAL AREAS OF THE PROVINCE OF PONTEVEDRA

In the context of contemporary environmental challenges, extreme weather events occupy a central position due to their capacity to generate high levels of hazard and impact on exposed territories. These events, whose recurrence and magnitude have intensified in recent decades, constitute tangible expressions of the alteration of global climate systems, closely linked to anthropogenic climate change. The interaction between atmospheric systems and pressure dynamics on a planetary scale plays a decisive role in shaping extreme weather patterns, with direct implications for ecological stability, human security, and territorial resilience.

Figure 5. Analysis of the Arctic Oscillation (NAO) indices in the study region.

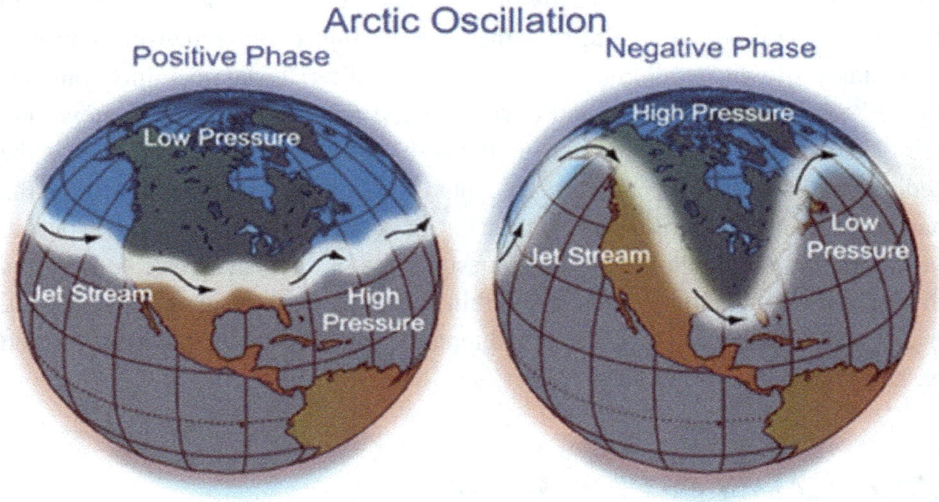

Source: National Oceanic and Atmospheric Administration (2024).

The coastal region of the Province of Pontevedra, located in the northwest of the Iberian Peninsula, represents a paradigmatic case of climate vulnerability. Its exposure to fluctuations in the Northern Hemisphere, particularly the North Atlantic Oscillation, determines the frequency and intensity of meteorological phenomena affecting its coastal and continental areas. This oscillation pattern, defined by variations in the pressure differential between the Icelandic low pressure center and the Azores high, modulates zonal atmospheric circulation and is closely linked to the formation of extratropical cyclonic low-pressure systems. The variability of the North Atlantic Oscillation, in correlation with the Arctic Oscillation indices, determines the alternation between periods of temperate and dry climate (associated with positive values), and episodes of greater instability, humidity, and temperature decline, associated with negative values.

Figure 6. Analysis of the North Atlantic Oscillation (NAO) indices in the study region.

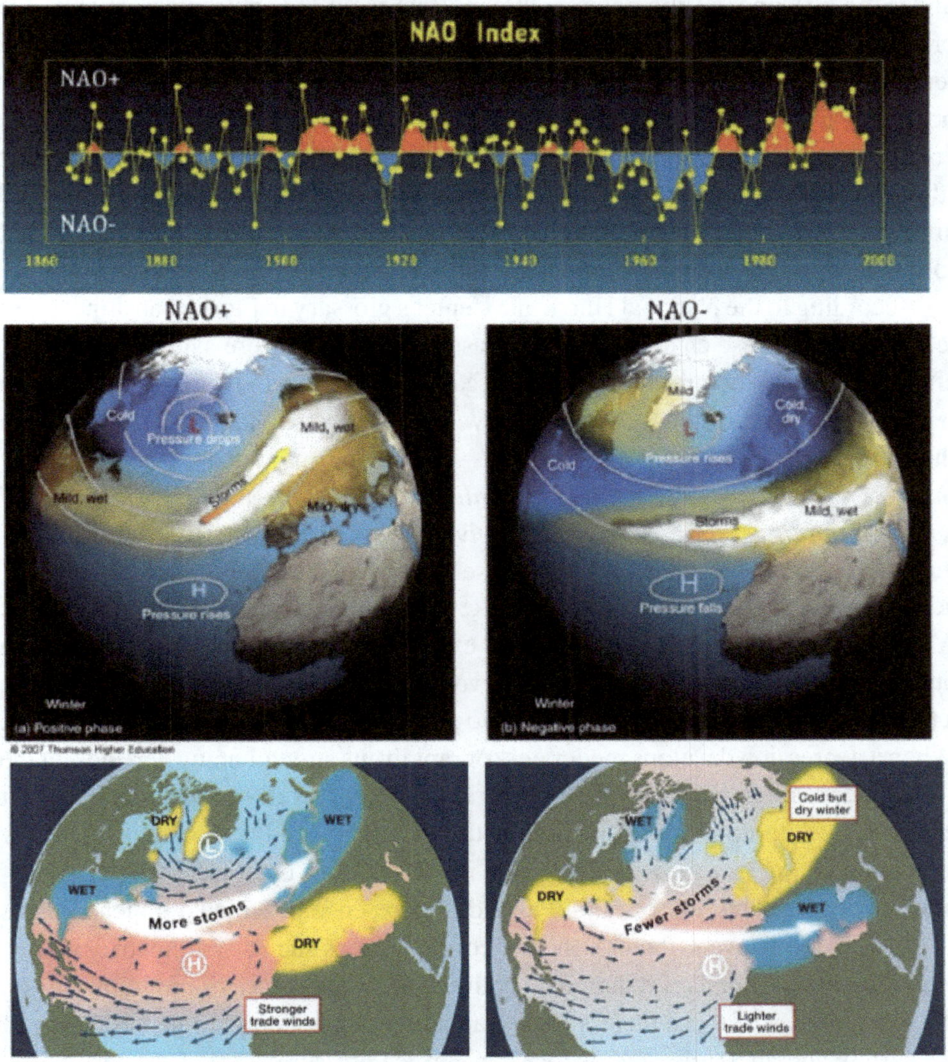

Source: Thomson Higher Education & Maggie Nelson (obtained from Cazatormentas.com, 2024).

In light of the above, if we analyse Figures 5 and 6, we can see how, in the Arctic Oscillation indices, the positive values indicate a more intense zonal current, associated with the strengthening of the Azores High, while the negative values reflect a weakening of said current, which translates into a loss of intensity of the anticyclone. On the other hand, if we analyse the North Atlantic Oscillation (which measures the variability of atmospheric circulation from North America to Europe and even to Northern Asia. Its value is estimated by the difference in sea level pressure that occurs between the low pressures of Iceland and the high pressures of the Azores), positive values indicate a zonal current displaced towards more northern latitudes, under the influence of the polar jet stream (that is, in the subtropical Atlantic the pressure is higher than usual and in Iceland the opposite occurs, thanks to this difference the winter storms that cross the Atlantic Ocean are more intense and their trajectory is located further north than during the negative phase), while, on the contrary, negative values imply a displacement towards the south of said current (during this phase both the subtropical high pressure and the low pressure of Iceland are weakened so that the pressure gradient is small so that the winter storms that follow a westerly trajectory or circulate further south than usual).

According to the National Hurricane Center's glossary in Florida, an atmospheric system with these characteristics is classified as a cold-core synoptic structure, characterized by the presence of well-organized frontal bands. In meteorological terms, it is an extratropical cyclone or low pressure system, that is, a phenomenon that forms once a tropical cyclone loses its essential properties. *The genesis of these storms is linked to baroclinic instability, a mechanism activated when marked temperature differences occur over relatively short distances* (Kodra, E. & Ganguly, A.R., 2014). This situation gives rise to a strong horizontal thermal gradient, accompanied by significant wind shear, key conditions for the system to intensify. All cyclones that develop in the Atlantic experience a phase of cyclogenesis, both during their formation and during their evolution and persistence, which results in a pronounced drop in surface atmospheric pressure, especially in the initial stages. When a depression develops moderately or rapidly, its strengthening depends on the simultaneous interaction of various factors at different levels of the atmosphere. At high levels, determining factors include deep troughs, cold depressions, the polar jet stream acting in zones of divergence in the outflow, cyclonic vorticity advection maxima around 300 or 500 hPa, and the presence of isolated upper-air lows (IADLs). At low levels, the presence of configurations such as barometric minima or depressional troughs is required, in addition to the influence of orographic lows. The coincidence of these factors in the same region favors rapid intensification of the system and an accelerated decrease in surface pressure.

When wind flows converge toward a specific point on the Earth's surface, an acceleration occurs proportional to the magnitude of the compression exerted on them. This phenomenon manifests itself with particular intensity in large-scale cyclonic systems, such as hurricanes and typhoons, where the strongest winds originate from the vigorous rise of warm, humid air from the ocean surface to upper levels of the atmosphere. This rise can even reach the stratopause, an extremely thin layer that marks the transition between the troposphere and the stratosphere, characterized by its thermal stability. This type of atmospheric dynamics takes on particular relevance when considering that, despite its limited vertical extension, the Earth's atmosphere constitutes an effective medium for the propagation of extreme weather phenomena.

These include events such as torrential rainfall, prolonged droughts, heat waves, intrusions of cold air from polar origin, and arid conditions similar to those of the Sahara Desert. In this context, Spain's geographic location places it in a zone of interaction between various meteorological systems in the Northern Hemisphere, which increases its vulnerability to these types of events. The figures analysed demonstrate the interaction between two air masses with contrasting characteristics: on the one hand, a significant advection of water vapor -precipitable water-, from the warm surface waters of the ocean; on the other, an intrusion of cold air carried by the polar jet stream, which circulates at altitudes close to eleven thousand meters. This interaction favors the formation of small-scale but high-intensity cyclonic systems with hurricane-like characteristics. An example of this can be seen in northwestern Galicia, where a system of this nature has been detected moving northward or northwestward, generating significant impacts due to both the strength of its winds and the associated rainfall. These phenomena should not be considered isolated events, but rather recurring manifestations within the same hydrological year. Their frequency has direct implications for territorial risk levels, especially in areas such as the aforementioned, which are highly exposed due to their status as tourist destinations. In this sense, the recurrence of extreme events affects not only the perception of safety, but also the local economy and the need to implement effective planning and risk management strategies in regions that can be defined as "at-risk regions" (Sotelo Navalpotro, J.A., 2000).

Thus, Spain presents itself as a profoundly heterogeneous territory, the result of the interaction between its peninsular status, its geostrategic location in southwestern Europe, its complex topography, its historical and cultural evolution, and the configuration of its demographic and settlement structure. These elements shape a unique territorial identity, but also place it at a high risk of exposure to a wide range of natural risks. This situation of vulnerability repeatedly translates into significant impacts in both human and economic terms, the severity of which is accentuated when considering the country's level of socioeconomic development. In this context, territorial development, due to its close dependence on environmental and climatic stability, becomes one of the realities most sensitive to natural risks. Within this analytical framework, the notion of risk must be understood as the result of the conflicting interaction between natural processes and human occupation of the territory. The concrete expression of this risk can range from hazardous events to disasters or catastrophes, depending on the magnitude of the phenomenon and the response capacity of the affected system. Particularly in the area under study, the territorial reality is conditioned by the recurrence and severity of extreme meteorological phenomena, with episodes associated with Extratopical Cyclonic Low Pressure Systems processes being especially relevant (see Figure 7).

Figure 7. Extratropical Cyclonic Low Pressure System "Justine", seen from satellite (2021).

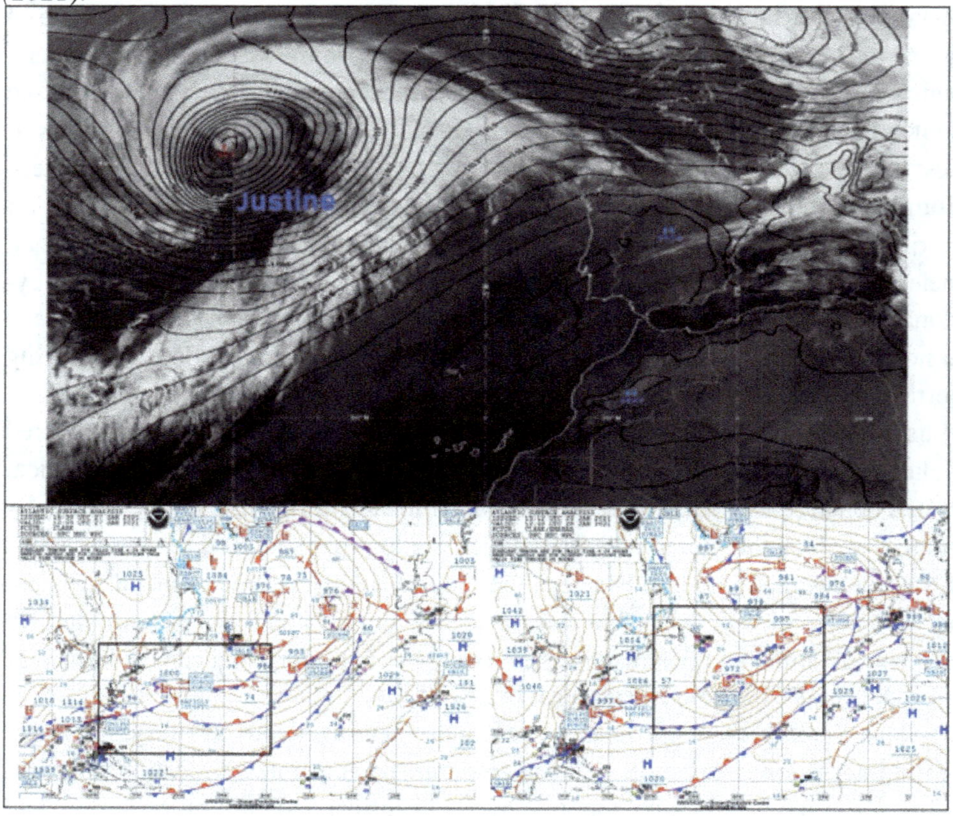

Source: National Oceanic and Atmospheric Administration (2021)

These events, characterized by rapid intensification of low-pressure systems, generate high-risk situations, with direct effects on the territorial foundation and sustainable development of the aforementioned study area. This structural vulnerability is closely linked to the characteristics of the physical environment and, in particular, to the high climatic variability that defines the surrounding area. Consequently, some regions can be classified as "at-risk territories," not only from a geographical or operational perspective, but also from the perception of visitors themselves, whose consumption decisions are influenced by the presence or absence of safe conditions. This condition poses significant challenges for territorial planning, risk management and resilience building in contexts where adverse natural phenomena are part of the recurring scenario of action.

It is in these contexts that the so-called Extratopical Low Pressure Systems develop, such as, for example, the aforementioned case of the extratropical cyclonic storm "Justine", which affected Galicia at the end of January 2021 with strong winds and intense rains, as a result of a meteorological phenomenon characterized by a rapid drop in atmospheric pressure (more than 30 hPa in less than 24 hours), this storm unleashed violent storms in the region, marked by the sudden development of low-pressure systems capable of generating severe storms. Justine caused significant damage, especially in the coastal areas of A Mariña and Costa da Morte, where waves of up to 8 meters were recorded. In the coastal areas of Pontevedra, winds exceeded 80 km/h, and in other areas, gusts caused trees to fall and affected traffic. The rainfall was intense throughout Galicia, with the greatest impact in the provinces of A Coruña and Pontevedra. The storm formed in the Atlantic and moved toward the Iberian Peninsula, particularly affecting the northwest. Upon arrival, red alerts were issued for strong waves and winds, in addition to other orange and yellow alerts for adverse weather conditions. *Galicia, due to its geographical location, is especially vulnerable to Atlantic storms, particularly between October and March, when storms can be intense and even fuelled by remnants of tropical cyclones, as was the case with Justine, which highlighted the impact these phenomena can have on the region* (Martín León, F., 2021). In fact, the storm left behind a landscape devastated by adverse conditions, with displaced street furniture, swollen rivers, and damaged vegetation. "Justine" once again highlighted Galicia's vulnerability to extratropical cyclones, which are common between October and March due to its geographical location in the path of Atlantic storms, which often arrive driven by strong southwesterly winds laden with moisture. In some cases, such as Justine's, these storms can be influenced by remnants of tropical cyclones that have lost their original structure at higher latitudes, further increasing their intensity. These types of phenomena not only compromise the physical safety of the population and coastal infrastructure, but also generate an economic impact on sectors such as fishing, agriculture, and transportation, in addition to disrupting the daily routines of thousands of people. The experience of Storm Justine in Pontevedra confirms the need to maintain effective early warning systems, emergency plans tailored to each municipality, and greater public awareness of the effects of climate change, which can contribute to making these extreme events more frequent and intense.

For its part, Hurricane "Kirk" surpassed the impacts generated by previous extratropical cyclones that hit the Galician coast, creating new meteorological risks, becoming "a historic hurricane in Galicia" (as described by the press). Let us pause to analyse it: on September 28, 2024, the United States National Hurricane Center began monitoring a large area of low pressure located west of the Cape Verde Islands, which initially presented limited convective activity. As the hours passed, the showers and thunderstorms associated with this system began to show more structure. Satellite images on September 29 confirmed an increasingly defined circulation, leading to its official classification as a tropical depression that evening. It was located approximately 950 kilometers west-southwest of Cape Verde.

During the following day, the system continued to intensify and was designated Tropical Storm "Kirk", now located more than a thousand kilometers west of the archipelago. Favored by favorable environmental conditions, "Kirk" underwent rapid intensification. In the early hours of October 1, the partial formation of an eyewall was observed, a clear indication of its strengthening. That same evening, it reached hurricane strength, and during the early hours of October 2, it rapidly evolved into a major Category 3 hurricane, while located nearly 2.000 kilometers east-northeast of the Leeward Islands. After fluctuating in intensity, "Kirk" reached Category 4 strength during the night of October 3, with maximum sustained winds of 230 kilometers per hour and a minimum central pressure of 934 hectopascals, marking its peak intensity. Subsequently, as it moved northward over cooler waters, the hurricane began a gradual weakening process. Accelerating northeastward, it completed its transition to an extratropical cyclone on the afternoon of October 7, while located approximately 1.055 kilometers west-northwest of the Azores archipelago. From there, the system continued northward over the archipelago, then moved toward the western coast of France and the rest of western Europe (see Figure 8).

Figure 8. Extratropical Cyclonic Low Pressure System "Kirk", seen from satellite (2024).

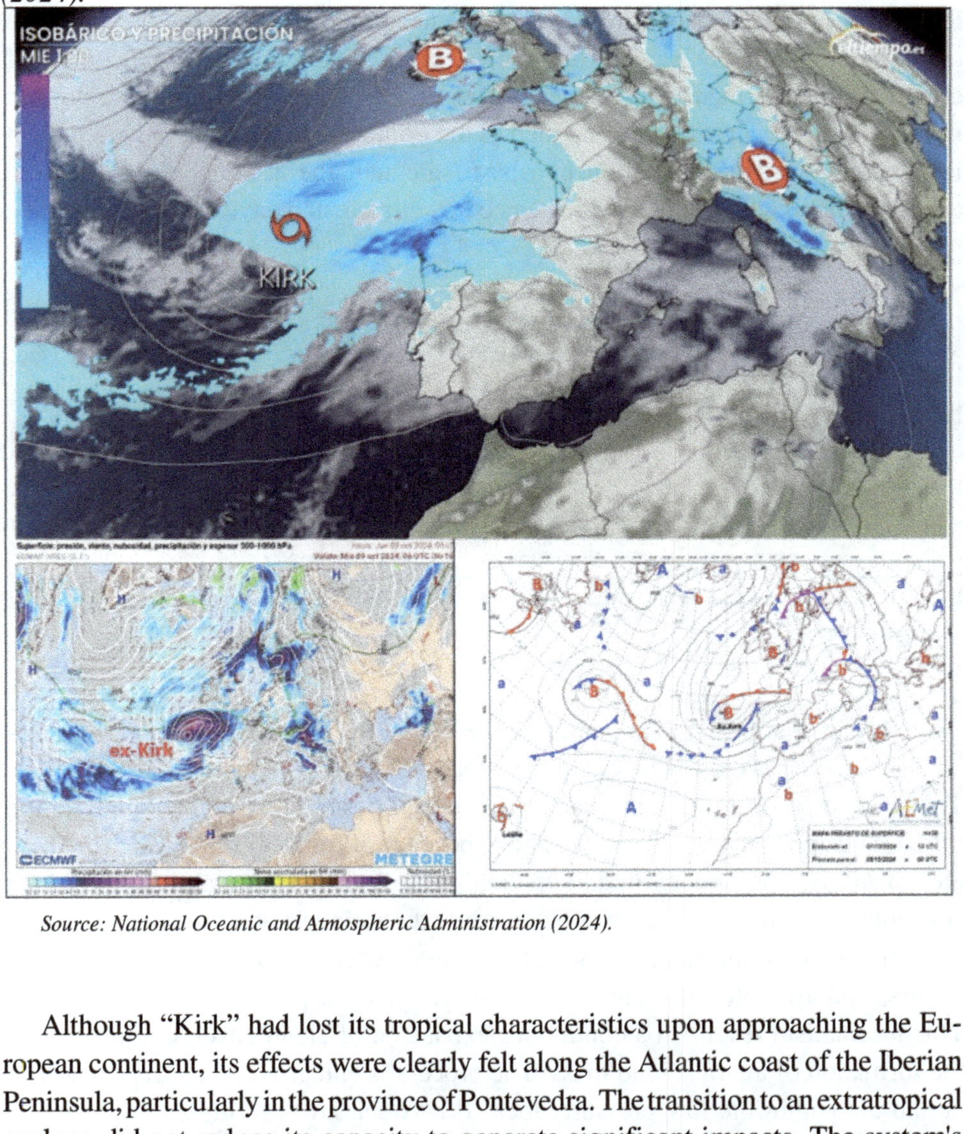

Source: National Oceanic and Atmospheric Administration (2024).

Although "Kirk" had lost its tropical characteristics upon approaching the European continent, its effects were clearly felt along the Atlantic coast of the Iberian Peninsula, particularly in the province of Pontevedra. The transition to an extratropical cyclone did not reduce its capacity to generate significant impacts. The system's interaction with other mid-latitude air masses intensified atmospheric instability, causing torrential rains, strong winds, and a significant deterioration in sea conditions.

On the Pontevedra coast, these conditions resulted in numerous adverse effects: power outages, fallen trees, damage to urban and coastal infrastructure, disruptions to transportation services, and the temporary halt of fishing and tourism activities. Economic sectors linked to tourism, especially those located in exposed coastal areas, were particularly vulnerable, suffering material and operational losses.

This episode highlights how phenomena originating in distant ocean basins can have direct repercussions in northwestern Spain due to their evolution into extratropical cyclones. Furthermore, it highlights the need to incorporate these events into territorial planning and risk prevention and management systems. In a context of climate change and increasing frequency of extreme events, it is essential to strengthen the resilience of fragile territories such as the Pontevedra coast, not only to protect infrastructure and the population, but also to safeguard the sustainability of the local tourism model, which is deeply influenced by climate and environmental stability.

Figure 9. Analysis of precipitation, temperatures and wind speed.

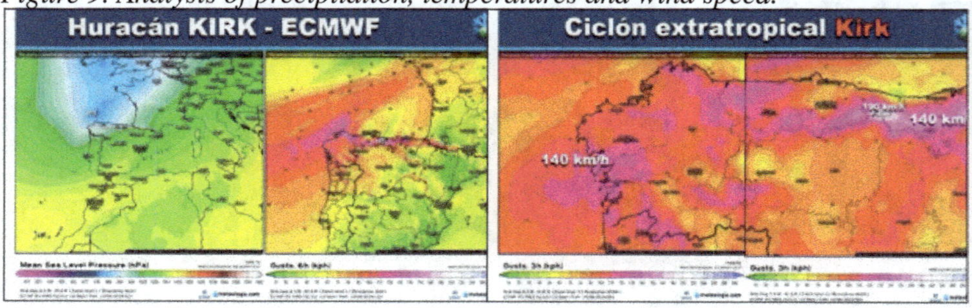

Source: Meteovigo (2024).

Analysis of atmospheric dynamics reveals that the entry of air into various tropospheric levels is closely linked to the direct influence of an extratropical cyclone. This system generates a particularly pronounced positive thermal anomaly in the middle and upper levels of the atmosphere, between 500 and 300 hPa, which contributes to an asymmetric distribution of cloudiness. As a result, precipitation tends to concentrate north of the cyclone's track, while the strongest winds are located on its southern flank. In the polar region of the system, where the maximum wind output occurs, updrafts intensify, favoring the progressive deepening of the depression. This mechanism results in a continuous expansion of cloud cover, culminating in the formation of a large cloud mass. The process is enhanced by the rise of warm, humid air within the low pressure, located between areas of higher atmospheric pressure dominated by cold, dense air masses that penetrate in a wedge-shaped formation. This configuration favors the development of convective cloud cover with significant vertical development, giving rise to a high-intensity storm.

Figure 10. Precipitation map and temperature map after the passage of extratropical cyclone "Kirk" (2024).

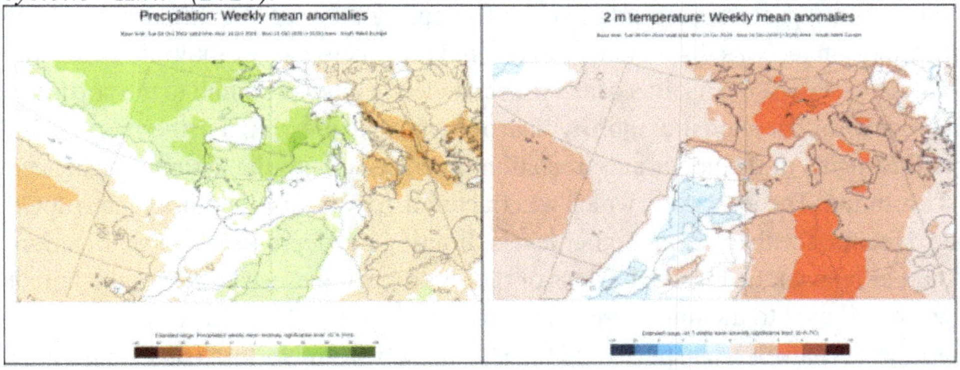

Source: Martín León, F. (2021), Meteored.

The virulence of the system, its rapid evolution, and its destructive potential make it a high-risk phenomenon. The cyclone's impact was not limited to intense rainfall but was amplified by persistent winds that reached hurricane-force speeds in a short period of time, especially in the coastal areas of the study area, where significant waves were recorded (see Figure 10). Unlike previous episodes, rainfall was widespread throughout the Autonomous Community of Galicia. However, the greatest severity was observed in the province of Pontevedra, with continuous, persistent, and high-intensity rainfall, demonstrating the capacity of these systems to generate large-scale hydrometeorological impacts.

Having analysed in detail the development of both episodes of adverse weather associated with the passage of an extratropical low with a marked process of explosive cyclogenesis, and, more specifically with the case study of Hurricane "Kirk", it is worth highlighting that, during its mature phase, the system experienced a process of warm seclusion. This evolution is often accompanied by the presence of very strong, even hurricane-force winds, especially on the periphery of the warm front surrounding the occlusion, in this case located south of the low-pressure core. However, the southerly wind gusts that had been forecast to be more intense failed to manifest themselves with the expected force in Galicia. This weakening can be explained by the cyclone's interaction with land, which altered its internal structure. Upon making landfall, the storm lost access to oceanic sources of energy and moisture, and friction with the land surface increased, preventing atmospheric pressure from dropping as sharply as it would have over the sea. This circumstance caused an increase in the extent of the wind field, expanding its radius of action but reducing its average intensity over Galician territory. If the system maintained a completely maritime path, skirting the coast of A Coruña, it would likely have reached greater depths, generating a more severe meteorological scenario, with stronger winds, although more localized to the northwest. The cyclone's somewhat erratic movement in the vicinity of the northwest of the peninsula can also be attributed to a partial filling process, resulting from a decoupling between the different levels of the atmosphere. This disorganization favored the emergence of multiple structures within the depression itself: a center associated with intense rainfall to the north and another with characteristics more typical of a typical depression. Furthermore, small readjustments in the upper-air circulation, interacting with the system, influenced its evolution. The jet stream, which was fractured, may have temporarily interrupted the dynamic feedback from higher levels, limiting the depression's intensification. Another relevant factor was the loss of the subtropical mesoscale environment that had contributed to the initial organization of cyclogenesis. The intrusion of dry air into the mid-levels, transported by the storm itself from North Africa, could have caused evaporation in the rising layers of warm, humid air, generating adiabatic cooling and partial stabilization of the atmosphere. This process ultimately inhibited further deepening of the system, reducing its destructive potential compared to the initially anticipated scenario.

Beyond their local impact, these phenomena are part of a system of transboundary environmental risks, the management of which requires a supranational and integrated approach. In an interconnected world, the effects of climate change ignore political or administrative borders, spreading through shared ecosystems, trade flows, financial networks, and migration dynamics. As outlined in the "National Climate Change Adaptation Plan", impacts generated outside our borders can significantly impact domestic vulnerability, requiring international cooperation mechanisms, multilevel governance, and intersectoral coordination. Cross-border risks include, among others, the disruption of supply chains due to extreme weather events in third countries, the spread of air or water pollutants, and pressure on shared natural resources. In this sense, the region under study not only faces challenges derived from its own climate exposure, but is also affected by global dynamics that impact its environmental and socioeconomic stability.

Thus, a detailed analysis of atmospheric oscillation patterns and their relationship to climate change allows for the identification of risk trends and the design of territorial adaptation strategies. These strategies must incorporate a cross-border perspective that recognizes the interdependence between regions and promotes equity in access to resources, technologies, and institutional capacities. Territorial resilience, understood as the capacity to anticipate, resist, and recover from climate shocks, thus becomes a priority objective of public policies, both nationally and internationally. An example of this can be seen in the coastal areas of the Province of Pontevedra, where it has become clear how the coastal areas of the province of Pontevedra are particularly exposed to extreme weather events, the economic impacts of which vary depending on the frequency and intensity of these natural events. In certain years, the losses caused by these events can represent up to one percent of the regional Gross Domestic Product, with the greatest impact on sectors whose activity depends on stable climatic conditions.

In the studied area, rapidly intensifying storms have generated emergency situations that temporarily disrupt economic and social activity. These episodes cause significant territorial disruption and test the response capacity of institutions. For this reason, it is essential to advance the development and implementation of effective risk prevention and management strategies to reduce the vulnerability of the territory and prevent negative perceptions about the affected region from taking root.

6. BY WAY OF CONCLUSIONS

After what has been said so far, we can conclude with the fundamental idea that climate change has ceased to be a hypothetical threat and has become an observable reality both locally and globally. Its transboundary nature makes it a phenomenon that transcends geographical and political limitations, simultaneously affecting multiple regions of the planet. Manifestations such as the accelerated retreat of polar glaciers, the intensification of tropical cyclones, and the growing instability of atmospheric patterns demonstrate the magnitude of this climate transformation. As the effects of global warming intensify, the environmental and social impacts that transcend borders, such as rising sea levels, forced species migration, and the spread of extreme events, all of which have direct implications for international governance and shared ecological security, are becoming more evident. At the local level, this crisis translates into significant alterations in the biological cycles of fauna and flora, with shifts in the geographic distribution of species and changes in their behavioural patterns. Although these changes may seem limited to specific territories, their impact extends to a global scale, affecting the interdependent ecological networks that link distant ecosystems. Daily climate change thus manifests itself in a network of interactions that connects the immediate environment with global processes, revealing the complexity of natural systems in the face of climate disturbances.

Answer to this environmental emergency requires a massive economic and social commitment. *Current estimates place the cost of the transition to a low-carbon economy in the order of several trillion euros, primarily directed toward reducing greenhouse gas emissions and transforming energy systems* (Oliver, J. & Peter, J., 2020). This effort must be accompanied by an international cooperation strategy that facilitates the development and implementation of clean technologies, such as advanced solar energy and nuclear fusion, capable of reducing dependence on fossil fuels and promoting a more sustainable energy model. The interconnectedness of climate systems also requires cross-border public policies that favor technology transfer to vulnerable regions and promote equity in climate adaptation.

In this context, geoengineering is presented as an emerging technological option, although not without controversy. Proposals such as ocean fertilization -which consists of adding nutrients like iron to stimulate phytoplankton growth and increase atmospheric CO_2 capture-, or solar radiation management through particle dispersion in the stratosphere seek to mitigate the effects of global warming on a large scale. Likewise, carbon capture and storage from industrial sources represents a way to reduce cumulative emissions and limit transboundary damage from air pollution. While these technologies require rigorous assessment of their ecological and ethical risks, they could play a significant role in global climate strategy.

Climate change is also generating structural transformations in development models across multiple territories. Phenomena such as floods, prolonged droughts, and extreme weather events have repercussions that go beyond the environmental sphere, affecting human security, social cohesion, and political stability. Regions that depend on shared natural resources -such as international watersheds or marine ecosystems- face heightened risks of conflict and displacement, reinforcing the need for global climate governance that enables these challenges to be managed cooperatively and equitably. Therefore, Climate change demands a comprehensive response that recognizes the transboundary nature of environmental risks and articulates mitigation, adaptation, and technological innovation efforts. Only through effective international collaboration, supported by sustainable policies and a shared scientific vision, will it be possible to address the climate challenges of the present and build a resilient future for the planet's societies and ecosystems.

In addition, it can be concluded that this research highlights how extreme weather events and environmental risks derived from climate change are intertwined with territorial development, especially in coastal areas such as the province of Pontevedra. As global warming affects ecosystems, communities, and human activities, there is an urgent need to understand and anticipate the impacts these risks have on the natural environment and the societies that inhabit it. In this sense, it is essential to understand the dynamics of extreme weather events, such as intense storms and hurricanes, and to analyse their transboundary effects, since changes in the climate of one region can impact surrounding territories, generating a global risk. Climate change has profoundly altered global weather patterns, leading to an increase in the frequency and intensity of meteorological phenomena such as extratropical cyclonic low-pressure systems, which severely affect the coastal areas of the province of Pontevedra. These events not only have direct consequences for social well-being, but also put the region's socioeconomic structure at risk. In this context, the vulnerability of communities and their infrastructure is accentuated by the lack of adequate management of the risks associated with climate change. It is essential that development and economic growth policies be adapted to integrate an ecological and sustainable vision that mitigates these impacts. Throughout the research, it has been observed that the interaction between sociohistorical factors and natural phenomena generates a situation of "socionatural" vulnerability. In the coastal areas of Pontevedra province, the high population density and the transition to an economic model that has promoted the expansion of industrial activity and the service sectors (despite being in its infancy) has increased exposure to extreme weather events, such as storms or destructive winds generated by cyclogenesis. Furthermore, the shift in the development model, which moves away from traditional activities such as fishing or agriculture, reinforces this vulnerability, as new productive structures are not always prepared to cope with the magnitude of natural disasters.

The impacts of climate change on coastal ecosystems are not limited to the modification of natural habitats, as exemplified in the case study, but can also trigger economic and social crises. Environmental deterioration, coupled with the failure to adequately manage cross-border risks arising from phenomena such as extratropical cyclonic low-pressure systems, compromises the stability of the region. The fragility of the environment in Pontevedra's coastal areas, a result of changing climate patterns, requires the urgent implementation of effective conservation measures and the integration of sustainability into the regional economic development model.

Risk management, therefore, must focus on territorial planning and the protection of natural resources in a manner that is equitable and accessible to all citizens. The economic growth model must be rethought, abandoning the traditional concept of development solely oriented toward increased production and moving toward an approach that prioritizes community resilience and adaptation to environmental risks. It is also essential that the regional economy diversify and focus on sectors that not only contribute to economic growth but also promote the protection and restoration of local ecosystems.

So, the impact of climate change on cross-border coastal regions underscores the urgent need to adopt more effective measures to combat pollution and environmental risks. The lack of prevention and response plans for environmental disasters in areas with high maritime traffic, such as the Galician coast, is surprising and alarming. Despite a history of disasters, such as maritime accidents along the coast, public administrations have failed to implement adequate protection mechanisms, generating an amplified environmental risk affecting local, national, and even international cross-border regions. This lack of emergency response, coupled with the lack of effective contingency protocols, has led to widespread pollution, as occurred with oil spills, which dispersed across vast maritime areas. Despite theoretical and legislative advances on paper, the lack of practical implementation highlights ineffective environmental risk management. At the same time, increasing diffuse pollution and the poor practices of some companies and ships, which take advantage of the lack of protection in certain areas to dump pollutants into the sea, have been a constant in recent years. These actions are not isolated cases, but rather a recurring problem that exacerbates environmental degradation in the affected regions, where environmental risks amplify their vulnerabilities, impacts, and damage. Therefore, it is imperative to establish a robust surveillance and prevention system that not only deters these illegal practices but also forms part of a comprehensive approach to environmental risk management. In this regard, international cooperation and the strengthening of cross-border policies are essential to mitigate the impacts of climate change and pollution, protecting both marine ecosystems and coastal communities from dangers that extend beyond national borders. Failure to address these challenges leads to increasing vulnerability, affecting not only local coasts but also the environmental security of surrounding regions.

In conclusion, environmental risks and cross-border environmental risks not only affect the economies of territories, but also profoundly alter the quality of life of populations. Hence, the Pontevedra region, and especially its coastal areas, require more effective integration between economic development policies and climate change management strategies to ensure that natural resources are conserved and protected from the threats posed by extreme weather events and global climate change. Creating a sustainable and resilient development model will be key to protecting future generations from the most serious impacts of the climate crisis.

REFERENCES

Berkeley Earth System. (2025). *Global Temperature Report for 2024*. Written by Robert Rohde.

ESA. (2019). *Applications Copernicus*. The European Space Agency.

Hoylman, ZH, Bocinsky, RK & Jencso, KG. (2022). "La evaluación de la sequía ha sido superada por el cambio climático: argumentos empíricos para un cambio de paradigma". Comunicaciones de la naturaleza volumen 13, Nº 2715.

Kodra, E., & Ganguly, A. R. (2014). Asymmetry of projected increases in extreme temperature distributions. *Scientific Reports*, *4*(1), 5884. DOI: 10.1038/srep05884 PMID: 25073751

Martín León, F. (2021). *Justine hacia la Península después de una ciclogénesis explosiva*. Meteored.

National Oceanic and Atmospheric Administration (2024).

NSIDC. (2019). *National Snow and Ice Data Center "About the data". Images of Thomas Mote*. University of Georgia.

NSIDC. (2022). *National Snow and Ice Data Center "Current conditions and data tools". Images of Thomas Mote*. University of Georgia.

Oliver, J., & Peter, J. (2020). *Trends in global CO_2 and total Greenhouse Gas Emissions*. PBL Netherlands Environmental Assessment Agency.

Sotelo Navalpotro, J. A. (2000). *Regional Development Models*. Oxford University Press.

Sotelo Pérez, I., & Sotelo Pérez, M. (2023). "Urban Planning and the Third Landscape in the Field of Local Development: Case Study", in AA.VV. In Global, I. G. I. (Ed.), *Geoinformatics in Support of Urban Politics and the Development of Civil Engineering*. DOI: 10.4018/978-1-6684-6449-6.ch001

Sotelo Pérez, M., & Sotelo Pérez, I. (2017). Una aproximación al marco teórico de los riesgos. *Observatorio Medioambiental, 20*, 9–36. DOI: 10.5209/OBMD.57944

Thomson Higher Education & Maggie Nelson (2004). Images obtained from *Cazatormentas.com*

Waters, C. et al. (2016). "The Anthropocene is functionally and stratigraphically distinct from the Holocene". Science, nº. 351.

ENDNOTE

1 This research is part of the MINECO Research Project (2024), PID2023-148597NB-I00. PYROLANDSCAPE. Ministry of Science, Innovation and Universities.</bok>

Chapter 13
The Impact of Regional Climate Cooperation in the EU in the Green Transition

Şükrü C. Demirtaş
https://orcid.org/0000-0002-6953-3599
Yildiz Technical University, Turkey

Senem Çakmak Şahin
https://orcid.org/0000-0003-3395-9122
Yildiz Technical University, Turkey

ABSTRACT

This chapter examines the role of regional climate cooperation in the European Union's Green Transition process, with a particular focus on subnational governance mechanisms. Drawing on multi-level and polycentric governance theories as well as transnational municipal networks, the study examines the strategic importance of local and regional actors in shaping and implementing EU climate policies. In this study we examine climate governance frameworks and four empirical case studies (Vienna, Transdanube.Pearls, Eurometropolis Lille-Kortrijk and Ghent). The chapter assesses how regional initiatives contribute to innovation, participatory governance and cross-border policy coherence. With IPCC recommendations, we explain that effective climate governance in the EU requires institutional coherence, democratic legitimacy and the integration of resilience-sensitive strategies in line.

DOI: 10.4018/979-8-3373-3972-6.ch013

INTRODUCTION

Climate change is one of the most important environmental problems that needs to be paid attention globally. It has profound effects on natural systems, socio-economic structures and urban life all over the world. The increase in heat experienced, endangered species and endangered animals appear as an effect of global warming. Even these effects alone can lead to consequences that cannot be avoided in the future. On top of this, studies have been conducted at the state level. These studies have led to the signing of the European Green Deal. This agreement consists of multifaceted strategies that include a just transition, such as becoming carbon neutral, a sustainable economy and the protection of wildlife. EU countries aim to make the climate carbon neutral by 2050 (European Commission, 2019).

The success of the EU climate policy implementation depends largely on the effective cooperation of subnational actors and their openness to innovative approaches (Bulkeley & Betsill, 2005). Local networks and alliances have gained momentum in advancing the goals of the European Green Deal. Their proximity to regional contexts allows them to design region-specific solutions and influence policy through bottom-up participation.

Transnational municipal networks, such as the Convention of Mayors, ICLEI and Energy Cities, support information sharing and the dissemination of practical solutions. These platforms facilitate joint climate projects and increase the policy capacity of local and regional governments. In addition to supporting local actions, they also encourage cross-border cooperation. Thanks to the contracts, a joint movement plan is being implemented. However, the limited academic literature evaluating the actual effectiveness of these cooperation mechanisms in advancing EU climate goals shows that more empirical research is needed in this area (Kern & Bulkeley, 2009).

Although sub-national governance actors are recognized as key components of multi-level climate governance in the EU, their institutional integration into formal decision-making processes remains unclear (Hooghe & Marks, 2019). These actors contribute significantly to democratic participation, local innovation, adaptation and institutional resilience (Ostrom, 2017). EU climate governance is still largely seen as Brussels-centric. This structural imbalance raises ongoing questions about how to more effectively incorporate regional and local contributions into a broader framework (Kern, 2019).

Climate change is a multidimensional crisis that profoundly affects not only natural systems but also multi-level governance structures, institutional capacities and decision-making processes. Current assessments (Dodman et al., 2022) that if the current trend in greenhouse gas emissions continues, keeping global warming below the 1,5 °C threshold will become extremely unlikely, and that this situation will exponentially increase economic, social and administrative risks. Particularly at the EU level, combating climate change is no longer an area that can be addressed solely through technology-based interventions in the energy, transport or agriculture sectors; it requires the evolution of governance mechanisms into a multi-level, multi-level and multi-disciplinary institutional framework (Hooghe & Marks, 2019).

The European Union has presented an ambitious framework that will transform existing governance paradigms by aiming for a climate-neutral economy by 2050 with the European Green Deal announced in 2019 (European Commission, 2019). The Deal is not limited to carbon reduction but is built on a multi-level strategy including energy transition, circular economy, protection of natural capital, sustainable financing, and a just transition mechanism (Hickel, 2021). However, while central policies play a major role in the European Green Deal, sub-national actors in governance-local governments, regions, metropolitan areas, and cross-border governance platforms-are critical to the success of the process (Bulkeley & Betsill, 2005; Kern & Bulkeley, 2009).

Climate change is recognized as a complex issue that deepens social and economic vulnerabilities. The bad effects that may occur if not controlled will primarily affect disadvantaged communities. These communities are already experiencing inequalities at the global and regional levels (Birkmann et al., 2022). The vulnerability of disadvantaged communities is often due to structural inequalities that determine their capacity to respond and adapt. The possible climate change impacts are primarily environmental, but not limited to this. Institutional and socio-economic impacts arise through difficulties in housing, health services, employment and access to basic services. In response to these challenges, the European Union has taken on the role of a leading actor in the global green transition by developing a climate governance model that supports multi-level and multi-actor cooperation. The climate governance model combines regulatory tools, financing mechanisms and policy integration aimed at aligning environmental goals with social justice.

EU regional climate governance examines the institutional dynamics of climate initiatives and their impact on decision-making within a multi-level governance framework. It highlights the role of sub-national actors, their capacity for innovation and their contribution to democratic participation. In line with the EU climate targets, cross-border cooperation platforms such as the Convention of Mayors, ICLEI and Energy Cities are evaluated in terms of their performance and limitations. These platforms are evaluated not only in terms of their ability to harmonize policies, but also in terms of how representative and institutionally consistent. The long-term effectiveness of EU climate governance is based on maintaining institutional diversity, improving transparency, ensuring multi-level representation and adapting institutions to emerging challenges.

When the literature is examined, it is seen that academic studies on the role of subnational governance in EU climate policy focus on participation in decision-making processes of multilateral governance mechanisms, institutional coherence, transparency and effectiveness. Therefore, it is necessary to integrate subnational actors into EU governance processes. In this way, the effectiveness of climate policy and the development of institutional pluralism and democratic legitimacy are ensured, and this situation emerges as a strategic necessity (Ostrom, 2017).

The IPCC, without adaptation policies, both economic and non-economic losses disproportionately affect the poorest and most vulnerable regions warns about that all the participants (Birkmann et al., 2022). Participants face high risks due to having limited resources and weak institutions. The existence of risks shows that there is an urgent need for climate justice-based approaches in EU policy-making. The European Green Deal and the Fair Transition Mechanism aim to reduce gender inequalities, combat structural poverty and achieve this adaptation by giving priority to those who are least able to cope with climate impacts. Successful adaptation requires more than financing. In general, an adequate level of adaptation depends on the creation of institutional environments that support inclusive and resilient development. The long-term effects of climate change make it clear that EU climate policies must address both emission reductions and the promotion of cross-border social and economic resilience. Today, when climate risks intersect with the ongoing legacies of inequality, migration and marginalization, cooperation at the EU level is becoming more important. The EU's multi-level governance model, globally just and inclusive climate action is increasingly being cited as a reference (Birkmann et al., 2022).

The European Union climate cooperation initiatives operate using various governance structures in order to ensure effectiveness towards the objectives of the European Green Deal. Regional initiatives are included in the EU climate frameworks by shaping policy agendas, participating in implementation and influencing institutional processes. This cooperation increases democratic participation and strengthens community resilience. Local and regional initiatives often serve as drivers of innovation. Cross-border cooperation contributes to policy coherence, especially when national legal systems differ. These dynamics show that sub-national efforts, i.e. regional initiatives, can align with the EU climate goals by addressing the challenges. EU climate cooperation reveals the dimensions of the fragility of the initiatives identified by the IPCC, especially those related to poverty, exclusion and weak institutional structures. It takes more than carbon reduction to prevent this vulnerability. When climate action is linked to poverty reduction, social inclusion and institutional development, the Green Transition can contribute not only to mitigation, but also to long-term resilience and justice. This orientation allows the restructuring of climate policies with a more just and resilient development approach. Thus, while ensuring that the achievements of the Green Transformation also reach the most vulnerable segments, it becomes possible to proactively address structural obstacles.

Climate change is already affecting nature and people in Europe, exacerbating existing social and economic vulnerabilities. This situation exacerbates spatial and sectoral inequalities. The IPCC AR6 WGII report reveals that temperatures are rising faster in Europe than the global average. The southern and central regions are especially at risk due to rising temperatures below 2-3 °C Global Warming Level (GWL), intensifying droughts and increasing water scarcity. It is seen that there is an urgent need for regional climate cooperation within the EU in order to address the multidimensional effects of climate change on various socio-economic and ecological landscapes. Especially in Southern Europe, it is estimated that more than a third of the population will be affected by water shortages at 2 °C GWL. It is estimated that this figure could double to 3 °C GWL, causing significant economic impacts on water-dependent sectors. It is estimated that most of Europe will experience a two- to three-fold increase in heat stress deaths and crop losses due to the combined effects of heat and drought. This makes the need for significant adaptation measures a priority. On the other hand, benefits such as long growing seasons are expected to be realized in Northern Europe (Bednar-Friedl et al., 2022).

The IPCC findings point to the critical role of regional climate cooperation in addressing these unequal impacts and in mobilizing cross-border governance and resources. The EU Green Deal, the Just Transition Mechanism, and the Adaptation Strategy already provide institutional platforms for aligning regional efforts. IPCC evidence confirms that adaptation within the EU still largely focuses on incremental, technical measures, such as flood defences and early warning systems, with limited transformational approaches. In many instances, existing adaptation strategies remain confined by institutional path dependencies and the economic interests of dominant actors, making it challenging to implement long-term, systemic changes (Bednar-Friedl et al., 2022).

More robust and inclusive regional cooperation can reduce residual risk across the EU. By aligning adaptation planning with IPCC recommendations for long-term climate resilience, EU institutions can:

- Integrate climate risk and vulnerability data across borders for joint planning;
- Mainstream nature-based and ecosystem-based adaptation across EU policies;
- Establish multi-level governance platforms for managing transboundary water, energy, and ecosystem services;
- Invest in resilience for highly exposed southern and central EU areas while facilitating knowledge and technology transfers from less affected northern regions.

The IPCC AR6 WGII Chapter 17 mark the central role of institutional design, multi-level governance, and well-targeted financing as critical enablers for effective climate risk management and adaptation across global and regional situations. In the EU, regional climate cooperation operates within a highly structured multi-level governance framework, making it a pivotal testbed for achieving long-term climate resilient development. Yet, as the IPCC stresses, climate risk decision-making must evolve from incremental, fragmented interventions towards a more transformational approach if it is to effectively reduce vulnerability and adapt to deep uncertainties (New et al., 2022).

At present, the EU has implemented a broad range of adaptation options addressing climate hazards across sectors - from flood protection and resilient infrastructure design, to changes in agricultural practice, water efficiency, and early warning technologies. However, IPCC assessments emphasize that much of the current adaptation is still largely incremental and focused on maintaining existing economic and institutional patterns, making it insufficient for long-term climate resilient development. In contrast, transformational adaptation - including systemic shifts such as re-localization of agriculture, nature-based restoration of degraded ecosystems, and equitable economic redistribution - emerges as increasingly necessary in a warming world (New et al., 2022).

For the EU, IPCC findings show an urgent need for multi-level and multi-level governance at national, regional and local scales. Inclusive governance means that institutional reforms should be implemented in the EU that prioritize participatory planning and use scientific and local knowledge together. The inequality, historical responsibility and the right to sustainable development brought about by climate risk policies have been addressed through inclusive governance. Global climate justice should reflect the principles of inclusive governance. It is also a key possibility for reducing vulnerability and aligning climate risk management with long-term sustainability goals (New et al., 2022).

CONCEPTUAL FRAMEWORK

Understanding the form and application of climate policies at different levels of government in the EU requires a theoretical basis. The theory of Multi-Level Governance (MLG) comes across here. Theories such as Multi-Level Governance and Transnational Municipal Networks (TMN) offer valuable insights into the dynamics of shared authority, overlapping jurisdictions, and distributed responsibilities. This information includes the method by which regional and local actors contribute to climate action beyond the challenges of national policymaking. When considered in terms of global challenges such as climate change, it helps to explain why cross-border cooperation is necessary.

MLG theory is a theory developed on the sharing and coordination of authority at national, regional and local levels. In contrast to traditional state-centered models, MLG includes non-state actors such as non-governmental organizations, municipalities and private organizations in the model while public policy is being shaped and emphasizes the increasing role of these actors (Hooghe & Marks, 2003). It is a valid theory, especially for the EU, where the climate targets set at the supranational level are implemented through national legislation and localized actions. MLG theory examines the situations in which broad climate goals can be adapted to various regional realities and develops a common method. The cities where the theory is applied are leading the way in EU climate governance. Thanks to this leadership, the ability of smaller-scale actors to create innovative solutions is gradually increasing (Di Gregorio et al., 2019).

Today, the necessity of an alternative to centralized and hierarchical decision-making processes in public administration is being discussed. It is possible that there may be more than one decision center in a governance area, these centers can interact with each other, and institutional diversity can improve the governance quality. Multi-level governance constitutes the decision centers in public policy in a multi-level structure. The ideal "metropolitan" management model should have a multi-level structure that can provide diversity, flexibility and efficiency in decision-making (Ostrom, Tiebout & Warren, 1961). It has been shown that the governance of natural partners can be achieved much more effectively with multi-level, rule-based governance mechanisms (Ostrom, 1990).

The theory of MLG provides a theoretical framework for integrating global climate goals into local conditions and institutionalizing multi-level cooperation processes in governance. Considering the effectiveness of sub-scale actors in decision-making mechanisms, multi-level governance is not just an institutional guidance system. It is a governance model that facilitates the implementation of multi-scale solutions with various disciplines involving multiple actors. It increases innovation and participation at the local level by revealing the centrality of decisional processes between different government departments (Ostrom, 2017). Instead of a single central authority, there are multiple decision-making centers that are interconnected and partially autonomous. These multi-level structures in the fight against climate change enable local actors to develop solutions adapted to their specific conditions. Since methods and results such as trial and error are shared, the ability to learn together and gain flexibility is developing. This sharing feature is quite evident in the EU's regional climate cooperation initiatives. However, since different regions face different climate risks, they need different adaptation abilities. This diversity and flexibility in needs enables the implementation of climate policies in a more effective and decentralized way. MLG increases the diversity of information, creates innovative solutions and ensures that local needs are reflected in governance. Multi-level structures are especially effective in the management of natural common areas, energy, climate and urban policies.

The MLG process has been criticized as creating serious problems in terms of transparency, accountability and speed of decision-making. Some researchers argue that multi-level governance cannot go beyond the realm of normative ideals, because institutional and social resistance prevents the establishment of rapid governance mechanisms. By causing diversity in decision-making, multi-level governance can reduce accountability when corporate roles are not clearly defined (Hooghe & Marks, 2003). If there are design flaws, powerful actors can exploit their structural advantages, leading to imbalances in participation and influence. The performance of such systems largely depends on how clearly the roles are distributed and how transparent the rules of interaction are.

In short, multi-level governance helps explain how climate goals defined at higher levels are shaped by decisions made closer to the ground. It draws attention to the influence of local governments, civil society and private actors in climate policy, especially in systems such as the EU, where decision-making is shared between different layers of authority. It decries cooperation between actors operating with varying degrees of autonomy, rather than a top-down chain of command.

Transnational municipal networks (TMNs) focus on how cities and regions participate in cross-border cooperation and shape policy outcomes through shared action (Kern & Bulkeley, 2009). These networks help cities to exchange information, implement effective practices more widely and harmonize their goals. In this way, solution-oriented feedback is provided. Initiatives such as the Convention of Mayors, ICLEI and Energy Cities enable local governments to work together on climate issues that cross national borders. By doing so, they strengthen local capacity and contribute to more coordinated climate strategies across Europe. Thanks to these networks, local actors are playing an active role in the process of 'Europeanization' by giving a stronger voice to EU-level policy-making processes.

Transnational Municipal Networks are recognized by many theorists as sites of “soft governance” in the fields of sustainable development, climate change or human rights (Bouteligier, 2012). However, a skeptical view may find the voluntary and goodwill-based nature of these processes insufficient to transform normative principles into institutional rules. Networks with expanding spheres of influence often have difficulty making decisions independently of central states, limiting the practical applicability of the theoretical claim.

When recent studies are examined, cities and regions are effective actors of multi-level climate governance at the EU level. The basic assumption of this theory is that municipalities are no longer just sub-units governed by nation states but can also act independently within multi-level governance. However, this claim requires careful examination. Many critics emphasize that networks of relations between municipalities idealize the concept of polycentric governance, ignoring existing asymmetries (Curtis, 2016). There are doubts about the extent to which local actors with limited institutional capacity, financial power or geopolitical position can find representation in these networks.

While the theory of transnational municipal networks offers a critical space for understanding the involvement of local governance in international relations, the explanatory and directive nature of the theory is limited by institutional limitations, problems of representation and lack of transparency of decision-making processes.

Global urbanization is a critical process in which climate change-related risks are generated, accumulated and become multidimensional. Chapter 6 of the IPCC Sixth Assessment Report, emphasizes that climate change-related impacts are exacerbated in urbanization processes and that these impacts generate cascading risks at multiple scales, spanning multi-sectoral relationships. It is determined that the global urban population increased by 397 million between 2015 and 2020, with 90% of this increase taking place in low- and middle-income regions (Dodman et al., 2022). This rapid growth coincides with the spread of low-income, unorganized settlements, making urban risks multidimensional.

The report explains the interaction of climate change with urbanization through multiple risk mechanisms. Flash floods, droughts, sea level rise, heat waves and other extreme weather events generate risks of cascading collapse of energy, transportation, health, water and food infrastructure. The interconnectedness of multiple risks means that urban adaptation goes beyond just 'physical infrastructure' interventions. It requires comprehensive multi-level solutions to protect social infrastructure, corporate governance and finance and increase functionality. The existing adaptation strategies for building urban resilience are quite limited due to lack of institutional capacity and insufficient financing in terms of monitoring and evaluation mechanisms. This limit is related to the interconnectedness of the risks affecting urban adaptation. Flash floods, droughts or extreme heat waves generate multiple and compounded risks, triggering cascading collapses in energy, water, transportation and health infrastructure. It is understood that compliance strategies require multi-sectoral and multi-level governance mechanisms. According to IPCC, reducing adaptation to only physical infrastructure interventions is insufficient, and the lack of investments in social infrastructure (social resilience, health, education) limits adaptation (Dodman et al., 2022).

Emphasizing that climate change-related risks are particularly concentrated in low-income, informal settlements, IPCC points out that 56% of the urban population in Sub-Saharan Africa lives in slum conditions according to 2018 data. In addition, it is projected that the world's urban population will increase by 2.5 billion by 2050, with 90% of this increase concentrated in Asia and Africa. This means that droughts, floods, heat waves and other climate risks, which vary according to the level of global warming, will be spread over a very large urban area. For example, under a 2°C warming scenario, the number of people exposed to drought risk in urban areas would reach 410.7 million, while more than one billion people in low-altitude coastal cities would be at risk from sea level rise and tropical storms (Dodman et al., 2022).

The report emphasizes that institutional capacities, multi-level governance, financing mechanisms and investments in social infrastructure are critical for risk governance. Indeed, cities that improve institutional coordination through multi-level governance platforms have a much better chance of achieving the goals of risk reduction, sustainable development and climate-resilient growth. However, the IPCC emphasize persistent imbalances in financing adaptation. Funding for large-scale gray infrastructure projects risks future “idle assets” due to lack of maintenance and deferral of risks (Dodman et al., 2022). Instead, multidisciplinary adaptation strategies combined with nature-based solutions should be preferred, and financing mechanisms should be diversified accordingly.

In conclusion, IPCC AR6 WGII report reveals that global urbanization trends and fast-growing low-income settlements are undergoing a critical test of resilience to climate risks. The report emphasizes that sustainable and resilient urban development is possible through multi-level governance, institutional capacity building, diversifying financing mechanisms, and adaptation strategies that integrate social-nature-physical infrastructure components (Dodman et al., 2022).

IPCC report reveals that for future urban adaptation to succeed, institutional governance needs to be multi-level, risks need to be addressed in multiple ways, and institutional capacity and financing gaps in low-income cities need to be addressed. This perspective guides policies aimed at developing regional cooperation and multi-level governance mechanisms, not only for local governments. Table 1 summarizes the key findings of the IPCC on the challenges of urban adaptation (Birkmann et al., 2022) and aligns them with relevant EU governance responses. It emphasizes areas in which the EU's experience, such as multi-level governance, rural-urban integration and investment in social infrastructure, can inform global adaptation strategies.

Table 1. IPCC Finding and Relevance Themes

Theme	IPCC Finding	Relevance
State of Adaptation Plans	Although adaptation plans are being prepared in many cities, only a limited number are being implemented.	The EU's examples of good governance can provide a good model for other regions.
Social Infrastructure Gaps	Adaptation measures often focus on physical infrastructure, while social infrastructure (community, health, education) is neglected.	Good examples in the EU can provide guidance for adaptation in low-income cities.
Institutional Capacity	In fast-growing cities, institutional deficiencies (financing, governance, monitoring) limit adaptation.	The EU's experience with multi-level governance can be a good model.
Global Climate Impact	Infrastructure decisions affect not only cities but also rural areas, with knock-on effects on food, energy and water risks.	EU examples of rural-urban cooperation can be useful in integrating adaptation.

Source: (Birkmann et al., 2022)

Drylands comprise approximately 45-47% of the global land area and host roughly three billion people[1], many of whom face growing climate-related threats (Mirzabaev et al., 2022). The IPCC AR6 WGII Chapter 3 shows that these environments have already been significantly impacted by climate change, exhibiting surface warming of 1,2-1,3 °C between 1920-2015 -a rate higher than the global humid average (0,8-1,0 °C). The resulting intensification of drought shifts in vegetation patterns, and rising sand and dust storm activity have reshaped both ecosystems and human livelihoods across drylands, with profound implications for economic and social stability.

For the European Union, the findings underscore the urgency of aligning climate and land policies within the framework of regional cooperation. In Southern and Mediterranean Europe - already prone to increasing aridity and prolonged drought - IPCC findings project further intensification of water scarcity and desertification-like conditions unless robust, multi-level governance and nature-based adaptation measures are implemented (Mirzabaev et al., 2022). These impacts are compounded by rising temperatures, decreasing surface water availability, and the degradation of agricultural lands, reinforcing the critical role of EU-wide collaboration across borders, sectors, and disciplines.

The IPCC emphasize the disproportionate impacts of desertification and dryland degradation on vulnerable populations, many of whom operate with limited access to economic, infrastructural, and institutional resources (Mirzabaev et al., 2022). Similar dynamics arise within the EU itself, where southern and rural regions risk long-term socio-ecological disruption if climate and land-degradation policies fail to integrate equity and resilience. The IPCC emphasizes that successful climate adaptation and resilient development hinge upon multi-level governance, inclusive planning, and mobilization of financial and institutional resources (Mirzabaev et al., 2022).

METHODOLOGY

This chapter is based on a qualitative research approach aimed at understanding how regional and local climate governance initiatives contribute to the European Green Deal objectives. Rather than focusing on a single case or relying on primary data collection, the research uses document analysis to explore a range of institutional practices and governance models across Europe.

The selection of case studies -Vienna, Transdanube.Pearls, Eurometropolis, and Ghent- was guided by their relevance to key themes in EU climate governance, such as technological innovation, cross-border cooperation, participatory democracy, and policy integration. These cases were chosen deliberately to reflect geographical diversity, sectoral variety (e.g. energy, mobility, tourism), and differing governance structures. This approach provides a comparative view of how subnational actors interpret and implement climate objectives in their respective contexts.

The materials reviewed include academic publications, European Commission policy papers, IPCC Assessment Reports, and publicly available reports from municipal and regional institutions. The analysis followed a thematic framework, identifying recurring patterns related to governance practices, local innovation, institutional coherence, and social inclusion.

While this study does not include primary fieldwork, efforts were made to cross-reference official documents with peer-reviewed literature to ensure consistency and credibility. The research acknowledges certain limitations, particularly in terms of context-specific insights that might be gained through interviews or on-site observation. However, given the policy-oriented scope of the chapter, the use of secondary sources was considered both appropriate and enough to support the analysis.

LITERATURE REVIEW

We have adopted a comprehensive and systematic approach using qualitative research methods to study the subject in depth. A multi-faceted methodological review was required to understand the governance mechanisms of regional climate cooperation initiatives within the EU and their effectiveness in achieving the European Green Deal goals. This approach consists of a literature review; an analysis of EU policy documents; and a detailed review of selected regional and local case studies. This approach aims to provide a holistic analysis of the problem by integrating theoretical background and empirical applications. The literature on green transformation demonstrates the importance of cities as key players in climate management and sustainable development.

Bulkeley and Betsill (2005) argue extensively for multi-level governance structures in their work. They emphasize that successful green transformations require coordinated efforts at different scales, integrating municipal actions within broader regional, national, and transnational networks. This approach is supported by a later study by Andonova, Betsill, and Bulkeley (2009). The study examines the role of transnational climate governance in global environmental policies the interaction between various stakeholders and institutional structures.

Kern and Bulkeley (2009) emphasize the evolving role of European cities within multi-level governance. Details their active and passive participation across regional and global climate networks.

Bouteligier (2012) describes cities as incubators of innovation in environmental governance. The study acknowledges the vulnerability of the theory to capture by powerful economic interests. This tension between innovative governance and economic pressures is also reflected in the nuanced perspective offered by Hickel (2021), who examines the principles of degrowth and challenges traditional economic growth paradigms to promote truly sustainable urban development.

Acuto (2013) argues for local-level initiatives that directly contribute to global climate solutions, emphasizing the emerging role of global cities as active actors in diplomacy and governance.

Barber (2013) argues that city mayors should have the authority to use their agility and pragmatic management styles. The study states that cities should position themselves as a viable alternative to often dysfunctional national frameworks.

The European Commission's 2019 Green Deal report provides an example of a structured initiative to harmonize local climate efforts. A comprehensive vision for sustainable transformation within Europe is provided.

Kern (2019) explains how the EU leadership facilitates local studies by incorporating successful practices into broader climate policy architectures.

Barbie and Macedo (2019) show how transnational municipal networks strengthen urban communities in both the Global South and the North. In the study, the dual dynamics of being involved in global environmental policy were also revealed.

Kitzmann et al. (2022) investigates the dynamics of climate innovation decoupling among global urban networks. They examined the dissemination of sustainability policies and technologies using a complex model. They explain the basic role of decoupling between global cities. The study shows that the adoption of climate-related innovations does not depend on local factors, and the decoupling between cities also affects innovations. As a result, he suggests that global connectivity can trigger local policy change. This approach demonstrates how global network dynamics can shape urban sustainability outcomes. It also improves the understanding of multi-level climate governance.

The OECD (2023) report emphasizes the importance of citizen participation in climate governance in EU countries. It states that consultation and participation are necessary to inform effective policymaking. He also emphasizes the importance of hydropolitical analysis, multi-basin case studies in understanding transboundary water and climate management. This refers to the complexity and interconnectedness of environmental problems.

These studies highlight the complexity of the green transformation process, advocating governance that is simultaneously localized, multi-level, and integrative. They also position green transformation as dependent on multi-level governance, inter-city learning, democratic participation, and adaptive regulatory structures. Such governance can respond to urgent urban challenges while also working in line with overarching global environmental goals. They also promote systemic changes that enable global climate action based on local innovation, collaborative practice, and institutional trust.

CLIMATE GOVERNANCE FRAMEWORKS AND REGIONAL PRACTICES IN CLIMATE GOVERNANCE

In order to understand the EU's climate policies, policy priorities and support mechanisms for sub-national actors, it is necessary to examine climate governance frameworks at the global and EU levels. In addition, examining practices that represent various aspects of regional climate cooperation in different geographical and sectoral situations will be useful for understanding how regional cooperation and governance mechanisms work in practice, the challenges encountered, and the factors contributing to success.

4.1. Global and EU-Level Climate Governance Frameworks

Table 2 highlights the central role of international policy frameworks in shaping urban climate adaptation and risk reduction. Collectively, these agreements emphasize the interplay between global climate goals and local action, underscoring the importance of multi-level governance, targeted financing, and inclusive planning. Together, they reflect a shift toward aligning sustainable urbanization, disaster risk management, and climate resilience, reinforcing the critical role of cities and settlements in achieving long-term, equitable climate and development goals.

Table 2. International policy agreements with implications for urbanization and climate adaptation

Agreement (Date)	Scope	Relevance for Cities, Settlements, and Infrastructure	Relevance for Addressing Climate Change Risk
Sendai Framework for Disaster Risk Reduction (March 2015)	Global agreement focused on disaster risk reduction across all levels of governance.	Identifies rapid urbanization as a core risk driver and emphasizes risk reduction through cooperation between national and local authorities.	Highlights systemic, compound, and cascading risks from climate hazards and promotes a shift from reactive disaster response to proactive risk management.
Addis Ababa Action Agenda (July 2015)	International agreement underscoring the financing needs for sustainable development, especially for local and sub-national actors.	Emphasizes the role of local actors and the importance of financing for municipal and local government capacity in addressing climate and disaster risk.	Identifies financing as critical for effective risk reduction in urban settings, although highlights institutional and capacity constraints.
Transforming Our World: The 2030 Agenda for Sustainable Development (September 2015)	Includes the 17 Sustainable Development Goals (SDGs).	Defines SDG11 ("Make cities inclusive, safe, resilient and sustainable") with extensive linkages to climate resilience, access to basic services, and inclusive urban planning.	Provides a framework for aligning climate action (SDG13) with other SDGs and addressing root drivers of climate risk.
The Paris Agreement (December 2015)	Global climate agreement adopted by 194 nations.	Acknowledges the role of local and sub-national authorities and encourages city-level climate action (both mitigation and adaptation).	Supports the integration of urban actors into climate policies, with a focus on aligning national and sub-national efforts.
The World Humanitarian Summit (May 2016)	Summit addressing disaster and climate-related displacements and vulnerabilities.	Outlines core responsibilities for urban areas and recognizes the role of NGOs and local authorities in addressing displacement and crisis.	Links climate change to migration and displacement, highlighting the increasing vulnerability of urban centers and underscoring the need for effective data and collaborative governance.
The New Urban Agenda (October 2016)	United Nations agreement focusing on sustainable urbanization and housing.	Provides a global framework for resilient urban planning, aligning national policies and local implementation for climate mitigation and adaptation.	Establishes cities and sub-national authorities as critical actors in achieving global climate and sustainability objectives.

Source: (Birkmann et al., 2022)

4.2. Regional Practices in Climate Governance

The four case studies included in this chapter were chosen to represent a cross-section of climate governance practices within the European Union. Each case is examined not only for its innovative contributions, but also for the practical or institutional barriers it reveals in the pursuit of a just and effective green transition. Each case highlights a distinct aspect of regional climate action and demonstrates the diversity of institutional, sectoral, and geographical contexts in which the EU Green Deal objectives are being implemented.

- *Vienna* was selected as a leading example of metropolitan-level innovation, particularly in the integration of smart technologies and energy efficiency in urban governance.
- *Transdanube.Pearls* represents a macro-regional, tourism-centered initiative that exemplifies sustainable mobility and cross-border cooperation in the Danube corridor.
- *Eurometropolis Lille-Kortrijk-Tournai* illustrates institutional innovation across national boundaries and provides insight into joint planning for climate adaptation.
- *Ghent* serves as a model of participatory governance, with strong emphasis on local democracy, citizen assemblies, and energy cooperatives.

Together, these cases provide a comparative framework for analyzing the multiple dimensions of regional climate cooperation -technological, organizational, institutional, and social- within the broader EU governance structure.

This section examines four key case studies selected to illustrate how regional climate cooperation and governance mechanisms work in practice. These cases represent various dimensions of regional climate action in different geographical and sectoral situations. The analyses focus on key themes such as policy integration, participatory governance, cross-border cooperation and innovative practices, demonstrating their potential in achieving EU climate goals.

4.2.1. Vienna: Technological Innovation and Smart Urbanism

With one of Europe's leading Smart City initiatives, Vienna is pioneering the integration of digital and social innovations with climate goals at the local level. With the Smart City initiative, it is aimed to develop integrated solutions to the challenges. These developments cover the areas of energy, mobility, buildings, infrastructure and digital services, and include urban sustainability. An important goal of Vienna's Smart City strategy is to increase energy efficiency with renewable energy sources and reduce dependence on fossil fuels. In the field of efficiency, smart grids, passive house standards and renewable heating systems are being installed to reduce energy use in buildings. Decarbonization efforts are based on two foundations: essential part is the strong cooperation between the public and private sectors and the active participation of citizens in the process. When open participation in planning and implementation is examined, neighborhood residents become a part of the transformation of the city. In this way, it strengthens the public's trust in the initiative and its results, and therefore makes it more sensitive to climate policies and more open to cooperation.

When the initiatives are examined, the Aspern Smart City Research (ASCR) project, which is one of the quite ambitious initiatives, was launched in 2013. This initiative is among the largest digital twin projects in Europe. It involves the integration of enterprise technology networks developed in partnership with Siemens, Wien Energie, Wiener Netze and the City of Vienna. These networks are provided by the integration of technologies such as Building Information Modeling (BIM), Geographic Information Systems (GIS) and Internet of Things (IoT) sensor networks. It is being developed simultaneously with artificial intelligence-oriented analysis. Systems are used throughout the city that allow real-time monitoring, simulation and control of energy flows.

Another initiative is Siemens' LV Insights X platform, a SAAS created for distribution network transparency by digitally mimicking low-voltage power grids. This increases the flexibility of intervention, simplifies the integration of renewable energy sources and minimizes system failures. According to field reports, such applications increase energy efficiency by up to 30%. In addition, it can reduce operating costs and minimize unplanned outages. Another important initiative at this level is the circular financing mechanism embedded in the Vienna model. Rather than treating efficiency gains as one-off savings, the city reinvests them into further digital upgrades. This reinvestment cycle transforms climate policy from a financial burden into a strategic investment tool—yielding long-term returns and reinforcing innovation over time (European Commission, 2023).

The integrated energy systems developed represent a multidimensional transformation in grid management. These systems provide several critical functions. The model enables instantaneous monitoring of the carbon footprint at every energy consumption point, from industrial facilities to residential buildings. It also provides decision-makers with an active control mechanism to monitor emissions and implement reduction strategies. In addition, forecasting energy demand with artificial intelligence algorithms and integration with renewable resources allows the system to prepare for future requirements.

The KlimaTicket—introduced nationwide on October 26, 2021—costs €1,179.30 for the standard annual pass and €884.20 for the reduced fare available to youth, seniors, and individuals with disabilities (25% discount) (KlimaTicket, 2025). A further price increase to €1,300 (standard) and €975 (reduced) is scheduled from August 1, 2025 (KlimaTicket, 2025).

By mid-2022, approximately 170,000 KlimaTicket passes had been sold, representing roughly 3% of Austria's adult population (Follmer & Justin, 2023). Including regional variants, the estimated penetration rose to roughly 15% (OECD, 2021).

Survey data collected within the first year of implementation indicate that approximately 66% of holders increased their use of public transportation and about 20% replaced trips that would otherwise have been made by private car (Follmer & Justin, 2023). Econometric analyses employing synthetic control and difference-in-differences methods estimate a net rise of 3.3-6.8 percentage points in rail ridership attributable to the KlimaTicket, relative to counterfactual trends. The initial effects were statistically significant in 2021, although the trend stabilized by 2022 (Wallimann, 2024).

The transportation sector constitutes approximately 39.8% of Austria's total CO_2 emissions (OECD, 2021). According to official reporting, the KlimaTicket's implementation is estimated to have saved roughly 64,631 t CO_2e in 2022, based primarily on user-reported shifts from car to public transport (Follmer & Justin, 2023).

The energy rehabilitation of heritage buildings in Vienna, particularly from the early 20th century, showcases an integrated approach to preserving architectural authenticity while achieving high energy-performance gains. A case study on simulated student housing indicates that deep retrofit measures-with envelope upgrades and renewable integration-reduced annual energy demand from approximately 249 kWh/m^2 to 54 kWh/m^2, corresponding to a 78% decrease (Elnagar et al., 2021).[2]

This comprehensive strategy combines advanced envelope technologies and renewables. Triple-glazed windows and aerogel-enhanced insulation have proven effective in historic contexts, significantly improving thermal performance without altering facades. Additionally, small photovoltaic systems generated nearly 22.5% of the reduced energy demand (Elnagar et al., 2021).

The digitalization of energy infrastructure in Vienna has demonstrated significant economic benefits and improved operational efficiency, aligning with the city's long-term sustainability and climate policies. According to the Austrian Research and Technology Report (European Commission, 2023), investments in smart grid and digital monitoring technologies have an average payback period of approximately seven years. In addition, the benefits that can be achieved are mainly due to several factors. These are operational efficiency gains, improved predictive maintenance and the use of climate finance. When the gains are examined, it has been shown that advances in IoT-based monitoring and predictive maintenance have reduced energy losses by about 22% (Siemens, 2023). It shows that predictive maintenance can reduce maintenance-related costs by 30% and unplanned interruptions by 70-

75% (Aberdeen Group, 2022; IoT Analytics, 2023). These gains justify investments in digital infrastructure, leading to significant savings in the energy sector. Digitalization provides improved carbon management and cost optimization within the framework of the European Union Emissions Trading Scheme (EU ETS). The EU ETS has become an effective mechanism for monetizing carbon reduction efforts and is in line with Austria's national climate goals (European Commission, 2023). The combination of improved operational efficiency, predictive maintenance and optimized emission management provides a solid foundation for long-term cost reduction and sustainability in the energy sector.

The energy poverty surveys conducted in Austria decelerated from about 12.4% to 8.6%, representing a reduction of 31% (between 2019 and 2023). It is linked to policies aimed at reducing the energy cost burden for low-income households (Statistics Austria, 2023). As a result of the investments made in Vienna's green economy (from 2020 to 2023), climate-oriented policies have been developed, economic growth and social equality have been supported and approximately 2,150 new jobs have been created (Vienna Enterprise Agency, 2023). Indirect additional benefits of the developed climate policies to the economy are also emerging. The improvement in air quality has increased labor productivity by reducing medical expenses. In this way, annual cost savings are achieved in the field of healthcare and contributions are made to the budget. Based on these results, it can be said that it has been concluded that the digital and systemic implementation of climate governance provides environmental benefits as well as improves social welfare and the economy.

The climate policies implemented in Vienna provide a valuable experience for other EU cities. This experience shows how digital climate tools can support social well-being without compromising financial or environmental goals. The policy approach applied proves that investments in digital infrastructure can meet EU climate targets within a seven-year turnaround time. As a result, digitalization, rather than being a technical upgrade, reaches a structural step of sustainable urban development and economic resilience (European Commission, 2023).

4.2.2. Transdanube.Pearls: Tourism and Sustainable Mobility Governance

Another major project initiated within the framework of climate policies is The Transdanube.Pearls project. The project, which was created within the scope of the EU's Interreg Danube Transnational Program, was launched in 2017 by bringing together 15 organizations from 9 countries. The project aims to promote low carbon mobility and regional tourism. With a total budget of €2.9 million, it focused on strengthening cross-border cooperation along the Danube corridor (Interreg Danube, 2018).

The main goal was to develop a network of "mobility pearls"—selected destinations accessible through sustainable transport such as cycling, electric mobility, river cruises, and integrated public systems. In 2018, a strategic plan was prepared based on SWOT analysis, stakeholder workshops, and shared goals for sustainable travel along the EuroVelo 6 route, particularly in Upper Austria, Bulgaria, Hungary, Serbia, and Romania (Interreg Danube, 2018; EuroVelo, 2020; Danube.Pearls, 2020).

Governance was structured around transnational mechanisms promoting knowledge exchange, common standards, staff training, and shared implementation among regions such as Bavaria (Germany), Lower Austria (Austria), Duna-Dráva NP (Hungary), Belgrade (Serbia), and the Danube Delta (Romania) (Interreg Danube, 2018; Danube.Pearls, 2020). Outcomes included installing bike carriers on buses in identified risk zones and deploying a multilingual app, downloaded more than 10,000 times, facilitating seamless cycling and mobility between pearls (Danube. Pearls, 2020).

The project's governance model was built upon a multi-level cooperation structure, combining strategic coordination with local implementation. At the strategic level, a Steering Committee comprised of representatives from partner institutions oversaw the project, ensuring its alignment with EU policies and priorities (Interreg Danube, 2018). The Transdanube.Pearls project is managed by a special Project Management Unit. This unit oversees pilot implementations, infrastructure improvements and communication activities in the participating countries (Danube Pearls, 2020).

The project aims to bring together public institutions, NGOs, transport operators and local communities for a broad participation. This collaborative structure aims to support low-emission transport, integrate bicycles with public transport and promote sustainable tourism along the Tuna corridor. The wide participation also contributes to the creation of a long-term cooperation framework that promotes cross-border and climate-resistant mobility. The initiative demonstrates the adaptability between EU-level policy objectives and regional governance practices (Interreg Danube, 2018).

4.2.3. Lille-Kortrijk-Tournai Eurometropolis: Institutional Innovation across Borders

The Lille-Kortrijk-Tournai Eurometropolis was established in 2008 as a transnational administrative formation on the Franco-Belgian border as the first European Grouping of Territorial Cooperation (EGTC). It covers 147 municipalities and about 2.1 million people in Lille (France), Kortrijk and Tournai (Belgium). The EGTC operates under a dual structure, with France's "Code Général des Collectivités Territoriales" harmonizing Belgium's Flemish regional legislation with each other.

The collective budget of the municipalities is approximately €42 million per year. This budget corresponds to about 0.3% of the total GDP of the countries.

Eurometropolis stands out in areas such as water management, flood risk reduction and energy efficiency. By coordinating the planning between the countries, it ensures the implementation of common policy decisions. These efforts are in order to be able to respond to the challenges related to climate changes that cross-national borders. The climate agenda has been updated and expanded to be compatible with the European Green Deal and the EU Climate Adaptation Strategy (Eurometropolis, 2021). Thanks to this structure, Eurometropolis has overcome the institutional disintegration. Corporate design allows for resource sharing, cross-border expertise and coordinated planning. Thanks to its climate adaptation design, it has presented strategies suitable for both urban and rural areas. Eurometropolis, one of the most prominent cases of EU cross-border climate cooperation, supports participants for infrastructure resilience of coordinated governance. It also promotes sustainable regional development.

The central component of Eurometropolis' adaptation strategy is the climate resistant housing initiative. This initiative consists of planning principles that reduce flood vulnerability and heat stress and support long-term habitability. Priority is given to sustainable design, low-carbon construction and spatial strategies that are compatible with EU climate targets. These priorities include urban planning, water risk management and decarbonization of existing housing stock (Eurometropolis, 2021). The Eurometropolis serves as a practical reference for EU-wide climate and housing policy. Technical standards and institutional supports show that local and regional actors can play meaningful roles in the EU's broader climate transition.

4.2.4. Ghent: Energy Democracy and Local Climate Assemblies

Within the framework of the climate policy in the city of Ghent (Belgium), the governance model of energy democracy is being implemented through the Sustainable Energy and Climate Action Plan (SECAP) adopted in 2020 (City of Ghent, 2020). The model, developed in cooperation with local governments, non-governmental organizations and experts, aligns national climate policy instruments with local needs and supports developments in renewable energy distribution.

With this approach, democratic participation is ensured, and all actors are integrated into local climate governance. The cooperation that develops through consultative democracy is connected to the solidarity of scientific expertise and local knowledge (Dryzek and Niemeyer, 2019). Ghent's climate governance practices aim to strengthen local climate action in achieving the EU's broader climate neutrality goals. The model includes practical recommendations, such as setting minimum standards for local climate councils, providing tax exemptions for green bonds, and creating an

EU-level technical office for energy cooperatives. Energy justice is given priority in the city by addressing issues such as energy poverty in a sensitive way to income inequality. In the Ghent model, the transformation of economic returns into social value and the contribution to a circular economy are also encouraged. In terms of supporting the broader climate goals of inclusive governance and local innovation, Ghent is an example for other European cities. As summarized in Table 3, Ghent also addresses common structural and social barriers through several participatory and financially innovative mechanisms.

The success of Ghent is proof of how nationally set energy and climate goals can be implemented in innovative ways that are appropriate to the local context. The participation of citizens in these goals has played a central role in the success of Ghent. Climate projects have reached a wide scale with citizen-oriented initiatives such as energy cooperatives and innovative financial mechanisms. It shows that local governments can put forward more sustainable and citizen-oriented climate solutions by linking national policies with local needs.

A major challenge across EU municipalities is the lack of sustained citizen engagement in climate planning. Ghent has responded through the institutionalization of Climate Citizen Assemblies, which enhance deliberative inclusivity and provide structured input into the city's SECAP. The high transferability of this mechanism stems from its low-cost, adaptable design and alignment with EU participatory governance standards.

Ghent's use of revenue-sharing mechanisms with local citizens and cooperatives has mitigated opposition and fostered community buy-in, a strategy with high applicability in contexts requiring social license for energy transitions.

Finally, institutional inertia, often tied to centralized planning structures and legacy systems, has been countered in Ghent by supporting energy cooperatives that decentralize implementation and distribute agency to non-state actors. This model is not only scalable but has proven particularly effective in enhancing both legitimacy and innovation in local energy transitions.

Table 3. Difficulties and Ghent Solutions

Challenge	Ghent's Solution	Transferability
Lack of citizen participation	Climate Citizen Assemblies	High
Limited financing capacity	Municipal Green Bonds	Medium
Social resistance to renewables	Revenue-sharing with local cooperatives	High
Institutional inertia	Energy cooperatives and decentralization	High

Source. Created by authors.

Table 4 provides dramatic examples of the diversification of climate governance at different geographical scales, sectors, governance types and innovation spaces, as illustrated by Vienna, Transdanube.Pearls, Eurometropolis and Ghent.

Table 4. Comparative Case Selection Criteria

Criteria	Vienna	Transdanube.Pearls	Eurometropolis	Ghent
Geographic Scale	Metropolitan	Macro-regional	Cross-border	Urban
Sector	Energy – Mobility	Tourism – Transportation	Water Management – Planning	Energy – Social Innovation
Governance Type	Single-centered	Network-based	Shared governance	Participatory / Cooperative
Innovation Type	Technological (digital twin)	Organizational (mobility networks)	Institutional (joint planning)	Social (energy democracy)
Stakeholder Structure	Public-private & citizenry	Multi-stakeholder	Municipal platform	Citizen assemblies & cooperatives

Source. Created by authors.

Vienna represents a metropolitan scale, concentrated in the energy and mobility sectors and adopting a mono-centric governance system. Its innovation style is technology-driven, with digitalization and infrastructure solutions at the forefront.

Transdanube.Pearls operates in the tourism and transportation sectors with a macro-regional approach. The networked governance system is characterized by multi-actor decision-making mechanisms, while the innovation style is focused on organizational solutions. With this, it is possible to see how governance is mainstreamed in the institutional framework through level cooperation.

Eurometropolis, as previously stated, is a model operating in a cross-border framework. Focusing on water management, this formation stands out with its shared governance system while implementing institutional innovations. In this way, it emphasizes the importance of multi-stakeholder cross-border governance in the development of institutional infrastructure.

When examined in this situation, Ghent offers a model for the energy sector that represents the urban scale. Ghent stands out with its participatory governance style and prioritizes social governance and social innovation in the field of innovation. This shows that climate governance is a multi-faceted, multi-level and multi-stakeholder area shaped by a wide range of issues, from institutional innovations to technology, from organizations to social solutions.

5. LIMITATIONS AND FUTURE RESEARCH DIRECTIONS

While this chapter provides a comparative overview of regional climate governance in the EU, several limitations must be acknowledged. First, the analysis relies exclusively on secondary sources, including policy documents, academic literature, and publicly available case reports. The absence of primary field data such as stakeholder interviews or on-site observations may limit the contextual depth of some case evaluations. In addition, the selected case studies -though diverse- cannot fully represent the heterogeneity of regional governance experiences across all EU member states.

The study mainly reflects developments up to 2023 and does not capture evolving dynamics in EU climate policy implementation after major geopolitical or economic disruptions (e.g. the Russian invasion of Ukraine, energy price shocks). These factors may have significantly influenced regional energy transitions, especially in Eastern and Southern Europe.

Future research could explore several directions. First, comparative fieldwork could assess how citizens, municipal actors, and private stakeholders perceive and interact with local climate policies. Second, deeper investigation into underrepresented regions -such as Central and Eastern Europe or Mediterranean drylands- may offer more balanced insights into EU-wide governance capacities. Third, future studies should examine the long-term effects of digital and financial innovations (e.g. green bonds, carbon pricing platforms) on inclusive urban adaptation. Lastly, the integration of climate justice metrics into regional governance assessments may provide a more robust framework for evaluating equity and resilience in future EU strategies.

An examination of the EU climate governance literature reveals a structural bias that tends to prioritize Western and Northern European countries, particularly those with high institutional and technological capacity. While our selected cases reflect this pattern, they do not fully represent the diversity of governance experiences across the Union. Eastern and Southern European regions, which often operate under different institutional constraints and sociopolitical dynamics, are understudied in both academic analyses and EU policy discourse. Addressing this imbalance requires expanded empirical research.

6. CONCLUSION

There are examples of multi-level governance in EU, Vienna and Ghent. These cooperations are important. These cooperations have significant potential to achieve the climate goals of the European Union (EU). Regional cooperation mechanisms increase the effectiveness of combating climate change by enabling the development and dissemination of innovative local policies. The implementation of innovative technologies and financing methods in cities has been successful locally. as predicted by the theory of multicenter governance, it refers to the capacity of different local units to develop and implement solutions adapted to their own conditions. It is thought that these innovations developed at the local level can later spread to other regions. These innovations may contribute to accelerating climate action across the EU in the future. It shows that 'bottom-up' approaches combined with central planning can produce stronger results.

The importance of democratic participation and the role of civil society are clearly stated in the cases we have examined. The participation of citizens in governance plays a critical role in the effective implementation of policies. This is also an element that increases local resilience. Initiatives such as the energy cooperatives in Ghent show that active participation of the public in climate action not only improves policy outcomes, but also supports community ownership and long-term sustainability. This highlights the importance of developing more participatory and inclusive governance models. Urban climate policies become more effective with the participation of local actors and civil society (Bulkeley & Betsill, 2005).

Transdanube.Pearls project and the Lille-Kortrijk Eurometropolis examples have facilitated integration with EU-level policy objectives by increasing policy coherence at the regional and local levels. Such collaborations show how transnational municipal networks strengthen climate management through information sharing (Kern & Bulkeley, 2009). There are important opportunities for efficient use of resources in the coordination of cross-border cooperation projects and climate policies. Common risk management, integrated infrastructure planning and common financing mechanisms are of critical importance in climate change. The theory of multi-level governance that cooperation and coordination between different levels of governance are important in dealing with complex environmental challenges.

The effects of cross-border cooperation on policy coherence are seen in the Eurometropolis (Lille-Kortrijk). It can be assumed that the regions of countries with different legal and administrative frameworks can be managed with integrated policies around common climate goals. Thanks to cooperation, it prevents the disintegration of politics and allows for the development of more coherent strategies at the regional level. Since the effects of climate change cross national borders, integration is an important factor in achieving the EU's overall climate goals. Integrated policies facilitate the achievement of environmental goals by ensuring more efficient use of resources between different sectors.

EU regional climate cooperation could make IPCC recommendations useful in the following way:

- Investing in nature-based and land-based adaptation measures tailored to southern EU drylands, addressing desertification and preserving critical ecosystem services for agriculture, water availability, and rural livelihoods;
- Strengthening institutional frameworks that enable multi-level governance across EU borders, aligning rural policies, agricultural incentives, and climate resilience strategies;
- Incorporating Indigenous and local knowledge from EU dryland communities, aligning IPCC-endorsed climate-resilient development pathways with EU rural revitalization policies.

Several chapters of the IPCC AR6 report position drylands as both hotspots for climate vulnerability and repositories of resilience knowledge (Mirzabaev et al., 2022). EU climate cooperation plays a key role in transforming it into actionable policies that promote resilience in global arid regions. It is an exemplary model for other global arid areas.

EU institutions are talking about the importance of data sharing and digital cooperation for climate change. They aim to create an infrastructure that can provide resources for data based on the decoupling of climate change data between sectors and borders. Mechanisms are being developed for common digital ecosystems that allow joint planning, risk assessment and resource allocation with infrastructure. The EU has also moved towards cooperative and multi-level approaches to overcoming traditional administrative and judicial barriers. The establishment of common technical standards for water quality, such as EN16941-1, harmonizes environmental management practices throughout the Member States. Risk pooling mechanisms and cross-border cooperation platforms have become increasingly important for managing climate-related hazards that cross national borders. It is also seen that climate management is an institutional and technological effort as well as an environmental one. The EU aims to facilitate cross-border cooperation by promoting standardized approaches. By using digital infrastructure, it is being tried to create the basis for a more coordinated climate policy. In the event of a possible climate crisis, the gradual transformation of traditional administrative borders is encouraged. It offers valuable theoretical and practical information for the future of climate management across Europe.

REFERENCES

Aberdeen Group. (2022). *The impact of unplanned downtime in the energy sector*. Aberdeen Group.

Acuto, M. (2013). *Global cities, governance and diplomacy: The urban link*. Routledge. DOI: 10.4324/9780203073810

Andonova, L. B., Betsill, M. M., & Bulkeley, H. (2009). Transnational climate governance. *Global Environmental Politics*, *9*(2), 52–73. DOI: 10.1162/glep.2009.9.2.52

Barber, B. R. (2013). *If mayors ruled the world: Dysfunctional nations, rising cities*. Yale University Press.

Barbi, F., & de Macedo, L. V. (2019). Transnational municipal networks and cities in climate governance: Experiments in Brazil. In van der Heijden, J., Bulkeley, H., & Certomà, C. (Eds.), *Urban climate politics: Agency and empowerment* (pp. 59–79). Cambridge University Press. DOI: 10.1017/9781108632157.004

Bednar-Friedl, B., Biesbroek, G. R., Schmidt, D. N., Alexander, P., Børsheim, K. Y., Carnicer, J., . . . Rama, B. (2022). Europe: From chapters and cross-chapter papers. In *Climate change 2022: Impacts, adaptation and vulnerability* (pp. 1817–1927). Cambridge University Press.

Birkmann, J., Liwenga, E., Pandey, R., Boyd, E., Djalante, R., Gemenne, F., Leal Filho, W., Pinho, P. F., Stringer, L., & Wrathall, D. (2022). Poverty, livelihoods and sustainable development. In Pörtner, H.-O. (Eds.), *Climate change 2022: Impacts, adaptation and vulnerability* (pp. 1171–1274). Cambridge University Press., DOI: 10.1017/9781009325844.010

Bouteligier, S. (2012). *Cities, networks, and global environmental governance: Spaces of innovation, spaces of cooptation*. Routledge. DOI: 10.4324/9780203106488

Bulkeley, H., & Betsill, M. (2005). Rethinking sustainable cities: Multilevel governance and the 'urban' politics of climate change. *Environmental Politics*, *14*(1), 42–63. DOI: 10.1080/0964401042000310178

City of Ghent. (2020). *2020–2025 Climate Plan*. Office of the Deputy Mayor for Environment, Climate, Housing and North–South.

Curtis, S. (2016). *Global cities and global order*. Oxford University Press. DOI: 10.1093/acprof:oso/9780198744016.001.0001

Danube.Pearls. (2020). *Transdanube.Pearls: Network for sustainable mobility along the Danube*. Interreg Europe. https://www.interregeurope.eu/good-practices/transdanube-pearls-network-for-sustainable-mobility-along-the-danube

Di Gregorio, M., Fatorelli, L., Paavola, J., Locatelli, B., Pramova, E., Nurrochmat, D. R., May, P. H., Brockhaus, M., Sari, I. M., & Kusumadewi, S. D. (2019). Multi-level governance and power in climate change policy networks. *Global Environmental Change*, *54*, 64–77. DOI: 10.1016/j.gloenvcha.2018.10.003

Dodman, D., Hayward, B., Pelling, M., Castán Broto, V., Chow, W., Chu, E., & Ziervogel, G. (2022). Cities, settlements and key infrastructure. In *Climate change 2022: Impacts, adaptation and vulnerability* (pp. 907–1040). Cambridge University Press., DOI: 10.1017/9781009325844.008

Dryzek, J. S., & Niemeyer, S. (2019). Deliberative democracy and climate governance. *Nature Human Behaviour*, *3*(5), 411–413. DOI: 10.1038/s41562-019-0591-9 PMID: 30962616

Elnagar, E., Munde, S., & Lemort, V. (2021). Energy efficiency measures applied to heritage retrofit buildings: A simulated student housing case study in Vienna. *Heritage*, *4*(4), 3919–3937. DOI: 10.3390/heritage4040215

European Commission. (2019). *The European Green Deal*. https://ec.europa.eu/info/strategy/priorities-2019-2024/european-green-deal_en

European Commission. (2023). *Austrian research and technology report 2023*. European Union. https://ec.europa.eu/

EuroVelo. (2020). *Transdanube.Pearls – EuroVelo 6 [Project summary]*. https://pro.eurovelo.com/projects/2020-03-03_transdanube-pearls---eurovelo

Follmer, J., & Justin, S. (2023). KlimaTicket–Report 2022. In *Republic of Austria – Green Investor Report 2023*. Austrian Government Printing Office.

Hickel, J. (2021). What does degrowth mean? A few points of clarification. *Globalizations*, *18*(7), 1105–1111. DOI: 10.1080/14747731.2020.1812222

Hooghe, L., & Marks, G. (2003). Unraveling the central state, but how? Types of multi-level governance. *Reihe Politikwissenschaft*, Institut für Höhere Studien, 87. https://nbn-resolving.org/urn:nbn:de:0168-ssoar-246200

Hooghe, L., & Marks, G. (2019). Grand theories of European integration in the twenty-first century. *Journal of European Public Policy*, *26*(8), 1113–1133. DOI: 10.1080/13501763.2019.1569711

Interreg Danube. (2018). *Sustainable regional tourism mobility plan for the Danube Region (Transdanube.Pearls).* Interreg Danube Transnational Programme https://www.interreg-danube.eu/

IoT Analytics. (2023). *Predictive maintenance market: 5 highlights for 2023 and beyond.* https://iot-analytics.com/

Kern, K. (2019). Cities as leaders in EU multilevel climate governance: Embedded upscaling of local experiments in Europe. *Environmental Politics, 28*(1), 125–145. DOI: 10.1080/09644016.2019.1521979

Kern, K., & Bulkeley, H. (2009). Cities, Europeanization and multi-level governance: Governing climate change through transnational municipal networks. *Journal of Common Market Studies, 47*(2), 309–332. DOI: 10.1111/j.1468-5965.2009.00806.x

Kitzmann, N. H., Romanczuk, P., Wunderling, N., & Donges, J. F. (2022). Detecting contagious spreading of urban innovations on the global city network. *The European Physical Journal. Special Topics, 231*(9), 1609–1624. DOI: 10.1140/epjs/s11734-022-00470-4

KlimaTicket. (2025). *Current prices and conditions.* https://www.klimaticket.at/en/home/

Mirzabaev, A., Stringer, L. C., Benjaminsen, T. A., Gonzalez, P., Harris, R., Jafari, M., & Zakieldeen, S. (2022). Cross-chapter paper 3: Deserts, semiarid areas and desertification. In *Climate change 2022: Impacts, adaptation and vulnerability* (pp. 2195–2231). Cambridge University Press., DOI: 10.1017/9781009325844.020

New, M., Reckien, D., & Viner, D.. (2022). Decision-making options for managing risk. In *Climate change 2022: Impacts, adaptation and vulnerability* (pp. 2539–2654). Cambridge University Press., DOI: 10.1017/9781009325844.026

OECD. (2021). *Austria's KlimaTicket to promote low-carbon mobility.* OECD International Policy Action Committee. https://www.oecd.org/

Ostrom, E. (1990). *Governing the commons: The evolution of institutions for collective action.* Cambridge University Press., DOI: 10.1017/CBO9780511807763

Ostrom, E. (2017). Polycentric systems for coping with collective action and global environmental change. In *Global justice* (pp. 423–430). Routledge. DOI: 10.4324/9781315254210-18

Ostrom, V., Tiebout, C. M., & Warren, R. (1961). The organization of government in metropolitan areas: A theoretical inquiry. *The American Political Science Review, 55*(4), 831–842. DOI: 10.2307/1952530

Siemens. (2023). *Readiness for predictive maintenance at scale report.* Siemens AG. https://www.siemens.com/

Statistics Austria. (2023). *Energy poverty in Austria: Annual report.* https://www.statistik.at/

UNCCD. (2020). *Global land outlook: Thematic report on drylands.* United Nations Convention to Combat Desertification. https://www.unccd.int

Vienna Business Agency. (2023). *Green economy and employment in Vienna.* https://www.viennabusinessagency.at/

Wallimann, H. (2024). Austria's KlimaTicket: Assessing the short-term impact of a cheap nationwide travel pass on demand. *Transport Policy, 159*, 201–214. DOI: 10.1016/j.tranpol.2024.10.008

ENDNOTES

1 As of 2020, drylands cover approximately 40.6% of the global land surface (excluding Antarctica) and are home to about 2.3 billion people—roughly 31% of the world's population (UNCCD, 2020)

2 Based on documented retrofit of student-housing prototypes in Vienna, where envelope measures and PV integration reduced energy consumption from 248.9 kWh/m^2-year to 54.3 kWh/m^2-year (Elnagar et al., 2021).

Chapter 14
The Role of Local and Regional Governments in Fostering Cross-Border Climate Cooperation

Ananya Pandey
https://orcid.org/0000-0002-2419-6314
Christ University, Bangalore, India

ABSTRACT

With climate change increasingly crossing political boundaries, the importance of local and regional governments (LRGs) in promoting cross-border climate co-operation has grown significantly. LRGs are often better positioned than national governments to tackle localized climate challenges through context-specific collaboration. This chapter examines the growing importance of LRGs in transboundary climate governance, emphasizing their ability to innovate, involve communities, and carry out adaptive solutions. It investigates mechanisms like cross-border networks, inter-municipal forums, joint environmental initiatives, and digital platforms that facilitate collaboration among subnational actors across borders. Through the examination of global case studies and policy scrutiny, this chapter pinpoints essential obstacles—like financial shortfalls and legal constraints—and proposes tactical policy suggestions, such as bolstering legal acknowledgment, establishing regional climate funds, and fostering knowledge sharing.

DOI: 10.4018/979-8-3373-3972-6.ch014

INTRODUCTION

Climate change is a fundamentally global issue that surpasses national borders, socio-economic boundaries, and governance jurisdictions. Though the causes of climate change are varied and its impacts unequal, its consequences—including rising sea levels, extreme weather events, biodiversity loss, and food insecurity—touch nearly every part of the planet. Even after years of collaborative attempts—including the UNFCCC and the Paris Agreement—global climate initiatives have not been able to match the necessary speed and magnitude to restrict warming to 1.5°C above pre-industrial levels (IPCC, 2023). In light of this context, there is an increasing amount of research and policy discussion that highlights the necessity of multi-level governance. This approach involves climate solutions being propelled not just by national governments, but also necessitates the substantial involvement of local and regional governments (LRGs) across different territories.

Because climate systems are complex and interdependent, they require a response that is decentralized but also coordinated. Conventional models that focus on the state and are based on principles of national sovereignty and hierarchical decision-making often do not adequately address the complexities of local climate vulnerabilities or stimulate innovation and action across borders. For example, while central governments may concentrate on wide-ranging regulatory frameworks and macroeconomic tools, they may lack the detail and flexibility needed to carry out adaptation and mitigation strategies tailored to specific regions. Moreover, climate policies at the national level often face obstacles due to political inertia, bureaucratic complexity, and competing geopolitical interests (Bulkeley & Betsill, 2013). This has led to a growing shift in the focus of significant climate action toward subnational actors, who are in a unique position to connect disparate policies and carry out solutions that are customized to fit local conditions.

Over the last 20 years, a major development has been the rise of local and regional governments (LRGs) as climate leaders. LRGs are taking on a more significant role in climate diplomacy, planning, and implementation, from cities like Amsterdam and Cape Town to regional coalitions such as the Under2 Coalition and the Covenant of Mayors for Climate & Energy. These actors have a distinctive blend of attributes: they are geographically close to communities impacted by climate change, they have specialized knowledge in land use and infrastructure planning, and they operate within robust participatory governance systems that facilitate more inclusive decision-making (Setzer & Higham, 2021). Furthermore, LRGs frequently serve as "first responders" to the effects of climate change—overseeing flood defenses, public health emergencies, and emergency services—which positions them at the forefront of climate resilience and adaptation efforts. Significantly, numerous LRGs are not only acting independently but also participating in cross-border cooperation, creating transnational networks and collaborative platforms designed to promote coordinated climate responses. These collaborations transcend mere symbolic diplomacy; they act as conduits for knowledge sharing, alignment of policies, execution of joint initiatives, and mobilization of financial resources (Acuto et al., 2017). In areas that share ecological systems—like river basins, mountain ranges, and coastal zones—cross-border climate cooperation is particularly vital. For example, the Great Lakes–St. Lawrence Cities Initiative promotes cooperative water management among U.S. and Canadian municipalities, whereas Euroregions such as the Basque Eurocity advocate for decarbonization and sustainable transportation strategies across national boundaries (Selleslaghs, 2017). These initiatives demonstrate how LRGs can go beyond the constraints of national policies and serve as laboratories for climate innovation and diplomacy.

The shortcomings of approaches focused on the state become particularly clear when one takes into account the inertia that is frequently seen in negotiations of an international nature. The Paris Agreement was a historic turning point for global climate governance, but because it relies on nationally determined contributions (NDCs) and lacks binding enforcement mechanisms, much depends on voluntary ambition and domestic political will. Conversely, particularly in democratic settings where local constituencies hold LRGs directly accountable, subnational actors often exhibit greater policy experimentation and political continuity (Ostrom, 2017). Additionally, city and regional leaders frequently work within more adaptable institutional settings, enabling them to test green infrastructure, climate-smart transportation, and community-based adaptation initiatives without the limitations imposed by national-level gridlock.

In addition, innovation at the local level and regional coordination are essential for attaining climate justice and guaranteeing that marginalized communities do not get overlooked in the green transition. Numerous vulnerable populations, including indigenous groups, coastal communities, and smallholder farmers, are governed by municipal or regional authorities rather than national ministries. These authorities are more capable of incorporating indigenous knowledge, tackling localized risks, and crafting context-specific interventions (Nagendra & Ostrom, 2012). From this perspective, the empowerment of LRGs is a question of equity and inclusivity as much as it is a question of efficiency. Therefore, in the light of the above perspective, chapter examines the role that LRGs play in promoting cross-border climate cooperation, given their increasing importance in climate governance. It aims to elucidate the institutional, political, and operational factors that facilitate or hinder such collaboration, using theoretical frameworks of multilevel governance and polycentricity (Jordan et al., 2015). Additionally, the chapter highlights effective strategies and insights gained from practical case studies, thus offering a guide for policymakers, researchers, and members of civil society who wish to utilize subnational efforts for transboundary climate solutions.

CONCEPTUAL FRAMEWORK

Cross-border climate cooperation involves collaborative efforts to address climate-related challenges through environmental actions taken by various governmental actors, especially at subnational levels, across administrative or national boundaries. For dealing with transboundary environmental problems like air and water pollution, biodiversity decline, and climate-driven migration—issues that disregard political borders—this type of collaboration is crucial (Bulkeley et al., 2013). Cross-border cooperation, as opposed to national or supranational strategies that function at a macro-policy level, is frequently based on local and regional initiatives that take into account shared geographies, ecosystems, and socio-economic interdependencies. Cooperation can take many forms, including joint mitigation initiatives, coordinated adaptation strategies, shared environmental monitoring systems, and institutional partnerships between cities and regions. Such arrangements challenge the conventional view that international climate diplomacy is exclusively the domain of nation-states, suggesting a growing "diplomatic agency" of cities and regions (Acuto, 2013).

THEORETICAL FOUNDATIONS

Several interrelated theoretical frameworks provide the basis for cross-border climate cooperation at the subnational level: multilevel governance, climate polycentricity, and subnational diplomacy. These frameworks elucidate the increasing authority and ability of local and regional governments (LRGs) to shape global climate agendas.

Multilevel Governance (MLG)

According to Marks & Hooghe (2004), multilevel governance describes a system in which decision-making power is shared among various interconnected layers of government, including local, regional, national, and international levels. In the context of climate change, MLG acknowledges that solutions are more effective when they involve input from stakeholders at all levels, especially those nearest to the point of impact. MLG stresses that local governments should not be viewed merely as enactors of policies from higher levels; they are also active policy-makers in their own right. A relevant instance is the EU, where regional cohesion policies integrate climate aims via cooperative funding methods and cross-border collaborations through tools like INTERREG (Bache & Flinders, 2004). MLG provides the necessary institutional flexibility for LRGs to create climate alliances that go beyond national borders, thus promoting innovative and context-specific responses.

Climate Polycentrism

The theory of polycentric governance, brought to prominence by Ostrom (2017), advocates for a framework comprising multiple, overlapping decision-making centers that are formally autonomous from one another. Within the framework of polycentric climate governance, local and regional authorities possess both the power and motivation to try out context-sensitive policies, learn from one another, and expand successful models. The complexity and urgency of climate change necessitate a variety of approaches, which is acknowledged by climate polycentricity. It promotes cross-border collaboration as a means of horizontal learning and mutual adaptation, rather than through the top-down imposition of standardized solutions. Transboundary river basin organizations, such as the Rhine Commission, facilitate polycentric responses to flooding and water pollution by involving multiple regional actors from various countries (Moss & Newig, 2010).

Subnational Diplomacy

Subnational diplomacy, which is also known as paradiplomacy, involves the international relations carried out by local and regional governments to promote their interests and values (Tavares, 2016). LRGs, though traditionally excluded from formal diplomacy, are increasingly acknowledged as influential actors in global climate forums, including UNFCCC processes and city networks like C40 and ICLEI. Subnational diplomacy offers LRGs a platform to form climate partnerships that transcend borders, especially in situations where national governments lack the will or capacity for decisive action. The diplomatic potential of LRGs in bridging geopolitical divides and sustaining transnational climate ambitions is illustrated by the proactive roles of cities like New York, Paris, and Shenzhen in forming climate pacts with counterparts in other countries (Acuto & Rayner, 2016).

DISTINCTION BETWEEN NATIONAL, REGIONAL, AND MUNICIPAL ROLES IN CLIMATE GOVERNANCE

Grasping the differentiated functions of national, regional, and municipal actors is crucial for recognizing the unique worth of cross-border initiatives spearheaded by LRGs.

National Governments

As the primary signatories of global climate accords, national governments usually establish the broad policy and regulatory foundations for climate action. They create nationally determined contributions (NDCs) as per the Paris Agreement and distribute funding and responsibilities to subnational governmental tiers. Nonetheless, national governments might encounter political inertia, diplomatic constraints, or administrative inefficiencies that restrict their capacity for a rapid or innovative response to climate threats (Jordan et al., 2015). Additionally, there may be a misalignment between national priorities and the localized nature of numerous climate impacts. In federal or decentralized systems like those of the USA, Germany, or India, state or regional governments may hold substantial legislative power concerning environmental issues, which highlights the importance of subnational governance.

Regional Governments

Regional governments (such as states, provinces, and autonomous communities) frequently act as intermediaries between the national and municipal levels. They may have the authority to legislate on land use, infrastructure, transportation, and public health—all essential areas for climate mitigation and adaptation. Regional authorities have a unique opportunity to foster cross-border collaboration by consolidating local needs and organizing regional responses. They have the capability to mobilize resources, establish standardized regulatory approaches, and create large-scale initiatives like regional emissions trading systems or biodiversity corridors that cross multiple jurisdictions. The involvement of regional governments in the EU's Committee of the Regions and global initiatives such as the Under2 Coalition exemplifies their increasing influence in climate diplomacy (Happaerts et al., 2011).

Local Governments

Municipal or city governments are nearest to citizens and ecosystems, giving them unique insights into the immediate effects of climate change. They carry out the implementation of most climate-related policies, such as those on waste management, energy efficiency, building regulations, and urban transport. Cities have shown extraordinary ability to initiate green technologies, implement climate action plans, and establish cross-border collaborations. As an example, cities along the border such as San Diego (USA) and Tijuana (Mexico) have worked together on initiatives for cross-border air quality monitoring and wastewater management (Mayer, 2020). Even when national policies are stagnant or regressive, municipalities' nimbleness and capacity for innovation make them powerful agents of cross-border cooperation.

THE UNIQUE POSITION OF LOCAL AND REGIONAL GOVERNMENTS

As the fight against climate change grows ever more pressing, local and regional governments (LRGs) have become essential players in this struggle. These subnational entities—municipalities, provinces, districts, and states—have distinct advantages that place them at the leading edge of climate mitigation and adaptation efforts. Their closeness to impacted communities, ability to innovate, participatory governance structures, and nimbleness in policy execution provide them with an advantage in developing effective and context-sensitive climate responses. LRGs, in contrast to national-level entities that are frequently burdened by bureaucracy or geopolitical interests, are rooted in localized realities. This allows them to promote inclusive, responsive, and cooperative cross-border climate action.

Proximity to Communities and Ecosystems

Impacted by Climate Change LRGs possess a significant advantage due to their geographic and sociocultural closeness to communities and ecosystems impacted by climate change. This proximity enables a detailed comprehension of local vulnerabilities and socio-environmental dynamics. As an example, municipalities located on the coast face the direct impacts of rising sea levels, whereas those situated in dry areas grapple with desertification. Consequently, LRGs serve as both first responders and primary decision-makers in the development of localized adaptation strategies. Due to their embeddedness, they can incorporate indigenous knowledge systems, historical land-use practices, and social needs into climate planning (Bulkeley & Betsill, 2013). Additionally, LRGs often oversee local infrastructure, waste management systems, urban development, and public transit—domains that are all closely linked to climate outcomes. Their management of these sectors amplifies their ability to implement policies that lower emissions or bolster resilience. Cities such as Rotterdam and Singapore have, for instance, developed innovative urban water management systems that are adapted to their specific topographies. These systems combine engineering with ecological design to address flooding and improve resilience (Mees et al., 2013).

Flexibility and Innovation in Policy Experimentation

Another characteristic of LRGs is their relative ability to experiment with new policy tools and governance arrangements. Subnational actors, liberated from the political inertia that frequently hampers national action, have led the way in climate innovations that are subsequently scaled up or replicated across borders. Due to their smaller scale and closeness to stakeholders, they can test out experimental policies that involve lower risks and quicker feedback cycles (Anguelovski & Carmin, 2011).

As an example, LRGs have pioneered the implementation of green infrastructure, low-emission zones, and circular economy models. Both Freiburg (Germany) and Portland (USA) have adopted innovative urban sustainability strategies that combine renewable energy systems with land-use planning and community involvement. The experimental ethos prevalent among LRGs is demonstrated by these efforts, which enable them to create novel solutions tailored to their specific environmental and socio-economic contexts (Kern & Alber, 2009). This ability to innovate extends beyond just technological solutions. LRGs have created institutional innovations like participatory budgeting, climate citizens' assemblies, and transboundary councils that have enhanced the inclusivity and transparency of climate governance. Because of their bottom-up approaches, they often enjoy enhanced legitimacy, effectiveness, and social buy-in—factors that promote the longevity and success of policies.

Responsibility and Participatory Governance

Local and regional authorities frequently benefit from greater public trust and involvement than national authorities. This proximity nurtures a participatory form of governance, allowing citizens and civil society to actively influence climate priorities and oversee their execution. Participatory mechanisms are essential for guaranteeing democratic accountability, particularly when it comes to making tough trade-offs regarding climate adaptation and mitigation strategies (Pierre, 2011). The legitimacy and effectiveness of climate policies are enhanced by the participatory nature of LRGs. Inclusion of community members in the decision-making process enhances the congruence of policies with local needs and boosts public support, which in turn diminishes the chances of resistance to implementation. For example, in cities such as Copenhagen and Medellín, participatory urban planning initiatives have effectively involved communities in rethinking mobility systems and the design of green spaces, taking into account climate issues and social justice needs (Aylett, 2015).

Furthermore, the principle of subsidiarity—established in numerous governance frameworks like the European Charter of Local Self-Government—guarantees that decisions are made at the closest or most local level that is consistent with their resolution. This principle bolsters democratic oversight of climate policy and fosters resilience by strengthening local agency and stewardship (EU Committee of the Regions, 2020). For instance, the city of Durban in South Africa developed a municipal climate protection program based on localized scientific research and stakeholder consultations, resulting in tailored biodiversity and adaptation projects (Roberts & O'Donoghue, 2013). Furthermore, the institutional closeness of LRGs to public and private service providers—such as utilities, transit authorities, and waste management companies—facilitates the coordinated delivery of climate-related services. Their ability to form coalitions with local stakeholders—businesses, NGOs, universities, and indigenous groups—fosters collaborative ecosystems where innovation and accountability can thrive.

Capacity for Swift Execution and Local Adjustment

Due to the close administrative and operational relationship of LRGs, policies can be implemented and adjusted more rapidly. LRGs can often bypass bureaucratic delays and quickly implement climate initiatives, in contrast to national bureaucracies that may be hindered by hierarchical decision-making and political compromise. In emergency situations like floods, wildfires, or heatwaves, this agility becomes especially vital, as timely decisions can reduce damage and preserve lives (OECD, 2022). Moreover, LRGs are frequently more skilled at customizing solutions to fit the socio-ecological context of their areas. They can create context-specific interventions thanks to localized climate modeling, vulnerability assessments, and risk mapping. As an example, Durban (South Africa) created a municipal climate protection program that was informed by localized scientific studies and consultations with stakeholders. This led to the development of customized biodiversity and adaptation projects (Roberts & O'Donoghue, 2013). Moreover, the institutional proximity of LRGs to public and private service providers—like utilities, transit authorities, and waste management companies—enhances the coordinated provision of climate-related services. Their capacity to establish alliances with local stakeholders—including businesses, NGOs, universities, and indigenous groups—promotes collaborative ecosystems conducive to innovation and accountability.

MECHANISMS FOR CROSS-BORDER CLIMATE COOPERATION

With the urgency of tackling climate change going beyond national borders, local and regional governments (LRGs) have increasingly assumed proactive roles in promoting cross-border collaboration. LRGs function near the communities and environments that climate change directly impacts, which allows for responses that are more immediate and tailored to the context—unlike those of national governments, which are centralized. Even though they are often constrained by jurisdictional limits, these governments are nonetheless creating innovative mechanisms for cross-border collaboration. These mechanisms can be classified into four categories: formal agreements, informal networks, joint initiatives, and digital platforms for data sharing and coordination.

FORMAL MECHANISMS

Agreements between Sister Cities

Agreements for sister cities, which are also referred to as twin town partnerships, act as both symbolic and practical collaborations between municipalities from various countries. Initially created for the purposes of cultural exchange and peace-building after World War II, these agreements have developed to include objectives related to climate change mitigation and sustainability. Cities utilize these connections to share best practices regarding renewable energy, sustainable transportation, urban planning, and disaster resilience. As an illustration, the sister city relationship between San Francisco (USA) and Bangalore (India) has involved collaboration in clean technology and sustainability strategies (Sister Cities International, 2023).

Climate Compacts at the Regional Level

The Under2 Coalition is a significant formal framework that links more than 270 subnational governments dedicated to maintaining the global temperature increase below 2°C. This coalition enables coordinated climate action across regional boundaries and provides a platform for member governments to establish joint decarbonization goals, exchange policy templates, and monitor progress together (Under2 Coalition, 2022). Additionally, regional compacts bolster subnational diplomacy and provide a certain level of autonomy to local and regional governments (LRGs), allowing them to chase climate objectives even in instances of national government shortcomings.

Organizations for Transboundary River Basins

Natural ecosystems like rivers, forests, and mountain ranges extend beyond political boundaries, making transboundary governance structures essential. River Basin Organizations (RBOs), including the International Commission for the Protection of the Danube River (ICPDR), serve as formal institutional mechanisms that enable multiple countries and their regional entities to collaboratively manage shared water resources. These RBOs assist local governments in working together on pollution control, flood risk management, and ecosystem protection (Schmeier, 2013). Another instance of LRGs in Southeast Asia serving in advisory capacities for flood forecasting and climate adaptation is the Mekong River Commission.

INFORMELLE NETZWERKE

ICLEI – Lokale Regierungen für Nachhaltigkeit

ICLEI is an international network comprising more than 2,500 local governments striving for sustainability and climate resilience. Though it does not have the force of law, ICLEI facilitates transnational collaboration through the provision of policy advice, technical assistance, and venues for conversation. Its initiatives, like the Cities for Climate Protection campaign, enable city-to-city learning and the development of cooperative strategies (ICLEI, 2021). Informal networks such as ICLEI often address the shortcomings of formal diplomacy, enabling more agile and trust-based collaboration.

Climate Leadership Group of C40 Cities

C40 comprises 96 megacities, accounting for one-twelfth of the global population and a quarter of the worldwide economy. C40 Cities (2022) reports that cities such as Paris, Tokyo, Johannesburg, and São Paulo exchange climate action plans and assist each other in lowering carbon emissions by improving energy efficiency, reforming transportation, and minimizing waste. C40 is characterized by its focus on peer-to-peer engagement and measurable outcomes, with numerous cities openly publishing their GHG emissions data to promote accountability.

UCLG (United Cities and Local Governments)

UCLG plays an important role in promoting decentralized international cooperation. It acts as a worldwide representative for democratic local self-government and climate justice. UCLG strengthens the ability of municipalities to express common environmental interests and affect global climate negotiations through policy dialogues, workshops for capacity building, and missions in climate diplomacy (UCLG, 2022). Informal networks such as UCLG play a crucial role in enhancing the representation of localities in intergovernmental forums like COP summits.

GEMEINSAME INITIATIVEN

Grüne Infrastrukturprojekte über die Grenze hinweg

LRGs are becoming more involved in developing cross-border green infrastructure initiatives. This encompasses community parks, initiatives for reforestation, and habitat linkages that fulfill ecological and societal roles. For example, the Green Belt of Fennoscandia represents a collaboration among Finland, Norway, and Russia aimed at safeguarding biodiversity across national borders via a system of transboundary protected areas (UNECE, 2020). These projects serve a dual purpose: they help reduce climate change effects and bolster transboundary governance and ecological solidarity.

Collaborative initiatives for reducing disaster risk

Climate change increases how often and how serious natural disasters are, making shared early warning systems, emergency preparedness protocols, and mutual aid agreements essential. The Alpine region of Europe has seen the establishment of integrated landslide monitoring and avalanche control systems through cross-border partnerships involving Italy, Switzerland, Austria, and Slovenia (Alpine Convention, 2021). These collaborative initiatives are vital for bolstering regional resilience and averting displacement and infrastructure damage caused by climate change.

Emission Trading Systems

Emissions trading schemes (ETS) have been put into action in certain areas, enabling local organizations across various jurisdictions to exchange carbon credits. An exemplary case is the Western Climate Initiative (WCI), which comprises states and provinces from both Canada and the U.S. Local and regional actors can use ETS to provide economic incentives for reducing emissions, while also generating funds for additional climate initiatives (Haites, 2018). Such mechanisms demonstrate how subnational cross-border cooperation can scale market-based solutions.

Digital Platforms and Climate Data Sharing

Digital technology has emerged as a potent facilitator of international climate governance. To coordinate responses and harmonize policies, LRGs are adopting shared data platforms and real-time climate monitoring systems. Instruments like Open Climate, Climate Watch, and CDP Cities provide municipalities the means to report their emissions, climate risks, and resilience measures in standardized formats. This facilitates benchmarking and mutual learning (World Resources Institute, 2023). Moreover, Geographic Information Systems (GIS) and satellite-based climate monitoring facilitate shared situational awareness across borders. As an example, the Copernicus Climate Change Service (C3S), run by the European Union, offers localized and transboundary climate data that aids decision-making at the municipal level. These platforms enable cities and regions to enhance their forecasting of climate impacts, evaluate vulnerabilities, and devise collaborative interventions.

CHALLENGES TO CROSS-BORDER CLIMATE COOPERATION

Local and regional governments (LRGs) can tackle common environmental issues through joint efforts, thanks to cross-border climate cooperation. Nevertheless, even though their role is increasingly acknowledged, many institutional, legal, political, and economic obstacles persist that impede effective transboundary collaboration. These challenges become particularly evident in complex geopolitical contexts, where subnational actors must navigate across national boundaries with varying priorities and governance systems.

Jurisdictional Fragmentation and Legal Incompatibilities

A major obstacle to cross-border climate cooperation is the fragmentation of jurisdiction and incompatibility of laws. Subnational governments function within national legal frameworks, which may not be harmonized across borders. Legal and administrative frameworks differ regarding environmental standards, planning processes, and the competencies assigned to regional and local authorities. Such a lack of coherence results in coordination difficulties and leads to inconsistent or conflicting climate policies (Piattoni, 2010). For instance, in the transboundary water management among countries in the Mekong River Basin, varying national water laws and institutional arrangements have restricted the effectiveness of joint environmental initiatives, even when mutual interests are present (Sneddon & Fox, 2006). Furthermore, in federal systems where states are decentralized, the degree of local autonomy can vary greatly, further complicating matters. In the absence of legal mechanisms for cross-border mandates, subnational cooperation is informal and susceptible to changes in national politics.

Resistance or Absence of Support from National Governments

A further significant barrier is the resistance or absence of support from national governments, which frequently perceive subnational international engagement—often called "paradiplomacy"—as infringing upon their foreign policy authority (Keating, 2013). Despite the formal acknowledgment of the necessity for multilevel climate governance by numerous national governments, they may, in practice, refrain from providing the resources or legal authority required for cross-border collaboration. This dynamic is apparent in numerous developing nations, where the centralization of climate finance and foreign relations limits LRGs' ability to access international funding or form partnerships without national oversight (Betsill & Bulkeley, 2006). Even in developed contexts, national-level decisions, such as the withdrawal of the United States from the Paris Agreement during the Trump administration, can significantly undermine the credibility and continuity of local climate partnerships (Romero-Lankao et al., 2016). National policies that prioritize economic or security concerns over climate action may leave subnational actors without support or even actively undermine their efforts.

Differences in Capacity and Limits on Funding

It takes considerable technical, administrative, and financial resources for cross-border cooperation to happen; however, numerous local and regional authorities function with restricted budgets and limited institutional capacity. This disparity in abilities, particularly among urban, rural, or economically disparate areas, obstructs fair involvement and collective ownership of climate initiatives. For instance, although bigger cities such as Amsterdam or San Francisco may possess the infrastructure and staff to facilitate international environmental collaboration, smaller municipalities frequently do not have the necessary expertise or funding to establish or sustain such partnerships (Kern & Bulkeley, 2009). Furthermore, accessing climate finance continues to be a crucial obstacle. A number of international funds, including the Green Climate Fund, necessitate that applicants fulfill intricate fiduciary and environmental standards. These requirements are frequently unattainable for smaller or under-resourced local governments (Tanner et al., 2010). Moreover, disparities in capacity across borders can intensify tensions and erode trust, particularly when one party feels that it bears a disproportionate share of responsibility or costs. When one area possesses a more developed monitoring or implementation system, disparities can impede advancement because of coordination inefficiencies.

Political and Ideological Disparities Between Adjacent Areas

Ultimately, political and ideological disparities between adjacent jurisdictions can interrupt or impede climate cooperation across borders. Variations in leadership vision, public priorities, or governance styles can greatly influence dedication to common objectives. In areas where politics are highly polarized, climate cooperation can get caught up in partisan disputes, complicating efforts at long-term planning and execution. The challenge is apparent within the European Union, where political disagreements have often weakened cohesive climate strategies, especially in Central and Eastern Europe, despite the existence of institutional frameworks for regional collaboration (Jordan & Huitema, 2014). Moreover, nationalist or protectionist agendas can deter collaboration by portraying climate cooperation as a threat to sovereignty or economic competitiveness. Moreover, the significance of climate change may vary across borders. A locality might place a higher priority on economic recovery or job creation than on environmental action, resulting in differing climate agendas. To close these gaps, it is necessary to have strong dialogue mechanisms and a political commitment to address the ideological and developmental disparities.

ENABLING CONDITIONS AND SUCCESS FACTORS

It is becoming ever clearer that cross-border climate cooperation among LRGs is vital for effective, place-based climate responses. The success of these collaborations, however, depends on several facilitating conditions. These factors encompass decentralization and supportive national policies, access to technical and financial resources, robust data along with shared metrics, trust-building and ongoing leadership, as well as active involvement from citizens and stakeholders.

Decentralization and Supportive National Policies

The extent of political and fiscal decentralization is one of the fundamental factors that enables local and regional authorities to participate in international and inter-regional collaboration for cross-border climate cooperation. Cooperation becomes more organized and impactful when national governments offer legal and institutional recognition for LRGs to engage in transboundary climate actions (Peters & Pierre, 2016). To enable local and regional governments (LRGs) to sign agreements, share data, and collaboratively implement adaptation and mitigation strategies, national frameworks that legitimize subnational diplomacy—often called "paradiplomacy"—are crucial. Germany, the Netherlands, and Canada exemplify multi-level governance structures that enable regional authorities to work together across borders with federal oversight (Betsill & Bulkeley, 2006). Furthermore, national policies that support and align with global climate objectives—like the Nationally Determined Contributions (NDCs) established by the Paris Agreement—can furnish LRGs with strategic frameworks and benchmarks for their local initiatives. Subnational actors are more likely to be agile and committed when national policy environments are flexible and supportive of regional innovation (OECD, 2019).

Access to Financial Resources and Technical Expertise

The availability of financial resources and technical expertise is another essential element for maintaining cross-border collaboration. Numerous successful initiatives are supported by international development agencies, multilateral climate funds, or regional integration organizations like the European Union (EU). As an example, the EU's INTERREG programme has aided a variety of cross-border climate and environmental initiatives through funding and capacity-building resources for regional stakeholders (European Commission, 2020). Local governments can ground their decisions in empirical evidence thanks to technical support that encompasses scenario modeling, emissions inventories, and GIS-based climate risk assessments. Such assistance is often facilitated by agencies like the UNDP and the Global Environment Facility (GEF), which helps ensure consistency across regional efforts and alignment with national and international standards (UNDP, 2022). Cross-border projects may risk stagnation due to differences in regional resources or technical expertise if they lack sufficient funding and technological capacity.

Resilient Data Gathering and Collective Climate Metrics

To cooperate internationally, we need to share a common measurement language and mutual comprehension of climate risks, impacts, and opportunities. This requires strong data collection methods and standardized metrics that promote comparability, transparency, and accountability. Platforms for shared climate data and open-access monitoring tools can facilitate cross-regional tracking of greenhouse gas emissions, temperature variations, land use patterns, and vulnerability assessments (Franzke et al., 2020). As an illustration, the Great Lakes–St. Lawrence Cities Initiative employs shared data systems to collaboratively track water levels, pollution, and flood risks across U.S. and Canadian cities (Gallagher et al., 2020). It is essential for coordinated planning and reporting that methodologies are aligned across borders so that partners are working from the same assumptions and standards.

Establishing Trust and Ongoing Leadership

Trust is essential for any collaboration that spans a long duration, with its importance heightened in cross-border contexts characterized by potential cultural, linguistic, or political differences. According to Ansell & Gash (2008), mutual trust and credibility among partner governments are enhanced by regular communication, transparent decision-making processes, and joint governance mechanisms. Successful collaborations often rely on sustained leadership, especially from local leaders who are politically stable and visionary. Local or regional climate champions can generate momentum, obtain funding, and uphold the political commitment needed to withstand shifts in national policy or public opinion (Barber, 2013). When leadership remains consistent, it helps ensure the retention of institutional memory and the development of long-term strategies required for climate governance.

Involvement of Citizens and Stakeholders

Another factor that contributes to success is the inclusive involvement of civil society, businesses, indigenous communities, and other local stakeholders. Cross-border projects that embody the interests and insights of a diverse range of community members tend to exhibit greater social resilience and political legitimacy. Public involvement fosters ownership and adherence, particularly in fields such as land use planning, water management, and the implementation of renewable energy (Patterson et al., 2017). Methods that involve participation — like community co-design workshops, transboundary stakeholder forums, and deliberative assemblies — assist in closing the gaps between policy and practice. Moreover, when local communities comprehend and endorse collaborative endeavors, it often bolsters transborder cooperation and enhances democratic accountability.

CONCLUSION AND POLICY RECOMMENDATIONS

Given that climate change is increasingly surpassing political and geographic boundaries, it becomes essential for local and regional governments to foster cross-border cooperation. This chapter has underscored the unique ability of LRGs, due to their proximity to communities and ecosystems, to carry out climate actions tailored to their specific contexts that align with national and global initiatives. Cross-border climate cooperation spearheaded by LRGs offers a means to tackle transboundary challenges—like watershed degradation, air pollution, and migratory impacts—through decentralized yet coordinated action. It allows for reciprocal learning and collaborative innovation among communities encountering comparable dangers. To fully realize this potential, it is essential to empower subnational actors through strong legal recognition, sufficient financial backing, institutional mechanisms for ongoing dialogue, and access to data and technology. It is crucial to note that the inclusion of local perspectives in global climate policy is still lacking development. While international frameworks such as the Paris Agreement refer to multilevel governance, practical methods for incorporating LRGs into decision-making processes are still being developed. This necessitates a recalibration of global climate governance that recognizes local and regional actors as co-creators of climate solutions, rather than just implementers. In the future, as climate impacts accelerate, the necessity of expanding regional collaborations will increase. This necessitates a more integrated architecture in which LRGs function within supportive institutional and financial ecosystems that empower them to connect national divides and create sustainable partnerships. A change of this kind would represent a crucial advance toward a genuinely inclusive and effective worldwide reaction to the climate emergency.

Policy Recommendations

Due to their closeness to local challenges and ability for innovation, LRGs are increasingly acknowledged as crucial players in climate governance. To enhance their role in cross-border climate cooperation, targeted policy support is crucial. Although international relations have traditionally been the domain of national governments, the increasing involvement of LRGs in transboundary environmental action makes it necessary to acknowledge "subnational climate diplomacy." To accommodate and legitimize the climate actions of local and regional actors, national legal systems and international frameworks need to be reformed. With legal recognition, LRGs can enter into climate cooperation agreements, establish international partnerships, and utilize global climate finance mechanisms (Keohane & Victor, 2011). As an illustration, giving formal support to city networks such as the Global Covenant of Mayors boosts their credibility in international negotiations.

One of the primary limitations that LRGs encounter when it comes to participating in cross-border climate initiatives is access to sufficient financing. Setting up regional climate funds aimed at bolstering local cross-border initiatives—particularly in areas with shared ecosystems or climate vulnerabilities—would assist in stimulating action. These resources can be gathered from public, private, and multilateral sources and applied to joint risk evaluations, shared infrastructure, and cooperative mitigation initiatives (Bulkeley & Betsill, 2013). Technical assistance can be provided by climate funds, which helps to lessen the gap in administrative capacities between municipalities with resources and those lacking them. Cross-border climate cooperation flourishes through mutual learning. It is essential to empower LRGs to exchange best practices, innovations, and lessons learned via co-learning platforms. Training programs, exchanges, and conferences can be organized by regional institutions and international agencies to fulfill a mediating role. These activities promote the spread of policy and establish networks of practice that go beyond national boundaries (Biermann et al., 2009). Initiatives such as ICLEI's Resilient Cities Network and C40's knowledge hubs demonstrate the potential of horizontal learning for local climate governance. Utilizing digital resources like geographic information systems (GIS), climate data platforms, and artificial intelligence (AI) can improve the effectiveness and accuracy of cross-border climate planning. With these technologies, it is possible to monitor the environment in real time, create predictive risk models, and visualize common dangers like increasing sea levels or air pollution. Digital platforms can facilitate participatory planning by enabling cross-border communities to share their viewpoints and work together on adaptation strategies (Krause, 2011). To allow local actors to fully utilize these tools, governments need to invest in digital infrastructure and capacity development.

REFERENCES

C40 Cities. (2022). *About C40*. Retrieved from https://www.c40.org

Acuto, M. (2013). *Global Cities, Governance and Diplomacy: The Urban Link*. Routledge. DOI: 10.4324/9780203073810

Acuto, M., Morissette, M., & Tsouros, A. (2017). City diplomacy: Towards more strategic networking? *Learning from the WHO Healthy Cities Network. Global Policy*, *8*(1), 14–22. DOI: 10.1111/1758-5899.12382

Acuto, M., & Rayner, S. (2016). City networks: Breaking gridlocks or forging (new) lock-ins? *International Affairs*, *92*(5), 1147–1166. DOI: 10.1111/1468-2346.12700

Anguelovski, I., & Carmin, J. (2011). Something borrowed, everything new: Innovation and institutionalization in urban climate governance. *Current Opinion in Environmental Sustainability*, *3*(3), 169–175. DOI: 10.1016/j.cosust.2010.12.017

Ansell, C., & Gash, A. (2008). Collaborative governance in theory and practice. *Journal of Public Administration: Research and Theory*, *18*(4), 543–571. DOI: 10.1093/jopart/mum032

Aylett, A. (2015). Institutionalizing the urban governance of climate change adaptation: Results of an international survey. *Urban Climate*, *14*, 4–16. DOI: 10.1016/j.uclim.2015.06.005

Bache, I., & Flinders, M. (2004). *Multi-level Governance*. Oxford University Press. DOI: 10.1093/0199259259.001.0001

Barber, B. R. (2013). *If Mayors Ruled the World: Dysfunctional Nations, Rising Cities*. Yale University Press.

Betsill, M. M., & Bulkeley, H. (2006). Cities and the multilevel governance of global climate change. *Global Governance*, *12*(2), 141–160. DOI: 10.1163/19426720-01202004

Biermann, F., Pattberg, P., Van Asselt, H., & Zelli, F. (2009). The fragmentation of global governance architectures: A framework for analysis. *Global Environmental Politics*, *9*(4), 14–40. DOI: 10.1162/glep.2009.9.4.14

Bulkeley, H., & Betsill, M. M. (2013). *Cities and climate change: Urban sustainability and global environmental governance*. Routledge. DOI: 10.4324/9780203077207

Convention, A. (2021). *Climate Action in the Alpine Region*. Retrieved from https://www.alpconv.org

EU Committee of the Regions. (2020). *The role of local and regional authorities in the implementation of the European Green Deal.* https://cor.europa.eu/

European Commission. (2020). *INTERREG: European Territorial Cooperation.* https://ec.europa.eu

Gallagher, G. E., Duncombe, R. K., & Steeves, T. M. (2020). Establishing climate change resilience in the Great Lakes in response to flooding. *Journal of Science Policy & Governance*, *17*(01), 2–7. DOI: 10.38126/JSPG170105

Haites, E., ... 2018: Experience with Carbon Taxes and Greenhouse Gas Emissions Trading Systems. SSRN *Electron. J.*, XXIX, 109, DOI: 10.2139/ssrn.3119241

Happaerts, S., Van den Brande, K., & Bruyninckx, H. (2011). Subnational governments in transnational networks for sustainable development. *International Environmental Agreement: Politics, Law and Economics*, *11*(4), 321–339. DOI: 10.1007/s10784-010-9128-4

ICLEI. (2021). *Global Review 2021*. ICLEI – Local Governments for Sustainability. Retrieved from https://www.iclei.org

IPCC. (2023). *Sixth Assessment Synthesis Report: Climate Change 2023*. Intergovernmental Panel on Climate Change. https://www.ipcc.ch/report/ar6/syr/

Jordan, A., & Huitema, D. (2014). Policy innovation in a changing climate: Sources, patterns and effects. *Global Environmental Change*, *29*, 387–394. DOI: 10.1016/j.gloenvcha.2014.09.005

Keating, M. (2013). Regions and international affairs: Motives, opportunities and strategies. In *Paradiplomacy in action* (pp. 1–16). Routledge.

Keohane, R. O., & Victor, D. G. (2011). The regime complex for climate change. *Perspectives on Politics*, *9*(1), 7–23. DOI: 10.1017/S1537592710004068

Kern, K., & Alber, G. (2009). Governing climate change in cities: Modes of urban climate governance in multi-level systems. *OECD Conference on Cities and Climate Change*. https://www.oecd.org

Kern, K., & Bulkeley, H. (2009). Cities, Europeanization and multi-level governance: Governing climate change through transnational municipal networks. *Journal of Common Market Studies*, *47*(2), 309–332. DOI: 10.1111/j.1468-5965.2009.00806.x

Krause, R. M. (2011). Policy innovation, intergovernmental relations, and the adoption of climate protection initiatives by US cities. *Journal of Urban Affairs*, *33*(1), 45–60. DOI: 10.1111/j.1467-9906.2010.00510.x

Marks, G., & Hooghe, L. (2004). Contrasting visions of multi-level governance. *Multi-level governance*, 15-30.

Mees, H. L., Driessen, P. P., Runhaar, H. A., & Stamatelos, J. (2013). Who governs climate adaptation? Getting green roofs for stormwater retention off the ground. *Journal of Environmental Planning and Management*, *56*(6), 802–825. DOI: 10.1080/09640568.2012.706600

Moss, T., & Newig, J. (2010). Multilevel water governance and problems of scale: Setting the stage for a broader debate. *Environmental Management*, *46*(1), 1–6. DOI: 10.1007/s00267-010-9531-1 PMID: 20640851

Nagendra, H., & Ostrom, E. (2012). Polycentric governance of multifunctional forested landscapes. *The International Journal of the Commons*, *6*(2), 104. DOI: 10.18352/ijc.321

OECD. (2019). *Enhancing Climate Change Mitigation through Decentralised Government*. OECD Publishing.

OECD. (2022). *Decentralised governance for sustainable development: Reshaping the climate agenda at the local level*. https://www.oecd.org

Ostrom, E. (2017). Polycentric systems for coping with collective action and global environmental change. In *Global justice* (pp. 423–430). Routledge. DOI: 10.4324/9781315254210-18

Patterson, J., Schulz, K., Vervoort, J., Van Der Hel, S., Widerberg, O., Adler, C., & Barau, A. (2017). *Exploring the governance and politics of transformations towards sustainability. Environmental Innovation and Societal Transitions, 24, 1-16.Piattoni, S. (2010). The theory of multi-level governance: Conceptual, empirical, and normative challenges*. Oxford University Press.

Pierre, J. (2011). *The politics of urban governance*. Bloomsbury Publishing. DOI: 10.1007/978-1-137-28555-3

Roberts, D., & O'Donoghue, S. (2013). Urban environmental challenges and climate change action in Durban, South Africa. *Environment and Urbanization*, *25*(2), 299–319. DOI: 10.1177/0956247813500904

Romero-Lankao, P., Gnatz, D. M., Wilhelmi, O., & Hayden, M. (2016). Urban sustainability and resilience: From theory to practice. *Sustainability (Basel)*, *8*(12), 1224. DOI: 10.3390/su8121224

Schmeier, S. (2013). *Governing International Watercourses: River Basin Organizations and the Sustainable Governance of Internationally Shared Rivers and Lakes*. Routledge.

Selleslaghs, J. (2017, September). EU environmental cooperation with Latin America: A critical assessment. In *Paper presented at the 11th Pan-European conference on International Relations* (Vol. 13, p. 17).

Setzer, J., & Higham, C. (2021). Global trends in climate change litigation: 2021 snapshot. *Grantham Research Institute on Climate Change and the Environment.* https://www.lse.ac.uk/granthaminstitute/publication/global-trends-in-climate-change-litigation-2021/

Sister Cities International. (2023). *Sister City Partnerships for Climate Action.* Retrieved from https://sistercities.org

Sneddon, C., & Fox, C. (2006). Rethinking transboundary waters: A critical hydropolitics of the Mekong basin. *Political Geography*, *25*(2), 181–202. DOI: 10.1016/j.polgeo.2005.11.002

Tanner, T., Mitchell, T., Polack, E., & Guenther, B. (2010). Urban governance for adaptation: Assessing climate change resilience in ten Asian cities. *IDS Working Paper*, *2010*(315), 1–47. DOI: 10.1111/j.2040-0209.2009.00315_2.x

Tavares, R. (2016). *Paradiplomacy: Cities and States as Global Players.* Oxford University Press.

UCLG. (2022). *Climate and Environmental Action by Local Governments.* United Cities and Local Governments. Retrieved from https://www.uclg.org

Under2 Coalition. (2022). *Annual Report.* The Climate Group. Retrieved from https://www.under2coalition.org

UNDP. (2022). *Strengthening local governance for climate resilience.* https://www.undp.org

UNECE. (2020). *Green Belt of Fennoscandia: Cross-border Biodiversity Cooperation.* Retrieved from https://unece.org

World Resources Institute. (2023). *Climate Watch: Tracking Progress Toward Climate Goals.* Retrieved from https://www.climatewatchdata.org

Chapter 15
Why the World Fails to Act?
The Silent Collapse of International Climate Governance

Apoorva Misra
https://orcid.org/0009-0009-0476-8125
Alliance University, India

Keerthiraj
https://orcid.org/0000-0002-2536-8579
GFGC Punjalakatte, Mangalore University, India

ABSTRACT

This chapter is not a critique of failures; it is a reflection on why those failures persist despite repeated attempts. International climate governance has not lacked diplomacy, nor has it suffered from ignorance of science or absence of public concern. What it lacks, we argue, is intelligibility across worlds. The frameworks imposed in the name of global cooperation presume a world populated by rational actors, seeking utility, governed by secular institutions, and pursuing infinite growth. But such assumptions do not travel well. They do not describe most human communities; they describe a particular cultural experience made universal. Drawing on comparative insights and case studies, we show how this epistemological imposition produces alienation, not alignment; mistrust, not participation. Compliance becomes performative. To move beyond this impasse, climate governance must be reimagined—not as a global consensus on Western terms, but as a dialogue between diverse lifeworlds. Without plural epistemologies, transformation remains a metaphor.

DOI: 10.4018/979-8-3373-3972-6.ch015

INTRODUCTION

International climate governance is at a crossroads. In the three decades since the first Earth Summit and the creation of the UN Framework Convention on Climate Change (UNFCCC) in 1992, governments have convened dozens of high-profile meetings and signed multiple landmark agreements – from the Kyoto Protocol (1997) to the Paris Agreement (2015) (Dagnet et al., 2021; Karybekova, 2018). Yet the tangible outcomes remain disheartening. Global greenhouse gas emissions reached a record high of 37.4 billion tons of CO_2 in 2023, despite the pledges made under these accords (Mravcová, 2025). Year after year, Conference of the Parties (COP) summits conclude with grand statements, but atmospheric carbon concentrations keep climbing and climate impacts – heatwaves, floods, wildfires – intensify unabated. The world is off-track to meet the Paris goal of limiting warming to 1.5°C, facing a projected warming well above 2°C unless dramatically stronger action is taken (Thakur, 2021; Held & Roger, 2018).

Why does global action consistently lag behind global rhetoric? The routine explanations are by now familiar refrains in political discourse: corruption and mismanagement, conflicting national interests, short-term politics, lack of enforcement etc. Each of these factors no doubt contributes to the paralysis. However, these conventional explanations remain unsatisfying in one crucial sense: they do not fully account for the persistence and universality of the failure. After all, if the core problem were simply corruption or lack of political will, one might expect that as climate dangers became more visibly dire (with public concern rising globally), leaders would eventually overcome these hurdles. Yet even as scientific warnings turn into observable catastrophes, meaningful action still falls short. Something more fundamental appears to be amiss – a deeper systemic flaw that underlies and perpetuates the ineffectiveness of international climate governance. In this chapter, we contend that the stalemate goes beyond faulty politics or economics. Instead, the root of the problem may lie in the way contemporary institutions conceptualize human beings and human societies. Specifically, modern international governance – heavily influenced by Western intellectual history – operates on a set of hidden assumptions about human nature, society, and motivation. These assumptions shape how climate solutions are framed and pursued. If they are significantly flawed or culturally biased, even the best-intentioned policies will repeatedly miss their mark. In short, we may have been formulating answers to the wrong questions, due to an epistemological lens that obscures more than it reveals.

One striking observation is how climate change is predominantly framed as a technical and economic problem in global forums. At its core, climate change is seen as an "emissions" problem – essentially a matter of managing and allocating the allowable carbon dioxide output within the finite budget dictated by physics.

The discourse revolves around carbon metrics, cost-benefit analyses, market-based mechanisms, and finance (Ekardt, 2012; Tallarita, 2021). Human beings in this framework appear mainly as rational economic actors driven by material self-interest: if emissions are costly (through carbon pricing or regulations), people and states will reduce them; if green technologies are made cheaper, societies will adopt them. The underlying model is the familiar homo economicus of modern economics – a world of individuals (or nations conceived as unitary actors) maximizing their utility and competing for resources. Needs and desires are treated as unlimited, so governance is about incentivizing restraint through clever design of institutions (markets, treaties, enforcement regimes). But is this portrayal of human motivation universally accurate? Are human societies simply aggregations of self-interested individuals responding predictably to rewards and punishments? Or could this view be a culturally specific artifact – a product of Western intellectual and theological history that has been mistakenly universalized? These questions rarely surface in mainstream climate policy debates, yet they are critical (Taylor, 2013; Rickards et al., 2014). If the architects of international climate agreements have mischaracterized the nature of human aspirations and societal values, then the solutions built atop those assumptions will continually misfire. The "silent collapse" of climate governance may in fact be the result of a profound epistemological mismatch: global institutions keep proposing frameworks that make sense within a Western paradigm, but fail to resonate with many cultures' lived realities – and perhaps even fail to capture essential aspects of Western publics' own motivations beyond economic calculus (Cabello & Gilbertson, 2012; Engen, 2025).

This chapter will explore that possibility in depth. We begin by outlining our conceptual framework and methodology, on how Western theological ideas became embedded in secular social sciences. We then present the key findings of our research, examining both empirical data on climate governance outcomes and qualitative evidence of cultural conceptions that diverge from the dominant paradigm. In the discussion, we interpret these findings through the lens of epistemological critique: how the "common sense" of current climate policy might itself be the culprit behind policy failure. Finally, we conclude with reflections on the way forward – arguing that meaningful international cooperation on climate change requires a shift toward pluralistic epistemologies. In practice, this means embracing diverse cultural understandings of nature and well-being, rather than imposing a one-size-fits-all blueprint drawn from a single (historically specific) worldview.

METHODOLOGY

What does it mean to investigate a failure that hides in plain sight? Our approach to this question was not methodological in the conventional sense of assembling tools from various disciplines, but conceptual: it began by acknowledging that the failure of climate governance is not a mere shortfall of action, but a misrecognition of the problem itself. Thus, to examine such a misrecognition, we required a methodology capable of detecting epistemological misalignments—where institutions speak in one register while societies inhabit another. To this end, we combined empirical analysis of international climate agreements and emissions data with a hermeneutic investigation into the conceptual vocabularies animating these frameworks. Policy records and data repositories (e.g., Global Carbon Project, UNEP) were not treated as self-evident facts, but as artifacts constructed within a particular worldview. They told us what had happened—but not why it failed to matter.

RESULTS AND FINDINGS

Repeated Failures of International Climate Agreements

A sobering pattern emerges when examining the history of international climate agreements: ambitious targets on paper, but insufficient action in reality. Since the early 1990s, countries have established numerous treaties and held annual negotiations with the stated aim of curbing greenhouse gas emissions. Yet by nearly every hard metric, these efforts have not altered the upward trajectory of emissions and global warming. Key findings include:

Global Emissions Continue to Rise: Carbon dioxide emissions have climbed steadily throughout the era of climate diplomacy. After the 1997 Kyoto Protocol – which obligated developed countries to cut emissions – global emissions actually increased by roughly 32–38% from 1990 to 2010, largely due to surging output from rapidly developing nations. Even in the wake of the 2015 Paris Agreement, which saw virtually every country pledge some form of climate action, emissions have not peaked (Milinski & Marotzke, 2022). In 2023, energy-related CO_2 emissions reached a record high of 37.4 Gt, up 1.1% from the previous year (IEA, 2024). The world remains on a path of slowly increasing or plateauing emissions at historically high levels – far from the sharp decline needed (a 45-50% cut by 2030) to meet the 1.5°C goal (UNEP, 2024). The slight dip in emissions during the COVID-19 pandemic (2020) proved temporary; by 2022-2023 emissions rebounded to new highs (Stanford Sustainability, 2024).

Figure 1. Annual CO_2 emissions by region (1750-2023) . Source: Global Carbon Budget 2024

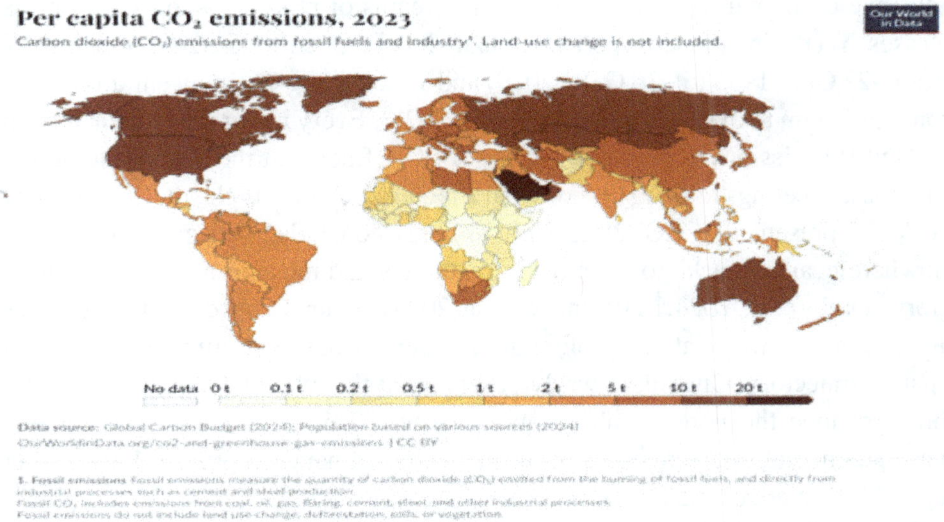

Despite international accords like Kyoto and Paris, global emissions have risen sharply in recent decades. Figure 1 shows that the growth is driven primarily by Asia and other developing regions industrializing, and also some developed regions (Ritchie et al., 2023; Our World in Data, 2024).

Limited Impact of Specific Treaties: The Kyoto Protocol's first commitment period (2008–2012) provides a revealing case. Participating industrialized countries did, in aggregate, reduce their emissions by around 20% below 1990 levels – exceeding Kyoto's target of 5% reduction *(Mulhern, 2020)*. However, this achievement was overshadowed by the growth in emissions from outside the treaty's obligations. The United States (then the world's largest emitter) never ratified Kyoto, and major developing economies like China and India had no binding targets under Kyoto. Consequently, global emissions rose by ~38% during the same period Kyoto was in force *(Mulhern, 2020)*. In essence, cuts by a subset of countries were negated by increases elsewhere. Similarly, the Paris Agreement, while hailed as a diplomatic breakthrough for its inclusivity, relies on voluntary national pledges (NDCs) that are collectively inadequate. Current country pledges put us on course for about 2.4–2.7°C of warming by century's end, far above the Paris goals *(Milinski & Marotzke, 2022)*. There is also widespread non-compliance with NDCs: many nations are not on track to fulfill the emissions cuts they promised for 2025 or 2030. The "ratcheting up" mechanism of Paris – where countries increase ambition every few years – has so far failed to close the emissions gap (UNEP, 2024).

Emissions vs. Conferences – an Inverse Correlation: The sheer number of meetings and declarations underscores the paradox. Nearly 30 COP summits have been held (one each year since 1995), producing reams of reports, accords, and press releases. Yet during the same period, annual global CO_2 emissions rose by over 60% (from ~22 Gt in 1995 to ~36 Gt by 2019) and atmospheric CO_2 concentration leapt from ~360 ppm to over 415 ppm (Nyirenda, 2023). Every major assessment – from the UNEP Emissions Gap Reports to the IPCC – finds that the actions stemming from these meetings fall far short of what's required. For example, the 2023 Global Stocktake presented at COP28 in Dubai bluntly concluded that governments are "nowhere near on track" to meet the Paris targets and must dramatically ramp up efforts (Darby et al., 2024; Maizland & Fong, 2025). Figure 1 above vividly illustrates this disconnect: negotiations proliferated, but emissions kept climbing, as if on a separate trajectory. One might wryly observe that the more we talk about solving climate change, the more greenhouse gases we emit. This inverse relationship points to a fundamental ineffectiveness of the current governance approach (Paiement et al., 2022; Depledge, 2024).

Adverse Trends in Climate Impacts: Not only are emissions rising, but real-world indicators of climate disruption are worsening, underscoring the human cost of inaction. The last decade was the hottest on record globally; unprecedented heatwaves and wildfires have struck every inhabited continent. Climate-related disasters have nearly doubled in frequency compared to 20 years ago, by some estimates, and their economic damage is soaring (Akhtar et al., 2024). Low-lying island nations and coastal areas are increasingly inundated by sea-level rise and storm surges, while prolonged droughts threaten food security in parts of Africa and Asia. While not all these impacts can be laid at the feet of governance failure, the lack of sufficient mitigation certainly contributes to the severity of these outcomes. The disconnect between what science says is necessary and what policy has delivered led the UN Secretary-General in 2023 to lament, "we are in the midst of a climate emergency… the window to act is closing fast" *(Arasu, 2023)*. Yet despite such warnings from global leaders, the pace of emissions reduction remains glacial (Obergassel et al., 2022).

Limitations of Conventional Explanations

Before advancing our thesis about deeper assumptions, it is essential to scrutinize why the widely cited causes of climate governance failure, mentioned in the Introduction, do not fully account for the persistent stalemate. Each conventional explanation contains some truth but also serious limitations:

Corruption and Bureaucratic Inefficiency: Yes, instances of corruption or misuse of climate funds exist, and large bureaucracies (national or UN bodies) can be slow and prone to red tape. However, corruption alone doesn't explain a global phenomenon. Many countries with relatively low corruption (e.g. in Northern Europe) are also struggling to meet climate goals. Moreover, even honest and efficient administrations often fall short on climate targets. This suggests the problem is not simply malfeasance or waste – even a perfectly well-run international system might fail if it is conceptually misdirected. Focusing on corruption also doesn't explain why public supportfor strong climate action often remains tepid; voters are not holding leaders accountable for failure in the way one might expect if it were just a matter of poor governance. In short, cleaning up institutions, while necessary, is insufficient – we might end up with efficient bureaucracies doing the wrong things (Girardi & Preite, 2017).

Competing National Interests and Free-Riders: There is no doubt that unequal burdens and the temptation to free-ride plague climate agreements. Climate change is the textbook global commons problem – every ton of CO_2 contributes to the harm, but no single actor can solve it alone, leading to collective action dilemmas *(Milinski & Marotzke, 2022)*. However, this explanation, rooted in game theory, treats states as unitary rational actors and assumes each calculates costs and benefits similarly. It fails to consider that how nations define their interests is not static or uniform – interests are shaped by internal values, ideologies, and public pressures. The "national interest" in climate matters is often narrowly interpreted as short-term GDP growth or energy security, but why are those the overriding metrics for all nations? Some countries (like Bhutan, which pursues Gross National Happiness and remains carbon-negative) clearly weigh interests differently. The free-rider framing also presumes a lack of trust and global solidarity – yet one might ask, why has genuine solidarity been so hard to achieve despite common existential threat? Reducing it to "that's just how self-interest works" is unsatisfying; humans have overcome collective dilemmas in other cases (e.g. ozone layer protection via the Montreal Protocol). Thus, while free-riding is observed, it might itself be a symptom of a deeper issue: perhaps the proposed cooperative schemes do not align with many societies' internal priorities, making lukewarm participation rational (Schmidt et al., 2022; Magistro et al., 2024; Bernstein et al., 2020)..

Short-term Politics and Voter Apathy: Democratic cycles do incentivize short-termism; politicians facing election may avoid policies with upfront costs (carbon taxes, for example) even if they yield long-term gains. Additionally, the abstract, global nature of climate change can lead to public complacency – the "climate crisis"lacks the immediacy of economic recessions or pandemics, until disaster strikes. Still, this cannot be the whole story. Public opinion surveys show that majorities in many countries do worry about climate change and support action in principle. Why doesn't this translate into sustained political demand? Part of the answer could be the way climate action is framed to the public – often as a sacrifice or cost today to avert problems later. If citizens in different cultures have varying expectations of government or different levels of trust in technocratic solutions, the standard climate pitch may not inspire them. Moreover, even authoritarian regimes (with long-term horizons) have not necessarily excelled at climate action unless it aligns with other goals. So it's not simply democracy's fault; it may be democracy has not been mobilized effectively on this issue because the narrative and approach fail to connect with people's values deeply (Girardi & Preite, 2017; Magistro et al., 2024).

Lack of Enforcement in International Law: International agreements indeed lack the hard enforcement tools of domestic law – there is no global government or police to punish climate treaty violators. Consequently, some argue, countries default because they can. However, international law in other arenas (trade, human rights, etc.) sometimes holds even without a world government, through norms and reputational pressure. The Paris Agreement's architects knew it was non-binding yet hoped transparency and moral suasion would work. The deeper question is, why hasn't the normative pressure been stronger? If devastating climate change were truly seen as an unacceptable horror by all nations, one would expect diplomatic and public pressure to mount on laggards. Yet climate "naming and shaming" is relatively weak. Often, countries defend their poor performance by pointing to special circumstances or by questioning the fairness of expectations. This again hints that beneath the surface, there isn't a universal consensus on responsibility and obligation – instead, there are clashing viewpoints on how to balance development versus environment, present versus future, rich versus poor, etc. Enforcement is only as strong as the shared values underpinning it. The lack of enforcement may reflect a lack of genuine conceptual agreement on the problem (Saxena & Singh, 2024; Ostrom, 2010).

Assumptions of the Dominant Climate Governance Paradigm

International climate governance, as it stands today, is built upon a set of deeply embedded assumptions about human societies – assumptions so ingrained that they are seldom articulated or questioned. These include:

1. Universal Rationality and Incentives: Humans (and by extension nation-states) are presumed to respond primarily to material incentives and deterrents in a calculative manner. Design the right mix of costs and benefits – for example, carbon pricing mechanisms, subsidies for green tech, or penalties for non-compliance – and behavior will shift accordingly. This assumes a kind of universal economic rationality across cultures (Ekardt, 2012; Rengs et al., 2020). The Paris Agreement itself embodies this logic: it relies on transparency and peer pressure to incentivize compliance, assuming countries will not want to be seen as laggards if the reputational cost is high. Cap-and-trade systems, carbon

credits, and international climate finance likewise reflect faith that monetary incentives and market signals are the key drivers of change.

2. Infinite Needs, Finite Resources: At the core of modern economics (which heavily informs climate policy) is the idea that human wants are limitless, but resources are limited – hence allocation via markets or governance is necessary. Climate change is framed as the tragedy of too many wants (for energy, consumption, comfort) straining the global atmospheric commons. The solution thus becomes managing and allocating the carbon budget. Crucially, this viewpoint accepts the expansion of human desires as given. Nowhere in the mainstream climate dialogue is the possibility raised that societies might curb desires themselves; instead, all effort is on supplying those desires with cleaner technology (Deflorian, 2015). For instance, major energy outlooks predict growing energy demand and strategize how to meet it with renewables – they do not question the trajectory of demand. Underlying this is the assumption that economic growth (and the consumptive lifestyle it entails) is non-negotiable and universally aspired to. Climate solutions, therefore, must be win-win: allowing continued growth while cutting emissions (through efficiency, innovation, etc.).
3. One-Size-Fits-All Metrics of Progress: Concepts like GDP, Human Development Index, or even tonnes of CO_2 have become universal yardsticks for progress and responsibility. Countries are compared on these scales (e.g. per capita emissions, development status) as if these metrics capture their "place" in the climate change problem. While useful, these metrics carry implicit value judgements – for example, that increasing GDP is inherently good (development) even if it correlates with higher emissions, or that all low-GDP countries want to industrialize to a Western standard of living. The climate negotiations often frame issues in terms of carbon budgets and dollar amounts (cost of mitigation, climate finance needed), speaking a language of globalized technocracy. This framing assumes every society views abstract, quantitative indicators as valid guides for policy. It privileges a managerial view of the world, often neglecting qualitative values (like cultural integrity, spiritual relationship with land, etc.) that some communities prioritize over numeric growth (Hite, 2018).
4. Secular-Technical Approach to Nature: Modern governance typically approaches the natural world in a secular, utilitarian way – nature is a resource to be managed, a set of ecosystem services to be maintained for human benefit. Climate change is thus a technical problem of engineering (energy systems) and economics (resource allocation). Moral or spiritual dimensions are largely stripped away in official discourse. For example, UN climate documents rarely invoke ethical or spiritual responsibility to Mother Earth; they stick to terms like "sustainable development," "resilience," and "ecosystems." This betrays an assumption that only a secular, scientifically informed perspective is valid for policymaking.

Traditional or religious worldviews that sacralize nature or define human's role in non-utilitarian terms are sidelined as "local color," not part of serious global policy. In effect, the climate governance paradigm has internalized the Western historical trajectory where science and secular law superseded religion – forgetting that this trajectory is not universal. Many cultures never saw "religion" and "science" as separate domains in the first place, so the imposition of a purely technical rationality might not resonate at the community level (Spicer, 2011).

5. The Nation-State as the Primary Unit of Action: Climate agreements are negotiated by nation-states and assume that national governments can universally represent their societies' will and act cohesively. This Westphalian view overlooks the internal diversity in how communities within a country relate to climate issues. It also assumes that solutions must be top-down via state policies. However, in many places local governance or customary institutions hold more sway on land and resource use. The climate regime's state-centric approach may miss opportunities for grassroots action because it isn't designed to incorporate them except indirectly (e.g. via national plans). Furthermore, the concept of a "nation" with unified interests is itself a Western export – many post-colonial states are conglomerates of diverse ethnic and cultural groups with different relationships to their environment. The global process doesn't account for these nuances well, treating nations rather monolithically (Lane & Montgomery, 2014).

Cultural Conceptions of Human Needs and Nature: Diverse Realities

Contrary to what is often assumed in global governance discourses, empirical findings from anthropology, historical records, and the experiences of various communities across the globe indicate something else altogether. Societies, we now know, differ profoundly in how they conceptualize human needs, desires, and their relation to nature. These are not superficial variations, nor mere cultural idiosyncrasies. Rather, they are manifestations of distinct epistemologies—divergent ways of knowing and being-in-the-world. This recognition is not a moral plea for diversity but a conceptual prerequisite to understanding why a universal model of governance fails in a pluriversal world. Let us consider, then, a few salient examples.

Take the notion of desire and its moral appraisal. In many cultural traditions, the idea that desires ought to be cultivated endlessly is not merely absent—it is explicitly rejected. Buddhism, for instance, identifies desire (tanhā) as the source of suffering and holds that the cessation of craving is the path to liberation. Traditional Christian monasticism extolled asceticism and a virtue of "enoughness," even if Protestant theology later revalorized work and wealth. In both Hindu and Jain traditions, the figure who is santu a—content with what one has—is held in esteem, and excess accumulation is often viewed as a mark of spiritual degeneration. Gandhi's oft-cited dictum—"The world has enough for everyone's need, but not for everyone's greed" (Gandhi, 1947)—was not a slogan, but a philosophical stance against the metaphysics of industrial modernity. What emerges here is a contrast not of ethics, but of ontology. Modern economic theory, by contrast, normalizes greed through the language of utility maximization. Contentment is no longer a virtue; it is seen as stagnation. This divergence has real consequences. In societies shaped by such contentment-values, the logic of financial incentive may not move people. Policies that assume individuals are utility-maximizers might not just misfire; they may appear morally repugnant (Milanez et al., 2022).

A similar epistemic gap appears in how identity is constituted. In liberal theory, society is an aggregate of individuals or nuclear families, each with preferences and rights. Yet many cultures do not start with the individual. They begin with the community. The African notion of *ubuntu*—that a person becomes a person only through others—is not a sentimental idea but a framework for making decisions (Mbiti, 1969). Indigenous governance often relies on elders and consensus, not on individuated property rights. Consequently, climate policies designed around individual incentives (such as paying landowners for ecosystem services) can induce conflict in contexts where land is held communally or is considered sacred. There are numerous cases where cash-for-conservation schemes have disrupted cohesive communities by atomizing them—introducing fractures where none existed before. And yet, when the same goals are pursued through community-based institutions—such as *gram sabhas* or forest councils in India and Nepal—the outcomes differ (Byskov & Hyams, 2022; Misra & Keerthiraj, 2025). What changes is not merely the structure but the motivational framework: honor, belonging, and relational responsibility take precedence over monetary reward. Here, again, the divergence is epistemic: a moral universe oriented around *we* cannot be navigated using the compass of *I*.

This epistemic chasm extends to the relationship between humans and the natural world. Modern industrial society, inheriting from Judeo-Christian and Enlightenment worldviews, treats nature as an object—external, inert, to be studied, mastered, used, or protected. The Biblical notion of dominion morphs into scientific mastery. Yet many indigenous traditions do not draw such a line between human and non-human. Nature is not out there; it is kin, ancestor, spirit. In such worldviews, rivers are not resources but goddesses (as with the Ganga), mountains are not geological formations but deities (as with Arunachala), and forests are not carbon sinks but sacred groves (Zandvliet et al., 2023). The Maori concept of *kaitiakitanga*—which led to the Whanganui River being granted legal personhood—is not a legal innovation but a civilizational intuition (Herriman, 2017). Such conceptions do not emerge from environmentalism; they precede it. And so, when global negotiators speak of "Loss and Damage" or "carbon markets," they are often met with incomprehension, not because the indigenous actors are ignorant, but because they are not speaking the same conceptual language (Blue, 2016). One side talks in carbon, the other in kinship. One in service-provision, the other in sacredness. This is not a communication gap—it is an epistemological abyss.

Equally instructive is how different cultures measure well-being. GDP and life expectancy dominate global metrics today, but not all traditions have shared this calculus. The Sanskrit notion of *pūrna*—wholeness, fullness—does not imply accumulation but harmony. Bhutan's Gross National Happiness index, incorporating environmental, cultural, and spiritual health, offers a rare non-Western alternative. In such contexts, policies that promise economic gain might fail to persuade (Ura et al., 2012). Indeed, several indigenous groups have resisted climate-friendly development projects—like biofuel plantations or hydropower dams—because they tear into their social fabric (New Zealand Parliament, 2017; Hite, 2018). What appears rational to the economist appears profane to the community. A solar farm, offered as a gift, may be refused—not out of irrationality, but out of fidelity to a way of life. History, too, is not neutral. It sediments into memory. In much of the world, colonialism was not a distant past—it is a present inheritance. Development projects often echo the same grammar: this is for your own good. When climate governance arrives with that voice, it triggers not denial of science but distrust of power. African and Asian leaders often hear climate advice as yet another civilizing mission—only now clothed in green. And when promised support—such as the $100 billion climate fund—fails to materialize, it confirms their suspicions (Bertilsson & Soneryd, 2023). What Western countries trust—global markets, international law—are institutions they designed. What others trust—local knowledge, communal norms—are often dismissed as parochial. Once more, the issue is not technical. It is epistemological: trust is structured differently across histories.

All this points to one basic conclusion: the assumption that human motivations and social structures are globally uniform is false. It is this falsehood that haunts international climate governance. A framework that might have emerged under specific historical conditions in Europe gets projected as universally valid. But policies that work in Norway may fail in Nepal—not because of flawed design, but because of misplaced assumptions. Take, for example, the promotion of solar energy in rural areas. In some contexts, microgrids and solar cookers bring real benefits. In others, they fail dramatically. A much-cited case in Central America involved solar cookers that required women to stand and cook in a manner wholly alien to their practices (Parsons et al., 2016). The designers assumed energy-efficiency was the women's priority. It wasn't. What mattered more was the ability to sit, cook, watch the children, and converse. The project failed—not because the technology was faulty, but because the social body it encountered was misunderstood.

What emerges, then, is not a mystery but a pattern. When monocultural governance frameworks confront a pluricultural world, the result is friction, resistance, and failure. But this failure is not always heard. It is silent. Official reports speak of targets missed, funds delayed, or enforcement lapses. What they rarely speak of is the deeper misfit—the inability of a global governance apparatus to reckon with civilizational diversity. The next section turns to precisely this issue: how an epistemological mismatch generates institutional inertia, and why that inertia is not a technical glitch but a structural feature of the current regime.

Epistemological Blindness and Institutional Paralysis

We use the term "epistemological blindness" to describe the inability of international institutions to see outside their own knowledge framework. This blindness has concrete manifestations in climate governance, leading to what we call the silent collapse – institutions going through the motions of action (meetings, reports, pledges) while fundamentally unable to effect change on the ground. Several dynamics illustrate how the dominant paradigm's narrow vision causes paralysis:

Misreading of Problems, Mis-design of Solutions: When institutions view all societies through the lens of Western rationalism, they may misdiagnose the cause of climate-harming behaviors. For example, tropical deforestation is a major source of emissions. The dominant view assumes that locals deforest for maximum short-term profit (e.g., to sell timber or raise cattle for income). Hence solutions like REDD+ (Reducing Emissions from Deforestation and Forest Degradation) were created to pay communities to not cut trees – essentially, replace the foregone profit with carbon payments. In reality, deforestation drivers vary: some communities clear land for subsistence farming, not profit, and deeply value the forest for food, medicine, and culture. Offering them money not to use their land could even offend their sense of stewardship or provoke conflict over who gets the funds (Purdon, 2020; Reymond et al., 2020). In some cases, REDD+ projects failed or faced backlash from indigenous groups who resented external NGOs suddenly dictating local land use and treating their forest as a carbon offset for foreigners. The epistemological blindness was the assumption that a market solution (paying for carbon) would be universally welcomed as logical. Where it failed, it wasn't just due to poor execution; it was conceptually flawed by not seeing the forest through the locals' eyes, so to speak (Marino, 2024).

Absurd Incentive Outcomes: The reliance on incentives can backfire in perverse ways, especially when it collides with different value systems. A striking example emerged under the Kyoto Protocol's Clean Development Mechanism (CDM). Companies in some developing countries realized they could get paid (through carbon credits) to destroy certain industrial gases (HFCs) that are potent greenhouse contributors. The incentive was so lucrative that it apparently led some manufacturers to produce more of those harmful gases just so they could then be paid to destroy them *(Mulhern, 2020)*. This was a loophole exploitation, yes, but at a deeper level it shows how an incentive scheme, abstracted from local ethical context, can produce outcomes opposite to intention. In a governance system attuned to cultural context, such a bizarre cycle might have been avoided by incorporating normative checks (e.g., a stigma or regulation against gaming the system for profit). But a culturally neutral, technocratic mechanism had no immune system against such behavior – it assumed profit-seeking is fine as long as emissions go down somewhere. When that assumption met reality, it facilitated moral hazard. To be clear, this was not limited to non-Western actors; Western companies and traders have also gamed carbon markets. The point is that a framework that appeals only to extrinsic motivation (money) and not to any intrinsic sense of duty or morality can engender cynicism and opportunism across cultures. If people do not internalize the purpose of a climate measure – if it doesn't resonate with any of their intrinsic values – they may treat it as just another scheme, to be taken advantage of or ignored (Marino, 2024).

Communication Gaps and Trust Deficits: Epistemological mismatches lead to talking past each other in negotiations. One side might emphasize historical responsibility and climate justice (often voices from the Global South), essentially speaking in moral terms about righting past wrongs and equitable sharing of the atmospheric commons. The other side (often developed countries) responds in terms of forward-looking efficiency and participation of all major emitters, speaking the language of realism and pragmatism. Each side finds the other's arguments somewhat baffling or irritating: "Why are they harping on colonialism and reparations? We need everyone to cut now, or we all lose" versus "Why are they ignoring the fact that their centuries of emissions created this crisis? We have a right to develop." Both have valid points, yet negotiations bog down. This is not just stubbornness; it reflects different moral epistemologies. Western delegates often come from secular, utilitarian policy training – they focus on crafting the most effective global deal now. Non-Western delegates may be drawing on collective historical memory and ethical worldviews where justice and honor are foundational – no solution is acceptable if it doesn't acknowledge and rectify unfairness. When the climate regime fails to bridge these worldviews, the outcome is half-baked compromises or stalemates. The "loss and damage" fund debate is illustrative: it took decades for the idea of rich countries paying for climate damages in poorer ones to be accepted, because it was resisted by a worldview allergic to anything resembling guilt or liability. Only relentless pressure rooted in an ethical argument ("those suffering least caused it least") pushed it through. Even now, its implementation is uncertain. The trust deficit remains because the communities that needed an ethical acknowledgement got very late and begrudging satisfaction; meanwhile, those who prefer a no-fault, all-in-this-together narrative felt railroaded. Trust cannot deepen if each side thinks the other just doesn't "get it" (Reymond et al., 2020; Purdon, 2020).

Internalization vs. External Enforcement: In any effective governance, a norm ideally becomes internalized by society – consider how littering or smoking indoors, once common, became socially unacceptable in many places. If climate-friendly behavior can be internalized (people conserving energy or treasuring forests out of habit or principle), enforcement isn't the main driver. But internalization requires alignment with cultural values. The current global approach has mostly failed to achieve internalization outside of relatively small environmentalist subcultures. Instead, it relies on external measures – laws, fines, international monitoring. This is costly and brittle; people find ways around rules they don't truly believe in. Why hasn't low-carbon living become a moral norm globally akin to, say, not committing theft or caring for one's children? Perhaps because the climate message has been framed in alienating ways: telling people to value abstract future benefits over present needs, or to trust global institutions over their local wisdom, or to accept sacrifices without seeing reciprocal fairness. A telling case is personal consumption: Western campaigns often emphasize individual carbon footprints ("eat less meat, fly less"). But in some cultures, communal feasting or pilgrimage travel are deeply ingrained traditions – simply labeling them "bad for the climate" will not persuade. A more effective internalization might occur if climate action could be tied to existing virtues – e.g., linking it to religious stewardship teachings (as some faith leaders are attempting), or to traditional concepts of respect for ancestors/descendants. Lacking that, the burden falls on surveillance and enforcement, which is patchy globally. The epistemological blind spot is not recognizing that without cultural embedding, climate policies remain an external imposition – obeyed when convenient, discarded when not.

Colonial Consciousness within Nations: Earlier we noted how colonial-era descriptions of non-Western societies (like the notion of a "caste system" in India, or viewing Indian epics as degraded history versus purana) created distorted understandings (Marino, 2024). This legacy means that elites in post-colonial countries often share the Western epistemology more than their own populace's worldview. Thus, climate policies made by these elites might mimic Western templates and fail to engage local populations. For example, an English-educated bureaucrat in South Asia might design an urban climate resilience plan focusing on high-tech solutions and strict regulations, ignoring informal community networks or traditional water harvesting practices that locals trust more. When the plan falters due to lack of public buy-in, officials blame "ignorance" or "superstition" among people, rather than reflecting on their own imported framework. This is a form of cognitive enslavement – preferring foreign categories even when they misdescribe local reality (Marino, 2024). The result is often policies that exist on paper (to satisfy international donors or rankings) but do not take root. The silent collapse here is at the implementation stage: things look aligned with global goals in reports, but the actual situation on ground diverges. For instance, many developing countries' climate action plans cite Western concepts of sustainability, yet continue to pursue contradictory actions (like expanding coal power) because there isn't an internal consensus behind those plans – they were written in the language of international expectation, not domestic conviction.

DISCUSSION

Our findings compel a reconsideration of foundational questions: What is the climate crisis really asking of us as a global civilization? On the surface, it asks for emissions cuts, new technologies, resilient infrastructure – all true. But at a deeper level, it is asking for coordination among diverse peoples on an unprecedented scale, and for a reevaluation of how humans relate to each other and to the Earth. In this discussion, we integrate our insights with broader theoretical perspectives and address potential counterarguments, all while maintaining fidelity to the conceptual framework that guided this research.

Climate governance is a secular endeavor, yet it inherits a conceptual lineage: ideas of stewardship, sin, and salvation have their analogs in "sustainability," "carbon footprint guilt," and visions of techno-fixes or green transitions that "save" the world. The critique of the universality of religion is analogous to what we see with climate change: the West had a particular historical experience (industrial revolution fueled by a certain worldview) and now implicitly expects all to follow that narrative – first as sinners (polluters), then as converts to green growth doctrines. Non-Western traditions that never separated the sacred and secular, or never valorized dominating nature, find this script alien. Our analysis does not claim that Western culture is wrong and others are right; rather, it echoes the point that applying one culture's categories to another can be a grave cognitive error. In colonial times, this error led to misrepresenting Indian civilization through the lens of "religion" and "caste". Today, it leads to misrepresenting global climate action as solely a technical-economic project divorced from culture. Recognizing the topos – the conceptual ground – of Western terms like "progress" or "nature" is the first step in breaking out of the blindness (Jahnel, 2023; David, 2024).

One might ask, isn't climate change a scientific reality that demands a single, coordinated response? Yes, the science of climate (the physics, chemistry) is universal. But the human response need not – and perhaps cannot – be uniform, because it is mediated by culture. Consider an analogy: a disease like COVID-19 was biological, universal to humans, yet responses differed – some societies emphasized individual choice, others collective discipline, based on cultural values; some used high-tech solutions, others traditional herbs alongside vaccines, etc. Eventually, a mix of approaches helped. With climate, any effective unified response may actually have to be a mosaic of differentiated efforts (Parsons, 2023). The error is thinking that unity equals uniformity. Unity can also be achieved through harmony of different parts – like an orchestra with distinct instruments playing together. Right now, climate governance is more like everyone being asked to play the same instrument, and many either cannot or will not .

Another theoretical lens: postcolonial theory and decolonial thought provide insight into why power dynamics persist in knowledge. The climate regime, though well-meaning, often has a paternalistic tinge – wealthy nations setting rules and expecting poorer ones to follow, albeit with some financial aid. Decolonial scholars would point out that unless we de-center Euro-American expertise and elevate subaltern perspectives, we will repeat patterns of domination (Wilkens & Datchoua-Tirvaudey, 2022). Our findings support that: the failure to act is partly a rebellion (sometimes inchoate) of those who feel the agenda is externally imposed or hypocritical. Incorporating plural epistemologies is in essence a decolonization of climate action – recognizing multiple centers of knowledge and agency (Khan et al., 2021). This aligns with calls for "climate justice" which demand not just outcome fairness but procedural fairness – the right to define the problem and solutions, not just receive instructions.

A potential counterargument to our thesis is: Time is short, and we don't have the luxury to indulge cultural differences – we need urgent action through existing frameworks. Indeed, the climate emergency is pressing. Some might fear that accommodating many approaches could slow down collective action or weaken hard targets. However, our rebuttal is that the urgency itself is why we must fix the approach – the current one has not delivered fast enough. Continuing on the same path hoping for a different result is unlikely to suddenly succeed (as emissions data clearly show). Embracing cultural plurality is not about delaying action; it's about unlocking new action. It may seem messier than a single global mandate, but the latter isn't happening anyway beyond paper promises. History shows that during crises (world wars, natural disasters), people mobilize extraordinarily when they truly believe in the cause. Belief is culturally mediated. If climate action remains a technocratic mandate, it won't evoke the level of sacrifice and innovation required. We need it to become a people's movement in every nation, albeit taking different forms. That could actually accelerate action – imagine thousands of local initiatives reducing emissions and adapting in parallel, rather than waiting for instructions from above (Bratman & DeLince, 2022).

Another question: Could plural epistemologies lead to relativism or conflict, undermining the scientific consensus on climate? It's important to clarify that we are not advocating multiple truths about the physical reality – the consensus (that GHGs warm the climate, that we must reach net-zero, etc.) remains a common guiding star. Plural approaches are about different paths to the same goal, not different goals (Voisard & Wallimann-Helmer, 2023). There is a risk that opportunistic leaders could cloak inaction in "cultural differences" (e.g., "our culture prioritizes development over environment, so we'll keep emitting"). But note, our argument isn't to accept every claim at face value; it's to genuinely engage and see if alternatives exist that meet both cultural priorities and climate needs. Often, there are win-win options if creativity is applied. For instance, a country insisting on development could pursue leapfrog green technology that provides jobs (aligning with their priority) while cutting emissions – but to sell that domestically, one might frame it as restoring past civilizational glory through innovation, if that narrative resonates. Also, part of plural epistemology is cross-cultural dialogue – calling bluff where culture is misused as excuse, but in a respectful way that maybe offers a solution that saves face. It's delicate but doable. Ultimately, if a culture truly values something that causes harm (say, a tradition that involves environmental damage), a plural approach wouldn't just give a free pass – it would invite discussion on how to preserve the essence of that practice while mitigating the damage. This is a more nuanced negotiation than simply condemning it from an outside perspective.

We also tie our insights to classical thinkers in social theory: aside from Weber (1905) whom we cited, consider Antonio Gramsci and the idea of cultural hegemony (Gramsci, 1929). The current climate governance model is a form of hegemonic common sense – alternatives are hard to imagine because one worldview dominates. Yet, Gramsci also emphasized that hegemony is never total; counter-hegemonic ideas can emerge from the periphery. In climate activism, we see that with youth movements and indigenous protests injecting moral urgency and challenging business-as-usual (Keerthiraj, Misra, & Vang-Phu, 2025). They are basically saying the same: the old paradigm is failing, we need a new vision (Amo-Agyemang, 2021). Our work academically buttresses that sentiment with analysis. Hannah Arendt could be invoked on the power of collective action and "new beginnings" – plurality was central to her political thought (Arendt, 1958). She warned against treating humans as just labor or animal needs (the economic lens) and losing the higher capacity for action and natality (innovation, starting new things). In a way, the dominant climate paradigm reduced humans to homines economici; a more Arendtian approach would see humans as capable of principled action that transcends mere survival calculus. That is precisely what a value-driven climate movement would leverage.

Finally, we reflect on India (given the conceptual framework mentions India specifically) as a microcosm. India is a civilization where diverse epistemologies have coexisted – modern secular, Gandhian simplicity, Hindu ideas of dharma, etc. Its climate actions sometimes reflect this jumble: it champions renewables (modern technocratic approach) but also invokes lifestyle change (e.g., PM Modi's call for "Mission LiFE" – Lifestyle for Environment – which resonates with Gandhian ideas). However, often the narrative even domestically remains borrowed (talk of carbon trading, of India becoming a "developed" nation in Western sense). There's a rich opportunity for India to articulate a climate philosophy drawing from its own civilizational ethos – emphasizing balance (the middle path), respect for all life (as in ahimsa), and duty (dharma) to planet (Mbah & Ezegwu, 2024). If it did so, and aligned its policies accordingly, it could offer a powerful alternative model on the world stage. This is just one example; every culture could contribute a chapter to a new climate ethos compendium.

In closing this discussion, we acknowledge that the transformation we are talking about is profound. Epistemological shifts are not easy; they often happen in times of great stress or change. The climate crisis unfortunately provides that stress, but also an impetus for change. The silver lining is that climate change, being a global challenge, forces us to confront global plurality. It has inadvertently made us realize how interconnected we are and how different we are. As the saying goes, unity in diversity might be our only way out – leveraging the unity of purpose (saving our common home) through the diversity of approaches (honoring our many homes and traditions).

This perspective is optimistic: it assumes humans can learn and adapt not just technologically but culturally. History offers examples of cultural renaissance when old paradigms fail – the late antiquity shift to new religions, or the post-war creation of the United Nations out of wrecked empires. Perhaps climate change will spur a new cultural renaissance where wisdom from all corners is pooled. In that sense, solving climate change might also solve something internal to humanity: the estrangement caused by centuries of one worldview trying to flatten others. A plural epistemology climate movement could become a healing project, reconnecting modern humanity with both its ancestors' wisdom and its collective future.

CONCLUSION

The world is not failing to act because it does not know what to do. It is failing because it cannot recognize what action means across its many ways of life. What appears as high effort with low impact is, in truth, a symptom of a deeper blindness: institutions continue to operate within a single conceptual framework, mistaking it for universal reason. That framework casts the climate crisis as a technical problem to be solved by Homo economicus—through prices, incentives, and innovation. But this figure is not a representation of humanity; it is a cultural invention. When policies premised on this abstraction are deployed across diverse societies, they encounter not compliance but confusion, resistance, or indifference. What is mistaken for apathy is often incommensurability. The world is not refusing action; it is refusing a particular conception of action.

Our analysis suggests that governance collapses not from malice or inertia, but from epistemological misalignment. When treaties fail, and targets remain unmet, it is not simply because enforcement is weak. It is because the governance paradigm lacks intelligibility in worlds where nature is sacred, desire is to be curtailed, and well-being is not reducible to consumption. To move forward, we must not globalize a single model but pluralize the epistemologies we allow into the conversation. This does not mean abandoning science, but recognizing that science without cultural translation becomes silent authority, not shared knowledge. What we propose is not relativism, but resonance—policy that resonates with local meanings. Thus, the way ahead is not more ambition within the same frame, but a reimagining of the frame itself. If we dare to think plurally, then perhaps action—real, situated, meaningful action—can finally begin. Not as a global mandate, but as a chorus of worlds, each finding its way to care for the Earth.

REFERENCES

Akhtar, S., Shaima, S., Rita, G., Rashid, A., & Rashed, A. J. (2024). *Navigating the global environmental agenda: A comprehensive analysis of COP conferences, with a spotlight on COP28 and key environmental challenges*. Nature Environment and Pollution Technology., DOI: 10.46488/NEPT.2024.v23i03.057

Amo-Agyemang, C. (2021). Decolonising the discourse on resilience. *International Journal of African Renaissance Studies, 16*(1), 4–30. DOI: 10.1080/18186874.2021.1962725

Arasu, S. (2023, September 8). UN secretary-general urges G20 leaders to send a strong message on climate change. Associated Press. https://apnews.com/article/climate-change-antonio-guterres-g20-united-nations-8ba2f8cfc5d18e59a2eb728e015b8fed

Arendt, H. (1958). *The human condition*. University of Chicago Press.

Bernstein, S., Hoffmann, M., Weinthal, E., Mildenberger, M., & Aklin, M. (2020). Introduction. *Global Environmental Politics, 20*(1), 1–3. DOI: 10.1162/glep_e_00545

Bertilsson, J., & Soneryd, L. (2023). Indigenous peoples and inclusion in the Green Climate Fund. *Environmental Sociology, 9*(3), 233–242. DOI: 10.1080/23251042.2023.2177091

Blue, G. (2016). Framing climate change for public deliberation: What role for interpretive social sciences and humanities? *Journal of Environmental Policy and Planning, 18*(1), 67–84. DOI: 10.1080/1523908X.2015.1053107

Bratman, E. Z., & DeLince, W. P. (2022). Dismantling white supremacy in environmental studies and sciences: An argument for anti-racist and decolonizing pedagogies. *Journal of Environmental Studies and Sciences, 12*(2), 1–11. DOI: 10.1007/s13412-021-00739-5

Byskov, M. F., & Hyams, K. (2022). Epistemic injustice in climate adaptation. *Ethical Theory and Moral Practice, 25*(4), 613–634. DOI: 10.1007/s10677-022-10301-z

Cabello, J., & Gilbertson, T. (2012). A colonial mechanism to enclose lands: A critical review of two REDD+-focused special issues. *Environmental Science & Policy*.

Dagnet, Y., Simeonova, K., Cogswell, N., Holt, M., La Viña, T., & Warszawski, N. (2021). *Toward more effective implementation of the Paris Agreement: Learning from 30 years of experience*. World Resources Institute., DOI: 10.46830/wriwp.20.00069

Darby, S. J., Hammond, G. P., & Wu, J. (2024). Stocktaking global warming – the outcomes of the 2023 Dubai climate summit (COP28). Proceedings of the Institution of Civil Engineers - Energy. DOI: 10.1680/jener.24.00005

David, J. O. (2024). Decolonizing climate change response: African indigenous knowledge and sustainable development. *Frontiers in Sociology*, *9*, 1456871. Advance online publication. DOI: 10.3389/fsoc.2024.1456871 PMID: 39640009

Deflorian, M. (2015). Governing the "Enough" in a Warming World: The Discourse of Sufficiency from a Climate Governmentality Perspective.

Depledge, J. (2024). *The future of negotiations under the climate change COP: Implementation is not enough.* Dialogues on Climate Change., DOI: 10.1177/29768659241293212

Ekardt, F. (2012). A critical review of "efficiency ethics": The case of climate economics. In Climate Justice and Sustainability (pp. 181–210). DOI: 10.1007/978-94-007-1869-2_9

Engen, S. (2025). *On green swans and catastrophic futures: Climate change as risk and uncertainty in central banking*. Valuation Studies., DOI: 10.3384/VS.2001-5992.2025.12.1.171-194

Gandhi, M. K. (1947). *The Mind of Mahatma Gandhi* (Prabhu, R. K., & Rao, U. R., Eds.). Navajivan Trust.

Girardi, G. G. G., & Preite, G. (2017). Escaping the economist's straightjacket: Overcoming the free-rider mentality which prevents climate change from being effectively addressed. In Corporate Social Responsibility and Governance (pp. 561–575). DOI: 10.1007/978-3-319-58214-6_35

Gramsci, A. (1929). *Selections from the prison notebooks* (Hoare, Q., & Smith, G. N., Eds. & Trans.). International Publishers.

Held, D., & Roger, C. B. (2018). Three models of global climate governance: From Kyoto to Paris and beyond. *Global Policy*, *9*(4), 527–537. Advance online publication. DOI: 10.1111/1758-5899.12617

Herriman, N. (2017). Management of biodiversity: Creating conceptual space for Indigenous conservation. J3ea, 19, 42–52. DOI: 10.5038/2162-4593.19.1.1184

Hite, E. (2018). Political ecology of Costa Rica's climate policy: Contextualizing climate governance. *Journal of Environmental Studies and Sciences*, *8*(4), 469–476. DOI: 10.1007/s13412-018-0480-y

IEA. (2024). CO_2 emissions in 2023. International Energy Agency. https://www.iea.org/reports/co2-emissions-in-2023

Jahnel, C. (2023). *God dwells in flesh: Decolonial ecojustice and planetary ethics in the "Anthropocene"*. Scriptura., DOI: 10.7833/122-1-2135

Karybekova, G. (2018). International instruments in the field of climate change. *Science World, 21*(3), 108–111. DOI: 10.31433/1605-220X-2018-21-3(1)-108-111

Keerthiraj, M. A., & Vang-Phu, T. (2025). India's BRICS Engagement: A Strategic Lever in South Asia and the Indian Ocean. In A. Gedikli, S. Erdogan, & H. Çalışkan Terzioğlu (Eds.), Changing the Global Political Economy: BRICS Countries and Alternative Relations Strategies (pp. 189-214). IGI Global Scientific Publishing. DOI: 10.4018/979-8-3693-7393-4.ch006

Khan, M., Ruszczyk, H., Rahman, M. F., & Huq, S. (2021). Epistemological freedom: Activating co-learning and co-production to decolonise knowledge production. *Disaster Prevention and Management, 30*(6), 681–694. DOI: 10.1108/DPM-03-2021-0070

Lane, L., & Montgomery, W. (2014). An institutional critique of new climate scenarios. *Climatic Change, 122*(3), 447–458. DOI: 10.1007/s10584-013-0919-9

Magistro, B., Abramson, C., Ebanks, D., Debnath, R., & Alvarez, R. M. (2024). Identifying American climate change free riders and motivating sustainable behavior. *Scientific Reports, 14*(1), 6575. Advance online publication. DOI: 10.1038/s41598-024-57042-w PMID: 38503779

Maizland, L., & Fong, C. (2025, January 21). Global climate agreements: Successes and failures. Council on Foreign Relations. https://www.cfr.org/backgrounder/paris-global-climate-change-agreements

Marino, E. (2024). Rethinking carbon offsets: An ethnographic study of REDD+ implementation in indigenous contexts.

Mbah, M., & Ezegwu, C. (2024). The decolonisation of climate change and environmental education in Africa. *Sustainability (Basel), 16*(9), 3744. DOI: 10.3390/su16093744

Mbiti, J. S. (1969). *African religions and philosophy*. Heinemann.

Milanez, F., Menton, M., & Souza, J. M. R. D. (2022). Epistemological justice: Decoloniality, climate change, and ecological conditions for future generations. *IDS Bulletin, 53*(4). Advance online publication. DOI: 10.19088/1968-2022.140

Milinski, M., & Marotzke, J. (2022). Economic experiments support Ostrom's polycentric approach to mitigating climate change. *Humanities & Social Sciences Communications*, *9*(1), 442. DOI: 10.1057/s41599-022-01436-6 PMID: 36530541

Misra, A. & Keerthiraj. (2025). Integrating Sustainability in India's Tourism Sector: An Uphill Battle. In S. Poddar, B. Paul, & M. Luperi (Eds.), Sustainable Business Ecosystems and Social Perspectives (pp. 359-386). IGI Global Scientific Publishing. DOI: 10.4018/979-8-3693-8437-4.ch016

Mravcová, A. (2025). *Assessing the effectiveness of international climate agreements in mitigating global warming*. Studia Ecologiae et Bioethicae., DOI: 10.21697/seb.5835

Mulhern, O. (2020, October 8). A brief assessment of the Kyoto Protocol. Earth.Org. https://earth.org/data_visualization/a-brief-assessment-of-the-kyoto-protocol

New Zealand Parliament. (2017). Innovative bill protects Whanganui River with legal personhood. Retrieved from https://www.parliament.nz/en/get-involved/features/innovative-bill-protects-whanganui-river-with-legal-personhood/

Nyirenda, H. (2023). Fiddling at the conference of the parties? A review of GHG emissions and deforestation trends during COPs. *Environment, Development and Sustainability*. Advance online publication. DOI: 10.1007/s10668-023-04277-8

Obergassel, W., Bauer, S., Hermwille, L., Aykut, S. C., Boran, I., Chan, S., Fraude, C., Klein, R. J. T., Mar, K. A., Schroeder, H., & Simeonova, K. (2022). From regime-building to implementation: Harnessing the UN climate conferences to drive climate action. *Wiley Interdisciplinary Reviews: Climate Change*, *13*(6), e797. DOI: 10.1002/wcc.797

Ostrom, E. (2010). Polycentric systems for coping with collective action and global environmental change. *Global Environmental Change*, *20*(4), 550–557. DOI: 10.1016/j.gloenvcha.2010.07.004

Our World in Data. (2024). Global carbon budget; population-based emissions data. Retrieved from https://ourworldindata.org/co2-and-greenhouse-gas-emissions

Paiement, P., Webster, E., & Anderson, R. (2022). After COP26: Appraising the transnational climate regime. *Transnational Legal Theory*, *13*(2), 157–164. DOI: 10.1080/20414005.2023.2170758

Parsons, M. (2023). Governing with care, reciprocity, and relationality. *Dialogues in Human Geography*, *13*(2), 288–292. Advance online publication. DOI: 10.1177/20438206221144819

Parsons, M., Fisher, K., & Nalau, J. (2016). Alternative approaches to co-design: Insights from Indigenous/academic research collaborations. *Current Opinion in Environmental Sustainability, 20*, 99–105. DOI: 10.1016/j.cosust.2016.07.001

Purdon, M. (2020). The politics of evidence in climate governance. *Sustainability Science, 15*, 701–717. DOI: 10.1007/s11625-020-00793-1

Rengs, B., Scholz-Wäckerle, M., & van den Bergh, J. C. (2020). Evolutionary macroeconomic assessment of employment and innovation impacts of climate policy packages. *Journal of Economic Behavior & Organization, 169*, 332–368. DOI: 10.1016/j.jebo.2019.11.025

Reymond, M., Ossewaarde, R., & Peeters, M. (2020). Environmental justice and REDD+: Lessons from indigenous resistance. *Environmental Science & Policy, 112*, 319–326. DOI: 10.1016/j.envsci.2020.07.001

Rickards, L., Ison, R., Fünfgeld, H., & Wiseman, J. (2014). Opening and closing the future: Climate change, adaptation, and scenario planning. *Environment and Planning. C, Government & Policy, 32*(4), 587–602. DOI: 10.1068/c3204ed

Ritchie, H., Rosado, P., & Roser, M. (2023). CO_2 and greenhouse gas emissions. Our World in Data. https://ourworldindata.org/co2-and-greenhouse-gas-emissions

Saxena, K., & Singh, M. (2024). Investors reward countries for participating in climate agreements. *Nature Communications, 15*(1), 8908. Advance online publication. DOI: 10.1038/s41467-024-53012-y PMID: 39414786

Schmidt, R. C., Drupp, M., Nesje, F., & Hoegen, H. (2022). Testing the free-rider hypothesis in climate policy.

Spicer, M. (2011). Some reflections on rationality, argument, and conflict. *Administrative Theory & Praxis, 33*(3), 473–477. DOI: 10.2753/ATP1084-1806330314

Sustainability, S. (2024). Global carbon emissions from fossil fuels reached a record high in 2023. https://sustainability.stanford.edu/news/global-carbon-emissions-fossil-fuels-reached-record-high-2023

Tallarita, R. (2021). Portfolio primacy and climate change. *SSRN*. DOI: 10.2139/ssrn.3912977

Taylor, C. (2013). The discourses of climate change. In Governing Through Globalised Crime (pp. 17–31). DOI: 10.1057/9781137006127_2

Thakur, S. (2021). From Kyoto to Paris and beyond: The emerging politics of climate change. *India Quarterly, 77*(3), 366–383. DOI: 10.1177/09749284211027252

UNEP. (2024). Emissions Gap Report 2024: No more hot air … please! United Nations Environment Programme. https://www.unep.org/resources/emissions-gap-report-2024

Ura, K., Alkire, S., Zangmo, T., & Wangdi, K. (2012). *A short guide to Gross National Happiness Index.* Centre for Bhutan Studies. DOI: 10.35648/20.500.12413/11781/ii025

Voisard, A., & Wallimann-Helmer, I. (2023). Towards a practical climate ethics. *Ethics, Policy & Environment.* Advance online publication. DOI: 10.1080/21550085.2023.2236504 PMID: 39258256

Weber, M. (1905). *The Protestant ethic and the "spirit" of capitalism* (Baehr, P., & Wells, G. C., Trans.). Penguin Books.

Wilkens, J., & Datchoua-Tirvaudey, A. R. C. (2022). Researching climate justice: A decolonial approach to global climate governance. *International Affairs, 98*(4), 1187–1205. DOI: 10.1093/ia/iiab209

Zandvliet, D., Leddy, S., Inver, C., Elderton, V., Townrow, B., & York, L. (2023). Approaches to bio-cultural diversity in British Columbia. *Sustainability (Basel), 15*(8), 6422. Advance online publication. DOI: 10.3390/su15086422

Compilation of References

Abbass, K., Qasim, M. Z., Song, H., Murshed, M., Mahmood, H., & Younis, I. (2022). A review of the global climate change impacts, adaptation, and sustainable mitigation measures. *Environmental Science and Pollution Research International, 29*(28), 42539–42559. DOI: 10.1007/s11356-022-19718-6 PMID: 35378646

Abdallah, I., Alhosin, H., Belarabi, M., Chaouki, S., Mahmoud, N., & Tayah, J. (2024). A Pan-Asian energy transition? The new rationale for decarbonization policies in the world's largest energy exporting countries: A case study of Qatar and other GCC countries. *Energies, 17*(15), 3776. DOI: 10.3390/en17153776

Abdin, Z. (2024). Empowering the hydrogen economy: The transformative potential of blockchain technology. *Renewable & Sustainable Energy Reviews, 200*, 114572. DOI: 10.1016/j.rser.2024.114572

Aberdeen Group. (2022). *The impact of unplanned downtime in the energy sector.* Aberdeen Group.

Aburto-Oropeza, O., Johnson, A. F., Agha, M., Allen, E. B., Allen, M. F., González, J. A., Arenas Moreno, D. M., Beas-Luna, R., Butterfield, S., Caetano, G., Caselle, J. E., Gaytán, G. C., Castorani, M. C. N., Cat, L. A., Cavanaugh, K., Chambers, J. Q., Cooper, R. D., Arafeh-Dalmau, N., Dawson, T., & Taylor, J. E. (2018). Harnessing cross-border resources to confront climate change. *Environmental Science & Policy, 87*, 128–132. DOI: 10.1016/j.envsci.2018.01.001

Abuwaer, N., Ullah, S., & Al-Ghamdi, S. G. (2023). Building Climate Resilience Through Urban Planning: Strategies, Challenges, and Opportunities. In Al-Ghamdi, S. (Ed.), *Sustainable Cities in a Changing Climate* (1st ed., pp. 185–206). Wiley., DOI: 10.1002/9781394201532.ch12

Acuto, M. (2013). *Global cities, governance and diplomacy: The urban link.* Routledge. DOI: 10.4324/9780203073810

Acuto, M., Morissette, M., & Tsouros, A. (2017). City diplomacy: Towards more strategic networking? *Learning from the WHO Healthy Cities Network. Global Policy*, *8*(1), 14–22. DOI: 10.1111/1758-5899.12382

Acuto, M., & Rayner, S. (2016). City networks: Breaking gridlocks or forging (new) lock-ins? *International Affairs*, *92*(5), 1147–1166. DOI: 10.1111/1468-2346.12700

Addaney, M., Jegede, A. O., & Matinda, M. Z. (2019). The protection of climate refugees under the African human rights system: Proposing a value-driven approach? *African Human Rights Yearbook*, *3*, 242–259. DOI: 10.29053/2523-1367/2019/v3a12

Adegboyega, A. W., Ani, E. C., Oladunni, S. O., & Pramanik, A. S. (2024). Project management tools in renewable energy integration: A review of U.S. perspectives. Eng. Sci. Technol. j., 5(7), 2364–2378.

Adelphi & German Federal Foreign Office. (2015). *A new climate for peace: Taking action on climate and fragility risks*. Retrieved from https://www.newclimateforpeace.org/

Adeola, R. (2019). The impact of the African Union Convention on the Protection and Assistance of Internally Displaced Persons in Africa. *African Human Rights Law Journal*, *19*(2), 591–607. DOI: 10.17159/1996-2096/2019/v19n2a2

Adrot, A., Fiedrich, F., Lotter, A., Münzberg, T., Rigaud, E., Wiens, M., Raskob, W., & Schultmann, F. (2018). Challenges in Establishing Cross-Border Resilience. In Fekete, A., & Fiedrich, F. (Eds.), *Urban Disaster Resilience and Security* (pp. 429–457). Springer International Publishing., DOI: 10.1007/978-3-319-68606-6_25

African Court on Human and Peoples' Rights. (2017). *African Commission on Human and Peoples' Rights v. Republic of Kenya (Ogiek case)*, App. No. 006/2012, Judgment of 26 May 2017.

Agostini, C., Silva, C., & Nasirov, S. (2017). Failure of energy mega-projects in Chile: A critical review from sustainability perspectives. *Sustainability (Basel)*, *9*(6), 1073. DOI: 10.3390/su9061073

Agrawala, S., & Andresen, S. (1999). Indispensability and indefensibility? The United States in the climate treaty negotiations. *Global Governance*, *5*(4), 457–482. DOI: 10.1163/19426720-00504004

Ahmed, Z. S. (2013). Regionalism and regional security in South Asia: The role of SAARC. *Regional Studies*, *31*(1), 76–91. DOI: 10.4324/9781315604442

Akhtar, S., Shaima, S., Rita, G., Rashid, A., & Rashed, A. J. (2024). *Navigating the global environmental agenda: A comprehensive analysis of COP conferences, with a spotlight on COP28 and key environmental challenges*. Nature Environment and Pollution Technology., DOI: 10.46488/NEPT.2024.v23i03.057

Akiho, S., & Raita, M. (2017). An Agreement on Enhancing International Arctic Scientific Cooperation: Only for the Eight Arctic States and Their Scientists? In *The Yearbook of Polar Law Online* (8th ed., pp. 129–162). Brill Publishers.

Akomea-Frimpong, I., Jin, X., & Osei-Kyei, R. (2022). Managing financial risks to improve financial success of public—private partnership projects: A theoretical framework. *Journal of Facilities Management, 20*(5), 629–651. DOI: 10.1108/JFM-03-2021-0036

Albahri, A., Khaleel, Y. L., Habeeb, M. A., Ismael, R. D., Hameed, Q. A., Deveci, M., Homod, R. Z., Albahri, O., Alamoodi, A., & Alzubaidi, L. (2024). A systematic review of trustworthy artificial intelligence applications in natural disasters. *Computers & Electrical Engineering, 118*, 109409. DOI: 10.1016/j.compeleceng.2024.109409

Aleke, C. U., Ovie, E. O., Asere, J. B., Oriakhi, E. A., Igah, G. C., Ajibade, O. E., & Egbuna, I. K. (2025). Digital twins for climate-resilient infrastructure: Simulating environmental impact on buildings. *Asian J. Geo. Res., 8*(2), 132–141. DOI: 10.9734/ajgr/2025/v8i2270

Alessa, L., Kliskey, A., Gamble, J., Fidel, M., Beaujean, G., & Gosz, J. (2016). The role of Indigenous science and local knowledge in integrated observing systems: Moving toward adaptive capacity indices and early warning systems. *Sustainability Science, 11*(1), 91–102. DOI: 10.1007/s11625-015-0295-7

Alkhaldi, M., Moonesar, I. A., Issa, S. T., Ghach, W., Okasha, A., Albada, M., Chelli, S., & Takshe, A. A. (2023). Analysis of the United Arab Emirates' contribution to the sustainable development goals with a focus on global health and climate change. *International Journal of Health Governance, 28*(4), 357–367. 357-367. DOI: 10.1108/IJHG-04-2023-0040

Al-Sarihi, A., & Mason, M. (2020). Challenges and opportunities for climate policy integration in oil-producing countries: The case of the UAE and Oman. *Climate Policy, 20*(10), 1226–1241. DOI: 10.1080/14693062.2020.1781036

Alvarez, A. (2016). Borderlands, Climate Change, and the Genocidal Impulse. *Genocide Studies International, 10*(1), 27–36. DOI: 10.3138/gsi.10.1.03

Alvin, S. D. (2022). Effectiveness of the ASEAN Agreement on Transboundary Haze Pollution. *Environmental Law Review, 24*(4), 305–313. DOI: 10.1177/14614529221126335

Amin, M., Shah, H. H., Fareed, A. G., Khan, W. U., Chung, E., Zia, A., Rahman Farooqi, Z. U., & Lee, C. (2022). Hydrogen production through renewable and non-renewable energy processes and their impact on climate change. *International Journal of Hydrogen Energy, 47*(77), 33112–33134. DOI: 10.1016/j.ijhydene.2022.07.172

Amo-Agyemang, C. (2021). Decolonising the discourse on resilience. *International Journal of African Renaissance Studies, 16*(1), 4–30. DOI: 10.1080/18186874.2021.1962725

Andersen, D. J., & Prokkola, E.-K. (2021). *Borderlands Resilience: Transitions, Adaptation and Resistance at Borders* (1st ed.). Routledge., DOI: 10.4324/9781003131328

Andonova, L. B., Betsill, M. M., & Bulkeley, H. (2009). Transnational climate governance. *Global Environmental Politics, 9*(2), 52–73. DOI: 10.1162/glep.2009.9.2.52

Anguelovski, I., & Carmin, J. (2011). Something borrowed, everything new: Innovation and institutionalization in urban climate governance. *Current Opinion in Environmental Sustainability, 3*(3), 169–175. DOI: 10.1016/j.cosust.2010.12.017

Anika, O. C., Nnabuife, S. G., Bello, A., Okoroafor, E. R., Kuang, B., & Villa, R. (2022). Prospects of low and zero-carbon renewable fuels in 1.5-degree net zero emission actualisation by 2050: A critical review. *Carbon Capture Science & Technology, 5*, 100072. DOI: 10.1016/j.ccst.2022.100072

Ansell, C., & Gash, A. (2008). Collaborative governance in theory and practice. *Journal of Public Administration: Research and Theory, 18*(4), 543–571. DOI: 10.1093/jopart/mum032

Aprizal, A., Wiranatakusuma, D. B., Rizki, M., & Anugrah, R. A. (2025). The role of technology in climate resilience: A systematic literature review and mapping study approach. *Journal of Physics: Conference Series, 2989*(1), 12037. DOI: 10.1088/1742-6596/2989/1/012037

Aragani, V. M., Anumolu, V. R., & Selvakumar, P. (2025). Democratization in the Age of Algorithms: Navigating Opportunities and Challenges. In Wongmahesak, K., Wekke, I., Seftyono, C., & Nurdin, N. (Eds.), *Democracy and Democratization in the Age of AI* (pp. 39–56). IGI Global Scientific Publishing., DOI: 10.4018/979-8-3693-8749-8.ch003

Arasu, S. (2023, September 8). UN secretary-general urges G20 leaders to send a strong message on climate change. Associated Press. https://apnews.com/article/climate-change-antonio-guterres-g20-united-nations-8ba2f8cfc5d18e59a2eb728e015b8fed

Archer, M., Ravn, L., & Thylstrup, N. B. (2025). The political economy of platformed silos: Theorizing data storage reconfigurations in the age of interoperability capitalism. *Big Data & Society*, *12*(2), 20539517241303144. DOI: 10.1177/20539517241303144

Arendt, H. (1958). *The human condition*. University of Chicago Press.

Aronova, E. (2017). Geophysical Datascapes of the Cold War: Politics and Practices of the World Data Centers in the 1950s and 1960s. *Osiris*, *32*(1), 307–327. DOI: 10.1086/694094

Arvanatidis, C., Penev, L., López Lérida, J., Huertas Olivares, C., Sáenz Albanés, A. J., Basset, A., Montinaro, S., Vaira, L., Minadakis, N., Griniezakis, M., & López Paneque, J. (2025). Biodiversity Knowledge Hub: Addressing the impacts of environmental change by linking Research Infrastructures, Global Aggregators and community. *Networks*.

Ashwarya, S. (2023). India's national role conception and relations with GCC countries under Modi: A focus on Saudi Arabia. *Journal of Asian and African Studies*, *58*(4), 535–555. DOI: 10.1177/00219096231162104

Atalay, Y. (2018). Understanding input and output legitimacy of environmental policymaking in the gulf cooperation council states. *Environmental Policy and Governance*, *28*(1), 39–50. DOI: 10.1002/eet.1794

Atkinson, R. (1995). Climate debate ends on cool note greenhouse gas cutbacks stalled 2 years. The Washington Post, 8 April, p. 15.

Axon, S., Chapman, A., & Light, D. (2017). Stakeholder engagement in coastal sustainability transitions: An emerging research agenda. *Reg. Mag.*, *308*(4), 20–22. DOI: 10.1080/13673882.2017.11958673

Aylett, A. (2015). Institutionalizing the urban governance of climate change adaptation: Results of an international survey. *Urban Climate*, *14*, 4–16. DOI: 10.1016/j.uclim.2015.06.005

Bache, I., & Flinders, M. (2004). *Multi-level Governance*. Oxford University Press. DOI: 10.1093/0199259259.001.0001

Ballo, A., Valentin, K. K., Korgo, B., Ogunjobi, K. O., Agbo, S. N., Kone, D., & Savadogo, M. (2022). Law and policy review on green hydrogen potential in ECOWAS countries. *Energies*, *15*(7), 2304. DOI: 10.3390/en15072304

Balsalobre-Lorente, D., Shahbaz, M., Roubaud, D., & Farhani, S. (2018). How economic growth, renewable electricity and natural resources contribute to CO2 emissions? *Energy Policy*, *113*, 356–367. DOI: 10.1016/j.enpol.2017.10.050

Baomar, A. A. H. A. (2024). *The GCC'S Complex Interdependence with China and its Implications on Western Alliances*. Sultan Qaboos University.

Baral, N., & Heinen, J. T. (2007). Decentralization and people's participation in conservation: A comparative study from the Western Terai of Nepal. *International Journal of Sustainable Development and World Ecology*, *14*(5), 520–531. DOI: 10.1080/13504500709469751

Barber, B. R. (2013). *If mayors ruled the world: Dysfunctional nations, rising cities*. Yale University Press.

Barber, B. R. (2013). *If Mayors Ruled the World: Dysfunctional Nations, Rising Cities*. Yale University Press.

Barbi, F., & de Macedo, L. V. (2019). Transnational municipal networks and cities in climate governance: Experiments in Brazil. In van der Heijden, J., Bulkeley, H., & Certomà, C. (Eds.), *Urban climate politics: Agency and empowerment* (pp. 59–79). Cambridge University Press. DOI: 10.1017/9781108632157.004

Barnett, M., & Sikkink, K. (2008). From international relations to global society. In Reus-Smit, C., & Snidal, D. (Eds.), *The Oxford handbook of international relations* (pp. 62–83). Oxford University Press.

Basnet, D., Kandel, P., Chettri, N., Yang, Y., Lodhi, M. S., Htun, N. Z., Uddin, K., & Sharma, E. (2019). Biodiversity research trends and gaps from the confluence of three global biodiversity hotspots in the Far-Eastern Himalaya. *International Journal of Ecology*, *2019*(1), 1323419. DOI: 10.1155/2019/1323419

Baten, M. A., & Titumir, R. A. M. (2016). Environmental challenges of trans-boundary water resources management: The case of Bangladesh. *Sustainable Water Resources Management*, *2*(1), 13–27. DOI: 10.1007/s40899-015-0037-0

Batool, N., Wani, M. D., Shah, S. A., & Dada, Z. A. (2024a). Theory of planned behavior and value-belief norm theory as antecedents of pro-environmental behaviour: Evidence from the local community. *Journal of Human Behavior in the Social Environment*, *34*(5), 693–709. DOI: 10.1080/10911359.2023.2205912

Batool, N., Wani, M. D., Shah, S. A., & Dada, Z. A. (2024b). Tourists' attitude and willingness to pay on conservation efforts: Evidence from the west Himalayan eco-tourism sites. *Environment, Development and Sustainability*, *27*(8), 1–19. DOI: 10.1007/s10668-024-04679-2

Bednar-Friedl, B., Biesbroek, G. R., Schmidt, D. N., Alexander, P., Børsheim, K. Y., Carnicer, J., . . . Rama, B. (2022). Europe: From chapters and cross-chapter papers. In *Climate change 2022: Impacts, adaptation and vulnerability* (pp. 1817–1927). Cambridge University Press.

Benveniste, H., Oppenheimer, M., & Fleurbaey, M. (2020). Effect of border policy on exposure and vulnerability to climate change. *Proceedings of the National Academy of Sciences of the United States of America, 117*(43), 26692–26702. DOI: 10.1073/pnas.2007597117 PMID: 33046645

Benzie, M., Carter, T. R., Carlsen, H., & Taylor, R. (2019). Cross-border climate change impacts: Implications for the European Union. *Regional Environmental Change, 19*(3), 763–776. DOI: 10.1007/s10113-018-1436-1

Benzie, M., & Persson, Å. (2019). Governing borderless climate risks: Moving beyond the territorial framing of adaptation. *International Environmental Agreement: Politics, Law and Economics, 19*(4–5), 369–393. DOI: 10.1007/s10784-019-09441-y

Berbés-Blázquez, M., Mitchell, C. L., Burch, S. L., & Wandel, J. (2017). Understanding climate change and resilience: Assessing strengths and opportunities for adaptation in the Global South. *Climatic Change, 141*(2), 227–241. DOI: 10.1007/s10584-017-1897-0

Berkeley Earth System. (2025). *Global Temperature Report for 2024.* Written by Robert Rohde.

Berke, P. R., Quiring, S. M., Olivera, F., & Horney, J. A. (2021). Addressing Challenges to Building Resilience Through Interdisciplinary Research and Engagement. *Risk Analysis, 41*(7), 1248–1253. DOI: 10.1111/risa.13202 PMID: 30261118

Bernstein, S., Hoffmann, M., Weinthal, E., Mildenberger, M., & Aklin, M. (2020). Introduction. *Global Environmental Politics, 20*(1), 1–3. DOI: 10.1162/glep_e_00545

Berry, M. (2002). Validity and reliability issues in elite interviewing. *PS, Political Science & Politics, 35*(4), 679–682. DOI: 10.1017/S1049096502001166

Bertilsson, J., & Soneryd, L. (2023). Indigenous peoples and inclusion in the Green Climate Fund. *Environmental Sociology, 9*(3), 233–242. DOI: 10.1080/23251042.2023.2177091

Betsill, M. (2008). Environmental NGOs and the Kyoto Protocol negotiations: 1995 to 1997. In Betsill, M., & Corell, E. (Eds.), *NGO diplomacy: The influence of nongovernmental organizations in international environmental negotiations* (pp. 43–66). The MIT Press.

Betsill, M. M., & Bulkeley, H. (2006). Cities and the multilevel governance of global climate change. *Global Governance*, *12*(2), 141–160. DOI: 10.1163/19426720-01202004

Bexell, M., Tallberg, J., & Uhlin, A. (2010). Democracy in global governance: The promises and pitfalls of transnational actors. *Global Governance*, *16*(1), 81–101. DOI: 10.1163/19426720-01601006

Bhat, G. M. (2025). "Drill, Baby Drill" for Oil and Gas: COP29 New Climate Finance. *Journal of the Geological Society of India*, *101*(3), 287–290. DOI: 10.17491/jgsi/2025/174094

Bhatta, L. D., Aggarwal, P. K., & Wang, E. (2015). Climate-induced migration in South Asia: Evidence, future projections and implications. *Regional Environmental Change*, *15*(1), 43–55. DOI: 10.1007/s10113-014-0744-8

Bhat, W. A., Wani, M. D., Dada, Z. A., & Qureshi, R. A. (2025). Beyond Linear Models: A Complexity-Theoretic Approach to Understanding Tourist Behavior in Peacebuilding Contexts. *Journal of Quality Assurance in Hospitality & Tourism*, •••, 1–30. DOI: 10.1080/1528008X.2025.2466050

Biermann, F., Pattberg, P., Van Asselt, H., & Zelli, F. (2009). The fragmentation of global governance architectures: A framework for analysis. *Global Environmental Politics*, *9*(4), 14–40. DOI: 10.1162/glep.2009.9.4.14

Birkmann, J., Liwenga, E., Pandey, R., Boyd, E., Djalante, R., Gemenne, F., Leal Filho, W., Pinho, P. F., Stringer, L., & Wrathall, D. (2022). Poverty, livelihoods and sustainable development. In Pörtner, H.-O. (Eds.), *Climate change 2022: Impacts, adaptation and vulnerability* (pp. 1171–1274). Cambridge University Press., DOI: 10.1017/9781009325844.010

Biskaborn, B., Smith, S., Noetzli, J., Matthes, H., Vieira, G., Streletskiy, D. A., Schoeneich, P., Romanovsky, V. E., Lewkowicz, A., Abramov, A., Allard, M., Boike, J., Cable, W. L., Christiansen, H. H., Delaloye, R., Diekmann, B., Drozdov, D., Etzelmüller, B., Grosse, G., & Lantuit, H. (2019). Permafrost is warming at a global scale. *Nature Communications*, *10*(1), 264. DOI: 10.1038/s41467-018-08240-4 PMID: 30651568

Biyik, C., Allam, Z., Pieri, G., Moroni, D., O'fraifer, M., O'connell, E., Olariu, S., & Khalid, M. (2021). Smart parking systems: Reviewing the literature, architecture and ways forward. *Smart Cities*, *4*(2), 623–642. DOI: 10.3390/smartcities4020032

Björk, P. (2000). Ecotourism from a conceptual perspective, an extended definition of a unique tourism form. *International Journal of Tourism Research*, *2*(3), 189–202. DOI: 10.1002/(SICI)1522-1970(200005/06)2:3<189::AID-JTR195>3.0.CO;2-T

Blanc, N., Cannata, M., Collombin, M., Ertz, O., Giuliani, G., & Ingensand, J. (2022). Ogc api state of play – a practical Testbed for the national spatial data infrastructure in Switzerland. ISPRS - Int. Arch. Photogramm. Remote Sens. Spat. Inf. Sci., XLVIII-4/W1-2022, 59–65.

Blue, G. (2016). Framing climate change for public deliberation: What role for interpretive social sciences and humanities? *Journal of Environmental Policy and Planning*, *18*(1), 67–84. DOI: 10.1080/1523908X.2015.1053107

Boas, I., & Rothe, D. (2016). From conflict to resilience? Explaining recent changes in climate security discourses and practices. *Environmental Politics*, *25*(4), 613–632. DOI: 10.1080/09644016.2016.1160479

Boisson de Chazournes, L. (2016). Dispute settlement procedures and environmental protection: Refining rules and institutions. *Environmental Policy and Law*, *46*(1), 15–25. DOI: 10.3233/EPL-46103

Bostrom, A., Hayes, A. L., & Crosman, K. M. (2018). Efficacy, action, and support for reducing climate change risks. *Risk Analysis*, *39*(4), 805–828. DOI: 10.1111/risa.13210 PMID: 30368853

Bouteligier, S. (2012). *Cities, networks, and global environmental governance: Spaces of innovation, spaces of cooptation*. Routledge. DOI: 10.4324/9780203106488

Bouwer, K. (2018). The unsexy future of climate change litigation. *Journal of Environmental Law*, *30*(3), 483–506. DOI: 10.1093/jel/eqy017

Boyd, A., Gatewood, J., Thorson, S., & Dye, T. (2019). Data Diplomacy. *Science & Diplomacy*, *8*, 1. Retrieved June 2, 2021, from http://sciencediplomacy.org/article/2019/data-diplomacy PMID: 31598426

Boyle, , & Shneiderman, Boyle & Shneiderman. (2020). Redundancy, Resilience, Repair: Infrastructural Effects in Borderland Spaces. *Verge: Studies in Global Asias*, *6*(2), 112. DOI: 10.5749/vergstudglobasia.6.2.0112

Braithwaite, J., & Drahos, P. (2000). *Global business regulation*. Cambridge University Press. DOI: 10.1017/9780521780339

Bramwell, B., & Lane, B. (2011). Critical research on the governance of tourism and sustainability. *Journal of Sustainable Tourism*, *19*(4–5), 411–421. DOI: 10.1080/09669582.2011.580586

Bratman, E. Z., & DeLince, W. P. (2022). Dismantling white supremacy in environmental studies and sciences: An argument for anti-racist and decolonizing pedagogies. *Journal of Environmental Studies and Sciences, 12*(2), 1–11. DOI: 10.1007/s13412-021-00739-5

Bruch, C. (2020). Climate change and international environmental dispute settlement: The role of international arbitration. *Environmental Policy and Law, 50*(6), 413–422. DOI: 10.3233/EPL-200255

Brunet-Jailly, E. (2022). Cross-border cooperation: A global overview. *Alternatives, 47*(1), 3–17. DOI: 10.1177/03043754211073463 PMID: 35280936

Buch, E. (2019). Arctic In Situ Data Availability. In Copernicus Programme Services; European Environment Agency: Kobenhavn, Denmark; p. 86.

Bulkeley, H., & Betsill, M. (2005). Rethinking sustainable cities: Multilevel governance and the 'urban' politics of climate change. *Environmental Politics, 14*(1), 42–63. DOI: 10.1080/0964401042000310178

Bulkeley, H., & Betsill, M. M. (2013). *Cities and climate change: Urban sustainability and global environmental governance*. Routledge. DOI: 10.4324/9780203077207

Butler, R. W., & Suntikul, W. (Eds.). (2010). *Tourism and political change*. Goodfellow Publishers. DOI: 10.2307/jj.28209953

Byskov, M. F., & Hyams, K. (2022). Epistemic injustice in climate adaptation. *Ethical Theory and Moral Practice, 25*(4), 613–634. DOI: 10.1007/s10677-022-10301-z

C40 Cities. (2022). *About C40*. Retrieved from https://www.c40.org

Cabello, J., & Gilbertson, T. (2012). A colonial mechanism to enclose lands: A critical review of two REDD+-focused special issues. *Environmental Science & Policy*.

Cabré, A., Kaufmann, S., Marinov, I., & Weisberg, M. (2024). Introducing the global climate security atlas. *Glob. Sustain., 7*(e25), e25. DOI: 10.1017/sus.2024.18

Carlos Muñoz, J., & Oliva, H. (2024). Effect of combining carbon policies and price controls in cross-border trade of energy on renewable generation investments. *The Energy Journal (Cambridge, Mass.), 45*(1), 81–103. DOI: 10.5547/01956574.45.1.jmun

Carlson, C. J., Albery, G. F., Merow, C., Trisos, C. H., Zipfel, C. M., Eskew, E. A., Olival, K. J., Ross, N., & Bansal, S. (2022). Climate change increases cross-species viral transmission risk. *Nature, 607*(7919), 555–562. DOI: 10.1038/s41586-022-04788-w PMID: 35483403

Carpenter, C. (2001). Businesses, green groups and the media: The role of non-governmental organisations in the climate change debate. *International Affairs*, *77*(2), 313–328. DOI: 10.1111/1468-2346.00194

Causevic, S., & Lynch, P. (2011). Phoenix tourism: Post-conflict tourism role. *Annals of Tourism Research*, *38*(3), 780–800. DOI: 10.1016/j.annals.2010.12.004

Challinor, A. J., Adger, W. N., Benton, T. G., Conway, D., Joshi, M., & Frame, D. (2018). Transmission of climate risks across sectors and borders. *Philosophical Transactions. Series A, Mathematical, Physical, and Engineering Sciences*, *376*(2121), 20170301. DOI: 10.1098/rsta.2017.0301 PMID: 29712795

Chang, W. C. (2025). A grand plan for health equity. *International Journal for Equity in Health*, *24*(1), 1–15. DOI: 10.1186/s12939-025-02536-w PMID: 40542351

Charusombat, P. (2024). Enhancing transboundary haze pollution cooperation in Southeast Asia (2003–2020): External actors and ASEAN supporting mechanisms. *Asia Social Issues*, *18*(1), e270363. DOI: 10.48048/asi.2025.270363

Chen, M., Zhang, L., Teng, F., Dai, J., Li, Z., Wang, Z., & Li, Y. (2020). Climate technology transfer in BRI era: Needs, priorities, and barriers from receivers' perspective. *Ecosystem Health and Sustainability*, *6*(1), 1780948. DOI: 10.1080/20964129.2020.1780948

Chen, N. Sh., Hu, G. Sh., Deng, W., Khanal, N., Zhu, Y. H., & Han, D. (2013). On the water hazards in the trans-boundary Kosi River basin. *Natural Hazards and Earth System Sciences*, *13*(3), 795–808. DOI: 10.5194/nhess-13-795-2013

Chetail, V. (2023). The politics of soft law: Progress and pitfall of the Global Compact for Safe, Orderly and Regular Migration. *Frontiers in Human Dynamics*, *5*, 1243774. Advance online publication. DOI: 10.3389/fhumd.2023.1243774

Chileshe, M. (2023). International renewable energy finance mechanisms and the role of local stakeholders in project design and implementation: Perceptions from Zambia. *Open J. Bus. Manag.*, *11*(06), 3066–3080. DOI: 10.4236/ojbm.2023.116168

Chimni, B. S. (2004). TWAIL: A theoretical framework. *International Community Law Review*, *5*(3), 301–321. DOI: 10.1163/187197304X00130

Chowdhooree, I. (2019). Indigenous knowledge for enhancing community resilience: An experience from the south-western coastal region of Bangladesh. *International Journal of Disaster Risk Reduction*, *40*, 101259. DOI: 10.1016/j.ijdrr.2019.101259

Ciliberti, F. G., Berardi, L., Laucelli, D. B., & Giustolisi, O. (2023). Digital Water Services using Digital Twin paradigm. *IOP Conference Series. Earth and Environmental Science*, *1136*(1), 12002. DOI: 10.1088/1755-1315/1136/1/012002

City of Ghent. (2020). *2020–2025 Climate Plan*. Office of the Deputy Mayor for Environment, Climate, Housing and North–South.

Clarke, B., Otto, F., Stuart-Smith, R., & Harrington, L. (2022). Extreme weather impacts of climate change: An attribution perspective. *Environmental Research Climate*, *1*(1), 012001. DOI: 10.1088/2752-5295/ac6e7d

Columbia Center on Sustainable Investment. (2021). *ISDS and the environment: Trends and tensions*. Columbia University. https://ccsi.columbia.edu

Cons, J. (2018). Staging Climate Security: Resilience and Heterodystopia in the Bangladesh Borderlands. *Cultural Anthropology*, *33*(2), 266–294. DOI: 10.14506/ca33.2.08

Convention, A. (2021). *Climate Action in the Alpine Region*. Retrieved from https://www.alpconv.org

Cowls, J., Tsamados, A., Taddeo, M., & Floridi, L. (2021). The AI gambit: Leveraging artificial intelligence to combat climate change—opportunities, challenges, and recommendations. *AI & Society*, *38*(1), 283–307. DOI: 10.1007/s00146-021-01294-x PMID: 34690449

Curtis, S. (2016). *Global cities and global order*. Oxford University Press. DOI: 10.1093/acprof:oso/9780198744016.001.0001

Cvetkovic, V., & Grbic, L. (2021). Public perception of climate change and its impact on natural disasters. *Journal of the Geographical Institute Jovan Cvijic SASA*, *71*(1), 43–58. DOI: 10.2298/IJGI2101043C

D'Amore, L. J. (2009). Peace through tourism: The global coalition for peace. *Journal of Business Ethics*, *89*(4), 559–570. DOI: 10.1007/s10551-010-0407-3

Dada, Z. A., Wani, M. D., & Shah, S. A. (2022). The impact of pilgrim tourism on the bilateral diplomacy of India and Nepal and the mediation effect of people-to-people contact. *Asian Journal of Comparative Politics*, *7*(3), 495–506. DOI: 10.1177/20578911221076234

Dagher, L., Heeter, J., & Mansour, M. H. (2020). What can GCC countries learn from well-established green power markets in other countries? In *Gulf Conference on Sustainable Built Environment* (pp. 309-327). Cham: Springer International Publishing. https://doi.org/DOI: 10.1007/978-3-030-39734-0_19

Dagnet, Y., Simeonova, K., Cogswell, N., Holt, M., La Viña, T., & Warszawski, N. (2021). *Toward more effective implementation of the Paris Agreement: Learning from 30 years of experience*. World Resources Institute., DOI: 10.46830/wriwp.20.00069

Dalby, S. (2020). *Anthropocene geopolitics: Globalization, security, sustainability*. University of Ottawa Press., DOI: 10.2307/j.ctvx5w8dk

Danube.Pearls. (2020). *Transdanube.Pearls: Network for sustainable mobility along the Danube*. Interreg Europe. https://www.interregeurope.eu/good-practices/transdanube-pearls-network-for-sustainable-mobility-along-the-danube

Darby, S. J., Hammond, G. P., & Wu, J. (2024). Stocktaking global warming – the outcomes of the 2023 Dubai climate summit (COP28). Proceedings of the Institution of Civil Engineers - Energy. DOI: 10.1680/jener.24.00005

Dash, K. C. (2008). *Regionalism in South Asia: Negotiating Cooperation, Institutional Structures*. Routledge., DOI: 10.4324/9780203930366

Daugirdas, K. (2021). Reputation and the responsibility of international organizations. *European Journal of International Law*, *32*(1), 159–188. DOI: 10.1093/ejil/chab002

David, J. O. (2024). Decolonizing climate change response: African indigenous knowledge and sustainable development. *Frontiers in Sociology*, *9*, 1456871. Advance online publication. DOI: 10.3389/fsoc.2024.1456871 PMID: 39640009

De Brabandere, E. (2015). Human rights and transnational corporations: The limits of direct corporate responsibility. *Human Rights & International Legal Discourse*, *9*(1), 65–85.

De Sherbinin, A. (2014). Climate change hotspots mapping: What have we learned? *Climatic Change*, *123*(1), 23–37. DOI: 10.1007/s10584-013-0900-7

Debnath, P., Deshmukh, S. M., Laroia, M., Selvakumar, P., Manjunath, T. C., & Bhattacharya, S. (2025). Challenges Faced by Countries in Renewable Energy Adoption. In Özsungur, F., Chaychi Semsari, M., & Küçük Bayraktar, H. (Eds.), *Geopolitical Landscapes of Renewable Energy and Urban Growth* (pp. 295–320). IGI Global Scientific Publishing., DOI: 10.4018/979-8-3693-8814-3.ch010

Deflorian, M. (2015). Governing the "Enough" in a Warming World: The Discourse of Sufficiency from a Climate Governmentality Perspective.

Depledge, J. (2024). *The future of negotiations under the climate change COP: Implementation is not enough*. Dialogues on Climate Change., DOI: 10.1177/29768659241293212

Dewi, L. (2024). Exploring stakeholder perspectives on supply chain sustainability in renewable energy industries. In *Research Square*. DOI: 10.21203/rs.3.rs-4301842/v1

Di Gregorio, M., Fatorelli, L., Paavola, J., Locatelli, B., Pramova, E., Nurrochmat, D. R., May, P. H., Brockhaus, M., Sari, I. M., & Kusumadewi, S. D. (2019). Multi-level governance and power in climate change policy networks. *Global Environmental Change, 54*, 64–77. DOI: 10.1016/j.gloenvcha.2018.10.003

Diaz, S., Al Hammadi, N., El Nasr, A. S., Villasuso, F., Prakash, S., Baobaid, O., Gracias, D., & Mills, R. (2023, October). Green Corridor: A Feasible Option for the UAE Decarbonization Pathway, Opportunities & Challenges. In *Abu Dhabi International Petroleum Exhibition and Conference* (p. D011S017R003). SPE. DOI: 10.2118/216033-MS

Dieng, A. (2018). Protecting internally displaced persons: The value of the Kampala Convention as a regional example. *International Review of the Red Cross, 99*(904), 263–282. DOI: 10.1017/S1816383117000613

Dietrich, M., & Gutierrez, M. (2025). Empowering environmental research through the EOSC EU node: FAIR data and scalable services.

Do, T. N., Burke, P. J., & Lu, B. (2023). Harnessing solar and wind for sustainable cross-border electricity trade in the Greater Mekong Subregion. *Frontiers in Environmental Science, 11*, 11. DOI: 10.3389/fenvs.2023.1188335

Downie, C. (2012). Toward an understanding of state behavior in prolonged international negotiations. *International Negotiation, 17*(2), 295–320. DOI: 10.1163/157180612X651458

Downie, C. (2013). Transnational actors: Nongovernmental organizations, civil society and individuals. In Harris, P. (Ed.), *Routledge handbook of global environmental politics* (pp. 176–186). Routledge. DOI: 10.4324/9780203799055.ch14

Downie, C. (2014). *The politics of climate change negotiations: strategies and variables in prolonged international negotiations*. Edward Elgar. DOI: 10.4337/9781783472116

Dryzek, J. S., & Niemeyer, S. (2019). Deliberative democracy and climate governance. *Nature Human Behaviour, 3*(5), 411–413. DOI: 10.1038/s41562-019-0591-9 PMID: 30962616

Dsilva, J., Shaikh, M. I., & Usmani, F. Z. A. (2024). Net Zero Policies and Climate Resilience: A Focus on Emerging Technologies and Implementation Strategies in the Middle East. *Transition Towards a Sustainable Future: Net Zero Policies and Environmental Sustainability*, 119-138. DOI: 10.1007/978-981-97-5756-5_6

Dubey, B., Agrawal, S., & Sharma, A. K. (2023). India's renewable energy portfolio: An investigation of the untapped potential of RE, policies, and incentives favoring energy security in the country. *Energies, 16*(14), 5491. DOI: 10.3390/en16145491

Dutta, A. (2021). Unpacking the India-EU economic relationship. *A 2030 Vision for India's Economic Diplomacy*, 28. https://www.orfonline.org/public/uploads/posts/pdf/20230424134334.pdf#page=28

Ekardt, F. (2012). A critical review of "efficiency ethics": The case of climate economics. In Climate Justice and Sustainability (pp. 181–210). DOI: 10.1007/978-94-007-1869-2_9

Ekechukwu, D. E., & Simpa, P. (2024). A comprehensive review of renewable energy integration for climate resilience. Eng. Sci. Technol. j., 5(6), 1884–1908.

Elegbeleye, I., Oguntona, O., & Elegbeleye, F. (2025). Green Hydrogen: Pathway to Net Zero Green House Gas Emission and Global Climate Change Mitigation. *Hydrogen, 6*(2), 29. DOI: 10.3390/hydrogen6020029

Elnagar, E., Munde, S., & Lemort, V. (2021). Energy efficiency measures applied to heritage retrofit buildings: A simulated student housing case study in Vienna. *Heritage, 4*(4), 3919–3937. DOI: 10.3390/heritage4040215

Engen, S. (2025). *On green swans and catastrophic futures: Climate change as risk and uncertainty in central banking*. Valuation Studies., DOI: 10.3384/VS.2001-5992.2025.12.1.171-194

Erdoğdu, A., Dayi, F., Yanik, A., Yildiz, F., & Ganji, F. (2025). Innovative solutions for combating climate change: Advancing sustainable energy and consumption practices for a greener future. *Sustainability (Basel), 17*(6), 2697. DOI: 10.3390/su17062697

ESA. (2019). *Applications Copernicus*. The European Space Agency.

EU Committee of the Regions. (2020). *The role of local and regional authorities in the implementation of the European Green Deal*. https://cor.europa.eu/

European Commission. (2019). *The European Green Deal*. https://ec.europa.eu/info/strategy/priorities-2019-2024/european-green-deal_en

European Commission. (2020). *INTERREG: European Territorial Cooperation*. https://ec.europa.eu

European Commission. (2023). *Austrian research and technology report 2023*. European Union. https://ec.europa.eu/

European Court of Human Rights. (2024). *KlimaSeniorinnen v. Switzerland*, App. No. 53600/20, Judgment of 9 April 2024.

EuroVelo. (2020). *Transdanube.Pearls – EuroVelo 6 [Project summary]*. https://pro.eurovelo.com/projects/2020-03-03_transdanube-pearls---eurovelo

Eveloy, V., & Gebreegziabher, T. (2019). Excess electricity and power-to-gas storage potential in the future renewable-based power generation sector in the United Arab Emirates. *Energy*, *166*, 426–450. DOI: 10.1016/j.energy.2018.10.088

Fauchald, O. K., & Hunter, D. (2008). Science and environmental law: Cooperation across disciplines. *Yearbook of International Environmental Law*, *19*(1), 3–40. DOI: 10.1093/yiel/17.1.3

Ferris, E., & Bergmann, J. (2017). Soft law, migration and climate change governance. *Journal of Human Rights and the Environment*, *8*(1), 6–29. DOI: 10.4337/jhre.2017.01.01

Follmer, J., & Justin, S. (2023). KlimaTicket–Report 2022. In *Republic of Austria – Green Investor Report 2023*. Austrian Government Printing Office.

Forsyth, T. (2014). Public concerns about transboundary haze: A comparison of Indonesia, Singapore, and Malaysia. *Global Environmental Change*, *25*, 76–86. DOI: 10.1016/j.gloenvcha.2014.01.013

Fouilloux, A., & Magagna, B. (2025). FAIR2Adapt: Advancing Climate Change Adaptation Strategies through FAIR and open data sharing and Research Infrastructures.

Franck, S. D. (2018). Investor–state dispute settlement: Reform options for the future. *ICSID Review Foreign Investment Law Journal*, *33*(1), 1–35. DOI: 10.1093/icsidreview/siy001

Freeman, R. E. (1984). *Strategic management: A stakeholder approach*. Pitman.

Gallagher, G. E., Duncombe, R. K., & Steeves, T. M. (2020). Establishing climate change resilience in the Great Lakes in response to flooding. *Journal of Science Policy & Governance*, *17*(01), 2–7. DOI: 10.38126/JSPG170105

Gandhi, M. K. (1947). *The Mind of Mahatma Gandhi* (Prabhu, R. K., & Rao, U. R., Eds.). Navajivan Trust.

Gao, Y., Gao, X., & Zhang, X. (2017). The 2 C global temperature target and the evolution of the long-term goal of addressing climate change—From the United Nations framework convention on climate change to the Paris agreement. *Engineering (Beijing)*, *3*(2), 272–278. DOI: 10.1016/J.ENG.2017.01.022

Ghoddusi, H., Moghaddam, H., & Wirl, F. (2022). Going downstream–an economical option for oil and gas exporting countries? *Energy Policy*, *161*, 112487. DOI: 10.1016/j.enpol.2021.112487

Girardi, G. G. G., & Preite, G. (2017). Escaping the economist's straightjacket: Overcoming the free-rider mentality which prevents climate change from being effectively addressed. In Corporate Social Responsibility and Governance (pp. 561–575). DOI: 10.1007/978-3-319-58214-6_35

Gonda, N., Flores, S., Casolo, J. J., & Nightingale, A. J. (2023). Resilience and conflict: Rethinking climate resilience through Indigenous territorial struggles. *The Journal of Peasant Studies*, *50*(6), 2312–2338. DOI: 10.1080/03066150.2022.2161372

Gössling, S., & Hall, C. M. (2006). *Tourism and global environmental change: Ecological, social, economic and political interrelationships*. Routledge., DOI: 10.4324/9780203011911

Gough, C., & Shackley, S. (2001). The respectable politics of climate change: The epistemic communities and NGOs. *International Affairs*, *77*(2), 329–345. DOI: 10.1111/1468-2346.00195

Gramsci, A. (1929). *Selections from the prison notebooks* (Hoare, Q., & Smith, G. N., Eds. & Trans.). International Publishers.

Graveline, M.-H., & Germain, D. (2022). Disaster Risk Resilience: Conceptual Evolution, Key Issues, and Opportunities. *International Journal of Disaster Risk Science*, *13*(3), 330–341. DOI: 10.1007/s13753-022-00419-0

Grennfelt, P., Engleryd, A., Forsius, M., Hov, Ø., Rodhe, H., & Cowling, E. (2020). Acid rain and air pollution: 50 years of progress in environmental science and policy. *Ambio*, *49*(4), 849–864. DOI: 10.1007/s13280-019-01244-4 PMID: 31542884

Grossman, R. L., Boyles, R. R., Davis-Dusenbery, B. N., Haddock, A., Heath, A. P., O'Connor, B. D., Resnick, A. C., Taylor, D. M., & Ahalt, S. (2024). A framework for the interoperability of cloud platforms: Towards FAIR data in SAFE environments. *Scientific Data*, *11*(1), 241. DOI: 10.1038/s41597-024-03041-5 PMID: 38409183

Guo, L., Wei, J., Zhang, K., & Tian, F. (2021). Technical Note: Building a methodological framework and toolkit for news media dataset tracking of conflict and cooperation dynamics on transboundary rivers.

Guo, L., Wei, J., Zhang, K., Wang, J., & Tian, F. (2022). Building a methodological framework and toolkit for news media dataset tracking of conflict and cooperation dynamics on transboundary rivers. *Hydrology and Earth System Sciences*, *26*(4), 1165–1185. DOI: 10.5194/hess-26-1165-2022

Gupta, D., Kumar, P., Okano, N., & Sharma, M. (2025). Climate-induced migration in India and Bangladesh: A systematic review of drivers, impacts, and adaptation mechanisms. *Climate (Basel)*, *13*(4), 81. DOI: 10.3390/cli13040081

Guruparan, K., & Zerk, J. (2021, June 17). *Influence of soft law grows in international governance*. Chatham House. https://www.chathamhouse.org/2021/06/influence-soft-law-grows-international-governance

Guttikunda, S., Lodoysamba, S., Bulgansaikhan, B., & Batdorj, D. (2013). Particulate pollution in Ulaanbaatar, Mongolia. *Air Quality, Atmosphere & Health*, *6*(3), 589–601. DOI: 10.1007/s11869-013-0198-7

Hadjiyianni, I. (2023). The extraterritorial reach of environmental law: Legitimacy concerns and the role of domestic courts in controlling transnational regulatory power. In Parrish, A., & Ryngaert, C. (Eds.), *Research handbook on extraterritoriality in international law* (pp. 372–388). Elgar Publishing., DOI: 10.4337/9781800885592.00031

Hague Rules on Business and Human Rights Arbitration. (2020). *Final text and commentary*. Center for International Legal Cooperation.

Haites, E., ... 2018: Experience with Carbon Taxes and Greenhouse Gas Emissions Trading Systems. SSRN *Electron. J.*, XXIX, 109, DOI: 10.2139/ssrn.3119241

Hamblin, J. (2011). *Oceanographers and the Cold War*. University of Washington Press. DOI: 10.1515/9780295801858

Hamzah, A., Aqlan, F., & Baidya, S. (2024). Drone-based digital twins for water quality monitoring: A systematic review. *Digital Twins and Applications*, *1*(2), 131–160. DOI: 10.1049/dgt2.12021

Hannigan, J. (2018). Disasters Across Borders: Borderlands as Spaces of Hope and Innovation in the Geopolitics of Environmental Disasters. In Miller, M. A., Douglass, M., & Garschagen, M. (Eds.), *Crossing Borders* (pp. 79–94). Springer Singapore., DOI: 10.1007/978-981-10-6126-4_5

Hansen, A., Bi, L., Saniotis, A., & Nitschke, M. (2013). Vulnerability to extreme heat and climate change: Is ethnicity a factor? *Global Health Action*, *6*(1), 21364. DOI: 10.3402/gha.v6i0.21364 PMID: 23899408

Happaerts, S., Van den Brande, K., & Bruyninckx, H. (2011). Subnational governments in transnational networks for sustainable development. *International Environmental Agreement: Politics, Law and Economics*, *11*(4), 321–339. DOI: 10.1007/s10784-010-9128-4

Hare, W. L., Cramer, W., Schaeffer, M., Battaglini, A., & Jaeger, C. C. (2011). Climate hotspots: Key vulnerable regions, climate change and limits to warming. *Regional Environmental Change*, *11*(S1), 1–13. DOI: 10.1007/s10113-010-0195-4

Harris, K., Lager, F., Jansen, M. K., & Benzie, M. (2022). Rising to a new challenge: A protocol for case-study research on transboundary climate risk. *Weather, Climate, and Society*, *14*(3), 755–768. DOI: 10.1175/WCAS-D-21-0022.1

Hartmann, H., Bastos, A., Das, A. J., Esquivel-Muelbert, A., Hammond, W. M., Martínez-Vilalta, J., McDowell, N. G., Powers, J. S., Pugh, T. A., Ruthrof, K. X., & Allen, C. D. (2022). Climate change risks to global forest health: Emergence of unexpected events of elevated tree mortality worldwide. *Annual Review of Plant Biology*, *73*(1), 673–702. DOI: 10.1146/annurev-arplant-102820-012804 PMID: 35231182

Hartmann, J., Inkpen, A. C., & Ramaswamy, K. (2021). Different shades of green: Global oil and gas companies and renewable energy. *Journal of International Business Studies*, *52*(5), 879–903. DOI: 10.1057/s41267-020-00326-w

Hatano, A. (2021). *Emerging international norms to protect "climate refugees"?: Human Rights Committee's decision on Teitiota v New Zealand. Journal of Human Security Studies*, 10(2), 32–50. DOI: 10.34517/jahss.10.2_32

Heilmann, D. (2015). After Indonesia's ratification: The ASEAN Agreement on Transboundary Haze Pollution and its effectiveness as a regional environmental governance tool. *Journal of Current Southeast Asian Affairs*, *34*(3), 95–121. DOI: 10.1177/186810341503400304

Heiskanen, V. (2021). The liberal idea of transparency in international investment arbitration. *The Journal of World Investment & Trade*, *22*(4), 605–629. DOI: 10.1163/22119000-12340193

Held, D., & Roger, C. B. (2018). Three models of global climate governance: From Kyoto to Paris and beyond. *Global Policy*, *9*(4), 527–537. Advance online publication. DOI: 10.1111/1758-5899.12617

Helldén, D., Andersson, C., Nilsson, M., Ebi, K. L., Friberg, P., & Alfvén, T. (2021). Climate change and child health: A scoping review and an expanded conceptual framework. *The Lancet. Planetary Health*, *2021*(5), e164–e175. DOI: 10.1016/S2542-5196(20)30274-6 PMID: 33713617

Hempelmann, N., Ehbrecht, C., Plesiat, E., Hobona, G., Simoes, J., Huard, D., Smith, T. J., McKnight, U. S., Pechlivanidis, I. G., & Alvarez-Castro, C. (2022). Deployment of Ai-enhanced services in climate resilience information systems. ISPRS - Int. Arch. Photogramm. Remote Sens. Spat. Inf. Sci., XLVIII-4/W1-2022, 187–194.

Henderson, J. (2021). COP 26—Examining the balance between ambitious pledges and realistic expectations. In *Oxford Energy Forum: A Quarterly Journal for Debating Energy Issues and Policies* (No. 129). Oxford: The Oxford Institute for Energy Studies. https://www.econstor.eu/handle/10419/306991

Herriman, N. (2017). Management of biodiversity: Creating conceptual space for Indigenous conservation. J3ea, 19, 42–52. DOI: 10.5038/2162-4593.19.1.1184

Herzberg, J., Kerht, C., & Torma, F. (2018). *Ice and Snow in the Cold War: Histories of Extreme Climatic Environments.* Berghahn. DOI: 10.3167/9781785339868

Hettne, B., & Söderbaum, F. (2005). Civilian power or soft imperialism? The EU as a global actor and the role of interregionalism. *European Foreign Affairs Review, 10*(4), 535–552. DOI: 10.54648/EERR2005036

Hickel, J. (2021). What does degrowth mean? A few points of clarification. *Globalizations, 18*(7), 1105–1111. DOI: 10.1080/14747731.2020.1812222

Hilali, I. (2025). Accelerating the transition to a hydrogen economy: Achieving carbon neutrality. In *Accelerating the Transition to a Hydrogen Economy* (pp. 105–127). Elsevier., DOI: 10.1016/B978-0-443-24002-7.00009-4

Hilzenrath, D. (1993). Miscalculations, lobby effort doomed BTU tax plan. The Washington Post, 11 June, p. 1.

Hippe, S., Bertram, D., & Chilla, T. (2024). Convergence and resilience in border regions. *European Planning Studies, 32*(1), 186–207. DOI: 10.1080/09654313.2023.2170214

Hite, E. (2018). Political ecology of Costa Rica's climate policy: Contextualizing climate governance. *Journal of Environmental Studies and Sciences, 8*(4), 469–476. DOI: 10.1007/s13412-018-0480-y

Hodgkinson, D., Burton, T., Anderson, H., & Young, L. (2010). The Hour When the Ship Comes In: A Convention for Persons Displaced by Climate Change. *Monash University Law Review. Monash University. Faculty of Law, 36*(1), 69–120.

Hooghe, L., & Marks, G. (2003). Unraveling the central state, but how? Types of multi-level governance. *Reihe Politikwissenschaft*, Institut für Höhere Studien, 87. https://nbn-resolving.org/urn:nbn:de:0168-ssoar-246200

Hooghe, L., & Marks, G. (2003). Unraveling the central state, but how? Types of multi-level governance. *The American Political Science Review, 97*(2), 233–243.

Hooghe, L., & Marks, G. (2019). Grand theories of European integration in the twenty-first century. *Journal of European Public Policy, 26*(8), 1113–1133. DOI: 10.1080/13501763.2019.1569711

Horri, B. A., & Ozcan, H. (2024). Green hydrogen production by water electrolysis: Current status and challenges. *Current Opinion in Green and Sustainable Chemistry, 47*, 100932. DOI: 10.1016/j.cogsc.2024.100932

Hospers, G., & Beugelsdijk, S. (2002). Regional Cluster Policies: Learning by Comparing? *Kyklos, 55*(3), 381–402. DOI: 10.1111/1467-6435.00192

Hoylman, ZH, Bocinsky, RK & Jencso, KG. (2022). "La evaluación de la sequía ha sido superada por el cambio climático: argumentos empíricos para un cambio de paradigma". Comunicaciones de la naturaleza volumen 13, Nº 2715.

Hurlbert, M. A., & Akpan, J. (2023). Dialectic narratives, hostile actors, and Earth's resources in Saskatchewan, Canada. *Sustainability Science, 18*(1), 285–301. DOI: 10.1007/s11625-022-01214-y PMID: 36157091

Ibrahim, A. (2024). Transitioning Emerging Economies Toward Sustainability by Implementing Targeted Policy Interventions. *ESTIDAMAA, 2024*, 53-59. https://peninsula-press.ae/Journals/index.php/ESTIDAMAA/article/view/130

ICIMOD. (2021). *The Hindu Kush Himalaya assessment: Mountains, climate change, sustainability and people*. Springer Nature. https://lib.icimod.org/record/34383

ICLEI. (2021). *Global Review 2021*. ICLEI – Local Governments for Sustainability. Retrieved from https://www.iclei.org

IEA. (2024). CO_2 emissions in 2023. International Energy Agency. https://www.iea.org/reports/co2-emissions-in-2023

International Ecotourism Society (TIES). (2015). *What is ecotourism?* Retrieved from https://ecotourism.org/what-is-ecotourism/

Interreg Danube. (2018). *Sustainable regional tourism mobility plan for the Danube Region (Transdanube.Pearls)*. Interreg Danube Transnational Programme https://www.interreg-danube.eu/

IoT Analytics. (2023). *Predictive maintenance market: 5 highlights for 2023 and beyond*. https://iot-analytics.com/

IPCC. (2023). *Sixth Assessment Synthesis Report: Climate Change 2023*. Intergovernmental Panel on Climate Change. https://www.ipcc.ch/report/ar6/syr/

Ishola, A. O., Odunaiya, O. G., & Soyombo, O. T. (2024). Stakeholder communication framework for successful implementation of community-based renewable energy projects. *Int. J. Front. Eng. Technol. Res.*, *7*(2), 25–43. DOI: 10.53294/ijfstr.2024.7.2.0047

Jafari, G. (2024). Woman, Water, Freedom: Climate Justice, Women's Liberation, and Territorial Sovereignty in Iran-Iraq Borderlands. *Journal of Architectural Education*, *78*(1), 104–114. DOI: 10.1080/10464883.2024.2303926

Jahnel, C. (2023). *God dwells in flesh: Decolonial ecojustice and planetary ethics in the "Anthropocene"*. Scriptura., DOI: 10.7833/122-1-2135

Jakhar, T. C. A. T. R. (2024). Beyond Borders: CBAM's Revolutionary Potential and Challenges in Achieving Carbon Neutrality. *policy, 3*, 4.

Jasny, B. R. (2013). Realities of data sharing using the genome wars as case study—An historical perspective and commentary. *EPJ Data Science*, *2*(1), 1–15. DOI: 10.1140/epjds13

Jeelani, P., Ahad, F., Shah, S. A., & Rashid, H. (2025). *Site suitability evaluation for nature-based tourism using gis and ahp: a case study of Kashmir Valley*. Applied Geomatics.

Jeelani, P., & Shah, S. A. (2023). Tourism and the environmental components coexisting in the Himalayas: A trade-off between the two. *Journal of Quality Assurance in Hospitality & Tourism*, •••, 1–25.

Jeelani, P., Shah, S. A., Dar, S. N., Ahad, F., & Rashid, H. (2024). Managing mass tourism in mountain ecosystems through instrument of sustainable tourism indicators. *Environment, Development and Sustainability*, •••, 1–28. DOI: 10.1007/s10668-024-04915-9

Jeffery, M. I. (2009). Book review: *International law and the environment*, by Patricia Birnie, Alan Boyle, and Catherine Redgwell. *Osgoode Hall Law Journal*, *47*(3), 595–601. DOI: 10.60082/2817-5069.1143

Jha, V. (2021). 'Soft Law in a Hard Shell': India, International Rulemaking and the International Solar Alliance. *Transnational Environmental Law*, *10*(3), 517–541. DOI: 10.1017/S2047102520000400

Joachim, J. (2003). Framing issues and seizing opportunities: The UN, NGOs, and women's rights. *International Studies Quarterly*, *47*(2), 247–274. DOI: 10.1111/1468-2478.4702005

Jogdand, O. K. (2020). Study on the effect of global warming and greenhouse gases on environmental system. In *Green chemistry and sustainable technology* (pp. 275–306). Apple Academic Press. DOI: 10.1201/9780367808310-12

Jordan, A., & Huitema, D. (2014). Policy innovation in a changing climate: Sources, patterns and effects. *Global Environmental Change*, *29*, 387–394. DOI: 10.1016/j.gloenvcha.2014.09.005

Joseph, M. P. (2002). The Canadian Cryospheric Information Network: Facilitating Access to Sea Ice, Lake Ice, Snow Cover, Permafrost, and Glacier Data. In *Proceedings of the IEEE International Geoscience and Remote Sensing Symposium*, Toronto, ON, Canada, 24–28 June.

Joyce, C., & Stanford, J. (2023). Manufacturing the Energy Revolution. *Centre for Future Work, August,* https://futurework. org. *au/report/manufacturing-the-energy-revolution.*

Ju, Y. (2017). Tracking the $PM_{2.5}$ inventories embodied in the trade among China, Japan and Korea. *Journal of Economic Structures*, *6*(1), 27. DOI: 10.1186/s40008-017-0089-4

Kaack, L. H., Donti, P. L., Strubell, E., Kamiya, G., Creutzig, F., & Rolnick, D. (2022). Aligning artificial intelligence with climate change mitigation. *Nature Climate Change*, *12*(6), 518–527. DOI: 10.1038/s41558-022-01377-7

Kabir, M. E., & Kamruzzaman, P. (2022). Exploring the drivers of vulnerability among disadvantaged internal migrants in riverbank erosion prone areas in north-west Bangladesh. *Journal of South Asian Development*, *17*(1), 57–83. DOI: 10.1177/09731741221078741

Karybekova, G. (2018). International instruments in the field of climate change. *Science World*, *21*(3), 108–111. DOI: 10.31433/1605-220X-2018-21-3(1)-108-111

Keating, M. (2013). Regions and international affairs: Motives, opportunities and strategies. In *Paradiplomacy in action* (pp. 1–16). Routledge.

Keck, M., & Sikkink, K. (1998). *Activists beyond borders: advocacy networks in international politics*. Cornell University Press.

Keerthiraj, M. A., & Vang-Phu, T. (2025). India's BRICS Engagement: A Strategic Lever in South Asia and the Indian Ocean. In A. Gedikli, S. Erdogan, & H. Çalışkan Terzioğlu (Eds.), Changing the Global Political Economy: BRICS Countries and Alternative Relations Strategies (pp. 189-214). IGI Global Scientific Publishing. DOI: 10.4018/979-8-3693-7393-4.ch006

Kelly, I. (2006). Peace through tourism: A SWOT analysis. *Journal of Tourism Studies*, *17*(1), 36–46.

Kemp, L., Xu, C., Depledge, J., Ebi, K. L., Gibbins, G., Kohler, T. A., Rockström, J., Scheffer, M., Schellnhuber, H. J., Steffen, W., & Lenton, T. M. (2022). Climate Endgame: Exploring catastrophic climate change scenarios. *Proceedings of the National Academy of Sciences of the United States of America*, *119*(34), e2108146119. Advance online publication. DOI: 10.1073/pnas.2108146119 PMID: 35914185

Keohane, R. O., & Victor, D. G. (2011). The regime complex for climate change. *Perspectives on Politics*, *9*(1), 7–23. DOI: 10.1017/S1537592710004068

Kern, K., & Alber, G. (2009). Governing climate change in cities: Modes of urban climate governance in multi-level systems. *OECD Conference on Cities and Climate Change*. https://www.oecd.org

Kern, K. (2019). Cities as leaders in EU multilevel climate governance: Embedded upscaling of local experiments in Europe. *Environmental Politics*, *28*(1), 125–145. DOI: 10.1080/09644016.2019.1521979

Kern, K., & Bulkeley, H. (2009). Cities, Europeanization and multi-level governance: Governing climate change through transnational municipal networks. *Journal of Common Market Studies*, *47*(2), 309–332. DOI: 10.1111/j.1468-5965.2009.00806.x

Khagram, S. (2006). Possible future architectures of global governance: A transnational perspective/prospective. *Global Governance*, *12*(1), 97–117. DOI: 10.1163/19426720-01201008

Khan, M. I., & Al-Ghamdi, S. G. (2023). Hydrogen economy for sustainable development in GCC countries: A SWOT analysis considering current situation, challenges, and prospects. *International Journal of Hydrogen Energy*, *48*(28), 10315–10344. DOI: 10.1016/j.ijhydene.2022.12.033

Khan, M., Mfitumukiza, D., & Huq, S. (2021). Capacity building for implementation of nationally determined contributions under the Paris Agreement. In *Making climate action more effective* (pp. 95–106). Routledge., https://www.taylorfrancis.com/chapters/edit/10.4324/9781003162407-8/capacity-building-implementation-nationally-determined-contributions-paris-agreement-mizan-khan-david-mfitumukiza-saleemul-huq DOI: 10.4324/9781003162407-8

Khan, M., Ruszczyk, H., Rahman, M. F., & Huq, S. (2021). Epistemological freedom: Activating co-learning and co-production to decolonise knowledge production. *Disaster Prevention and Management*, *30*(6), 681–694. DOI: 10.1108/DPM-03-2021-0070

Kim, H. J., Choi, H. C., Agrusa, J., & Lee, H. (2007). The role of tourism in the inter-Korean relationship. *Journal of Tourism Studies*, *18*(2), 121–136.

Kitzmann, N. H., Romanczuk, P., Wunderling, N., & Donges, J. F. (2022). Detecting contagious spreading of urban innovations on the global city network. *The European Physical Journal. Special Topics*, *231*(9), 1609–1624. DOI: 10.1140/epjs/s11734-022-00470-4

KlimaTicket. (2025). *Current prices and conditions*. https://www.klimaticket.at/en/home/

Kodra, E., & Ganguly, A. R. (2014). Asymmetry of projected increases in extreme temperature distributions. *Scientific Reports*, *4*(1), 5884. DOI: 10.1038/srep05884 PMID: 25073751

Kowarsch, M., Garard, J., Riousset, P., Lenzi, D., Dorsch, M. J., Knopf, B., Harrs, J.-A., & Edenhofer, O. (2016). Scientific Assessments to Facilitate Deliberative Policy Learning. *Palgrave Communications*, *2*(1), 16092. DOI: 10.1057/palcomms.2016.92

Kraatz, J., Demmich, K., Schnell, J., Gräler, B., Bagli, S., & Mazzoli, P. (2024). A federated Data Fabric to enhance disaster resilience for extreme weather events.

Krause, R. M. (2011). Policy innovation, intergovernmental relations, and the adoption of climate protection initiatives by US cities. *Journal of Urban Affairs*, *33*(1), 45–60. DOI: 10.1111/j.1467-9906.2010.00510.x

Krzymowski, A. (2020). The European Union and the United Arab Emirates as civilian and soft powers engaged in Sustainable Development Goals. *Journal of International Students*, *13*(3), 41–58. https://www.ceeol.com/search/article-detail?id=979401. DOI: 10.14254/2071-8330.2020/13-3/3

Kularathna, A. H. T. S., Suda, S., Takagi, K., & Tabeta, S. (2019). Evaluation of co-existence options of marine renewable energy projects in Japan. *Sustainability (Basel), 11*(10), 2840. DOI: 10.3390/su11102840

Lake, D. (2008). The state and international relations. In Reus-Smit, C., & Snidal, D. (Eds.), *Oxford handbook of international relations* (pp. 41–61). Oxford University Press.

Lake, D. (2010). Rightful rules: Authority, order, and the foundations of global governance. *Studia Quaternaria, 54*(3), 587–613.

Lane, L., & Montgomery, W. (2014). An institutional critique of new climate scenarios. *Climatic Change, 122*(3), 447–458. DOI: 10.1007/s10584-013-0919-9

Lan, R., Eastham, S. D., Liu, T., Saikawa, E., & Barrett, S. R. H. (2022). Air quality impacts of crop residue burning in India and mitigation alternatives. *Nature Communications, 13*(1), 6537. DOI: 10.1038/s41467-022-34093-z PMID: 36376316

Larrinaga, F., Pérez, A., Aldalur, I., Hernández, J. L., Izkara, J. L., & Sáez de Viteri, P. (2021). A holistic and interoperable approach towards the implementation of services for the digital transformation of smart cities: The case of Vitoria-Gasteiz (Spain). *Sensors (Basel), 21*(23), 8061. DOI: 10.3390/s21238061 PMID: 34884064

Laurent, É. (2022). Air (ine)quality in the European Union. *Current Environmental Health Reports, 9*(2), 123–129. DOI: 10.1007/s40572-022-00348-6 PMID: 35347642

Leal-Arcas, R. (2024). Global Energy Transition: Reflecting on COP 28 and Case Studies of Qatar and Morocco. *Forthcoming in Florida Journal of International Law, 35*(1).

Leggett, J. (1999). *The carbon war: dispatches from the end of the oil century.* Allen Lane.

Lehmann, J., Bouillass, G., Fofack-Garcia, R., & Pérez-López, P. (2022). Towards social Life Cycle Assessment of Energy Systems: a case study on offshore wind farms from companies' perspective. E3S Web Conf., 349, 12002.

Lemos, M. C., & Agrawal, A. (2006). Environmental governance. *Annual Review of Environment and Resources, 31*(1), 297–325. DOI: 10.1146/annurev.energy.31.042605.135621

Leung, T. C. H., & You, C. S. X. (2023). Climate change in Hong Kong with the integration of UN SDG 13 climate action. In *SDGs in the Asia and Pacific Region* (pp. 1–26). Springer International Publishing., DOI: 10.1007/978-3-030-91262-8_18-1

Levy, D. (2005). Business and the evolution of the climate regime. In Levy, D., & Newell, P. (Eds.), *The business of global environmental governance* (pp. 73–104). The MIT Press.

Lim, A. B., Poelzer, G., & Noble, B. (2024). Social value of renewable energy in remote northern Indigenous communities. *Journal of Aboriginal Economic Development, 14*(1), 74–91. DOI: 10.29173/jaed8

Link, A. C., Oakes, R., Durand-Delacre, D., Thalheimer-Prezyna, L., & van der Geest, K. (2025). To what extent do climatic stressors drive human mobility in the world's drylands? *Population and Environment, 47*(16), 16. Advance online publication. DOI: 10.1007/s11111-025-00486-7

Liu, O. R., & Molina, R. (2021). The persistent transboundary problem in marine natural resource management. *Frontiers in Marine Science, 8*, 8. DOI: 10.3389/fmars.2021.656023

Liu, S., Xing, J., Wang, S., Ding, D., Chen, L., & Hao, J. (2020). Revealing the impacts of transboundary pollution on $PM_{2.5}$-related deaths in China. *Environment International, 134*, 105323. DOI: 10.1016/j.envint.2019.105323 PMID: 31759275

Liu, Y., Ma, J., Song, H., Qian, Z., & Lin, X. (2021). Chinese universities' cross-border research collaboration in the social sciences and its impact. *Sustainability (Basel), 13*(18), 10378. DOI: 10.3390/su131810378

Lizarralde, G., Bornstein, L., Robertson, M., Gould, K., Herazo, B., Petter, A., Páez, H., Díaz, J. H., Olivera, A., González, G., López, O., López, A., Ascui, H., Burdiles, R., & Bouchereau, K. (2021). Does climate change cause disasters? How citizens, academics, and leaders explain climate-related risk and disasters in Latin America and the Caribbean. *International Journal of Disaster Risk Reduction, 58*, 102173. DOI: 10.1016/j.ijdrr.2021.102173

Llambias, G., Bradach, B., Nogueira, J., González, L., & Ruggia, R. (2023a). Gateway-based interoperability for distributed ledger technology. CLEI Electron. J., 26(2).

Llambias, G., Bradach, B., Nogueira, J., González, L., & Ruggia, R. (2023b). Gateway-based Interoperability for DLT. In *TechRxiv*.

Lucatello, S. (2022). Climate Resilient Development Pathways in the US-Mexico Border Region: The Case of the El Paso del Norte Metropolitan Area. In Giorgi, E., Cattaneo, T., Flores Herrera, A. M., & Aceves Tarango, V. D. S. (Eds.), *Design for Vulnerable Communities* (pp. 361–375). Springer International Publishing., DOI: 10.1007/978-3-030-96866-3_19

Lund, E. (2013). Environmental diplomacy: Comparing the influence of business and environmental NGOs in negotiations on reform of the clean development mechanism. *Environmental Politics*, *22*(5), 739–759. DOI: 10.1080/09644016.2012.737253

Madgule, M., Vinayaka, N., Sonkhaskar, Y. M., Patel, D. R., Karthikeyan, R., & Selvakumar, P. (2023). Mechanical properties and microstructure of activated TIG welded similar joints of Inconel alloys by desirability approaches. *Materials Today: Proceedings*, *77*, 528–533.

Magistro, B., Abramson, C., Ebanks, D., Debnath, R., & Alvarez, R. M. (2024). Identifying American climate change free riders and motivating sustainable behavior. *Scientific Reports*, *14*(1), 6575. Advance online publication. DOI: 10.1038/s41598-024-57042-w PMID: 38503779

Magnani, N., & Cittati, V.-M. (2022). Combining the multilevel perspective and Socio-technical imaginaries in the study of community energy. *Energies*, *15*(5), 1624. DOI: 10.3390/en15051624

Mahadeo, S., & Del Savio, L. (2023). Transboundary maritime cooperation: The case of the Eastern Caribbean Region. *Frontiers in Marine Science*, *10*, 10. DOI: 10.3389/fmars.2023.1251911

Maizland, L., & Fong, C. (2025, January 21). Global climate agreements: Successes and failures. Council on Foreign Relations. https://www.cfr.org/backgrounder/paris-global-climate-change-agreements

Makarov, I., Chen, H., & Paltsev, S. (2020). Impacts of climate change policies worldwide on the Russian economy. *Climate Policy, 20*(10), 1242-1256. ttps://DOI: 10.1080/14693062.2020.1781047

Makki, M., Butt, F. A., Akash, S. A., Petrova, K., & Naeem, S. A. (2025). Fragile Geographies and the Climate-Conflict Nexus: Investigating Climate-Induced Security Risks, Migration, and Inequality in Balochistan, Pakistan. *Alternatives*, *50*(2), 350–375. DOI: 10.1177/03043754241291728

Mallah, A. Q., Gayen, M. H., & Basu, P. P. (2024). India's Environmental Diplomacy: In Pursuit of a Safer Planet. In *75 Years of India's Foreign Policy: Bilateral, Conventional and Emerging Trends* (pp. 463-483). Singapore: Springer Nature Singapore. DOI: 10.1007/978-981-97-6054-1_20

Manohar, H. (2024). Exploring the nexus of renewable energy and international investment law in India: Opportunities and issues. *Jindal Global Law Review*, *15*(2), 635–670. DOI: 10.1007/s41020-024-00239-x

Maqbool, R., Deng, X., & Rashid, Y. (2020). Stakeholders' satisfaction as a key determinant of critical success factors in renewable energy projects. *Energy, Sustainability and Society*, *10*(1), 28. DOI: 10.1186/s13705-020-00259-0

Marino, E. (2024). Rethinking carbon offsets: An ethnographic study of REDD+ implementation in indigenous contexts.

Marks, G., & Hooghe, L. (2004). Contrasting visions of multi-level governance. *Multi-level governance*, 15-30.

Martín León, F. (2021). *Justine hacia la Península después de una ciclogénesis explosiva*. Meteored.

Martínez-Zarzoso, I., Nowak-Lehmann, F., & Paschoaleto, R. D. L. (2022). Climate change, natural disasters, and international migration: A country-level analysis considering climatic zones. *Frontiers in Climate*, *4*, 986088. Advance online publication. DOI: 10.3389/fclim.2022.986088

Matar, A., Fareed, Z., Magazzino, C., Al-Rdaydeh, M., & Schneider, N. (2023). Assessing the co-movements between electricity use and carbon emissions in the GCC area: Evidence from a wavelet coherence method. *Environmental Modeling and Assessment*, *28*(3), 407–428. DOI: 10.1007/s10666-022-09871-0

Matias, D. M. S. (2020). Climate humanitarian visa: International migration opportunities as post-disaster humanitarian intervention. *Climatic Change*, *160*(1), 143–156. DOI: 10.1007/s10584-020-02691-9

Matz, C. A. (2018). State-sponsored sustainability within the Emirate of Abu Dhabi. *The Palgrave Handbook of Sustainability: Case Studies and Practical Solutions*, 803-833. DOI: 10.1007/978-3-319-71389-2_44

Maulu, S., Hasimuna, O. J., Haambiya, L. H., Monde, C., Musuka, C. G., Makorwa, T. H., Munganga, B. P., Phiri, K. J., & Nsekanabo, J. D. (2021). Climate Change Effects on Aquaculture Production: Sustainability Implications, mitigation, and adaptations. *Frontiers in Sustainable Food Systems*, *5*, 609097. Advance online publication. DOI: 10.3389/fsufs.2021.609097

Mbah, M., & Ezegwu, C. (2024). The decolonisation of climate change and environmental education in Africa. *Sustainability (Basel)*, *16*(9), 3744. DOI: 10.3390/su16093744

Mbiti, J. S. (1969). *African religions and philosophy*. Heinemann.

McAdam, J. (2011). Swimming against the tide: Why a climate change displacement treaty is not the answer. *International Journal of Refugee Law*, *23*(1), 2–27. DOI: 10.1093/ijrl/eeq045

McAdam, J., & Ferris, E. (2015). Planned relocations in the context of climate change: Unpacking the legal and conceptual issues. *Cambridge Journal of International and Comparative Law*, *4*(1), 137–166. DOI: 10.7574/cjicl.04.01.137

Mccall, C. (2013). European Union Cross-Border Cooperation and Conflict Amelioration. *Space and Polity*, *17*(2), 197–216. DOI: 10.1080/13562576.2013.817512

McDougal, M. S. (1960). Some basic theoretical concepts about international law: A policy-oriented framework of inquiry. *The Journal of Conflict Resolution*, *4*(3), 337–354. DOI: 10.1177/002200276000400305

McLean, B., & Barton, J. (2008). U.S.-Canada cooperation: The U.S.-Canada Air Quality Agreement. *Journal of Toxicology and Environmental Health. Part A.*, *71*(9–10), 564–569. DOI: 10.1080/15287390801997567 PMID: 18569627

McMichael, C., Katonivualiku, M., & Powell, T. (2019). Planned relocation and everyday agency in low-lying coastal villages in Fiji. *The Geographical Journal*, *185*(3), 325–337. DOI: 10.1111/geoj.12312

Mees, H. L., Driessen, P. P., Runhaar, H. A., & Stamatelos, J. (2013). Who governs climate adaptation? Getting green roofs for stormwater retention off the ground. *Journal of Environmental Planning and Management*, *56*(6), 802–825. DOI: 10.1080/09640568.2012.706600

Melious, J. (2014). The Canada-US Air Quality Agreement and its impact on air quality management in Canada. In Taylor, E., & McMillan, A. (Eds.), *Air quality management* (pp. 357–376). Springer., DOI: 10.1007/978-94-007-7557-2_16

Mengisteab, K., & Hagg, G. (Eds.). (2017). *Traditional Institutions in Contemporary African Governance* (1st ed.). Routledge., DOI: 10.4324/9781315227948

Menon, S. (2024). Fuzzy cooperative game model for Renewable Energy advancement: A case study of offshore wind farms. In *Research Square*. DOI: 10.21203/rs.3.rs-5720102/v1

Meus, J., den Bergh, K., Delarue, E., & Proost, S. (2019). On international renewable cooperation mechanisms: The impact of national RES-E support schemes. *Energy Economics*, *81*, 859–873. DOI: 10.1016/j.eneco.2019.05.016

Michener, W. K. (2015). Ecological Data Sharing. *Ecological Informatics*, *29*, 33–44. DOI: 10.1016/j.ecoinf.2015.06.010

Mike Campbell (Pvt) Ltd. and Others v. Republic of Zimbabwe. (2010). SADC Tribunal, Case No. SADC (T) 2/2007, Judgment of 28 November 2008.

Milanez, F., Menton, M., & Souza, J. M. R. D. (2022). Epistemological justice: Decoloniality, climate change, and ecological conditions for future generations. *IDS Bulletin*, *53*(4). Advance online publication. DOI: 10.19088/1968-2022.140

Miles, K. (2013). *The origins of international investment law: Empire, environment and the safeguarding of capital*. Cambridge University Press. DOI: 10.1017/CBO9781139600279

Milinski, M., & Marotzke, J. (2022). Economic experiments support Ostrom's polycentric approach to mitigating climate change. *Humanities & Social Sciences Communications*, *9*(1), 442. DOI: 10.1057/s41599-022-01436-6 PMID: 36530541

Misra, A. & Keerthiraj. (2025). Integrating Sustainability in India's Tourism Sector: An Uphill Battle. In S. Poddar, B. Paul, & M. Luperi (Eds.), Sustainable Business Ecosystems and Social Perspectives (pp. 359-386). IGI Global Scientific Publishing. DOI: 10.4018/979-8-3693-8437-4.ch016

Montville, J. V. (1991). Track two diplomacy: The work of healing history. *Whitehead Journal of Diplomacy and International Relations*, *2*(1), 15–25.

Morley, J. W., Selden, R. L., Latour, R. J., Frölicher, T. L., Seagraves, R. J., & Pinsky, M. L. (2018). Projecting shifts in thermal habitat for 686 species on the North American continental shelf. *PLoS One*, *13*(5), e0196127. DOI: 10.1371/journal.pone.0196127 PMID: 29768423

Moss, T., & Newig, J. (2010). Multilevel water governance and problems of scale: Setting the stage for a broader debate. *Environmental Management*, *46*(1), 1–6. DOI: 10.1007/s00267-010-9531-1 PMID: 20640851

Mravcová, A. (2025). *Assessing the effectiveness of international climate agreements in mitigating global warming*. Studia Ecologiae et Bioethicae., DOI: 10.21697/seb.5835

Mulhern, O. (2020, October 8). A brief assessment of the Kyoto Protocol. Earth.Org. https://earth.org/data_visualization/a-brief-assessment-of-the-kyoto-protocol

Müller, C., Cramer, W., Hare, W. L., & Lotze-Campen, H. (2011). Climate change risks for African agriculture. *Proceedings of the National Academy of Sciences of the United States of America*, *108*(11), 4313–4315. DOI: 10.1073/pnas.1015078108 PMID: 21368199

Nagendra, H., & Ostrom, E. (2012). Polycentric governance of multifunctional forested landscapes. *The International Journal of the Commons*, *6*(2), 104. DOI: 10.18352/ijc.321

Nag, T. (2021). Barriers to cross-border energy cooperation and implications on energy security: An Indian perspective with reference to energy trade in south Asia. *Global Business Review*, *22*(6), 1530–1552. DOI: 10.1177/0972150919826380

Najam, A., Huq, S., & Sokona, Y. (2003). Climate negotiations beyond Kyoto: Developing countries' concerns and interests. *Climate Policy*, *3*(3), 221–231. DOI: 10.3763/cpol.2003.0329

Nakahara, Y., Tabata, T., Ohno, T., Furukawa, F., Inokuchi, K., Katagiri, K., & Hirayama, Y. (2019). Discussion on regional revitalization using woody biomass resources as renewable energy. *International Journal of Energy and Environmental Engineering*, *10*(2), 243–256. DOI: 10.1007/s40095-019-0300-5

Narain, D., Teo, H., Lechner, A., Watson, J., & Maron, M. (2021). Biodiversity risks and safeguards of China's hydropower investments in Belt and Road Initiative (BRI) Countries. In *Research Square*. DOI: 10.21203/rs.3.rs-778318/v1

Nasr, A. K., Kashan, M. K., Maleki, A., Jafari, N., & Hashemi, H. (2020). Assessment of barriers to renewable energy development using stakeholders approach. *J. Entrep. Sustain. Issu.*, *7*(3), 2526–2541. DOI: 10.9770/jesi.2020.7.3(71)

National Oceanic and Atmospheric Administration (2024).

Naval, N., & Yusta, J.M. (2024). Technical versus socio-economic and environmental criteria in power transmission projects. Renew. Energy Power Qual. J., 21(2).

Nepal, S. K. (2000). Tourism in protected areas: The Nepalese Himalaya. *Annals of Tourism Research*, *27*(3), 661–681. DOI: 10.1016/S0160-7383(99)00105-X

New Zealand Parliament. (2017). Innovative bill protects Whanganui River with legal personhood. Retrieved from https://www.parliament.nz/en/get-involved/features/innovative-bill-protects-whanganui-river-with-legal-personhood/

Newell, P. (2000). *Climate for change: non-state actors and the global politics of the greenhouse*. Cambridge University Press. DOI: 10.1017/CBO9780511529436

NSIDC. (2019). *National Snow and Ice Data Center "About the data". Images of Thomas Mote*. University of Georgia.

NSIDC. (2022). *National Snow and Ice Data Center "Current conditions and data tools". Images of Thomas Mote*. University of Georgia.

Nüsser, M., & Schmidt, S. (2017). Nanga Parbat revisited: Evolution and challenges of tourism development in Pakistan's mountain areas. *Mountain Research and Development*, *37*(2), 153–162.

Nyirenda, H. (2023). Fiddling at the conference of the parties? A review of GHG emissions and deforestation trends during COPs. *Environment, Development and Sustainability*. Advance online publication. DOI: 10.1007/s10668-023-04277-8

Obergassel, W., Bauer, S., Hermwille, L., Aykut, S. C., Boran, I., Chan, S., Fraude, C., Klein, R. J. T., Mar, K. A., Schroeder, H., & Simeonova, K. (2022). From regime-building to implementation: Harnessing the UN climate conferences to drive climate action. *Wiley Interdisciplinary Reviews: Climate Change*, *13*(6), e797. DOI: 10.1002/wcc.797

Odell, J., & Sell, S. (2006). Reframing the issue: The WTO coalition on intellectual property and public health, 2001. In Odell, J. (Ed.), *Negotiating trade: developing countries in the WTO and NAFTA* (pp. 85–114). Cambridge University Press. DOI: 10.1017/CBO9780511491610.003

Odeyemi, O., Usman, F. O., Mhlongo, N. Z., Elufioye, O. A., & Ike, C. U. (2024). Sustainable entrepreneurship: A review of green business practices and environmental impact. *World Journal of Advanced Research and Reviews*, *21*(2), 346–358. DOI: 10.30574/wjarr.2024.21.2.0461

OECD. (2019). *Enhancing Climate Change Mitigation through Decentralised Government*. OECD Publishing.

OECD. (2021). *Austria's KlimaTicket to promote low-carbon mobility*. OECD International Policy Action Committee. https://www.oecd.org/

OECD. (2022). *Decentralised governance for sustainable development: Reshaping the climate agenda at the local level*. https://www.oecd.org

Ogino, K., Nakayama, M., & Sasaki, D. (2019). Domestic socioeconomic barriers to hydropower trading: Evidence from Bhutan and Nepal. *Sustainability (Basel)*, *11*(7), 2062. DOI: 10.3390/su11072062

OHADA. (2021). Arbitration and business law in Africa: Achievements and challenges. *OHADA Journal*, *18*(2), 45–67.

Olabi, A. G., Abdelkareem, M. A., Mahmoud, M. S., Elsaid, K., Obaideen, K., Rezk, H., Wilberforce, T., Eisa, T., Chae, K., & Sayed, E. T. (2023). Green hydrogen: Pathways, roadmap, and role in achieving sustainable development goals. *Process Safety and Environmental Protection*, *177*, 664–687. DOI: 10.1016/j.psep.2023.06.069

Olabi, V., & Jouhara, H. (2024). An assessment of current hydrogen supply chains in the Gulf Cooperation Council (GCC). *Energy, 131576*, 131576. Advance online publication. DOI: 10.1016/j.energy.2024.131576

Oliver, J., & Peter, J. (2020). *Trends in global CO_2 and total Greenhouse Gas Emissions*. PBL Netherlands Environmental Assessment Agency.

Omenge, P. M., Eshiamwata, G. W., Makindi, S. M., & Obwoyere, G. O. (2019, September). Public participation in environmental impact assessment and its substantive contribution to environmental risk management: Insights from eia practitioners and other stakeholders in Kenya's renewable energy sub-sector. Energy and Sustainability VIII.

Oshilalu, A. Z. (2024). Sustainability meets scalability: Transforming energy infrastructure projects into economic catalysts through supply chain innovation. *International Journal of Research Publication and Reviews*, *5*(12), 762–779. DOI: 10.55248/gengpi.5.1224.3417

Ostrom, E. (1990). *Governing the commons: The evolution of institutions for collective action*. Cambridge University Press., DOI: 10.1017/CBO9780511807763

Ostrom, E. (2010). Polycentric systems for coping with collective action and global environmental change. *Global Environmental Change*, *20*(4), 550–557. DOI: 10.1016/j.gloenvcha.2010.07.004

Ostrom, E. (2017). Polycentric systems for coping with collective action and global environmental change. In *Global justice* (pp. 423–430). Routledge. DOI: 10.4324/9781315254210-18

Ostrom, V., Tiebout, C. M., & Warren, R. (1961). The organization of government in metropolitan areas: A theoretical inquiry. *The American Political Science Review*, *55*(4), 831–842. DOI: 10.2307/1952530

Our World in Data. (2024). Global carbon budget; population-based emissions data. Retrieved from https://ourworldindata.org/co2-and-greenhouse-gas-emissions

Paiement, P., Webster, E., & Anderson, R. (2022). After COP26: Appraising the transnational climate regime. *Transnational Legal Theory*, *13*(2), 157–164. DOI: 10.1080/20414005.2023.2170758

Palacios-Abrantes, J., Sumaila, U. R., & Cheung, W. W. L. (2020). Challenges to transboundary fisheries management in North America under climate change. *Ecology and Society*, *25*(4), art41. DOI: 10.5751/ES-11743-250441

Parsons, M. (2023). Governing with care, reciprocity, and relationality. *Dialogues in Human Geography, 13*(2), 288–292. Advance online publication. DOI: 10.1177/20438206221144819

Parsons, M., Fisher, K., & Nalau, J. (2016). Alternative approaches to co-design: Insights from Indigenous/academic research collaborations. *Current Opinion in Environmental Sustainability, 20*, 99–105. DOI: 10.1016/j.cosust.2016.07.001

Patterson, J., Schulz, K., Vervoort, J., Van Der Hel, S., Widerberg, O., Adler, C., & Barau, A. (2017). *Exploring the governance and politics of transformations towards sustainability. Environmental Innovation and Societal Transitions, 24, 1-16.Piattoni, S. (2010). The theory of multi-level governance: Conceptual, empirical, and normative challenges*. Oxford University Press.

Paul, C., Oliver, K., & Wellstead, A. (2016). To Bridge the Divide between Evidence and Policy: Reduce Ambiguity as Much as Uncertainty. *Public Administration Review, 76*(3), 399–402. DOI: 10.1111/puar.12555

Perissi, I., & Jones, A. (2024). An emissions offset strategy to accomplish 2 °C long-term mitigation goals in the European Union. *Sustainability (Basel), 16*(11), 4783. DOI: 10.3390/su16114783

Permanent Court of Arbitration. (2001). *Optional rules for arbitration of disputes relating to natural resources and the environment*. https://pca-cpa.org

Pham, L., Palaneeswaran, E., & Stewart, R. (2018). Knowing maintenance vulnerabilities to enhance building resilience. *Procedia Engineering, 212*, 1273–1278. DOI: 10.1016/j.proeng.2018.01.164

Pierre, J. (2011). *The politics of urban governance*. Bloomsbury Publishing. DOI: 10.1007/978-1-137-28555-3

Pitzl, J., & Stewart, E. (2000). *The US position on the Kyoto Protocol: Senate ratification or not?* Georgetown University., Available from http://faculty.guisd.org/product.cfm?ProductID=193&MenuCategory=2

Poque González, A. B., Masip Macia, Y., Ferreira, L. D. C., & Valdes, J. (2023). Socio-ecological controversies from Chilean and Brazilian sustainable energy transitions. *Sustainability, 15*(3), 1861.

Porter, G.. (2000). *Global environmental politics*. Westview Press.

Porter, J. (2010). A Brief History of Data Sharing in the U.S. Long Term Ecological Research Network. *Bulletin of the Ecological Society of America, 91*(1), 14–20. DOI: 10.1890/0012-9623-91.1.14

Pradhan, S. (2010). India and the Gulf Cooperation Council (GCC): An economic and political perspective. *Strategic Analysis*, *34*(1), 93–103. DOI: 10.1080/09700160903351340

Prasad, S., Yadav, K. K., Kumar, S., Pandita, P., Bhutto, J. K., Alreshidi, M. A., Ravindran, B., Yaseen, Z. M., Osman, S. M., & Cabral-Pinto, M. M. (2024). Review on biofuel production: Sustainable development scenario, environment, and climate change perspectives– A sustainable approach. *Journal of Environmental Chemical Engineering*, *12*(2), 111996. DOI: 10.1016/j.jece.2024.111996

Prokkola, E.-K. (2019). Border-regional resilience in EU internal and external border areas in Finland. *European Planning Studies*, *27*(8), 1587–1606. DOI: 10.1080/09654313.2019.1595531

Purdon, M. (2020). The politics of evidence in climate governance. *Sustainability Science*, *15*, 701–717. DOI: 10.1007/s11625-020-00793-1

Putnam, R. (1988). Diplomacy and domestic politics: The logic of two-level games. *International Organization*, *42*(3), 427–460. DOI: 10.1017/S0020818300027697

Pu, Y., Li, Y., & Wang, Y. (2021a). Structure characteristics and influencing factors of cross-border electricity trade: A complex network perspective. *Sustainability (Basel)*, *13*(11), 5797. DOI: 10.3390/su13115797

Quamar, M. M., Hameed, S., & Kumaraswamy, P. R. (2023). United Arab Emirates. In *Persian Gulf 2023: India's Relations with the Region* (pp. 335-382). Singapore: Springer Nature Singapore. https://doi.org/DOI: 10.1007/978-981-99-6380-5_9

Rahayu, H. P., Zulfa, K. I., Nurhasanah, D., Haigh, R., Amaratunga, D., & Wahdiny, I. I. (2024). Unveiling transboundary challenges in river flood risk management: Learning from the Ciliwung River basin. *Natural Hazards and Earth System Sciences*, *24*(6), 2045–2064. DOI: 10.5194/nhess-24-2045-2024

Raina, S., Shandilya, S., Kothari, R., & Pathania, D. (2024). *Green Hydrogen: International Collaborators, Agreements, Opportunities, and Challenges, and Their Impact on India's Research and Development Sector. Green Hydrogen Economy for Environmental Sustainability* (Vol. 2). Applications, Challenges, and Policies., DOI: 10.1021/bk-2024-1474.ch014

Rajshree, M., Mehta, C., & Luhach, S. (2025). Displaced by climate change: Navigating legal framework for 'climate refugees' in international environmental law. *India Quarterly*, *81*(1), 9–24. DOI: 10.1177/09749284241307928

Ramakrishnan, T., Mohan Gift, M. D., Chitradevi, S., Jegan, R., & Subha Hency Jose, P. Nagaraja, Rajneesh Sharma H.N., Selvakumar P., Sintayehu Mekuria Hailegiorgis, (2022). "Study of Numerous Resins Used in Polymer Matrix Composite Materials", Advances in Materials Science and Engineering, vol. 2022, Article ID 1088926, 8 pages, 2022. https://doi.org/DOI: 10.1155/2022/1088926

Ramesh, S. (2025). Renewable Energy Technologies. In *The Political Economy of Contemporary Human Civilisation, Volume II: From Quantum Computing and Nuclear Fusion to War and Conflict* (pp. 183-218). Cham: Springer Nature Switzerland. DOI: 10.1007/978-3-031-84185-9_6

Ran, Y., Li, X., Cheng, G., Zhang, T., Wu, Q., Jin, H., & Jin, R. (2012). Distribution of permafrost in China: An overview of existing permafrost maps. *Permafrost and Periglacial Processes, 23*(4), 322–333. DOI: 10.1002/ppp.1756

Ran, Y., Xin, L., Guodong, C., Tingjun, Z., Qingbai, W., Huijun, J., Rui, J., Romanovsky, V. E., Smith, S. L., & Christiansen, H. H. (2010). Permafrost thermal state in the Polar Northern Hemisphere during the international polar year 2007–2009: A synthesis. *Permafrost and Periglacial Processes, 21*(2), 106–116. DOI: 10.1002/ppp.689

Rao, U., & Saad, M. (2025). Economic and financial aspects of renewable energy environment. In *Renewable Energy Projects and Investments* (pp. 243–258). Elsevier., DOI: 10.1016/B978-0-443-29869-1.00013-1

Rao, X., Ma, R., Zhang, L., & Liu, J. (2022). Research and design of a digital twin-driven smart river basin platform. *Metaverse, 3*(2), 11. DOI: 10.54517/m.v3i2.1815

Rasmussen, L. H., Zhang, W., Hollesen, J., Cable, S., Christiansen, H. H., Jansson, P.-E., & Bo, E. (2018). Modelling present and future permafrost thermal regimes in Northeast Greenland. *Cold Regions Science and Technology, 146*, 199–213. DOI: 10.1016/j.coldregions.2017.10.011

Readman, P., Radding, C., & Bryant, C. (2014). Introduction: Borderlands in a Global Perspective. In Readman, P., Radding, C., & Bryant, C. (Eds.), *Borderlands in World History, 1700–1914* (pp. 1–23)., DOI: 10.1057/9781137320582_1

Rengs, B., Scholz-Wäckerle, M., & van den Bergh, J. C. (2020). Evolutionary macroeconomic assessment of employment and innovation impacts of climate policy packages. *Journal of Economic Behavior & Organization, 169*, 332–368. DOI: 10.1016/j.jebo.2019.11.025

Rentschler, J., & Leonova, N. (2023). Global air pollution exposure and poverty. *Nature Communications*, *14*(1), 4432. DOI: 10.1038/s41467-023-39797-4 PMID: 37481598

Reymond, M., Ossewaarde, R., & Peeters, M. (2020). Environmental justice and REDD+: Lessons from indigenous resistance. *Environmental Science & Policy*, *112*, 319–326. DOI: 10.1016/j.envsci.2020.07.001

Reynolds, P. D. (2010). New Firm Creation: A Global Assessment of National, Contextual, and Individual Factors. *Foundations and Trends® in Entrepreneurship*, *6*(5–6), 315–496. DOI: 10.1561/0300000034

Reynolds, S. (2024). Stakeholder engagement and its impact on supply chain sustainability in the context of renewable energy. In *Preprints*.

Riaz, K., McAfee, M., Anton, I., & Gharbia, S. (2022). Conceptualising the management of climate extreme events through the GIS-based digital twin system.

Rickards, L., Ison, R., Fünfgeld, H., & Wiseman, J. (2014). Opening and closing the future: Climate change, adaptation, and scenario planning. *Environment and Planning. C, Government & Policy*, *32*(4), 587–602. DOI: 10.1068/c3204ed

Risse-Kappen, T. (1995). Bringing transnational relations back in: Introduction. In Risse-Kappen, T. (Ed.), *Bringing transnational relations back in: non-state actors, domestic structures and international institutions* (pp. 3–34). Cambridge University Press. DOI: 10.1017/CBO9780511598760.002

Ritchie, H., Rosado, P., & Roser, M. (2023). CO_2 and greenhouse gas emissions. Our World in Data. https://ourworldindata.org/co2-and-greenhouse-gas-emissions

Roberts, D., & O'Donoghue, S. (2013). Urban environmental challenges and climate change action in Durban, South Africa. *Environment and Urbanization*, *25*(2), 299–319. DOI: 10.1177/0956247813500904

Rockhopper Exploration Plc v. Italian Republic, ICSID Case No. ARB/17/14, Award (2022). *Chevron Corporation and Texaco Petroleum Corporation v. Republic of Ecuador*, PCA Case No. 2009-23, Final Award (2018). *Pacific Rim Cayman LLC v. Republic of El Salvador*, ICSID Case No. ARB/09/12, Award (2016). *Methanex Corporation v. United States of America*, UNCITRAL, Final Award (2005). *Glamis Gold Ltd. v. United States of America*, UNCITRAL, Award (2009).

Roldan-Fernandez, J. M., Gómez-Quiles, C., Merre, A., Burgos-Payán, M., & Riquelme-Santos, J. M. (2018). Cross-border energy exchange and renewable premiums: The case of the Iberian system. *Energies*, *11*(12), 3277. DOI: 10.3390/en11123277

Romero-Lankao, P., Gnatz, D. M., Wilhelmi, O., & Hayden, M. (2016). Urban sustainability and resilience: From theory to practice. *Sustainability (Basel), 8*(12), 1224. DOI: 10.3390/su8121224

Rosenzweig, B. R., McPhillips, L., Chang, H., Cheng, C., Welty, C., Matsler, M., Iwaniec, D., & Davidson, C. I. (2018). Pluvial flood risk and opportunities for resilience. *WIREs. Water, 5*(6), e1302. DOI: 10.1002/wat2.1302

Rountree, V., & Baldwin, E. (2018). State-level renewable energy policy implementation: How and why do stakeholders participate? *Frontiers in Communication, 3*, 3. DOI: 10.3389/fcomm.2018.00006

Saber, A. (2024). *Sovereign wealth funds in a post-pandemic landscape: a uk perspective* (Doctoral dissertation, University of Sussex). https://www.researchgate.net/profile/Ali-Saber-17/publication/384635695_Sovereign_wealth_funds_in_a_post-pandemic_landscape_a_UK_perspective/links/67006342869f1104c6c65638/Sovereign-wealth-funds-in-a-post-pandemic-landscape-a-UK-perspective.pdf

Sachseder, J., Stachowitsch, S., & Standke-Erdmann, M. (2024). Entangled Vulnerabilities: Gendered and Racialised Bodies and Borders in EU External Border Security. *Geopolitics*, •••, 1–29. DOI: 10.1080/14650045.2023.2291060 PMID: 39318502

Salacuse, J. W. (2010). *The law of investment treaties*. Oxford University Press. DOI: 10.1093/law/9780199206056.001.0001

Sambasivam, B., & Sarma, R. N. (2024). India's National Green Hydrogen Mission: An analysis of the strategies, policies for net-zero emissions and sustainability. *Environmental Research: Energy, 1*(4), 045015. DOI: 10.1088/2753-3751/ad8fb2

Sattar, U. (2022). A conceptual framework of climate action needs of the least developed party countries of the Paris agreement. *International Journal of Environmental Research and Public Health, 19*(16), 9941. DOI: 10.3390/ijerph19169941 PMID: 36011575

Savaresi, A., & Setzer, J. (2020). Mapping the whole of climate litigation: A proposed taxonomy of climate cases. *Climate Law, 10*(1), 91–120. DOI: 10.1163/18786561-01001006

Saxena, K., & Singh, M. (2024). Investors reward countries for participating in climate agreements. *Nature Communications, 15*(1), 8908. Advance online publication. DOI: 10.1038/s41467-024-53012-y PMID: 39414786

Schattschneider, E. E. (1960). *The semisovereign people: a realist's view of democracy in America*. Holt, Rinehart and Winston.

Scheyvens, R. (1999). Ecotourism and the empowerment of local communities. *Tourism Management*, *20*(2), 245–249. DOI: 10.1016/S0261-5177(98)00069-7

Scheyvens, R. (2000). Promoting women's empowerment through involvement in ecotourism: Experiences from the Third World. *Journal of Sustainable Tourism*, *8*(3), 232–249. DOI: 10.1080/09669580008667360

Schmeier, S. (2013). *Governing International Watercourses: River Basin Organizations and the Sustainable Governance of Internationally Shared Rivers and Lakes*. Routledge.

Schmidt, R. C., Drupp, M., Nesje, F., & Hoegen, H. (2022). Testing the free-rider hypothesis in climate policy.

School of Media, Communication Universities of Douala, C., Leicester, U. K., & Nyoh, I. (2021). Communicating climate change and energy in rural Africa: A case analysis to explain how participatory communication can support transition to renewables and adoption of solar technologies in rural Africa. Ciências e Pol\'\ iticas Públicas / Public Sciences &, 183–200.

Schoon, M., York, A., Sullivan, A., & Baggio, J. (2017). The emergence of an environmental governance network: The case of the Arizona borderlands. *Regional Environmental Change*, *17*(3), 677–689. DOI: 10.1007/s10113-016-1060-x

Schröter, K., Pia-Johanna Schweizer, Gräler, B., Cumiskey, L., Bharwani, S., Parviainen, J., Kropf, C., Hakansson Viktor Wattin and Drews, M., Irvine, T., Dondi, C., Apel, H., Löhrlein, J., Hochrainer-Stigler, S., Bagli, S., Huszti, L., Genillard, C., Unguendoli, S., & Steinhausen, M. (2024). Invited perspectives: Fostering interoperability of data, models, communication and governance for disaster resilience through transdisciplinary knowledge co-production.

Sebbagh, T., Şahin, M. E., & Beldjaatit, C. (2024). Green hydrogen revolution for a sustainable energy future. *Clean Technologies and Environmental Policy*, *26*(12), 4017–4040. DOI: 10.1007/s10098-024-02995-9

Secretariat, I. C. S. I. D. (2021). *Diversity in arbitration appointments: 2021 report*. International Centre for Settlement of Investment Disputes. https://icsid.worldbank.org

Seidel, D., Wurster, S., Jenks, J. D., Sati, H., Gangneux, J., Egger, M., Alastruey-Izquierdo, A., Ford, N. P., Chowdhary, A., Sprute, R., Cornely, O., Thompson, G. R.III, Hoenigl, M., & Kontoyiannis, D. P. (2024). Impact of climate change and natural disasters on fungal infections. *The Lancet Microbe*, *5*(6), e594–e605. DOI: 10.1016/S2666-5247(24)00039-9 PMID: 38518791

Selleslaghs, J. (2017, September). EU environmental cooperation with Latin America: A critical assessment. In *Paper presented at the 11th Pan-European conference on International Relations* (Vol. 13, p. 17).

Sell, S. (2003). *Private power, public law: the globalization of intellectual property rights*. Cambridge University Press. DOI: 10.1017/CBO9780511491665

Selvakumar, P., Seenivasan, S., Patel, S., Das, A., & Ravikumar, S. (2025). Environmental Sustainability. In *Renewable Energy and the Economic Welfare of Society* (pp. 103–132). IGI Global Scientific Publishing., DOI: 10.4018/979-8-3693-7580-8.ch005

Setzer, J., & Higham, C. (2021). Global trends in climate change litigation: 2021 snapshot. *Grantham Research Institute on Climate Change and the Environment*. https://www.lse.ac.uk/granthaminstitute/publication/global-trends-in-climate-change-litigation-2021/

Sewell, G. (1996). Conflicting beliefs: National implementation of the United Nations framework convention on climate change. *Environmental Impact Assessment Review*, *16*(3), 137–150. DOI: 10.1016/0195-9255(96)00031-5

Shah, G. M., Sapkota, K., & Spakota, M. (2025). *A Comprehensive Review of Theories, Models and Approaches Governing Climate Change Discourse and Global Climate Finance*. Models and Approaches Governing Climate Change Discourse and Global Climate Finance. DOI: 10.2139/ssrn.5287468

Shahnawaz, M., & Mir, R. Y. (2025). Strategic Convergence in the Arabian Gulf: Evaluating India-UAE Partnership Post 2014. [IJSSL]. *Indian Journal of Social Science and Literature*, *4*(4), 19–24. DOI: 10.54105/ijssl.C1166.04040625

Shahzad, S., Faheem, M., Muqeet, H. A., & Waseem, M. (2024). Charting the UK's path to net zero emissions by 2050: Challenges, strategies, and future directions. *IET Smart Grid*, *7*(6), 716–736. DOI: 10.1049/stg2.12185

Sharma, D., Singh, R., & Singh, R. K. (2020). Regional climate governance in South Asia: An assessment of cooperative initiatives. *Climate and Development*, *12*(1), 30–42. DOI: 10.1080/17565529.2019.1603734

Shee Weng, L. (2025). *The EU's Carbon Border Adjustment Mechanism (CBAM): Balancing Climate Action and Global Trade. The EU's Carbon Border Adjustment Mechanism (CBAM)*. Balancing Climate Action and Global Trade.

Shende, A. D., Tekade, S. A., Deshmukh, A. A., Tembhurkar, S. P., & Selvakumar, P. (2025). Water Quality Monitoring and Control With AI. In Ruiz-Vanoye, J., & Díaz-Parra, O. (Eds.), *Smart Water Technology for Sustainable Management in Modern Cities* (pp. 51–80). IGI Global Scientific Publishing., DOI: 10.4018/979-8-3693-8074-1.ch003

Shibata, Y., Nian, V., Bhandari, A., & Roychoudhury, J. (2024). Using hydrogen for decarbonization, industrial development, and energy security: Shared ambitions for Japan, ASEAN Member States, and India. In *The Clean Hydrogen Economy and Saudi Arabia* (pp. 329–373). Routledge., https://www.taylorfrancis.com/chapters/oa-edit/10.4324/9781003294290-14/using-hydrogen-decarbonization-industrial-development-energy-security-yoshiaki-shibata-victor-nian-amit-bhandari-jitendra-roychoudhury DOI: 10.4324/9781003294290-14

Shrestha, A., Souza, F. A. A., Park, S., Cherry, C., Garcia, M., Yu, D. J., & Mendiondo, E. M. (2022). Socio-hydrological modeling of the tradeoff between flood control and hydropower provided by the Columbia River Treaty. *Hydrology and Earth System Sciences*, *26*(19), 4893–4917. DOI: 10.5194/hess-26-4893-2022

Siemens. (2023). *Readiness for predictive maintenance at scale report*. Siemens AG. https://www.siemens.com/

Sim, L. C. (2024). The Arab Gulf states in the Asian energy market: Is the Russia-Ukraine war a game changer? *Policy Studies*, *45*(3-4), 633–652. DOI: 10.1080/01442872.2023.2279986

Simpson, N. P., Mach, K. J., Constable, A., Hess, J., Hogarth, R., Howden, M., Lawrence, J., Lempert, R. J., Muccione, V., Mackey, B., New, M. G., O'Neill, B., Otto, F., Pörtner, H., Reisinger, A., Roberts, D., Schmidt, D. N., Seneviratne, S., Strongin, S., & Trisos, C. H. (2021). A framework for complex climate change risk assessment. *One Earth*, *4*(4), 489–501. DOI: 10.1016/j.oneear.2021.03.005

Singh, A., Jamasb, T., Nepal, R., & Toman, M. (2018). Electricity cooperation in South Asia: Barriers to cross-border trade. *Energy Policy*, *120*, 741–748. DOI: 10.1016/j.enpol.2017.12.048

Singh, B. K., Delgado-Baquerizo, M., Egidi, E., Guirado, E., Leach, J. E., Liu, H., & Trivedi, P. (2023). Climate change impacts on plant pathogens, food security and paths forward. *Nature Reviews. Microbiology*, *21*(10), 640–656. DOI: 10.1038/s41579-023-00900-7 PMID: 37131070

Sister Cities International. (2023). *Sister City Partnerships for Climate Action*. Retrieved from https://sistercities.org

Siyech, M. S. (2020). Understanding India's increased counter terrorism relations with Saudi Arabia and the UAE. *India Review, 19*(4), 351–375. DOI: 10.1080/14736489.2020.1797316

Skendžić, S., Zovko, M., Živković, I. P., Lešić, V., & Lemić, D. (2021). The impact of climate change on agricultural insect pests. *Insects, 12*(5), 440. DOI: 10.3390/insects12050440 PMID: 34066138

Smith, D., & Fischbacher, M. (2009). The changing nature of risk and risk management: The challenge of borders, uncertainty and resilience. *Risk Management, 11*(1), 1–12. DOI: 10.1057/rm.2009.1

Smolak, A. (2010). Contextual factors influencing HIV risk behaviour in Central Asia. *Culture, Health & Sexuality, 12*(5), 515–527. DOI: 10.1080/13691051003658135 PMID: 20301020

Sneddon, C., & Fox, C. (2006). Rethinking transboundary waters: A critical hydropolitics of the Mekong basin. *Political Geography, 25*(2), 181–202. DOI: 10.1016/j.polgeo.2005.11.002

Soliev, I., & Theesfeld, I. (2020). Benefit sharing for solving transboundary commons dilemma in central Asia. *The International Journal of the Commons, 14*(1), 61–77. DOI: 10.5334/ijc.955

Sotelo Navalpotro, J. A. (2000). *Regional Development Models*. Oxford University Press.

Sotelo Pérez, I., & Sotelo Pérez, M. (2023). "Urban Planning and the Third Landscape in the Field of Local Development: Case Study", in AA.VV. In Global, I. G. I. (Ed.), *Geoinformatics in Support of Urban Politics and the Development of Civil Engineering*. DOI: 10.4018/978-1-6684-6449-6.ch001

Sotelo Pérez, M., & Sotelo Pérez, I. (2017). Una aproximación al marco teórico de los riesgos. *Observatorio Medioambiental, 20*, 9–36. DOI: 10.5209/OBMD.57944

Spenceley, A., & Goodwin, H. (2007). Nature-based tourism and poverty alleviation: Impacts of private sector and parastatal enterprises in and around Kruger National Park, South Africa. *Current Issues in Tourism, 10*(2–3), 255–277. DOI: 10.2167/cit305.0

Spicer, M. (2011). Some reflections on rationality, argument, and conflict. *Administrative Theory & Praxis, 33*(3), 473–477. DOI: 10.2753/ATP1084-1806330314

Spijkers, O. (2025, April 4). The status of the right to a clean, healthy and sustainable environment under customary international law. *Verfassungsblog*. https://verfassungsblog.de/ip-hr2he-customary-international-law/ https://doi.org/10.59704/b4113d1f4013cea9

Spiteri, J., & von Brockdorff, P. (2021). Transboundary air pollution and respiratory disease mortality: Evidence from European countries. *Journal of Economic Studies (Glasgow, Scotland)*, *48*(7), 1371–1387. DOI: 10.1108/JES-05-2020-0210

Sreekanth, K. J. (2024). Energy Policy Analyses, Energy Transition and Sustainability. In *Energy Efficiency Improvements with Emission Abatement for Energy Sustainability: Energy Statistics* (pp. 135-167). Singapore: Springer Nature Singapore. DOI: 10.1007/978-981-97-9480-5_8

Sreenivasulu, A., & Manimozhi, T. Satheesh kumar P., Selvakumar P. and Singh P.,(2022). "The Model of Smart Sensing Device For Sensitive Nanoclusters Modification in Sensing Properties," 2022 2nd International Conference on Advance Computing and Innovative Technologies in Engineering (ICACITE), Greater Noida, India, 2022, pp. 1043-1047, DOI: 10.1109/ICACITE53722.2022.9823846

Stanford Encyclopedia of Philosophy Stanford University. (2020). Models in Science. Available online: https://seop.illc.uva.nl/entries/models-science/

Statistics Austria. (2023). *Energy poverty in Austria: Annual report*. https://www.statistik.at/

Stern, N., & Lankes, H. P. (2022). *Collaborating and delivering on climate action through a climate club*. Grantham Research Institute on Climate Change and the Environment, London School of Economics & Political Science.

Stevens, R., Poterlot, C., Trieu, N., Rodriguez, H. A., & Hayes, P. L. (2024). Transboundary transport of air pollution in eastern Canada. *Environmental Science. Advances*, *3*(3), 448–469. DOI: 10.1039/D3VA00307H

Stojanov, R., Rosengaertner, S., de Sherbinin, A., & Nawrotzki, R. (2021). Climate mobility and development cooperation. *Population and Environment*, *43*(2), 209–231. DOI: 10.1007/s11111-021-00387-5 PMID: 34305224

Stroetmann, K. A. (2014). Health system efficiency and eHealth interoperability – how much interoperability do we need? In *Advances in Intelligent Systems and Computing* (pp. 395–406). Springer International Publishing.

Stronza, A., Hunt, C. A., & Fitzgerald, L. A. (2019). Ecotourism for conservation? *Annual Review of Environment and Resources*, *44*(1), 229–253. DOI: 10.1146/annurev-environ-101718-033046

Subraelu, P., Sefelnasr, A., Yagoub, M. M., Sherif, M., Ebraheem, A. A., Raj Sekhar, A., & Nageswara Rao, K. (2022). Global warming climate change and sea level rise: Impact on land use land cover features along uae coast through remote sensing and gis. *Journal of Ecosystem & Ecography*, *12*, 1–8. DOI: 10.4172/2157-7625.1000329

Sugumaran, K. M. D., Pandiyarajan V., Y. B S, Selvakumar P. and M. S, (2024). "Analysis of Electromyography Signals for Control of Mechanical Prosthesis using Machine Learning Techniques," 2024 2nd International Conference on Intelligent Data Communication Technologies and Internet of Things (IDCIoT), Bengaluru, India, 2024, pp. 1059-1067, DOI: 10.1109/IDCIoT59759.2024.10467707

Sumaila, U. R., Palacios-Abrantes, J., & Cheung, W. W. L. (2020). Climate change, shifting threat points, and the management of transboundary fish stocks. *Ecology and Society*, *25*(4), art40. DOI: 10.5751/ES-11660-250440

Suri, N., Ghosh, N., Taneja, K., Patil, S., & Mookherjee, P. (2024). India-Middle East-Europe Economic Corridor: Towards a New Discourse in Global Connectivity. *Observer Research Foundation, April, 9*. https://www.orfonline.org/research/india-middle-east-europe-economic-corridor-towards-a-new-discourse-in-global-connectivity?trk=public_post_comment-text

Sustainability, S. (2024). Global carbon emissions from fossil fuels reached a record high in 2023. https://sustainability.stanford.edu/news/global-carbon-emissions-fossil-fuels-reached-record-high-2023

Swagatika Priyadarshini, L. A., Kaushik, V., & Kataria, R. (2025). The Role of Anthropogenic Activities, Energy Demand, and Global Climate Change. In *The Intersection of Global Energy Politics and Climate Change: A Comprehensive Analysis of Energy Markets and Economics* (pp. 35–61). Springer Nature Singapore., DOI: 10.1007/978-981-96-0535-4_2

Tallarita, R. (2021). Portfolio primacy and climate change. *SSRN*. DOI: 10.2139/ssrn.3912977

Tallberg, J., & Jönsson, C. (2010). Transnational actor participation in international institutions: where, why, and with what consequences? In Jönsson, C., & Tallberg, J. (Eds.), *Transnational actors in global governance: patterns, explanations, and implications* (pp. 1–21). Palgrave Macmillan. DOI: 10.1057/9780230283220_1

Talukder, M. B. (2024). Implementing Artificial Intelligence and Virtual Experiences in Hospitality. In Innovative Technologies for Increasing Service Productivity (pp. 145-160). IGI Global. DOI: 10.4018/979-8-3693-2019-8.ch009

Talukder, M. B. (2025). Revolutionizing Hospitality: Leveraging Mobile Apps and Cutting-Edge Room Technologies in the Hotel Industry. In B. Varghese & S. H. (Eds.), Advancing Smart Tourism Through Analytics (pp. 315-332). IGI Global Scientific Publishing. DOI: 10.4018/979-8-3693-3715-8.ch017

Talukder, M. B., Hoque, M., & Kumar, S. (2024). Impact of Cybersecurity in the Aviation, Tourism, and Hospitality Industries. In *Corporate Cybersecurity in the Aviation, Tourism, and Hospitality Sector* (pp. 200–215). IGI Global Scientific Publishing. DOI: 10.4018/979-8-3693-2715-9.ch011

Talukder, M. B., Khan, M. R., Kumar, S., & Singh, K. (2025). Blockchain and Tourism: Enhancing Community Participation and Sustainability. In *Open Innovation and Technology in Tourism and Hospitality* (pp. 279–302). IGI Global Scientific Publishing. DOI: 10.4018/979-8-3693-8633-0.ch015

Talukder, M. B. (2025). Climate change adaptation strategies for a tourism destination: A study of Kuakata beach. *Acta Scientiarum Polonorum. Administratio Locorum*, *24*(1), 165–179. DOI: 10.31648/aspal.10073

Tanner, T., Mitchell, T., Polack, E., & Guenther, B. (2010). Urban governance for adaptation: Assessing climate change resilience in ten Asian cities. *IDS Working Paper*, *2010*(315), 1–47. DOI: 10.1111/j.2040-0209.2009.00315_2.x

Tanopir, C., Allard, S., Gouglass, K., Aydinoglu, A. U., Wu, L., Read, E., Manoff, M., & Frame, M. (2011). Data Sharing by Scientists: Practices and Perceptions. *PLoS One*, *6*(6), 6. DOI: 10.1371/journal.pone.0021101 PMID: 21738610

Tavares, R. (2016). *Paradiplomacy: Cities and States as Global Players*. Oxford University Press.

Taylor, C. (2013). The discourses of climate change. In Governing Through Globalised Crime (pp. 17–31). DOI: 10.1057/9781137006127_2

Thakur, S. (2021). From Kyoto to Paris and beyond: The emerging politics of climate change. *India Quarterly*, *77*(3), 366–383. DOI: 10.1177/09749284211027252

The Intergovernmental Panel on Climate Change (IPCC). (2019). United Nations Intergovernmental Panel on Climate Change's Special Report on the Ocean and Cryosphere. Available online: https://www.ipcc.ch/srocc/

Thompson, A. (2025). From rivals to partners: The cooptation of emerging powers into the climate regime. *Contemporary Security Policy*, *46*(1), 126–149. DOI: 10.1080/13523260.2024.2420477

Thomson Higher Education & Maggie Nelson (2004). Images obtained from *Cazatormentas.com*

Thornton, D. H., & Branch, L. C. (2019). Transboundary mammals in the Americas: Asymmetries in protection challenge climate change resilience. *Diversity & Distributions*, *25*(4), 674–683. DOI: 10.1111/ddi.12880

Thornton, F. (2022). The future is now: Climate displacement and human rights obligations—A note on recent developments in the UN Human Rights Committee. *Frontiers in Climate*, *4*, 1061474. Advance online publication. DOI: 10.3389/fclim.2022.1061474

Thuiller, W., Lavorel, S., Araújo, M. B., Sykes, M. T., & Prentice, I. C. (2005). Climate change threats to plant diversity in Europe. *Proceedings of the National Academy of Sciences of the United States of America*, *102*(23), 8245–8250. DOI: 10.1073/pnas.0409902102 PMID: 15919825

Tibni, H. (2024). Assessing the Effectiveness of the 10 United Nations Global and Multilateral Climate Treaties. *European Journal of Development Studies*, *4*(4), 41–55. DOI: 10.24018/ejdevelop.2024.4.4.366

Timothy, D. J. (2001). *Tourism and political boundaries*. Routledge., DOI: 10.4324/9780203214480

TransCanada Corporation v. United States of America, NAFTA/ICSID, Notice of Discontinuance (2017). *Lone Pine Resources Inc. v. Canada*, ICSID Case No. UNCT/15/2, Pending (as of 2024). Baxi, U. (2002). *The future of human rights*. Oxford University Press.

Trogal, K., Bauman, I., Lawrence, R., & Petrescu, D. (Eds.). (2018). *Architecture and Resilience: Interdisciplinary Dialogues* (1st ed.). Routledge., DOI: 10.4324/9781315159478

Turchetti, S., & Roberts, P. (2014). *The Surveillance Imperative: Geosciences during the Cold War and Beyond*. Palgrave. DOI: 10.1057/9781137438744

Tyson, L. D., & Weiss, D. (2025). Climate Finance: Progress, Challenges, and Opportunities on the Path to a Sustainable Planet. *California Management Review*, *67*(2), 5–39. DOI: 10.1177/00081256241296902

UCLG. (2022). *Climate and Environmental Action by Local Governments*. United Cities and Local Governments. Retrieved from https://www.uclg.org

UNCCD. (2020). *Global land outlook: Thematic report on drylands*. United Nations Convention to Combat Desertification. https://www.unccd.int

UNCITRAL. (2022). *Transparency in treaty-based investor-State arbitration: Practice guide*. United Nations. https://uncitral.un.org

Under2 Coalition. (2022). *Annual Report*. The Climate Group. Retrieved from https://www.under2coalition.org

Underwood, A., Marcantonio, R., Wood, D., & Crippa, P. (2024). Interstate air pollution governance in the United States: Exploring Clean Air Act Section 126. *Environmental Management*, *74*(3), 401–413. DOI: 10.1007/s00267-024-02002-3 PMID: 38896241

UNDP. (2022). *Strengthening local governance for climate resilience*. https://www.undp.org

UNECE. (2020). *Green Belt of Fennoscandia: Cross-border Biodiversity Cooperation*. Retrieved from https://unece.org

UNEP. (2024). Emissions Gap Report 2024: No more hot air … please! United Nations Environment Programme. https://www.unep.org/resources/emissions-gap-report-2024

United Nations Conference on Trade and Development. (2020). *World investment report 2020: International production beyond the pandemic*. United Nations.

United Nations Environment Programme. (2022). *Legal readiness for climate dispute resolution*. United Nations Environment Programme.

United Nations Framework Convention on Climate Change. (2015). *Paris Agreement* (adopted 12 December 2015, entered into force 4 November 2016). https://unfccc.int

United Nations. (2019). *Climate action and support trends 2019*. United Nations. https://unfccc.int

United States Senate. (1997). Senate resolution 98. 105th Congress. 1st Session. Available from: http://thomas.loc.gov/cgi-bin/bdquery/z?d105:S.RES.98: [Accessed 5 December 2013].

UNSDGs 17 (https://sdgs.un.org/goals)

UNWTO. (2017). *Tourism for Sustainable Development in Least Developed Countries*. United Nations World Tourism Organization., DOI: 10.18111/9789284418848

Ura, K., Alkire, S., Zangmo, T., & Wangdi, K. (2012). *A short guide to Gross National Happiness Index*. Centre for Bhutan Studies. DOI: 10.35648/20.500.12413/11781/ii025

Utelbayev, B. T., Azhigali, A. S., & Amangeldin, A. G. (2024). Environmental sustainability in cross-border energy trade: The case of the Kazakhstan – China pipeline. *Jour, 21*(1), 140–148. DOI: 10.55452/1998-6688-2024-21-1-140-148

Van Eerd, M. C. J., Wiering, M. A., & Dieperink, C. (2014). Exploring the Prospects for Cross-Border Climate Change Adaptation between North Rhine-Westphalia and the Netherlands. *Utrecht Law Review, 10*(2), 91. DOI: 10.18352/ulr.271

Van Harten, G. (2007). *Investment treaty arbitration and public law*. Oxford University Press.

Vanhala, L., & Calliari, E. (2022). Governing people on the move in a warming world: Framing climate change migration and the UNFCCC Task Force on Displacement. *Global Environmental Change, 76*, 102578. DOI: 10.1016/j.gloenvcha.2022.102578

Vattenfall AB and others v. Federal Republic of Germany, ICSID Case No. ARB/12/12, Decision on Jurisdiction (2013).

Verma, P., Sharma, B., & Mehta, P. (2024). Green Hydrogen Production: Impacts on Environmental Sustainability Scale. In *Green Hydrogen Economy for Environmental Sustainability. Volume 1: Fundamentals and Feedstocks* (pp. 29-48). American Chemical Society. DOI: DOI: 10.1021/bk-2024-1473.ch002

Vicedo-Cabrera, A. M., Scovronick, N., Sera, F., Royé, D., Schneider, R., Tobias, A., Astrom, C., Guo, Y., Honda, Y., Hondula, D. M., Abrutzky, R., Tong, S., De Sousa Zanotti Stagliorio Coelho, M., Saldiva, P. H. N., Lavigne, E., Correa, P. M., Ortega, N. V., Kan, H., Osorio, S., & Gasparrini, A. (2021). The burden of heat-related mortality attributable to recent human-induced climate change. *Nature Climate Change, 11*(6), 492–500. DOI: 10.1038/s41558-021-01058-x PMID: 34221128

Vienna Business Agency. (2023). *Green economy and employment in Vienna*. https://www.viennabusinessagency.at/

Vijayakumar, G., Rajkumar, M., Rajiv Chandar, N., Selvakumar, P., & Duraisamy, R. (2024). Muniyandi Rajkumar, Rajiv Chandar N, Selvakumar P and Ramesh Duraisamy, (2024). Environmentally friendly TDS removal from waste water by electrochemical ion exchange batch-type recirculation (EIR) technique. *Environmental Science. Water Research & Technology, 10*(4), 826–835. DOI: 10.1039/D3EW00793F

Vinogradov, S., & Wouters, P. (2020). Adaptation regulatory regimes to address climate change challenges in transboundary water basins: Can multilateral regionalism help? *Review of European, Comparative & International Environmental Law, 29*(3), 406–416. DOI: 10.1111/reel.12371

Visualizing Scientific Landscapes. VOSviewer. (n.d.). https://www.vosviewer.com/

Voisard, A., & Wallimann-Helmer, I. (2023). Towards a practical climate ethics. *Ethics, Policy & Environment*. Advance online publication. DOI: 10.1080/21550085.2023.2236504 PMID: 39258256

Voyer, M., Schofield, C., Azmi, K., Warner, R., McIlgorm, A., & Quirk, G. (2018). Maritime security and the Blue Economy: Intersections and interdependencies in the Indian Ocean. *Journal of the Indian Ocean Region*, *14*(1), 28–48. DOI: 10.1080/19480881.2018.1418155

Vulevic, A., Castanho, R. A., Gómez, J. M. N., Lausada, S., Loures, L., Cabezas, J., & Fernández-Pozo, L. (2021). Cross-Border Cooperation and Adaptation to Climate Change in Western Balkans Danube Area. In Berisha, E., Cotella, G., & Solly, A. (Eds.), *Governing Territorial Development in the Western Balkans* (pp. 289–308). Springer International Publishing., DOI: 10.1007/978-3-030-72124-4_14

Wakrim, A., & Fait, S. (2023). The future of hydropower: Civil engineering and renewable energy integration. *Int. J. Hydropower Civ. Eng.*, *4*(2), 12–15. DOI: 10.22271/27078302.2023.v4.i2a.30

Wallimann, H. (2024). Austria's KlimaTicket: Assessing the short-term impact of a cheap nationwide travel pass on demand. *Transport Policy*, *159*, 201–214. DOI: 10.1016/j.tranpol.2024.10.008

Walter, K. M. (2023). Vulnerable People or Vulnerable Borders?: EU External Migration Policies and Gendered Vulnerability. *Migration and Diversity*, *2*(1), 51–62. DOI: 10.33182/md.v2i1.2808

Wang, S., & Xu, O. (2024). Interoperability structure of smart water conservancy based on Internet of Things. *International Journal of Distributed Sensor Networks*, *2024*, 1–13. DOI: 10.1155/2024/7724783

Wang, Z., Kim, Y., Seo, H., Um, M.-J., & Mao, J. (2018). Permafrost response to vegetation greenness variation in the Arctic tundra through positive feedback in surface air temperature and snow cover. *Environmental Research Letters*, **•••**, 14.

Wani, M. D., Akhter, S., Shah, S. A., & Dada, Z. A. (2025f). Does Mountain Tourism Impact the Socio-Ecological Balance of the Indian Trans Himalayas, Ladakh? In *Balancing Mountain Tourism, Cultural Heritage, and Environmental Stability* (pp. 89-108). IGI Global Scientific Publishing. DOI: DOI: 10.4018/979-8-3693-8764-1.ch006

Wani, M. D., Baba, S. N., Shah, S. A., & Dada, Z. A. (2025e). Digital Nomads as Catalysts for Change: Enhancing Ethical Travel and Sustainability in Indian Tourism. In *Transforming Remote Work and Tourism in the Digital Nomad Era* (pp. 279-302). IGI Global Scientific Publishing.

Wani, M. D., Ahmad Shah, S., & Ahmad Dada, Z. (2025d). Eco-conscious tourism in India: Investigating the interplay between green hotel practices and traveler decision-making. *Tourism and Hospitality Research, 14673584251334238*, 14673584251334238. Advance online publication. DOI: 10.1177/14673584251334238

Wani, M. D., Batool, N., Dada, Z. A., & Shah, S. A. (2024c). Investigating the impact of community-based tourism on the residents' quality of life and their support for tourism. *Community Development (Columbus, Ohio), 55*(1), 138–159. DOI: 10.1080/15575330.2023.2272271

Wani, M. D., Dada, Z. A., Bhat, W. A., & Shah, S. A. (2025b). Community-driven ecotourism in the trans Himalayan region: A sustainable model for cultural and environmental preservation. *Community Development (Columbus, Ohio), 56*(4), 1–23. DOI: 10.1080/15575330.2025.2451358

Wani, M. D., Dada, Z. A., & Shah, S. A. (2023a). Behavioral intention and satisfaction levels among pilgrim tourists in South Asia. *Tourism Review, 78*(4), 1147–1163. Advance online publication. DOI: 10.1108/TR-09-2022-0432

Wani, M. D., Dada, Z. A., & Shah, S. A. (2023b). Evaluating the impact of cross Border tourism and people to people contact on the regional cooperation in South Asia: A study of India and Bangladesh. *Journal of Quality Assurance in Hospitality & Tourism*, •••, 1–22. DOI: 10.1080/1528008X.2023.2262137

Wani, M. D., Dada, Z. A., & Shah, S. A. (2023c). To utilize structural equation modelling to investigate the impact of cross-border tourism on regional cooperation: The mediating effect of community support. *The Review of Regional Studies, 53*(1), 80–99. DOI: 10.52324/001c.74889

Wani, M. D., Dada, Z. A., & Shah, S. A. (2024a). The impact of community empowerment on sustainable tourism development and the mediation effect of local support: A structural equation modeling approach. *Community Development (Columbus, Ohio), 55*(1), 50–66. DOI: 10.1080/15575330.2022.2109703

Wani, M. D., Dada, Z. A., & Shah, S. A. (2024b). Can consumption of local food contribute to sustainable tourism? Evidence from the perception of domestic tourists. *International Journal of Sustainable Development and World Ecology, 31*(2), 163–178. DOI: 10.1080/13504509.2023.2262950

Wani, M. D., Dada, Z. A., & Shah, S. A. (2024d). Divine Experiences: Exploring the Impact of Tourist Satisfaction on Religious Travel in Jammu and Kashmir, India. In Jermsittiparsert, K., Wekke, I., & Suanpang, P. (Eds.), *Sustainable Development, Humanities, and Social Sciences for Society 5.0* (pp. 387–414). IGI Global Scientific Publishing., DOI: 10.4018/979-8-3693-7989-9.ch019

Wani, M. D., Dada, Z. A., & Shah, S. A. (2024e). What factors influence consumers' intention and food waste reduction perceptions in Indian restaurants? *British Food Journal*. Advance online publication. DOI: 10.1108/BFJ-07-2024-0710

Wani, M. D., Dar, S. N., & Mohanty, P. P. (2025g). Integrating the Value Belief Norm Theory and Theory of Planned Behavior to Predict the Climate Change Mitigation and Adaption Behaviors in Agriculture Production. *Environmental Management*, *75*(10), 1–15. DOI: 10.1007/s00267-025-02217-y PMID: 40588689

Wani, M. D., Ganai, S. A., Wani, S. A., Dada, Z. A., & Shah, S. A. (2025c). Balancing Growth and Preservation: Sustainable Strategies for Overtourism Management. In Kumar, N., Zadeh, S., Naumov, N., & Sharma, S. (Eds.), *Solutions for Managing Overtourism in Popular Destinations* (pp. 1–22). IGI Global Scientific Publishing., DOI: 10.4018/979-8-3693-8347-6.ch001

Wani, M. D., Shah, S. A., & Dada, Z. A. (2025a). Fostering sustainable tourism in India: Harnessing cultural and heritage assets for enduring development. In *Integrating Architecture and Design Into Sustainable Tourism Development* (pp. 179–214). IGI Global Scientific Publishing., DOI: 10.4018/979-8-3693-7388-0.ch007

Wani, M. U. D., Dada, Z. A., & Shah, S. A. (2022). Building peace through tourism: The analysis of an ongoing Siachen Glacier dispute between India and Pakistan. *Asian Journal of Comparative Politics*, *7*(4), 836–848. DOI: 10.1177/20578911221118730

Wapner, P. (1995). Politics beyond the state: Environmental activism and world civic politics. *World Politics*, *47*(3), 311–340. DOI: 10.1017/S0043887100016415

Waris, M., Sufia Azlan, Z., Fadzline Muhamad Tamyez, P., Ullah, M., & Khan, A. (2019). *Analyzing the constructs of stakeholder engagement towards renewable energy projects success in Malaysia: A PLS approach*. KnE Soc. Sci. DOI: 10.18502/kss.v3i22.5090

Waters, C. et al. (2016). "The Anthropocene is functionally and stratigraphically distinct from the Holocene". Science, nº. 351.

Weber, M. (1905). *The Protestant ethic and the "spirit" of capitalism* (Baehr, P., & Wells, G. C., Trans.). Penguin Books.

Weissmann, M. (2008). Peacebuilding in East Asia: The role of Track 2 diplomacy, informal networks, and economic, social, and cultural regionalization. In *Conflict management, security and intervention in East Asia* (pp. 83-98). Routledge. https://doi.org/DOI: 10.4324/9780203928288-11

White, R. (1998). Kyoto and beyond. *Issues in Science and Technology*, *14*(3), 59–65.

Widdis, R. W. (2019). On globalization, borders, and borderlands: A historical geographical perspective. *Canadian Geographies / Géographies Canadiennes*, *63*(4), 573–593. DOI: 10.1111/cag.12580

Wieckowski, M. (2013). Eco-frontier in the mountainous borderlands of Central Europe. The case of Polish border parks. *Journal of Alpine Research*|. *Revue de Geographie Alpine*, •••(101-2), 101–102. DOI: 10.4000/rga.2107

Wilder, M., Scott, C. A., Pablos, N. P., Varady, R. G., Garfin, G. M., & McEvoy, J. (2010). Adapting Across Boundaries: Climate Change, Social Learning, and Resilience in the U.S.–Mexico Border Region. *Annals of the Association of American Geographers*, *100*(4), 917–928. DOI: 10.1080/00045608.2010.500235

Wilkens, J., & Datchoua-Tirvaudey, A. R. C. (2022). Researching climate justice: A decolonial approach to global climate governance. *International Affairs*, *98*(4), 1187–1205. DOI: 10.1093/ia/iiab209

Williams, J. M. (2018). Stagnant rivers: Transboundary water security in South and Southeast Asia. *Water (Basel)*, *10*(12), 1819. DOI: 10.3390/w10121819

Woodley, D. (2017). *Globalization and capitalist geopolitics: Sovereignty and state power in a multipolar world.* Taylor & Francis., DOI: 10.4324/9781315798165

Woodworth, F. (2023). Exclusion of climate migrants from the Global Compact on Refugees. *Geopolitics*, *29*(1), 118–147. DOI: 10.1080/14650045.2023.2225242

World Resources Institute. (2023). *Climate Watch: Tracking Progress Toward Climate Goals.* Retrieved from https://www.climatewatchdata.org

Wu, G. C., Deshmukh, R., Ndhlukula, K., Radojicic, T., & Reilly, J. (2015). *Renewable energy zones for the Africa clean energy corridor. Office of Scientific and Technical Information.* OSTI. DOI: 10.2172/1328753

Wu, J., Orlandi, F., O'Sullivan, D., & Dev, S. (2024). LinkedGeoClimate: An interoperable platform for climate data access within geographical context. *IEEE Journal of Selected Topics in Applied Earth Observations and Remote Sensing*, *17*, 10352–10366. DOI: 10.1109/JSTARS.2024.3391922

Xia, L., Li, Y., Zhan, B., & Wu, C. (2025). Are regional environmental policies always effective? Evidence from China. *Environmental Research Letters, 20*(7), 074067. Advance online publication. DOI: 10.1088/1748-9326/ade3f1

Yang, B. (2025). Current Development and Practice: The Economics of Climate Change. In *The Carbon Emission Liability Mechanism: A New Solution for Global Carbon Neutrality* (pp. 45-114). Singapore: Springer Nature Singapore. https://doi.org/DOI: 10.1007/978-981-96-3612-9_3

Yang, S., Chen, B., & Fath, B. (2015). Trans-boundary total suspended particulate matter (TSPM) in urban ecosystems. *Ecological Modelling, 318*, 59–63. DOI: 10.1016/j.ecolmodel.2014.10.006

Yin, R. (2009). *Case study research design and methods*. Sage Publications.

Zaiter, I. (2024). *Optimizing the UAE's Energy Sector through Hydrogen Integration for Cost-Effectiveness and Net-Zero Carbon Emissions* (Doctoral dissertation, Khalifa University of Science).

Zandvliet, D., Leddy, S., Inver, C., Elderton, V., Townrow, B., & York, L. (2023). Approaches to bio-cultural diversity in British Columbia. *Sustainability (Basel), 15*(8), 6422. Advance online publication. DOI: 10.3390/su15086422

Zargar, S., Kannangara, M., Gonzales-Calienes, G., Yang, J., Shadbahr, J., Decès-Petit, C., & Bensebaa, F. (2024). Data hub for life cycle assessment of climate change solutions—Hydrogen case study. *Data, 9*(11), 129. DOI: 10.3390/data9110129

Zhang, H., & Rajamani, L. (2022). Global climate litigation trends: Strengthening international environmental law through domestic courts. *Journal of Environmental Law, 34*(2), 227–251. DOI: 10.1093/jel/eqac002

Zidan, H., Malik, M. T., Rafique, U., Azeem, F., Manzoor, T., & Manzoor, H. U. (2025). Exploring the potential of cross-border energy trade in SAARC countries for achieving Sustainable Development Goals (SDGs). *Energy Science & Engineering, 13*(4), 2063–2081. DOI: 10.1002/ese3.70032

About the Contributors

Rui Alexandre Castanho holds an International Ph.D. on Sustainable Planning in Borderlands. Nowadays, he is a Vice-Dean for Sustainable Development at the WSB University, Poland, and a visitor Professor at the University of Johannesburg, South Africa. He is an European Climate Pact Ambassador. Besides, he completed a post-doc researcher on the GREAT Project, University of Azores, Ponta Delgada, Portugal. Furthermore, he collaborates with the ECIDES Research Group - University International of La Rioja (UNIR), Spain; Vice-Coordinator of VALORIZA - Research Center for the Enhancement of Endogenous Resources, Polytechnic Institute of Portalegre (IPP), Portugal; CITUR - Madeira - Centre for Tourism Research, Development and Innovation, Madeira, Portugal (Researcher); and AQUAGEO Research Group - University of Campinas (UNICAMP), Brazil. Moreover, he has Graduated in Green Spaces Engineering and Agricultural Engineering and holds a Master's Degree in Planning, Audit, and Control in Urban Green Spaces. He is also the author and/or editor of 20 books, more than a 130 indexed peer review papers, 90 indexed book chapters, 90 conference papers, and made several presentations worldwide academic events as keynote speaker. Also, he was awarded the Doctor Honoris Causa title in Lima-Peru by (UPCI).

***

Keerthiraj is currently with the Department of Political Science, GFGC Punjalakatte (affiliated with Mangalore University) and has previously served at the School of Law & Liberal Arts, Alliance University, Bengaluru. He completed his BA in History, Economics, and Political Science, followed by an MA in Political Science from Mangalore Universit. His doctoral research at Kuvempu University focused on Media, Politics, Colonial Consciousness, and Public Perceptions. His work spans political ideologies, diplomacy, public policy, strategy and international relations.

A. Ranjithkumar is currently serving as an Assistant Professor of Political Science at the School of Law, Vel Tech University, Avadi, Chennai, Tamil Nadu. With a distinguished academic and professional career, he holds a Ph.D. in Politics and International Studies from Pondicherry University, a Master of Population Studies (MPS) from the International Institute for Population Sciences, Mumbai, and an M.A. in Development Administration from Gandhigram Rural Institute – Deemed University. He has also completed a Postgraduate Diploma in Industrial Relations and Labour Law (PGDIR & LL) from Pondicherry University and a B.P.Ed. from Vinayaka Missions University, Salem. Dr. Ranjithkumar has qualified the National Eligibility Test (NET) for Assistant Professor in Human Rights and Duties, Political Science, and Public Administration, conducted by the University Grants Commission (UGC), and was awarded the Dr. S. Radhakrishnan Postdoctoral Fellowship in Humanities and Social Sciences for the year 2014-15. His professional experience includes roles as an Assistant Professor at DLR Arts and Science College, Arcot, and Gandhigram Rural Institute, as well as research positions at Alagappa University, the National Law School of India University, and the Institute of Grassroots Governance. Dr. Ranjithkumar's research focuses on the Indian Political System, Local Government, Political Thought and Theories, Dalit Studies, and Development Administration. He has published over 35 research papers in leading journals and authored/edited 10 books on topics such as social justice, gender, and caste. He has received an International Travel Grant from the Indian Council of Social Science Research (ICSSR) to participate in the "Democratization and Autocratization: IPSA 75th Anniversary Conference" in Lisbon, Portugal, and Financial Assistance from ICSSR, Hyderabad, for organizing a National Conference on "Empowering Vulnerable Communities: Strategies for Inclusive Social Development." Dr. Ranjithkumar is an active member of professional bodies such as the International Political Science Association (IPSA) and the Indian Political Science Association, and he has organized over 10 international webinars and seminars. His areas of interest include Indian politics, local governance, Dalit studies, gender studies, and human rights.

Tushar Bhatia, Research Scholar, pursuing PhD from Graphic Era University, India. I am pursuing research in India-ASEAN Bilateral Relations. As an IR scholar, beyond my research area. I am interested to write in other key fields related with International Politics and International foreign affairs like in South Asia, South-East Asia, West Asia, Maritime relations in Indo-pacific, climate change issue, Human rights problem, Blue Economy, AI and International diplomacy. Recently, three papers published as Review paper in maritime relations, India-Gulf relations and Oil diplomacy at global level. Despite this, I completed my graduation in Political

Science and post-graduation in International relation discipline of Political Science subject from Gautam Buddha University.

Kashvi Chaturvedi is presently enrolled at Ajeenkya DY Patil University to pursue a B.Tech in Artificial Intelligence and Data Science. Her technological skills are demonstrated by her knowledge on languages like Python, C++, C. Her Semiconductors Research Paper examined developments in microchip technology, while her work on the Comparative Analysis of BJT Configurations offered insights into electronic circuits. Kashvi is now researching on Hyperloop's potential to transform high-speed transportation. She recently took part in the IIT Madras-hosted Global Hyperloop Competition 2025, where she offered creative suggestions for this futuristic form of transportation. In addition, she is investigating the possibilities of Small Modular Reactors (SMRs) and Ethical dimensions of AI in sustainable energy solutions.

Susanta Das is a seasoned educator with a proven record of teaching and mentoring diverse students, working at multiple institutions, and providing service to the various initiatives of organizations. He received his Ph.D. and M.A. degrees from Western Michigan University (WMU), USA, and M.Sc. degree from Banaras Hindu University (BHU), India all in Physics. He continued his research as a Marie-Curie post-doctoral fellow at Stockholm University, Stockholm, Sweden on beam diagnostics for the DESIREE (Double ElectroStatic Ion Ring ExpEriment) in the project DITANET (Diagnostic Techniques for particle Accelerators – a Marie-Curie initial training NETwork), at the Indian Institute of Science Education and Research-Kolkata, India on high-pressure physics as a Project Scientist-B, and at the University of Electro-Communications, Tokyo, Japan on ion-surface interactions as a Post-doctoral fellow. Throughout his career, Dr. Das worked and collaborated with many researchers, Ph.D., master, and visiting students from different countries. He further visited many countries to discuss research and an international conference participant (UK, Italy, Germany, Greece, Belgium, Bulgaria, Romania, Brazil etc.). He worked at several institutions before assuming his current role at Ajeenkya D Y Patil University. He received the Marie-Curie post-doctoral fellowship in Sweden, Science Academies' Summer Research Fellowship in India, Gwen Frostic Doctoral Fellowship, Department Graduate Research and Creative Scholar Award by WMU, and Leo R. Parpart Doctoral Fellowship by Dept. of Physics, WMU, among many others, throughout his academic journey. He co-authored several research articles in WoS/Scopus indexed international journals, conference proceedings, and scholarly book chapters and co-edited several books published by renowned international publishing houses (IGI-Global, Springer Nature, Nova Science Publishers, Wiley, & Scrivener-Wiley, Notion Press). At present, Dr. Das continues to delve into cutting-edge fields, with a keen interest in nanotechnology, UN SDGs, and data science. His multifaceted contributions, spanning teaching, research, and administrative leadership, showcase a dedicated professional who is instrumental in advancing the vision and mission of the institutions he serves.

Şükrü C. Demirtaş is a computational social scientist, and economist. His research explores the intersection of data science, behavioral economics, and public policy, with a particular focus on the quantitative analysis of socio-economic systems and climate-related urban dynamics. He employs computational modeling and agent-based methods to study institutional cooperation, policy diffusion, and sustainable development processes in cross-border regions. In addition to his scholarly work, he contributes to international research collaborations on climate governance and regional planning, emphasizing evidence-based and interdisciplinary approaches to sustainability and economic transformation.

Gaurav Dimri Assistant Professor in Graphic Era (deemed to be) University, Dehradun. His PhD thesis explored the planning and governance ecosystem, Dehradun city. His academic interest in philosophhy, politics, cinema, arts and religin. Moreover, research interest areas in human settlements and planetary urbanization, Indian knowledge System, Research methodology

Kasagani Durga is a dedicated legal professional and an LLM student at NBM College of Law, under Andhra University, specializing in International Law. With a strong foundation in legal studies, she is passionate about global legal frameworks, human rights, and diplomatic relations. Durga is committed to deepening her expertise in cross-border legal issues, international treaties, and dispute resolution mechanisms. She aims to contribute to the evolving field of international law by engaging in research, advocacy, and legal practice. With a keen interest in legal reforms and justice, she strives to make a meaningful impact in the legal domain. Her academic pursuits and professional aspirations reflect her dedication to excellence in legal scholarship and international legal affairs.

E. Nirupama is a distinguished academician with over 22 years of experience in Commerce and Management Education. She is currently serving as a Professor in the MBA Program at Saveetha School of Management, bringing extensive expertise in Finance and Accounting. A multi-award-winning educator and researcher, she has published widely in reputed national and international journals. Dr. Nirupama has actively contributed to Academic Leadership, Curriculum Development, and Faculty Development initiatives.

G. Chitra is a dedicated educator in the Department of Mathematics at Karpaga Vinayaga College of Engineering, located in Chengalpattu, Tamil Nadu. With 19 years of teaching experience, she brings a wealth of knowledge and commitment to academic excellence, inspiring students through her expertise and passion for mathematics.

Shubhang Gomasta is currently pursuing Ph.D in Law and has done post-graduate LL.M specialisation in (Corporate and Commercial Law). He has pursued five years integrated under graduation (2014-19) from Tamil Nadu National Law University, Tiruchirappalli, Tamil Nadu. He has pursued PG Diploma in Environmental Laws & Human Rights from National Law School of India University, Bengaluru, Karnataka. He has also qualified competitive level examinations in India for the Assistant Professor. He began his career in academia in 2021 and has since served as an assistant professor of law at different Universities in India. He has been a part of numerous workshops, and has won awards for the best research papers he has presented at both national and international conferences and seminars. His work has been published in peer-reviewed journals, book chapters, the UGC Care Listed Journals and more than twenty-five research papers to his credits. Her primary areas of interest and expertise in Legal Profession are Arbitration Laws and Corporate and Commercial Laws.

Aakash Gupta is currently an Associate Professor of Law at Amity University Gwalior, India. He has obtained the degree of LL.M. from M.L.B. Government College of excellence Jiwaji University, Gwalior. He has obtained his PhD degree in Law from Jiwaji University and cleared UGC NET exmination in 2012. Prior to Amity University, he was working as an Assistant Professor of Law Prestige Institute of Management & Research Gwalior from 2018 to 2023 and in Madhav Law College Gwalior from 2014 to 2018. He has also worked as Guest Lecturer in SOS in Law, Jiwaji University till 2014. He has more than 13 years of experience in Academic and Litigation. His area of Interest in Constitutional Law and Criminal Law. He has 12 International / National Publications to his credit. He has attended various seminar, webinars and workshops and delivered special lectures in EML on numerous occasion. He has published 1 book and presented chapters in 5 books. He has also developed a unique pedagogy to teach law subjects in a poetic form.

Yadnyesh Khapekar is a student of Ajeenkya D. Y. Patil University, pursuing a Bachelor of Technology degree in Computer Science and Engineering. With a sharp analytical mind and a natural inclination for problem-solving, he thrives in tackling complex challenges with precision and creativity. His proficiency in C++ programming enables him to develop efficient algorithms and software solutions, making him an asset in the world of coding. Beyond programming, He has honed his skills in Computer-Aided Design (CAD) software, allowing him to conceptualize and create innovative designs with technical accuracy. His enthusiasm for 3D printing technology further complements his expertise, as he seamlessly translates digital models into tangible prototypes. His ability to design in three dimensions showcases his blend of technical prowess and artistic vision. An active participant in academic and technical competitions, he has proudly represented his university on multiple occasions, demonstrating his capability to perform under pressure and excel in competitive settings. His dedication to research and knowledge dissemination is evident in the numerous review papers he has authored and published in books and journals relating to concepts like artificial intelligence, lead nanoparticles, climate risks, and many more, contributing valuable insights to the fields. With a strong foundation in engineering principles and a drive to push boundaries, he aspires to innovate and lead in the realm of technology. His blend of coding expertise, design acumen, and research experience positions him as a future visionary in the ever-evolving field of computer science and engineering.

Sanjeev Kumar Is a Professor in Lovely Professional University Punjab India. He received Undergraduate in Hotel Management from Osmania University, Postgraduate in Tourism Management from Madurai Kamaraj University. He completed Doctor of Philosophy in Hospitality Management in 2019 from Amity University Jaipur India. His research areas lies in food Service Industry, his research focuses on Alcoholic beverages, Event management and Sustainable Management Practices, Metaverse, AI Machine Learning and Artificial Intelligence. He published more than 80 research papers, articles and chapters in Scopus Indexed, UGC Approved and peer reviewed Journals and books. Dr. Sanjeev Kumar participated and acted as resource person in various National and International conferences, seminars, session chair, research workshops and industry talks and his work has been widely cited. Dr. Sanjeev also edited more than 20 books with various International Publishers such as IGI Global, Cambridge Scholars, Emerald, CABI, Bentham Science, Taylor & Francis. Dr Sanjeev is an active researcher and has published numerous research articles in prestigious National and International journals and books.

Kannan Kunnathully, specialize in International Trade Law and related advertising infringing fundamental rights, now focusing my research on the use of technology in the infringement of fundamental. I am now serving as an Assistant Professor at Vinayaka Mission's Law School, Vinayaka Mission's Research Foundation, Chengalpattu, Tamil Nadu. Here, I teach International law, Legal Research and Manuscript Writing. My passion for teaching and in-depth subject knowledge have earned me recognition as a dedicated educator. In addition to my academic role, I actively engage with corporate contract vetting together with utilisation of CSR funds for legal awareness programs. Looking forward, I aim to expand my research into emerging legal issues in the digital age, especially in privacy rights, while guiding aspiring law students and future legal scholars.

T.C. Manjunath was born on 6th Feb. 1967 in Bangalore. He did his BE from RV College of Engg., Bangalore Univ. in 1989 with First Class, ME in Robotics and Automation from the Govt. LD College of Engg. in 1992 with First Rank – FCD from Gujarat Univ., Ahmedabad and PhD in Systems and Control Engg. from IIT Bombay in the year 2007. He is currently working as the Dean Research, R and D and as Professor in the Dept. of Computer Science and Engineering with specialization in IoT, Cyber Security and Blockchain Tech. in Rajarajeswari College of Engg., Bengaluru, Karnataka, India. He has got a teaching, academic, research and administrative experience of more than 35+ years with 10 years as Principal, 8 yrs as Professor - HoD and 17 yrs as faculty. He has published more than 1200+ research papers in various national and international journals, conferences, with 5000+ citations, 175+ patents, 50+ book-chapters, 25 awards, published 25 text books in various field of science, engineering and technology, completed funded research projects worth Rs. 9 Crores till date apart from having guided 10 PhD scholars, 25 PG Students, more than 200 UG students towards their project works. He is a member of 30 Professional Society Chapters and is a fellow of IE, IETE, AMIEE and a chartered engineer. He got the best research scholar award in 2006 in IIT Bombay amongst all the research scholars of the institute.

Apoorva Misra is a legal scholar specializing in international law, public policy, environmental law, and sustainable development. She holds a Doctorate of Philosophy in Law from Guru Gobind Singh Indraprastha University, Delhi (2023) for her thesis titled "Locating Environment in the Application of Sustainable Development: A Critical Analysis of Environmental Jurisprudence", and a Master of Law from TERI School of Advanced Studies, New Delhi (2018). She is serving as Assistant Professor of Law at Alliance University and has published widely in peer-reviewed journals and edited volumes.

P. Selvakumar is currently an Associate Professor in the Department of Science and Humanities at Nehru Institute of Technology, Coimbatore, Tamil Nadu, India. Prior to joining Nehru Institute of Technology, he held positions as an Assistant Professor in the Department of Science and Humanities various engineering colleges in Coimbatore, Tamil Nadu, India, spanning from 2011 to 2023. Dr.P. Selvakumar's academic and research interests lie primarily in the field of phytochemistry. He has made significant contributions to his field, with an impressive record that includes 30 journal publications, 12 patents, and 20 conference papers. Additionally, he has organized 5 seminars and conferences, further demonstrating his commitment to academic and scientific exchange. Throughout his career, Dr. P.Selvakumar has been recognized for his dedication and excellence in teaching, having been honored with several awards, including the Best Teacher Awards. He holds a prominent position in the academic community, contributing actively to research.

Ananya Pandey is a Research Scholar in School of Law at Christ University, Bangalore. She has her research interests in the field of Constitutional Law, Environment Law, Human Rights, Food Security, Intellectual Property Law and Jurisprudence.

Harita Ponnapalli is an Assistant Professor at the School of Architecture and Planning, Woxsen University. She is dedicated to fostering innovative solutions in the urban planning and architectural education, actively engaging in both academic and studio-based teaching to shape future architects.

Ravikumar R. Natarajan is an experienced academician with over 12+ years of teaching and research expertise across institutions in India, the UAE, and the Maldives. He currently serves as an Assistant Professor at Marwadi University, Rajkot, specializing in Machine Learning, Recommender Systems, and AI-driven applications in healthcare and security. He holds a Ph.D. in Machine Learning and Recommender Systems from Amity University, Jaipur, a Master of Engineering in CSE from VMRF University, Salem, and a Bachelor of Engineering in CSE from Anna University. He also has a MiniMBA from the Swiss eLearning Institute. Dr. Ravikumar has authored more than 8+ multiple SCIE/Scopus-indexed journal articles, 25+ IEEE/Springer conference papers, and co-authored a book on Artificial Intelligence. He has also published 9 patents and contributed 10+ book chapters in emerging domains such as Federated Learning, Blockchain, and Digital Twins for IGI Global, Wiley and Springer publisher. He actively mentors students through initiatives like the Research Activities Club (RAC) and Student Research Internship Training (SRIT), promoting innovation and research culture.

Aarthi S is a dedicated Assistant Professor in Computer Science with over six years of teaching experience. She is currently pursuing her Ph.D. at Marwadi University, Gujarat, and holds an M.Phil. and MCA from Bharathiar University, Tamil Nadu. Throughout her academic career, Aarthi has specialized in programming languages such as C++, Java, and C, and contributed to research in recommender systems and AI-related data privacy. She has published journal articles, contributed to book chapters, and actively participated in technical workshops. Known for her student mentorship and engagement in academic events, Aarthi remains committed to fostering innovation and excellence in computer science education.

Guru Basava Aradhya S is a seasoned academician with 17 years of experience in management education. He currently serves as the Director & Professor at Chandigarh School of Business, Mohali. He holds a Ph.D. in Management and has mentored five Ph.D. scholars. His expertise includes HR, marketing, and business analytics. A prolific researcher, he has published in SCOPUS, ABDC, and UGC CARE journals. He has organized numerous international conferences and contributed to NAAC and NBA accreditations. His industry collaborations include faculty exchange programs and consultancy projects. Passionate about academia, he actively fosters innovation, research, and student development.

Ravikumar S is a Civil Engineering academic and researcher specializing in Structural Engineering. He holds a B.E. in Civil Engineering, an M.Tech in Structural Engineering from JNTU Hyderabad, and a Ph.D. in Civil Engineering from Pondicherry University under the AICTE Doctoral Fellowship. His doctoral research, supported by the RPS Scheme and Institute R&D fund, focused on improving the seismic performance of reinforced concrete beam-column joints using a moment coupler technique. With over 10 years of teaching experience, Dr. Ravikumar has been honoured twice with the Best Faculty Award. He has published several research articles in international journals and is passionate about innovations in earthquake-resistant design and concrete technology.

Shivansu Sachan is Pursuing Ph.D. in Management, a Master's holder from Integral University, Lucknow, Bachelor's from IHM, Lucknow. Academic and professional accomplishments of 6 years with renowned organizations and institutes. His Teaching Career has been very satisfying and have continuously upgraded the skills and knowledge by attending many courses and training programs, seminars, conferences organized by different organizations and also delivered many lectures as a Hospitality sector. He is working as an Assistant Professor at Chhatrapati Shahu ji Maharaj University Kanpur

Ramesh Safare is an accomplished academic and researcher with over two decades of contributions to the fields of management education, research, and practice. Currently serving as an associate professor at Marwadi University in Rajkot, Gujarat, he has established himself as a thought leader in research methods, operations management, supply chain analytics, and international marketing. Holding a Ph.D. in Management, an MBA, and a B.Tech. in Mechanical Engineering, he brings a robust techno-managerial perspective to both teaching and research.

Sunil Sankathala is pursuing B. Tech Computer Science and Engineering at Ajeenkya D Y Patil University, with a diverse range of interests that bridge technology, sustainability, and innovation. He academic focus lies in computer science fundamentals like programming, data structures, algorithms, and software development. He actively supplements this with hands-on experience in web development and robotics. Beyond the core curriculum, He has a strong research inclination in emerging and impactful fields. He current research interests include Artificial Intelligence in Healthcare, bioplastics, nano bioplastics, e-waste management, and solid waste management—areas where he believes technology can make a meaningful difference in solving real-world problems. He has been publishing research articles and book chapters in reputed national and international journals and books. He is also part of an ambitious Hyperloop project, which involves interdisciplinary collaboration and cutting-edge engineering aimed at revolutionizing transportation. Working on this project has enhanced his skills in teamwork, innovation, and systems thinking. In the tech domain, he has developed proficiency in web development using technologies such as HTML, CSS, JavaScript, and React. He interests in robotics further reflects building smart, responsive systems that interact with the physical world. Throughout his course, he has helped strengthen problem-solving and collaborative skills. He proficient in languages such as Python, C, C++, MySQL, Ubuntu, Machine Learning and familiar with frameworks and tools like React, Git. He actively participates in hackathons, coding competitions, and tech events to challenge himself and learn from peers. He is also a quick learner and highly adaptable, with an eagerness to stay updated with the latest trends in the tech world. He aspires to become a skilled software engineer and contribute to building innovative solutions that have a positive impact on society. The long-term goal is to work in a dynamic environment where knowledge, grow professionally, and make meaningful contributions to the field of technology. Driven by a spirit of exploration and sustainability, he aims to contribute to technological solutions that address both human and environmental needs. The long-term goal is to become a versatile engineer and researcher, working at the intersection of computer science, environmental innovation, and human well-being.

Aditya Shrivastav is currently studying B.Tech. in Computer Engineering at the School of Engineering, Ajeenkya DY Patil University. His research interests are centered around nanoparticle biosynthesis, exploring ionic liquids' use in various applications, and applying Artificial Intelligence to different fields. Aditya is particularly focused on understanding how these technologies can be utilized together to solve practical problems. He aims to advance research in these areas and contribute to innovative solutions that address real-world challenges. He recently participated at the Global Hyperloop Competition (GHC) at IIT Madras as a member of the team Vac-Vectoor. The event took place from 21st February through 24th February 2025. He contributed significantly to the performance of his team. The team Vac-Vectoor achieved 2nd position in Business and Socio-Economic Impact of Hyperloop in the Innoquest category and Special Mention Certificate in Traction sub-system under Design X category.

Hemendra Singh is a Lecturer of Law at Jindal Global Law School, situated within O.P. Jindal Global University, Sonipat. He pursued his LL.M. in Intellectual Property Rights from National Law University, Jodhpur, following his graduation with a B.A. LL.B. (Hons.) from Rajiv Gandhi National University of Law, Punjab. With a focus on corporate law, intellectual property law, criminal law, and constitutional law, he has authored numerous research articles and book chapters published in both national and international journals and books.

Mohammad Badruddoza Talukder is an Associate Professor at the College of Tourism and Hospitality Management at the International University of Business Agriculture and Technology (IUBAT) in Dhaka, Bangladesh. Holding a PhD from Lovely Professional University, India, he is renowned for his innovative integration of cutting-edge technologies such as Artificial Intelligence (AI), the Internet of Things (IoT), and Blockchain into tourism and hospitality management. His research explores the transformative impact of these technologies on smart tourism, cloud kitchens, and the broader industry landscape. As a prolific author, Dr. Talukder has published numerous scholarly articles and books, with forthcoming titles that address AI innovations in hospitality, the evolution of cloud kitchens, and the enhancement of digital literacy within hospitality education. His work shapes contemporary tourism and hospitality practices, skillfully marrying technological advancements with practical industry applications. Since 2008, Dr. Talukder has imparted his knowledge through a diverse array of courses in tourism and hospitality across various universities in Bangladesh.

Divya J Thakur is serving as Professor at University School of Business Chandigarh University, Punjab. She has a total experience of 15+ years in academia and industry. She has completed her Doctorate of Philosophy (Ph.D.) in Management (HR). Her Research is on Responsible Leadership, Leadership Qualities. She also provides Student Development Programs, Personality Development Programs, Mock Interview Sessions. She has visited many international universities such as City University College Azmen Dubai, Kuala Lumpur University Malaysia, Ujbekistan and Indonesia. Prof. Thakur also has completed Coursera/LinkedIn courses. Prof. Thakur has successfully organised many webinars/conferences/workshops/FDPs. Also participated as speaker (Resource Person/Keynote Speaker/Guest Speaker) in many conference/workshops. She has published 32 research papers in reputed journals/Scopus Indexed, Thomson and Reuters, UGC care journals are worth mentioning. 22 Patents published in national and international, 6 are granted and three copyright has been published, Dr. Thakur has been published 30 Book Chapters with ISBN no. and organised international workshop and international conference as convener with conference proceedings. Dr. Thakur has received Best Citizen Award from IEEE Chandigarh subsection and also received Iconic Faculty of the Year 2021 award from Saksham Society Jaipur Rajasthan. Also organised one day ODOP fruitful workshop association with ministry of education, MGNCRE, govt of India and IPR workshop association with ministry of commerce and research, govt of India, Research Proposal workshop and 5days FDP association with Techsaksham under Microsoft, Edunet foundation and SAP. Also She is the Joint Secretary of the IPR cell. She has published four international and national books as first editor. On teacher's day she has received a award of appreciation of Best Researcher. She has received iconic award of Excellence in Research 2022. Received award as chief guest during national conference at Rajouri J & K. 5th International Scopus Indexed Book is coming soon as first editor.

Srinivasan Vasan is an Assistant Professor in the Department of Humanities and Social Sciences at Graphic Era (Deemed to be University), with 14 years of teaching and field expertise. He has served in the Ministry of Health and Family Welfare, as a Guest Faculty, and as a Mobilization Training Specialist with the State Government. Recently, he contributed to a tuberculosis project funded by GFATM and USAID. He has published 60 papers, with 20 more under peer review in UGC-CARE and SCOPUS journals.

Index

A

B

C

D

E

G

H

I

L

T

www.ingramcontent.com/pod-product-compliance
Lightning Source LLC
LaVergne TN
LVHW062339300926
849161LV00007B/1757
* 9 7 9 8 3 3 7 3 3 9 7 2 6 *